CHAPMAN & HALL/CRC APPLIED MATHEMATICS
AND NONLINEAR SCIENCE SERIES

QUANTUM COMPUTING DEVICES

PRINCIPLES, DESIGNS, AND ANALYSIS

Published Titles

Geometric Sturmian Theory of Nonlinear Parabolic Equations and Applications,
  Victor A. Galaktionov
Introduction to Fuzzy Systems, Guanrong Chen and Trung Tat Pham
Introduction to Partial Differential Equations with MATLAB®, Matthew P. Coleman
Mathematical Methods in Physics and Engineering with Mathematica,
  Ferdinand F. Cap
Optimal Estimation of Dynamic Systems, John L. Crassidis and John L. Junkins
Quantum Computing Devices: Principles, Designs, and Analysis, Goong Chen,
David A. Church, Berthold-Georg Englert, Carsten Henkel, Bernd Rohwedder,
Marlan O. Scully, and M. Suhail Zubairy

Forthcoming Titles

Mathematical Theory of Quantum Computation, Goong Chen and Zijian Diao
Stochastic Partial Differential Equations, Pao-Liu Chow

CHAPMAN & HALL/CRC APPLIED MATHEMATICS
AND NONLINEAR SCIENCE SERIES

QUANTUM COMPUTING DEVICES

PRINCIPLES, DESIGNS, AND ANALYSIS

Goong Chen
David A. Church
Berthold-Georg Englert
Carsten Henkel
Bernd Rohwedder
Marlan O. Scully
M. Suhail Zubairy

Chapman & Hall/CRC
Taylor & Francis Group
Boca Raton   London   New York

Chapman & Hall/CRC is an imprint of the
Taylor & Francis Group, an informa business

Chapman & Hall/CRC
Taylor & Francis Group
6000 Broken Sound Parkway NW, Suite 300
Boca Raton, FL 33487-2742

International Standard Book Number-10: 1-58488-681-1 (Hardcover)
International Standard Book Number-13: 978-1-58488-681-5 (Hardcover)

Library of Congress Cataloging-in-Publication Data

Quantum computing devices : principles, designs, and analysis / Goong Chen ... [et al.].
    p. cm. -- (Chapman & Hall/CRC applied mathematics and nonlinear science series ; #6)
Includes bibliographical references and index.
ISBN 1-58488-681-1 (alk. paper)
    1. Quantum computers. I. Chen, Goong, 1950- II. Title. III. Series.

QA76.889.Q832 2006
004.1--dc22   2006049559

Visit the Taylor & Francis Web site at
http://www.taylorandfrancis.com

and the CRC Press Web site at
http://www.crcpress.com

Goong Chen
Department of Mathematics and Institute for Quantum Studies
Texas A&M University
College Station, TX 77843, U.S.A.

David A. Church
Department of Physics
Texas A&M University
College Station, TX 77843, U.S.A.

Berthold-Georg Englert
Department of Physics
National University of Singapore
Singapore

Carsten Henkel
Institut für Physik, Universität Potsdam
D-14469 Potsdam, Germany

Bernd Rohwedder
Instituto de Física
Universidade Federal do Rio de Janeiro
Cx.P.68528,CEP 21941-972
Rio de Janeiro, Brazil

Marlan O. Scully
Departments of Physics, and Electrical and Chemical Engineering
and Institute for Quantum Studies
Texas A&M University
College Station, TX 77843, U.S.A.
and
Departments of Chemistry, and Mechanical and Aerospace Engineering
Princeton University
Princeton, NJ 08544, U.S.A.
and
Max-Planck Institut für Quantenoptik
D-85748 Garching, Germany

M. Suhail Zubairy
Department of Physics and Institute for Quantum Studies
Texas A&M University
College Station, TX 77843, U.S.A.

To Professor and Mrs. Pah (Paul) I. and May T. Chen

獻給陳伯儀教授及夫人吳美提

Contents

Preface

Quantum information science has a relatively short history, beginning during the 1980s through the pioneering work of P. Benioff, R. Feynman and D. Deutsch. Later on, P. Shor's mid-1990s "killer app", his quantum factoring algorithm, generated tremendous enthusiasm in the scientific community for quantum computing. His algorithm utilized quantum computers' peculiar properties of superposition and entanglement in quantum mechanics to achieve massive parallelism and unprecedented speedup. Such properties do not have any classical analogues and, thus, quantum computers have the potential to execute certain special tasks "exponentially faster" than classical computers.

On the hardware front, the computer industry will soon be facing one of its greatest challenges—the end of Moore's Law. The continued efforts to miniaturize microchips in electronics will hit a brick wall in 20 to 25 years. No longer will the industry be able to double the computing power of computer chips every 1.5 years at half the price, unless quantum effects can be integrated. Microelectronics is inevitably moving into *nanoelectronics*. Therefore, the development of quantum information science and quantum computing technology most surely constitutes the future of information science and technology. It has also become the major impetus to the development of new general quantum technologies, which, since the invention of the laser during the 1950s, have changed nearly every aspect of life for the modern human being.

As the theoretical foundation of quantum information science is largely in place, the key step in the design and construction of the quantum computer lies in the fabrication of *quantum logic gates* that can carry out computations reliably and faithfully. Many proposals have been made by scientists and engineers—literally in the hundreds—for such quantum logic gates. All proposed schemes have some promise, but experimental work has encountered various types of roadblocks and difficulties, including decoherence and dissipation, single photon sources, measurements, etc. Nevertheless, scientists and engineers have made tremendous progress, improving the design, understanding and operations of various types of quantum gates. As a result, the state of the art has improved dramatically over the past few years. Even though there is still no working quantum computer thus far, spin-offs from this research have lead to cutting edge technologies in computing and in areas other than computing important for the economy and defense. Most remarkably, of late we have noticed a dramatic increase of confidence—some leading researchers have now proclaimed that the year 2020 (or sooner) will see the inaugural run of a "real" quantum computer.

This book is aimed particularly at this *contemporary*, important aspect of quantum computation, namely *quantum electronic devices as quantum gates*. There are hundreds of proposals and thousands more papers in the literature. Obviously, it is not possible to cover them all. Even though the varieties are many, the fundamental ideas are similar. Presently, there is not yet any book dedicated to this area and we hope this book can fill the void. What's more, we hope this book will offer interesting and useful material to multidisciplinary researchers in physics, computer science, engineering and mathematics. Our organization and approach in writing this book is as follows:

(1) The first three chapters explain the basic ideas, fundamentals, and algorithms of quantum computing and information, quantum systems, operations and formalisms.

(2) Chapters 3 and 5–10 cover the most promising candidates for the future quantum computer, namely Cavity QED, ion traps, neutral atom traps, quantum dots, linear optics, superconducting quantum interference devices (SQUID), and nuclear magnetic resonance (NMR).

(3) For each device, we discuss the physical properties of the quantum system, the setup of qubits, the control actions that bring about the quantum gates which are universal for quantum computing, the measurements, and decoherence properties of the system.

The readers may first be astonished that there are seven coauthors for the book. No, this book is not an edited volume with seven people contributing disparate chapters. We wrote this book together and linked the material in a coherent way. Nevertheless, due to the different expertise of each coauthor, certain chapters are much heavily contributed to by an individual coauthor. (This also helps explain why there are seven coauthors, as this book is very cross-disciplinary, and we must confess that none of us could have written the book individually.) Among these chapters, some "inhomogeneities" may be spotted through different styles of structuring, usage of preferred words or notation, and occasional repetitions of items already introduced earlier or elsewhere. We tried to smooth out such things as consistently and coherently as possible. We trust that what has remained and not yet so rendered is mostly inevitable or innocuous. To achieve a truly unified presentation would require such effort and time that the book would lose its spontaneity and contemporaneousness, failing to do justice to this vivid field of research rapidly under development.

This book is written at a level suitable for an audience who have had some prior exposure to quantum computing and some maturity in atomic physics. But these are not the "absolute prerequisites" for reading the book. We have made a conscientious effort to make many parts of the book as self-contained and tutorial as possible. Several sections in Chapter 2 contain very condensed information, and terminologies from atomic physics are needed and used throughout, whose introductions cannot always be made with the best care without a long diversion from the main topics under discussion. Many readers will probably feel comfortable reading the book with some minor help using online search and internet resources, plus some occasional references to other relevant textbooks. Our own writing has benefitted from many online resources. For example, Wikipedia, the free internet encyclopedia (http://en.wikipedia.org), is a constant source of ready help and valuable information.

In the writing of this book, we have received tremendous help from Drs. Zijian Diao and Zhigang Zhang. We have also benefitted from the following people in many ways—collaborations, consulting, corrections, discussions and proofreading: Janet Anders, Andre Bandrauk, Geesche Boedecker, Michel C. Delfour, Jonathan Dowling, Jens Eisert, Stephen A. Fulling, David Yang Gao, Philippe Grangier, Philip R. Hemmer, Chia-Ren Hu, Jong Kim, Andreas Klappenecker, Antonio Negretti, Arup Neogi, Signe Seidelin (NIST) and Martin Wilkens. Several graphics and quantum device color photos included in the book are by courtesy of Michael Chapman, Arup Neogi, Pepijn Pinkse, Jean-Michel Raimond, and Gerhard Rempe. We are grateful to their generosity. All errors and inaccuracies are of course the responsibility of the authors.

The original contributions included in the book must be credited to the many frontier researchers in quantum computing and quantum devices. Their pioneering work is really to be admired and thanked for. It is not possible for us to cite the relevant literature and credit each individual's contributions exhaustively. We apologize in advance for any inadvertent omissions.

Financial support from the following agencies is gratefully acknowledged: DARPA QuIST Contract AFRL F306602-01-1-0594 and AFOSR FA9550-04-1-0206, NSF Grant PHY 9876899, Texas A&M University TITF No. 2001-055, Singapore A*STAR Temasek Grant No. 012-104-0040, and the European Commission under the contracts FASTNet (HPRN-CT-2002-00304), ACQP (IST-CT-2001-38863), and QUELE (EFP-003772), and a Robert A. Welch Foundation Grant No. A-1261.

We have tried to do the proofreading as carefully as possible, but undoubtedly there are still many errors not yet picked up. A website of Erratum for the book will be posted at (that of the first author's) http://www.math.tamu.edu/~gchen. The readers will also be welcome to communicate directly with us about any needed corrections. We hope to incorporate them in future revised editions.

Ms. Robin Campbell has provided us with the critical expertise of drafting, typing and editing. Mr. Bob Stern, Senior Executive Editor at Taylor & Francis, offered us continuous encouragement and expedited the book publishing process. To both of them we are deeply indebted.

Goong Chen
David A. Church
Berthold-Georg Englert
Carsten Henkel
Bernd Rohwedder
Marlan O. Scully
M. Suhail Zubairy

Chapter 1

Foundations of Quantum Informatics: Spins, the EPR Problem, and Landauer's Principle

- The Stern–Gerlach experiment and spins

- The EPR problem

- Bell's inequalities and locality

- Thermodynamics and information

- Landauer's principle on information erasure

The earliest ideas of simulating and utilizing quantum systems to do computation, i.e., quantum computation, can be attributed to P. Benioff [3] (1980) and R. Feynman [12] (1982). In his 1980 paper, Benioff introduced a quantum Turing machine model. Deutsch [6, 7] further developed more concrete proposals and introduced the quantum circuit model of computation. Later, Yao [34] showed that the quantum circuit model of computation is equivalent to the quantum Turing machine model.

But seeds for quantum computing were planted more than half a century earlier. The following three profound events actually constitute the most important preludes that paved the way for the development of a modern quantum computer (QC):

(1) The Stern–Gerlach experiment (1920s);

(2) The Einstein–Podolsky–Rosen (EPR) paper [8] (1935) and its reply from Schrödinger [26];

(3) The Landauer principle on information erasure [19] (1961).

Quantum computers execute tasks on a radically different principle from that based on Boolean logic in classical computers. Quantum machines process information and perform logic operations by exploiting the laws of quantum mechanics. A QC promises the power of massive parallelism and unprecedented speedup because of the peculiar properties of *superposition* and *entanglement* in quantum mechanics. Any ideal quantum logic operation is *reversible* and does not dissipate heat. How and why? In this chapter, we will motivate three of the giant pillars upholding quantum informatics, namely, *superposition, entanglement and reversibility*, by explaining the three important discoveries (1)–(3) above.

1.1   Spins: The Stern–Gerlach experiment and spin filter

Spin (throughout this book) refers to intrinsic angular momentum possessed by particles such as individual atoms, protons, or electrons. It is particularly important in quantum mechanics for systems at atomic length scales or smaller.

The fact that electrons have spins was first discovered in the celebrated Stern-Gerlach experiment, named after the German physicists Otto Stern and Walther Gerlach, in 1922 on deflection of silver atoms. (See the story in an interesting account by Friedrich and Herschbach [14].)

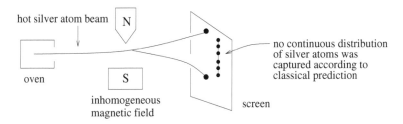

Figure 1.1: A beam of spin-1/2 particles is sent through an inhomogeneous magnetic field. The magnetic field is strongly increasing in one direction; let's call this the privileged "z" direction. Each such particle has a spin magnetic moment. The inhomogeneous magnetic field causes the particles to be deflected, 1/2 of them up and 1/2 of them down, totally separated rather than in a continuous distribution, and thus measures the magnetic moment of the particle and shows that the spin is discrete and (spatially) quantized.

The general concept of elementary particle spin was first proposed in 1925 by R. Kronig, G. Uhlenbeck, and S. Goudsmit (15, 16, 33). In quantum mechanics, the angular momentum is quantized. The magnitude $|J|$ of the angular momentum vector J can only take on the values

$$\hbar \sqrt{s(s+1)},$$

where $\hbar = 1.054 \cdot 10^{-34}$ Joule·sec is Planck's constant h divided by 2π, and s is a non-negative integer or half-integer (0, 1/2, 1, 3/2, 2, etc.). For example, electrons are "spin-1/2" particles because their intrinsic spin angular momentum has $s = 1/2$. Other elementary spin-1/2 particles include neutrinos and quarks. On the other hand, photons are spin-1 particles, i.e., $s = 1$.

Pauli formalized the theory of spin in 1927, using the new quantum theory of Schrödinger and Heisenberg being actively developed at that time. He devised Pauli spin matrices σ_x, σ_y and σ_z (see Section 2.6) as a representation of the spin operators, and introduced a 2-component (non-relativistic) spinor wave-function.

Particles with half-integer spin are *fermions* while those with integer spin are *bosons*. They are governed, respectively, by the Fermi-Dirac and Bose-Einstein statistics. Fermions are subject to the Pauli *exclusion principle*, which forbids them from sharing quantum states, and are described in quantum theory by "antisymmetric states" (which are antisymmetric with respect to particle exchange). In contrast, bosons are in "symmetric states".

Box 1.1 Spins of elementary particles

The physical setup of the Stern–Gerlach experiment as shown in Fig. 1.1 will be called a Stern–Gerlach apparatus (SGA) from now on. The particles whose trajectories are in the upper and lower split beams are said to have spin-up and spin-down, denoted by $|\uparrow\rangle$ and $|\downarrow\rangle$, respectively.

Thus, the SGA demonstrates a basic, *intrinsically quantum property of spins of elementary particles*. (Nevertheless, we need to say that there is *no* Stern–Gerlach experiment for electrons or protons because they carry electric charges and the Lorentz force will deflect them differently.) Textbooks that develop quantum kinematics by systematically exploiting the lessons of the Stern–Gerlach experiments are Schwinger [27] (graduate level) and Englert [9, Vol. I] (undergraduate level).

The quantum computing devices addressed in this book represent a number of important quantum systems in chemistry, physics and material science. Such quantum systems about which one has maximal knowledge are denoted by *kets* such as $|\uparrow\rangle$, $|\downarrow\rangle$, $|0\rangle$, $|1\rangle$, $|n, \ell, m\rangle$, etc., which are vectors in relevant Hilbert spaces $\mathcal{H}$. The interior of the symbol $|\ \rangle$ is labelled by spins $\uparrow, \downarrow$; or energy levels $0, 1$; or quantum numbers n, ℓ, m, depending on the quantum system under study. The dual space $\mathcal{H}^*$ of $\mathcal{H}$ is the space constituted by the *bra* (vectors) $\langle\uparrow|$, $\langle\downarrow|$, $\langle0|$, $\langle1|$, $\langle n,\ell,m|$, etc. Mathematically, such vectors are *linear functionals* on $\mathcal{H}$. In the physics jargon, such bras are related to the kets by Hermitian conjugation or by taking the adjoint as follows:

$$\langle\cdot| = |\cdot\rangle^\dagger, \qquad |\cdot\rangle = \langle\cdot|^\dagger.$$

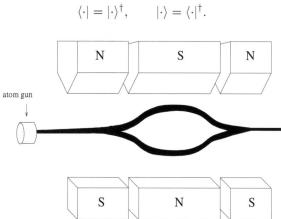

Figure 1.2: Three magnets are in serial connection. The middle magnet is longer and has opposite polarity. The particle beam is split first. But the strengths of the magnetic fields are such that the split beams can be re-combined at the end. (Adapted from [17].)

The theory of Hilbert spaces and linear operators is one of the major successes in mathematics of the 20th century. Its development first was aimed at providing a rigorous mathematical foundation for the flourishing field of quantum mechanics during the 1920s. Inner products (of two bras, or of two kets), normal operators, tensor products, unitary operators, spectral decompositions, evolution semigroups and groups and their infinitesimal generators, matrix representations of linear operators, etc., are thus built up or set up for the mathematical formulation, modeling and analysis of quantum systems, including all of those under study here for the purpose of quantum computation. This theory of Hilbert space and linear operators constitutes a bulk of the modern theory of *mathematical physics* [23, 24, 25]. In Chap. 2, we offer some rudiments of this material. Any readers who had no prior exposure to this may need to consult Sections 2.3 and 2.4 in Chap. 2 at this point and then return to continue the subsequent discussions of this chapter.

We now consider the concatenation of several SGAs to form a "spin filter" according to Feynman [11, Chap. 5] and Harrison [17]. We will first demonstrate that *measurement in quantum mechanics affects the particles measured.* Another profound result is *quantum states are necessarily described by a superposition with complex numbers as coefficients.* Whose demonstration must take up more space so here we will provide only some heuristics. But this will soon become self-evident in Chaps. 3–10.

The first concatenation is shown in Fig. 1.2, where three magnets are in serial connection, with the middle one having the opposite polarity. The overall magnetic field splits the particle (say, silver atoms) beam first, but then re-combines the split beams. The recombined beam is ideally assumed here to have the same strength and direction as the original beam, and also the same spin properties so that the device in Fig. 1.2 is meant to be equivalent to "doing nothing".

Next, consider the arrangement as shown in Fig. 1.3, where the spin-down beam is removed by some absorber in its path.

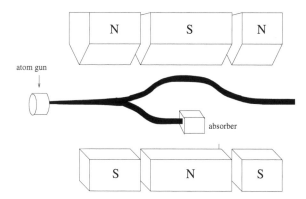

Figure 1.3: The magnets are aligned exactly as in Fig. 1.2, except that along the spin-down path, some particle absorption material is placed, which removes all the spin-down atoms. This device is meant to be performing a selective measurement. (Adapted from (17).)

We now enclose the apparatus in Fig. 1.3 in a box, constituting a *spin filter*. See Fig. 1.4.

The selection (or filtering) of spin-up electrons is independent of how the filter is oriented. Only half of the original beam will emerge. But what if we let the beam go through two spin filters? See Fig. 1.5.

The case (b) in Fig. 1.5 can be interpreted as follows: half of the particle beam will emerge from the first filter. But only 1-half of those atoms can pass through the second filter. The axis of the spin (filter) is now important as there are now two different axes. For the particle beam exiting the first filter in Fig. 1.5 (b), 1-half of it satisfies spin-up, and the other half satisfies spin-down, relative to the new axis of the second filter. There is no way of having a "spin-left/right" instead of a "spin-up/down" (relative to an up/down oriented SGA). Technically, the spin-left/right is a superposition state of spin up/down states with equal weights. See Eq. (1.8) below.

Consider the general situation of the alignment of two spin filters. If we slowly rotate the orientation of the second filter relative to the first one, from $0°$ to $180°$, then the percentage of the particles that finally exit both

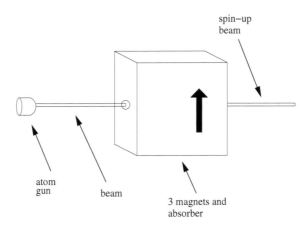

Figure 1.4: A spin filter box. Inside the box, three magnets and an particle-absorber are arranged as in Fig. 1.3. Only spin-up particles emerge after exiting the box. (Adapted from (17).)

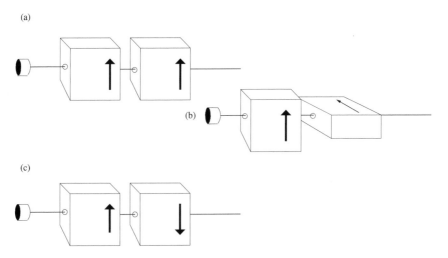

Figure 1.5: Put two spin filters in serial connection. Rotate the second filter so that its orientation is: (a) parallel, (b) perpendicular; and (c) anti-parallel, respectively, to that of the first filter. Then we observe the following consequences: (a) half of the beam emerges; (b) a quarter of the beam emerges; and (c) no beam emerges. (Adapted from (17).)

filters is

$$100\% \times \cos^2 \frac{\theta}{2}, \quad \theta = \text{relative angle of the two spin filter axes.} \qquad (1.1)$$

The spin-up $|\uparrow\rangle$ and spin-down $|\downarrow\rangle$ states will be denoted, respectively, as quantum bit (qubit) states $|0\rangle$ and $|1\rangle$ in subsequent chapters. It will become evident that atoms in the beam are representable in the form of superposition

$$\frac{1}{\sqrt{2}}(|0\rangle + |1\rangle), \qquad (1.2)$$

where the factor $1/\sqrt{2}$ is a normalization factor in order to achieve a unit norm.

Later on in subsequent chapters, we will soon see a general form of superposition

$$\frac{1}{\sqrt{2}}(\cos\beta|0\rangle + e^{i\alpha}\sin\beta|1\rangle), \qquad \alpha, \beta \in \mathbb{R},$$

where some "phase shift" α and weights $\cos\beta$ and $\sin\beta$ can be incorporated so that *complex numbers* enter as coefficients in a superposition.

The above discussions applying SGAs and spin filters to a particle beam can be reduced to the case where only a *single* particle (e.g., a silver atom) is emitted from the particle gun. Then what we observe is that the single particle has spin-up with probability 1/2, and spin-down with probability 1/2. As a matter of fact, in general depending on how this particle was "prepared", Eq. (1.1) shows that the probabilities of spin-up can range from 0% to 100%.

A much more complicated situation arises when we deal with two particle spins or other types of arrangements when they are *correlated*. This is the important *entanglement* case, a scenario arising, e.g., in an atomic decay problem where two (correlated) particles are emitted almost simultaneously. This we are going to address in the next section.

1.2 EPR and Bell's inequalities

In quantum cryptography, *entanglement* is used to transmit signals that, if eavesdropped, will leave a trace for alarm. In quantum computation, entanglement works together with superposition to allow certain calculations to be performed with fast speedup. In 1935, Einstein, Podolsky and Rosen (EPR) [8] drew on this special quantum entanglement phenomenon to show that measurements performed on spatially separated parts of a quantum system can have an instantaneous effect on each other, constituting a paradoxical so-called "nonlocal behavior". Indeed, EPR's elegant presentation of the entanglement (Schrödinger's word) is the basis for the Bell inequality, as well as much of the stuff of modern quantum mechanics such as quantum teleportation and quantum dense coding, etc. In the present section we shall present the EPR problem (not paradox), offer our resolution and make contact with quantum erasure. Our presentation is an adaptation of the EPR paper by Cantrell and Scully [5] as well as of Englert, Scully and Walther [10, Appendix].

1.2.1 The EPR problem

In order to set the stage, we paraphrase EPR as follows: consider two spin-$\frac{1}{2}$ particles (e.g., two silver (or mercury) atoms in a 2-atom molecule (i.e., a dimer)), in a bound state of zero angular momentum. Call the corresponding state vector as $|\psi_{1,2}\rangle$. At time $t = 0$, we disintegrate the dimer but do not disturb the spins in any way. The separate parts now move off to opposite sides of the laboratory. We consider two kinds of "experimental" arrangements I and II as follows.

"Experiment" I

Suppose we measure the z-component of spin 1 as indicated in Fig. 1.6.

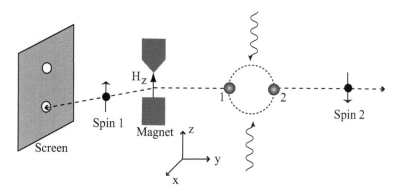

Figure 1.6: Measurement of the z-component of the spin of particle 1 in Experiment I, where "Spin 1" and "Spin 2" denote, respectively, the spins of particles 1 and 2.

Before measuring (at some time t_0) the z component of the spin of particle 1, we have

$$|\psi_{1,2}^<(I)\rangle = \frac{1}{\sqrt{2}} [|\uparrow_1, \downarrow_2\rangle - |\downarrow_1, \uparrow_2\rangle], \tag{1.3}$$

where $\psi^<$ means the state *before* ($t < t_0$) "looking" at particle 1 and where, the state $|\uparrow_1, \downarrow_2\rangle$ means that the spin of particle 1 is "up" and that of particle 2 is "down". If at ($t = t_0$) we find it to be, say, in the spin-up state $|\uparrow_1\rangle$, then for all later times ($t > t_0$) the state of particle 2 is given by

$$|\psi_2^>(I)\rangle = |\downarrow_2\rangle. \tag{1.4}$$

At this point EPR argue that: "Since at the time of measurement the two systems no longer interact, no real change can take place in the second system in consequence of any thing that may be done to the first system. This is, of course, merely a statement of what is meant by the absence of any interaction between the two systems". In other words, EPR argue that (in present notation)

$$|\psi_2^<(I)\rangle = |\psi_2^>(I)\rangle = |\downarrow_2\rangle. \tag{1.5}$$

Expressed in the density-matrix language, cf. Box 1.2, the description of particle 2 just before looking at particle 1, according to EPR, would be given by

$$\rho_2^<(I) = |\psi_2^<(I)\rangle\langle\psi_2^<(I)| = |\downarrow_2\rangle\langle\downarrow_2|, \tag{1.6}$$

which in matrix notation is

$$\rho_2^<(I) = \begin{pmatrix} 0 & 0 \\ 0 & 1 \end{pmatrix}_2, \tag{1.7}$$

if we use $\{|\uparrow\rangle_2, |\downarrow\rangle_2\}$ as the ordered basis for the spin of particle 2.

"Experiment" II

Next we consider the case where we measure the x-component of spin 1 as in Fig. 1.7. We now express our spins in turns of $|\pm x\rangle$ states as:

$$|\pm x\rangle \equiv |\pm\rangle = \frac{1}{\sqrt{2}}\left[|\uparrow\rangle \pm |\downarrow\rangle\right]. \tag{1.8}$$

This is a key quantum mechanics "magic" of constructing a "spin-left/right" state out of "spin-up/down", the superposition principle and we now write

$$|\psi_{1,2}^<(II)\rangle = \frac{1}{\sqrt{2}}\left[|+_1, -_2\rangle - |-_1, +_2\rangle\right]. \tag{1.9}$$

Now after finding spin 1 to be in, say, the state $|-_1\rangle$, we have

$$|\psi_2^>(II)\rangle = |+_2\rangle, \tag{1.10}$$

which implies

$$|\psi_2^<(II)\rangle = |+_2\rangle. \tag{1.11}$$

The density matrix for particle 2 just before t_0 is then

$$\rho_2^<(II) = |\psi_2^<(II)\rangle\langle\psi_2^<(II)|, \tag{1.12}$$

which in matrix notation is

$$\rho_2^<(II) = \frac{1}{2}\begin{pmatrix} 1 & 1 \\ 1 & 1 \end{pmatrix}_2. \tag{1.13}$$

Thus the density matrices (1.7) and (1.13) *ascribed* to spin 2 by EPR before the measurement are not equal:

$$\rho_2^<(II) \neq \rho_2^<(I). \tag{1.14}$$

Paraphrasing EPR: "Thus, *it is possible to assign two different state vectors* (in our notation $|\uparrow_2\rangle$ and $|+_2\rangle$) *to the same reality*."
Thus we regard (with EPR) the *inequality*

$$\rho_2^<(II) \neq \rho_2^<(I) \tag{1.15}$$

as the fundamental statement of the EPR paradox. Therefore, EPR argue that quantum mechanics is not "complete" or self consistent and does not provide an adequate picture of "physical reality". In other words, EPR argue that the state for particle 2 is the same just before $(t < t_0)$ looking at particle 1 as after $(t > t_0)$ since we haven't affected (that is, haven't applied any bumps, grinds, pushes, pulls, etc., to) particle 2.

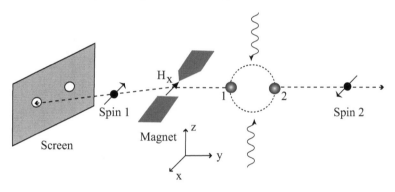

Figure 1.7: Measurement of the x-component of the spin of particle 1 in Experiment II.

1.2.2   Resolving the EPR conundrum

Based on this logic one would have to conclude that $\rho_2^> = \rho_2(t > t_0) = \rho_2(t < t_0) = \rho_2^<$. But now we are in trouble because we have (at least) four *different* states for particle 2 to choose from, namely

$$\rho_2^< = \left\{ \begin{pmatrix} 1 & 0 \\ 0 & 0 \end{pmatrix} \text{ or } \begin{pmatrix} 0 & 0 \\ 0 & 1 \end{pmatrix} \right\} \tag{1.16}$$

for up/down, type I, measurement and

$$\rho_2^< = \left\{ \frac{1}{2} \begin{pmatrix} 1 & 1 \\ 1 & 1 \end{pmatrix} \text{ or } \frac{1}{2} \begin{pmatrix} 1 & -1 \\ -1 & 1 \end{pmatrix} \right\} \tag{1.17}$$

for left/right, type II, measurements. How to resolve this problem? Simple, just be careful to note that any predictions quantum mechanics can make about the statistics of measurements on particle 2 alone are characterized by the reduced density matrix (cf. Box 1.2) $\rho_2 = \mathrm{tr}_1[|\psi\rangle\langle\psi|]$. At times $t < t_0$, we get

$$\rho_2^< = \frac{1}{2} \times \begin{pmatrix} 1 & 0 \\ 0 & 0 \end{pmatrix} + \frac{1}{2} \times \begin{pmatrix} 0 & 0 \\ 0 & 1 \end{pmatrix} = \frac{1}{2} \begin{pmatrix} 1 & 0 \\ 0 & 1 \end{pmatrix} \qquad (1.18)$$

if we employ $|\psi\rangle_I$ equivalently,

$$\rho_2^< = \frac{1}{2} \times \frac{1}{2} \times \begin{pmatrix} 1 & 1 \\ 1 & 1 \end{pmatrix} + \frac{1}{2} \times \frac{1}{2} \times \begin{pmatrix} 1 & -1 \\ -1 & 1 \end{pmatrix} = \frac{1}{2} \begin{pmatrix} 1 & 0 \\ 0 & 1 \end{pmatrix} \qquad (1.19)$$

if we make use of $|\psi\rangle_{II}$. (The weight factors of $\frac{1}{2}$ are the respective probabilities for finding particle 1 in either one of the states $|\downarrow_1\rangle$, $|\uparrow_1\rangle$, $|\leftarrow_1\rangle$, or $|\rightarrow_1\rangle$.) So we have the *same* state for particle 2 "before looking" at particle 1, irrespective of how we are going to look at it.

Then this is the lesson drawn from the EPR problem: Statements such as "particle 2 is in the spin-up state $|\uparrow_2\rangle$" already *before* particle 1 is found in the spin-down state $|\downarrow_1\rangle$ are meaningless. The property of being a pure-state spin-up particle, say, is not possessed by particle 2 before an up/down measurement on particle 1 has found it spin-down. This is quite analogous to the situation discussed in the quantum eraser; see Sections 3.5 and 3.6, Chap. 3. The property of being a first-slit atom is not possessed by the atom until the photon is found in the first cavity.

The density matrix as introduced by von Neumann provides a description of systems whose initial states are not completely specified. If our system is in a pure state $|\psi\rangle$, then the expectation value of the observable $\hat{Q}$ is

$$\langle \hat{Q} \rangle = \langle \psi | \hat{Q} | \psi \rangle = \sum_j \langle \psi | \hat{Q} | j \rangle \langle j | \psi \rangle. \qquad (1.20)$$

where we have expressed the state $|\psi\rangle$ in terms of a basis $\{|j\rangle\}$. The above may be written (in a more formal, mathematical notation) as

$$\langle \hat{Q} \rangle = \mathrm{tr}\{\hat{Q} |\psi\rangle\langle\psi|\}, \qquad (1.21)$$

where "tr" stands for the *trace operation* defined first for any ket-bra $|1\rangle\langle 2|$ by

$$\mathrm{tr}\{|1\rangle\langle 2|\} = \langle 2 | 1 \rangle \qquad (1.22)$$

and then extended by linearity. This trace operation satisfies the properties

$$\mathrm{tr}\{ABC\} = \mathrm{tr}\{BCA\} = \mathrm{tr}\{CAB\},$$

in particular,

$$\mathrm{tr}\{AB\} = \mathrm{tr}\{BA\}.$$

If A is interpreted as a matrix, then $\operatorname{tr} A$ is equivalent to taking the trace of the matrix A, i.e., the sum of all the diagonal entries of A. According to (1.21) and (1.22), for a quantum system in a pure system $|\psi\rangle$, we have the expectation value

$$\langle\widehat{Q}\rangle = \operatorname{tr}\{\widehat{Q}|\psi\rangle\langle\psi|\} = \langle\psi|\widehat{Q}|\psi\rangle,$$

so the above yields exactly (1.20).

If a system is not in a pure state but rather is an ensemble of states, described by a collection of kets $|j\rangle$ with probability p_j, such that $\sum_j p_j = 1$, then in taking the expectation of any observable $\widehat{Q}$, we need to take this *mixture* into account and get

$$\langle\widehat{Q}\rangle = \sum_j p_j \operatorname{tr}\{Q|j\rangle\langle j|\} = \operatorname{tr}\{Q \sum_j |j\rangle p_j \langle j|\} \tag{1.23}$$

$$= \operatorname{tr}\{Q\rho\},$$

where

$$\rho \equiv \sum_j p_j |j\rangle\langle j|, \quad \text{satisfying} \ \operatorname{tr}\rho = 1 \tag{1.24}$$

is called the *statistical operator* (or, density matrix), summarizing what we know about the statistical properties of the quantum system.

For a qubit, the statistical operator has the form

$$\rho = \frac{1}{2}(1 + \boldsymbol{s} \cdot \boldsymbol{\sigma}), \tag{1.25}$$

where $\boldsymbol{s} \in \mathbb{R}^3$ is a vector and $\boldsymbol{\sigma} = (\sigma_x, \sigma_y, \sigma_z)$ is a "3-vector" whose components are spin-1/2 operators σ_x, σ_y and σ_z (cf. (2.42) and (2.43), Chap. 2) with $|\boldsymbol{s}| \leq 1$. If $|\boldsymbol{s}| = 1$, we have a *pure* state, for which $\rho^2 = \rho$ holds (cf. (10.4)–(10.6) in Chap. 10, e.g., for an explicit representation of (1.25)). Otherwise we have a mixed state, for which $\operatorname{tr}\{\rho^2\} < 1$.

In a *composite system* composed of two subsystems A and B, the state vectors are, of course, *not* always direct products of the form

$$|a, b\rangle = |a\rangle \otimes |b\rangle. \tag{1.26}$$

For example, a superposition

$$|\psi\rangle = c_1|a_1, b_1\rangle + c_2|a_2, b_2\rangle \tag{1.27}$$

can not in general be expressed in direct-product form. The states (1.26) nevertheless contain a basis for the composite system, if $\{|a\rangle\}$ and $\{|b\rangle\}$ are complete sets for A and B, respectively. An observable $\widehat{Q}_A$ which refers to system A alone has the form

$$\widehat{Q} = \widehat{Q}_A \otimes 1_B \tag{1.28}$$

when considered as an observable of the composite system.

Because the states of the composite system are not generally direct products such as (1.26), the density matrix is also not generally a direct product:

$$\rho_{AB} \neq \rho_A \otimes \rho_B. \tag{1.29}$$

However, we can define a *reduced density matrix* which can be used to calculate the expectation value of an observable which refers to system A alone. Using the complete set (1.26), we have

$$\langle \hat{Q} \rangle = \mathrm{Tr}_{A,B}(\rho_{AB}\hat{Q}) = \sum_{a,b} \langle a, b | \rho_{A,B}\hat{Q} | a, b \rangle \tag{1.30}$$

$$= \sum_a \langle a | \sum_b \langle b | \rho_{AB} 1_B | b \rangle \hat{Q}_A | a \rangle = \mathrm{Tr}_A(\rho_A^{(r)}\hat{Q}_A).$$

The reduced density matrix for system A is then

$$\rho_A^{(r)} = \sum_b \langle b | \rho_{AB} 1_B | b \rangle = \mathrm{Tr}_B(\rho_{AB}), \tag{1.31}$$

which is, an operator on the states of system A.

See a further formal definition of the density matrix in Section 4.2 of Chap. 4.

Box 1.2   Expectation value and the density matrix concept

1.2.3   Bell's inequality

The EPR paper suggested that quantum mechanics was incomplete or only a kind of partial theory. That is, just as thermodynamics is a gross (macroscopic) theory of the underlying statistical mechanics, perhaps there are deeper hidden variables for which quantum mechanics plays the role of a kind of "macro-cover theory" (in the spirit of thermodynamics). In particular, Einstein hoped that quantum theory could be supplemented by some additional "hidden" variables. John S. Bell (1928–1990) then showed that a description of the EPR correlation, based on a local hidden variable theory, moves the problem from a philosophical discussion into the realm of experimental physics. We offer an argument adapted from [9, Vol. I] in the following, which is a variant of the one given by Bell [2] in the 1960s.

Consider a scenario where a particle-pair (e.g., pair of silver atoms) source always emits two particles, one moving to the left and the other to the right, as depicted in Fig. 1.8.

In the last row, the product +1 signifies that the two particles have the same outcome (i.e., both are ↑, or both are ↓), while the product −1 signifies the contrary. Using these products, we now define the *Bell correlation* $C(a, b)$ for the chosen setting specified by parameters a and b:

$$C(a, b) = \frac{(\text{number of } +1 \text{ pairs}) - (\text{number of } -1 \text{ pairs})}{\text{total number of observed pairs}}. \tag{1.32}$$

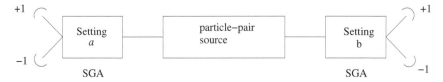

Figure 1.8: A pair of particles, one moving to the left and the other moving to the right, and each goes through an SGA. The exiting particle is then detected either as spin-up $(+1)$, or spin-down (-1). The two SGAs are assumed to have a number of parameter settings, denoted by a on the left SGA and by b on the right SGA. (Details of a and b are not important here; all we need is that different settings are possible.)

For an arbitrary setting, the following is a sample of collected experimental data; see Table 1.1.

particle-pair number	1	2	3	4	5	6	7	8	9	$\cdots$
detection on the left	$+1$	-1	-1	$+1$	$+1$	-1	-1	$+1$	$+1$	$\cdots$
detection on the right	$+1$	$+1$	-1	-1	-1	-1	-1	$+1$	$+1$	$\cdots$
product	$+1$	-1	$+1$	-1	-1	$+1$	$+1$	$+1$	$+1$	$\cdots$

Table 1.1: A sample of experimental data for the experimental setup as in Fig. 1.8.

Obviously, $C(a, b)$ satisfies

$$-1 \leq C(a, b) \leq 1,$$

and its lower bound -1 above can be attained when and only when all the measurement values -1 (resp., $+1$) on one side are matched with that of $+1$ (resp., -1) on the other side. The upper bound $+1$ of $C(a, b)$ can be attained when and only when the signs of measurement values on both sides are the same.

Following Bell [2], we now suppose that there were a (hidden) mechanism at work that determines the outcome on each side (in a *deterministic*, rather than *probabilistic*, way). We conceive each pair as being characterizable by a set of parameters collectively called λ, and that the source realizes various λ and obeys the hidden mechanism with different relative frequencies. Thus, there is a probability density function $\rho(\lambda)$ such that $\rho(\lambda)d\lambda$ is the probability of having the occurrence of a λ value within a $d\lambda$ volume around λ, satisfying

$$\rho(\lambda) \geq 0, \qquad \int \rho(\lambda)d\lambda = 1. \tag{1.33}$$

Here we see (some of) the beauty of Bell's arguments: λ can be any general set of parameters, and no specific details of λ are needed.

We denote by $A_\lambda(a)$ and $B_\lambda(b)$, respectively, as the measurement result on, respectively, the left setting a and the right setting b when the hidden control parameter has value λ. Since all measurement results are either $+1$ or -1, we have

$$A_\lambda(a) = \pm 1, \quad B_\lambda(b) = \pm 1, \quad \text{for all} \quad a, b, \lambda,$$

and, thus,

$$A_\lambda(a)B_\lambda(b) = \pm 1, \quad \text{for all} \quad a, b, \lambda. \tag{1.34}$$

The above can then be filled into each entry in the fourth row of Table 1.1. Upon substituting (1.34) into (1.32) and summing over all pairs, we obtain

$$C(a,b) = \int A_\lambda(a)B_\lambda(b)\rho(\lambda)d\lambda \tag{1.35}$$

for the Bell correlation, under the supposition that there were a set of hidden mechanism λ.

At this point, we remark that an important assumption has entered: we take for granted that the measurement result on the left does not depend on the setting of the apparatus on the right, and vice versa. This is an expression of *locality* as a natural consequence accepted by us due to Einstein's observation that spatially well separated events cannot be connected by *causality* if such events are *simultaneous* in a common reference frame. This being understood, we then don't need to consider the more general possibility of having the functional dependences $A_\lambda(a,b)$ and $B_\lambda(a,b)$ on, respectively, the left and the right. Such a b dependence of A_λ and an a dependence of B_λ are physically ruled out due to causality, although it may remain a mathematical possibility that cannot be excluded on purely logical grounds.

We are now in a position to state the main result of this subsection, a variant of the renowned Bell inequality.

Theorem 1.2.1 (Bell's inequality). *Let $A_\lambda(a), B_\lambda(b), \rho(\lambda)$ and $C(a,b)$ be as specified above. Then for any settings a and a' on the left and any settings b and b' on the right, the Bell correlation $C(\cdot,\cdot)$ satisfies*

$$|C(a,b) - C(a,b')| + |C(a',b) + C(a',b')| \leq 2. \tag{1.36}$$

Proof. From (1.35), we have

$$\begin{aligned}
C(a,b) - C(a,b') &= \int [A_\lambda(a)B_\lambda(b) - A_\lambda(a)B_\lambda(b')]\rho(\lambda)d\lambda \\
&= \int A_\lambda(a)B_\lambda(b)[1 \pm A_\lambda(a')B_\lambda(b')]\rho(\lambda)\,d\lambda \\
&\quad - \int A_\lambda(a)B_\lambda(b')[1 \pm A_\lambda(a')B_\lambda(b)]\rho(\lambda)\,d\lambda,
\end{aligned}$$

when the terms after "$\pm$" reconcile provided that we take either both upper signs or both lower signs. From (1.32) and (1.34), we have

$$\rho(\lambda) \geq 0, \quad |A_\lambda(a)B_\lambda(b)| = 1, \quad 1 \pm A_\lambda(a)B_\lambda(b) \geq 0$$

for both signs in "$\pm$" and all values of λ, a and b, repeated applications of the triangle inequality give

$$|C(a,b) - C(a,b')| \leq \int [1 \pm A_\lambda(a')B_\lambda(b')]\rho(\lambda)d\lambda$$
$$+ \int [1 \pm A_\lambda(a')B_\lambda(b)]\rho(\lambda)d\lambda$$
$$= 2 \pm [C(a',b) + C(a',b')].$$

Therefore,

$$|C(a,b) - C(a,b')| \leq \min\{2 \pm [C(a',b) + C(a',b')]\}$$
$$= 2 - [C(a',b) + C(a',b')],$$

which is (1.36) that we set out to prove. $\square$

Bell's inequality's derivation follows from intuitive, heuristic arguments and natural assumptions flowing seamlessly in the proof above. One would be totally convinced that it must be obeyed by the correlation $C(\cdot, \cdot)$ in any experiment as described in Fig. 1.8. Anything else would defy common sense, wouldn't it? However, the brutal fact is that rather *strong violations* are observed in actual physics experiments where the left-hand side of (1.36) substantially exceeds 2, getting close in fact to $2\sqrt{2}$, the maximal value allowed for Bell correlations in quantum mechanics.

Since we cannot possibly give up our convictions about locality, and thus about Einsteinian causality, the logical conclusion must be that *there just is no such hidden deterministic mechanics* (characterized by $\rho(\lambda)$ as well as $A_\lambda(a)$ and $B_\lambda(b)$).

Even though the experimental arrangement in Fig. 1.8 used pairs of spin-1/2 particles such as silver atoms, what is true for such correlated pairs of particles, by inference, is also true for individual particles, and for analogous physical properties (such as polarizations). There is no mechanism that decides whether the particle is spin-up or spin-down. Rather, it is a *truly probabilistic* phenomenon.

As another consequence, we have deduced that there are nonlocal correlations in the real world. Hence there are no local hidden variable theories.

Let us argue a little bit more about "locality" and "causality" from the EPR lesson. Consider the classical correlations of two balls in colors of white and black, respectively. You take one at random, but don't look at it and travel to the Moon, and leave the other one on Earth. The very moment you look at the ball on the Moon, you know for sure that the other on

Earth has the opposite color! This does not imply that information is flowing instantaneously from the Moon to the Earth, of course. The tricky thing about spin-1/2 is that this argument holds for *any* orientation of the SGA's. Correlated measurements along up/down could equivalently be described by the above classical correlations.

For more discussions on Bell's inequality, EPR, locality and reality, we refer the readers to Appendix A.

1.3 The Landauer principle

The design and operation of the modern electronic computer engenders serious heating of the machine which is a major problem. Such heat dissipation often causes malfunctioning of the machine and is an obstacle to computer circuits miniaturization; it has to be countered by expensive and sophisticated cooling devices and facilities. During the early 1960s, researchers at IBM, led by Rolf Landauer, took on this problem in the hope of designing computers that are energy efficient and not plagued by overheating.

Historically, many scientists (including Brillouin and von Neumann) believed that there is an intrinsic cost involved in information processing, such as the execution of a logic operation by CPU, the copying of information from one memory medium to another, and the measurements of outputs from information flow. For example, there is a so-called Szilard's Principle, which states that it is information gathering that requires an increase in entropy. The finding of the IBM researchers was astounding: that belief was actually a *misconception*. There is *no intrinsic irreducible thermodynamic cost required of information processing, acquisition and measurement*. However, the *logical reversibility or irreversibility* determines whether there is a minimum cost associated with the information processing. This is Landauer's principle, often regarded as *the basic principle of the thermodynamics of information processing*. It will be stated more precisely in the following paragraphs.

To see why this is true, we first recall some history of thermodynamics, information theory and the second law of thermodynamics.

The second law of thermodynamics, first formulated by Rudolph Clausius in 1850, states that the *entropy* of any totally isolated system not at a thermal equilibrium will tend to increase with respect to time, approaching a maximum value. Here, the entropy S is defined through its increment dS by

$$dS = \frac{dQ}{T}, \quad \text{satisfying} \quad dS \geq 0,$$

where dQ is the transfer in energy in the form of heat, and T is the temperature.

For macroscopic systems, the statistical formulation of entropy implies that the second law is overwhelmingly likely to be accurate. However, a non-zero probability is allowed for it to be measurably inaccurate. The great physicist J.C. Maxwell, in his study of thermodynamics, posed an intelligent being, later referred to as Maxwell's Demon, to demonstrate that there could be situations in which the second law is violated. He wrote the following (quoted in [20, p. 4], see also [21, 30]):

"...if we conceive of a being whose faculties are so sharpened that he can follow every molecule in its course, such a being, whose attributes are as essentially finite as our own, would be able to do what is impossible to us. For we have seen that molecules in a vessel full of air at uniform temperature are moving with velocities by no means uniform, though the mean velocity of any great number of them, arbitrarily selected, is almost exactly uniform. Now let us suppose that such a vessel is divided into two portions, A and B, by a division in which there is a small hole, and that a being, who can see the individual molecules, opens and closes this hole, so as to allow only the swifter molecules to pass from A to B, and only the slower molecules to pass from B to A. He will thus, without expenditure of work, raise the temperature of B and lower that of A, in contradiction to the second law of thermodynamics."

The violation of the second law of thermodynamics is a big controversy and Maxwell's Demon attracted the attention of many scientists. Leo Szilard [32] (1929) suggested a simplified version. We show it schematically in Fig. 1.9 by following Plenio and Vitelli [22].

In Maxwell's supposition, an element of *consciousness* or *intelligence* entered into the realm of physical science. The role of such intelligence of the Demon has generated significant debates as to how and whether one ought to characterize it physically. This issue of vagueness involving intelligence can be avoided by the substitution of an automatic mechanism which can perform Demon's functions. (More on this at the end of this section.) With this, the paradox of Maxwell's Demon was still not easy to resolve. A satisfactory exorcism of the Demon was finally made by Charles H. Bennett in [4] (1982) by applying Landauer's principle. We describe it succinctly next.

In Landauer's study [19] (1961), he had an important insight about a fundamental asymmetry in information processing. For example, the negation of a binary variable or the copying of classical information can be done *reversibly* without expending any energy. But when information is erased, or if an irreversible binary logic operation such as $a \wedge b = c$ is executed, where $\wedge$ denotes the "AND" operation (see Section 2.1, Chap. 2), there is always an energy cost of $kT \ln 2$ to be consumed, where k is the Boltzmann constant and T is the absolute temperature. In information theory, consider a system with n events (or states), each with probability of occurrence p_i, for $i = 1, 2, \ldots, n$, where $0 < p_i \leq 1$ and $p_1 + p_2 + \cdots p_n = 1$. Then the total information "value" of the system is (defined by) the Shannon entropy

$$H_S = -\sum_{i=1}^{n} p_i \log p_i, \quad (\log = \text{logarithm with base 2}). \tag{1.37}$$

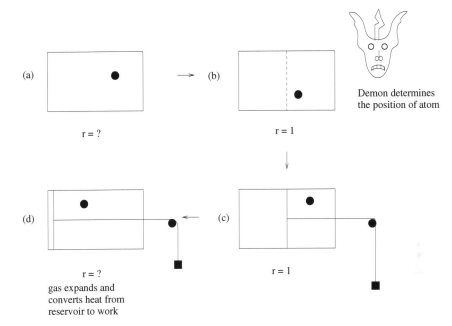

Figure 1.9: Schematic for a simplified version of Szilard's engine, adapted from Ref. (22). Here, r represents a register recording the position of a gas atom, $r = 0$ and $r = 1$ corresponds, respectively, to the location of the atom in the left half or right half of the engine chamber. Initially (a), that position is unknown. Then the Demon measures the position of the atom and inserts a piston if $r = 1$, in (b) to (c). Then the single atom gas expands and does work on the load attached to the piston, which is pushed to the left in (d). This process is repeated and work is apparently done by extracting heat from one reservoir only, violating the second law of thermodynamics.

The above entropy can be linked to the Boltzmann entropy

$$H_B = -k \ln 2 \sum_{i=1}^{n} p_i \log p_i, \quad (\ln = \text{the natural logarithm}) \tag{1.38}$$

by considering (from statistical mechanics) how many arrangements there are in the assemblage of matter and energy in a physical system. The term $k \ln 2$ is a scaling factor in the conversion from the classical Shannon information entropy (1.37)) to the Boltzmann entropy (1.38). For a (1-bit) binary system, the single 1-bit 0 and 1 occur with equal probability $p_1 = p_2 = 1/2$, so we obtain

$$H_B = -k \ln 2 \left[\frac{1}{2} \log \left(\frac{1}{2} \right) + \frac{1}{2} \log \left(\frac{1}{2} \right) \right] = k \ln 2. \tag{1.39}$$

Landauer argues that in any irreversible computing where information is lost, forgotten or erased, the entropy so obtained is the amount of thermodynamical entropy you will generate in erasing the information. The least of which is one bit of information. Thus, the Boltzmann entropy is (1.39), which after multiplying by T, gives the minimum of expended energy $kT \ln 2$ (dissipated into the environment). This is the Landauer principle of information erasure. From this principle, one can actually derive the second law of thermodynamics [22].

Bennett [4] applied Landauer's principle to Szilard's engine (Fig. 1.9) by first agreeing with Szilard that the Demon's mind could be viewed as a 2-state system where information about the particle is recorded as a single-bit register $r = 0$ or 1. After the particle in the engine is returned to the initial state, r remains unchanged. But the information about r needs to be erased by doing work that is lost as heat into the reservoir. By Landauer's principle, this erasure has an energy cost

$$W_{\text{erasure}} = -kT \ln 2$$

while the work done by the engine through isothermal expansion is

$$W_{\text{done}} = kT \ln 2.$$

Therefore, there is energy conservation and no net flow of heat or decrease of entropy, and the second law of thermodynamics is saved. The crux of Bennett's "exorcism" argument lies in his penetrating observation that information erasure has taken place that expends energy.

The discussions in this section so far are all classical. But now let us be equipped with quantum mechanics and revisit Maxwell's Demon. What Maxwell envisioned in his Demon's function having the ability to sort out molecules with faster and slower speeds is no longer as impossible or far-fetched as Maxwell originally thought. Indeed, an SGA can certainly achieve that in principle, where higher energy and lower energy atoms in an atom beam are separated (i.e., sorted out) by an SGA. The tracking and detection of their trajectories can be done, e.g., by cavity-QED and photo-detectors. However, in the quantum realm, some unexpected, amazing phenomena happen. When we observe and determine the path of an atom in a double-slit Young's experiment, the situation is as follows:

(i) If the path information is fully available (irrespective of whether there is any human knowledge about it), then there are no interference fringes;

(ii) If we have interference fringes of perfect visibility, then there cannot be any path knowledge available.

But it does not follow that whenever there are no fringes, then path knowledge must be available. Rather, it is quite easy to have a situation where neither path knowledge nor fringes are available. Even then, it may be possible to restore fringes by a "postselection" procedure called "quantum erasure". The nomenclature of "quantum erasure"—like so many other terms—could be somewhat misleading. Key studies of the quantum eraser process are [28, 29], and experimentally confirmed in [18, 35]. This puzzle of quantum eraser turns out to be a quantum mechanical analogue of Maxwell's Demon. Since information erasure has taken place in this process, Landauer's principle applies to the quantum eraser.

In order to discuss the concept and setup of the quantum eraser, some prerequisite on cavity QED is necessary. We thus defer the discussions to Sections 3.5 and 3.6 in Chap. 3.

An elementary introduction to Maxwell's Demon, Szilard's engine and quantum eraser can be found in Scully and Scully [30].

As a final note of this chapter, we mention that all quantum processes are unitary and reversible. Therefore, the logic operations in quantum computing, based on unitary operations, are always reversible (in "idealistic environments"). This is a major distinction between classical computing and quantum computing. Thus, ideally, a QC can carry out computations without any energy cost and problem associated with heating. This, as well as other fundamental differences involving quantum superposition and entanglement, will be expounded in the next Chap. 2. Nevertheless, we need to keep in mind that in reality, the environment is mostly not idealistic. If the situation is such that there is strong coupling between the system and reservoir, then information and energy loss occurs in between in a rather non neglegible way. This will be treated in Chap. 4.

References

[1] F. Belinfante, *A Survey of Hidden Variable Theories*, Pergamon, New York, 1973.

[2] J.S. Bell, *Physics* **1** (1964), 195.

[3] P. Benioff, The computer as a physical system: a microscopic quantum mechanical Hamiltonian model of computers as represented by Turing machines, *J. Stat. Phys.* **22** (1980), 563–591.

[4] C.H. Bennett, *Int. J. Theor. Phys.* **21** (1982), 905.

[5] C.D. Cantrell and M.O. Scully, *Phys. Rep. C* **43** (1978), 499.

[6] D. Deutsch, Quantum theory, the Church–Turing principle, and the universal quantum computer, *Proc. Royal Soc. London* **A400** (1985), 97–117.

[7] D. Deutsch, Quantum computational networks, *Proc. Royal Soc. London* **A425** (1989), 73.

[8] A. Einstein, B. Podolsky, and N. Rosen, *Phys. Rev.* **47** (1935), 777.

[9] B.-G. Englert, *Lectures on Quantum Mechanics*, Vol. I: Basic Matters; Vol. II: Simple Systems; and Vol. III: Perturbed Evolution, World Scientific, Singapore, 2006.

[10] B.-G. Englert, M.O. Scully, and H. Walther, *Sci. Am. (Int. Ed.)* **271** (1994), 56–61.

[11] R.P. Feynman, R.B. Leighton, and M. Sands, *The Feynman Lectures on Physics*, Vol. III, Addison-Wesley, Reading, Pennsylvania, 1965.

[12] R.P. Feynman, Simulating physics with computers, *Int. J. Theor. Phys.* **21** (1982), 467–488.

[13] S.J. Freedman and J.F. Clauser, *Phys. Rev. Lett.* **28** (1972), 938; E.S. Fry and R.C. Thompson, *Phys. Rev. Lett.* **36** (1976), 1223; A. Aspect, P. Grangier, and G. Roger, *Phys. Rev. Lett.* **49** (1982), 91; A. Aspect, J. Dalibard, and G. Roger, *Phys. Rev. Lett.* **49** (1982), 1804.

[14] B. Friedrich and D. Herschbach, Stern and Gerlach: How a bad cigar helped reorient atomic physics, *Phys. Today Dec. Issue* (2003), 53–59.

[15] S. Goudsmit, *Z. Physik* **32** (1925), 111.

[16] S. Goudsmit and R. de L. Kronig, *Naturwissenschaften* **13** (1925), 90; *Verhandelingen Koninklijke Akademie van Wetenschappen* **34** (1925), 278.

[17] D.M. Harrison, http://www.upscale.utoronto.ca/GeneralInterest/ Harrison/SternGerlach/SternGerlach.html (Department of Physics, University of Toronto, March 1998).

[18] Y.-H. Kim, R. Yu, S.P. Kulik, Y. Shih, and M.O. Scully, *Phys. Rev. Lett.* **84** (2000), 1.

[19] R. Landauer, *IBM J. Res. Develp.* **5** (1961), 183.

[20] H. Leff and A. Rex, Overview, in "Maxwell's Demon: Entropy, Information, Computing", H. Leff and A. Rex, (eds.), Princeton University Press, Princeton, New Jersey, 1990.

[21] E.H. Neilsen, http://users.ntsource.com/~neilsen/papers/demon/ node3.html

[22] M.B. Plenio and V. Vitelli, The physics of forgetting: Landauer's erasure principle and information theory, *Contemporary Physics* **42** (2001), 25–60.

[23] M. Reed and B. Simon, *Methods of Modern Mathematical Physics*, Vol. 1: Functional Analysis, Revised Edition, Academic Press, New York, 1980.

[24] M. Reed and B. Simon, *Methods of Modern Mathematical Physics*, Vol. 2: Fourier Analysis, Self-Adjointness, Academic Press, New York, 1975.

[25] M. Reed and B. Simon, *Methods of Modern Mathematical Physics*, Vol. 2: Analysis of Operators, Academic Press, New York, 1978.

[26] E. Schrödinger, *Naturwissenschaften* **23** (1935), 807 & 823 & 844; English translation by J.D. Trimmer, *Proc. Amer. Philos. Soc.* **124** (1980), 323.

[27] J. Schwinger, *Quantum Mechanics : Symbolism of Atomic Measurements*, B.-G. Englert (ed.), Springer-Verlag, Berlin, 2003.

[28] M.O. Scully and K. Drühl, *Phys. Rev. A* **25** (1982), 2208.

[29] M.O. Scully, B.-G. Englert, and H. Walther, *Nature (London)* **351** (1991), 111.

[30] R. Scully and M.O. Scully, *The Demon and the Quantum*, book to appear, 2006.

[31] M.O. Scully and M.S. Zubairy, *Quantum Optics*, Cambridge Univ. Press, Cambridge, U.K., 1997.

[32] L. Szilard, *Z. Phys.* **53** (1929), 840.

[33] G.E. Uhlenbeck and S. Goudsmit, *Naturwissenschaften* **47** (1925), 953.

[34] A. Yao, Quantum circuit complexity, *Proc. the 34th IEEE Symposium on Foundations of Computer Science*, IEEE Computer Society Press, Los Alamitos, CA, 1993, 352–360.

[35] X.Y. Zou, L.J. Wang, and L. Mandel, *Phys. Rev. Lett.* **67** (1991), 318; see also P.G. Kwait, A.M. Steinberg, and R.Y. Chiao, *Phys. Rev. A* **47** (1992), 7729.

Chapter 2

Quantum Computation and Quantum Systems

- Classical Turing machines and binary logic

- Quantum logic operations

- Universal quantum gates

- Major quantum algorithms

- Quantum circuits for addition and multiplication

- Quantum error correction

- DiVincenzo's criteria

2.1 Introduction: Turing machines and binary logic gates

The foundation of modern computer science is built on the theory of Turing machines. Alan Turing (1912–1954) [44] published his ideas of an abstract computing machine (now called a "Turing machine" (TM)) in 1936, which moved from one state to another using a precise finite set of rules, given by a finite table, and depending on a single symbol it read from a tape [46]. A TM, at the time Alan Turing invented it in his 1936 paper [61], was a hypothetical computer. It consists of the following:

(i) An infinite *tape* on which symbols may be read and written.

(ii) The machine travels right or left along the tape, following a *program*.

(iii) At each step the machine writes to the tape, travels either left or right and changes state, according to a set of *internal states*.

27

(iii) The set of symbols and the set of internal states are both finite sets.

Turing wrote [61]:

> "... Some of the symbols written down will form the sequences of figures which is the decimal of the real number which is being computed. The others are just rough notes to "assist the memory". It will only be these rough notes which will be liable to erasure. ..."

Further, he established that a universal Turing machine existed [61]:

> "... which can be made to do the work of any special-purpose machine, that is to say to carry out any piece of computing, if a tape bearing suitable "instructions" is inserted into it. ..."

The tape referred above in his description of a universal Turing machine was the "computer memory" and the instructions on the tape constituted the "computer program."

We may formalize the notion of a TM as follows.

Definition 2.1.1. A (classical) Turing machine (TM) is an abstract machine consisting of a finite control, an infinitely long tape divided into cells holding a finite set of symbols, and a tape head which reads/writes on the tape.

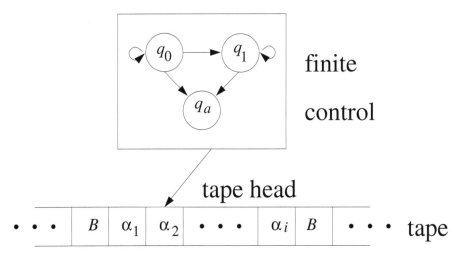

Figure 2.1: Components of a turing machine.

It can be described by a 6-tuple $(Q, A, B, \delta, q_0, q_a)$, where
$Q = \{q_0, q_1, q_2, \ldots, q_m\}$ is a finite set of control states;
$A = \{\alpha_1, \alpha_2, \alpha_3, \ldots, \alpha_n\}$, the alphabet, is a finite set of distinct symbols;
$B \in A$ is the blank symbol;

$\delta : Q \times A \to Q \times A \times \{L, R\}$ is the transition function which describes the rules for the moves of the TM, where "L" and "R" stand for the tape head moving left and right by one cell, respectively;

$q_0 \in Q$ is the initial state;

$q_a \in Q$ is the accepting state.

Initially, a finite string of symbols from A (the input) is written on the tape. All other cells on the tape hold the blank symbol B. The finite control is in the initial state q_0. The tape head is at the first cell of the input. The TM moves according to the transition function δ. For example, if the current state is q_i, the symbol in the cell at the tape head reads α_s, and $\delta(q_i, \alpha_s) = (q_j, \alpha_t, L)$, then the state of the finite control changes to q_j, the tape head writes α_t in the current cell and moves to the left (L). When the state of the finite control becomes q_a, the TM accepts the input string and halts. □

A TM is said to compute a function f if, started with input x on the tape, it halts with $f(x)$ on the tape. A simple TM is exemplified below.

Example 2.1.1. Consider

$$Q = \{q_0, q_1, q_a\},$$
$$A = \{0, 1, B\}.$$

Define

$$\delta(q_0, 0) = (q_0, 0, R), \qquad \delta(q_1, 0) = (q_a, 1, R),$$
$$\delta(q_0, 1) = (q_0, 1, R), \qquad \delta(q_1, 1) = (q_1, 0, L),$$
$$\delta(q_0, B) = (q_1, B, L), \qquad \delta(q_1, B) = (q_0, 1, L).$$

Then the above defined TM computes the increment of 1, i.e., it takes a string of 0 and 1's (an integer in binary form) and increases it by 1, e.g.,

$$1011 \to 1100.$$

It functions as follows. On the tape, initially the entries are:

	1	2	3	4	5		
$\cdots$	B	1	0	1	1	B	B	$\cdots$

Start from the state q_0, it first encounters 1, as

$$\delta(q_0, 1) = (q_0, 1, R),$$

so the TM remains in state q_0, writes 1 in cell 1, and moves right. At cell 2, as

$$\delta(q_0, 0) = (q_0, 0, R),$$

it keeps moving right with no change in the state and no modification on the tape. It does the same at cell 3 and cell 4. At the 5th cell, because

$$\delta(q_0, B) = (q_1, B, L),$$

the TM sees the end of the input, changes its state to q_1, and returns to cell 4. Now, because

$$\delta(q_1, 1) = (q_1, 0, L),$$

the TM rewrites cell 4 to 0, moves left to cell 3. For the same reason, it rewrites cell 3 to 0, and moves left to cell 2. Since

$$\delta(q_1, 0) = (q_a, 1, R),$$

the TM rewrites cell 2 to 1, enters the accepting state q_a, and finishes the computation. The above steps are shown sequentially as follows (where arrow indicates the tape head):

B 1 0 1 1 B B ···	Step 1: $\delta(q_0,1) = (q_0,1,R)$
B 1 0 1 1 B B ···	Step 2: $\delta(q_0,0) = (q_0,0,R)$
B 1 0 1 1 B B ···	Step 3: $\delta(q_0,1) = (q_0,1,R)$
B 1 0 1 1 B B ···	Step 4: $\delta(q_0,1) = (q_0,1,R)$
B 1 0 1 1 B B ···	Step 5: $\delta(q_0,B) = (q_1,0,L)$
B 1 0 1 0 B B ···	Step 6: $\delta(q_1,1) = (q_1,0,L)$
B 1 0 0 0 B B ···	Step 7: $\delta(q_1,1) = (q_1,0,L)$
B 1 1 0 0 B B ···	Step 8: $\delta(q_1,0) = (q_a,1,R)$

What is written on the tape is:

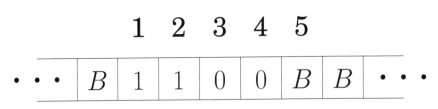

and the outcome is the result of increasing 1011 by 1.   □

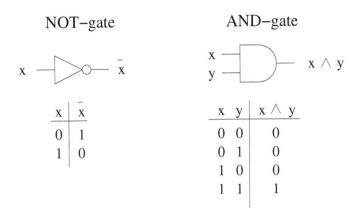

Figure 2.2: The NOT-gate and the AND-gate, and their truth tables. Note that the NOT-gate is reversible but the AND-gate is irreversible.

The execution of the transition function δ and the computation on a TM is done by logic gates. Such gates are based on binary logic. For example, we have the NOT-gate and AND-gate, as shown in Fig. 2.2

In Table 2.1, we provide a list of classical logic gates and notation for performing Boolean operations.

A gate operation is said to be *universal* if all the other logic operations can be expressed by this gate. The NAND-gate $\uparrow$ in Table 2.1 is universal, e.g., as we can carry out negation, $\wedge$ (and) and $\vee$ (or) by NAND:

$$\bar{x} = x \uparrow 1, \quad x \wedge y = (x \uparrow y) \uparrow 1, \quad x \vee y = \bar{x} \uparrow \bar{y} = (x \uparrow 1) \uparrow (y \uparrow 1).$$

The dissipation of heat generated by the computer has always been an important issue in the design of computers. As we have noted in Section 1.3 of Chap. 1, in the early 1960s, R. Landauer of IBM identified the (main) source of heat generation as due to the *irreversibility* of the computing process [36]. For example, the AND-gate "$\wedge$" is irreversible because if $x \wedge y = z$, then we cannot determine x and y from the value of z. This can be easily

AND: $\wedge$	$x \wedge y = x \cdot y$
OR: $\vee$	$x \vee y = x + y - x \cdot y$
NOT: $\neg$	$\neg x = 1 - x, \quad \bar{x} = 1 - x$
COPY	COPY $x = xx$
NAND: $\uparrow$ (NOT AND)	$x \uparrow y = \neg(x \wedge y) = (\neg x) \vee (\neg y)$
NOR: $\downarrow$ (NOT OR)	$x \downarrow y = \neg(x \vee y) = (\neg x) \wedge (\neg y)$

Table 2.1: Classical gates for Boolean operations, where $x, y \in \{0, 1\}$.

seen from the truth table for "$\wedge$" given in Fig. 2.2. For example, $0 \wedge 0 = 0$ and $1 \wedge 0 = 0$. So from $z = 0$ in $x \wedge y = z$, we cannot determine whether $(x, y) = (0, 0)$ or $(x, y) = (1, 0)$. Similarly, the NAND-gate is irreversible, albeit it is universal as mentioned in the preceding paragraph. *If all the computing gates are reversible, then heat generation by the computer can be dramatically curtailed.* (Ideally, such heat generation can be reduced to 0.)

Even though the 1-bit, $\wedge$, is irreversible AND-gate, an "equivalent" reversible 3-bit AND-gate, called the Toffoli gate, can be constructed, as shown in Fig. 2.3.

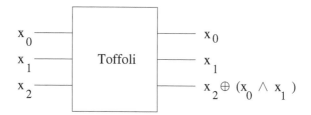

Figure 2.3: The reversible 3-bit Toffoli AND-gate. Here $\oplus$ means the usual addition, but modulo 2. By choosing $x_2 = 0$, the output on the lowest wire gives us $x_0 \wedge x_1 (= x_0 x_1)$.

In order to have a gate operation that is both reversible and universal, the number of bits involved must be at least 3. A famous theorem due to Fredkin and Toffoli [20] states the following:

> "There exists a 3-bit "universal gate" for reversible computation, i.e., a gate which, when applied in succession to different triplets of bits in a gate array, could be used to simulate any arbitrary reversible computation."

This universal, reversible gate is exactly the 3-bit Toffoli AND-gate as shown in Fig. 2.3.

Example 2.1.2. The Toffoli gate $T(x_0, x_1, x_2)(x_0, x_1, x_2 \oplus (x_0 \wedge x_1))$ as defined in Fig. 2.3 satisfies the property that $T(T(x_0, x_1, x_2)) = T^2(x_0, x_1, x_2) = (x_0, x_1, x_2)$. So T has itself as the inverse and T is reversible. □

For the classical gates and circuits, the following operations are also allowed:

(i) fan-out: (multi-copy, i.e., the output of gate G is wired

directly to several other gates downstream without requiring any additional interfacing circuitry);

$$(2.1)$$

(ii) fan-in: (the inputs from several gates feed directly into

gate G, the reverse of (i) above);   (2.2)

(iii) setting the input to a constant.   (2.3)

All the quantum computing gates to be studied in this book will soon be known to be reversible, as quantum operations are unitary transformations and thus are reversible. A *quantum Toffoli gate* will be given in Fig. 2.5. It can be represented by an 8×8 unitary matrix. Since the classical Toffoli gate is universal, we see that all the classical logic circuits can be simulated by quantum circuits.

We will also see that the three operations (i)–(iii) above in general are not allowable in quantum computing due to the No-Cloning Theorem and the reversibility of quantum computing. See Remark 2.9.1 subsequently.

2.2 Quantum mechanical systems; basics of atoms and molecules

We assume that the reader has rudimentary knowledge of modern physics and multivariate calculus. Quantum mechanics was developed in the 20th Century by the work of great physicists Planck, Einstein, Heisenberg, Schrödinger and Dirac. But here we begin our discussion by first introducing a quantum register and its probabilistic nature, and then presenting the Schrödinger equation as a fundamental postulate of quantum mechanics.

2.2.1 Quantum probability

We start with a sketch of the fundamental features of quantum computers and quantum systems in general, comparing them to usual (or "classical") computers. We restrict ourselves to a quantum device where the "data register" is "quantum". The program remains "classical" and is specified by a list of operations that are performed on the quantum register.

The basic difference between a classical and a quantum register is that the quantum one can "contain" many different numbers "simultaneously". This strange property will be exemplified below. It has the important consequence that the quantum computer can perform calculations in parallel, by performing them "simultaneously", in a single run, for many different input numbers. This is at the heart of the potential speed-up brought about by quantum computing. There is also a drawback: in many cases, the result

of a single run cannot be predicted with certainty.[1] This is another generic
feature of a quantum-mechanical process. Relative abundances of different
results can be found by performing several runs.

So how is a quantum register described? Assume that the register entries are characterized by binary numbers $x = x_0 x_1 \ldots x_{N-1}$ that characterize the register "states" of the classical Turing machine (Def. 2.1.1). In the
classical case, all possible values of x form the set Q of possible states of the
register. A quantum register with the same number of bits, on the other
hand, has much more possible "states".[2] The symbol $|\psi\rangle$ is conventionally
used for one of these states, and one has:

$$|\psi\rangle \;=\; \sum_x c_{x \in Q} |x\rangle, \tag{2.4}$$

$$\text{with } c_x \in \mathbb{C}, \qquad \sum_{x \in Q} |c_x|^2 = 1, \tag{2.5}$$

where the complex numbers c_x are called "probability amplitudes". A simple example for a quantum state of the register is the "classical state"

$$|\psi\rangle = |x\rangle, \tag{2.6}$$

where all but one probability amplitudes are nonzero. But states where
more than one classical entries are involved are possible as well, for example

$$|\psi\rangle = \frac{1}{\sqrt{2}} \left(|00 \ldots 0\rangle + e^{i\varphi} |11 \ldots 1\rangle \right) \tag{2.7}$$

with an arbitrary phase $\varphi \in [0, 2\pi)$. This type of state is called a "superposition state". Note that the addition in (2.7) differs conceptually from the
binary addition $00 \ldots 0 \oplus 11 \ldots 1$. We are dealing with the structure of a
complex vector space where each of the classical states corresponds to one
basis vector. The mathematical background is explained in more detail in
Section 2.4.

The "classical states" are important for the measurement of the quantum register. In fact, when a measurement is performed, the outcome is
one of the binary numbers $x = x_0 x_1 \ldots x_{N-1}$. If, before the measurement,
the register is in the state given by Eq. (2.4), one gets the outcome x with
a probability $p_x = |c_x|^2$. The normalization of the probability amplitudes
shown in Eq. (2.5) is essential to comply with this probabilistic interpretation.

The states of the form of Eq. (2.6) are special in the sense that the measurement outcome can be predicted with certainty to be x. This is not true
for superposition states: different outcomes are possible, and one cannot

---

[1]There are quantum algorithms that avoid this intrinsic uncertainty by producing a final
state of the quantum register that "contains" a single number. See examples below.

[2]From now on, we use the word "state" in the sense usually employed in quantum mechanics.

predict with certainty which one will be measured. For the state of Eq. (2.7), one gets the results $x = 00 \ldots 0$ and $x = 11 \ldots 1$ with probability $1/2$, respectively. So even if the quantum computer performs the same sequence of operations, starting from the same initial state, the "result" (outcome of the measurement) cannot be predicted with certainty. Only probabilistic statements are possible. This strange behavior offended, for example, A. Einstein until the end of his life, see Section 1.2 of Chap. 1, but it has been reproduced many times by experiments.

In the language of quantum mechanics, the state $|\psi\rangle$ is called a "state vector" or "abstract state vector". The notation $|\,\rangle$ is sometimes used for a typical state vector as well. One also uses $|\text{some string}\rangle$ where "some string" is a shorthand to describe the state at hand. (This avoids lengthy subscripts like $|\psi_{\text{some string}}\rangle$.) The vector of probability amplitudes $(c_x, x \in Q)$ in Eq. (2.4) is called the "representation of the state vector in the basis $\{|x\rangle, x \in Q\}$". Note that for other physical systems, other basis vectors are used that may form, for example, a continuous rather than discrete set. Examples are given below (Subsection 2.2.2) for the constituents of atoms and ions (electrons and atomic nuclei).

What happens to a state vector during quantum computation? One way to specify this is to go back to a physical implementation of the quantum register and to track its evolution in time. This is specified by the following differential equation, the Schrödinger equation:

$$i\hbar \frac{\partial}{\partial t}|\psi(t)\rangle = H(t)|\psi(t)\rangle, \qquad (2.8)$$

where $\hbar$ is the reduced Planck constant and the state vector $|\psi(t)\rangle$ is now time-dependent. It specifies the state of the quantum register at time t. The object $H(t)$ is called the "Hamiltonian operator" or in short "Hamiltonian": it is a linear mapping on the vector space of quantum states.[3] The time-dependence of $H(t)$ specifies the sequence of operations that are performed on the quantum register.[4]

The solution to the Schrödinger equation at time t, given an initial state $|\psi(0)\rangle$ at $t = 0$, is the result of a linear mapping in the state space,

$$|\psi(0)\rangle \mapsto |\psi(t)\rangle = U(t)|\psi(0)\rangle, \qquad (2.9)$$

where $U(t)$ is called the "time evolution operator". With respect to the basis of the register states $\{|x\rangle, x \in Q\}$, it is represented by matrix elements $U_{xy}(t)$, i.e.,

$$U(t)|x\rangle = \sum_{y \in Q} U_{xy}(t)|y\rangle. \qquad (2.10)$$

---

[3]The Hamiltonian has certain mathematical properties that ensure that the probability amplitudes remain normalized at all times, i.e., $\sum_{x \in Q} |c_x(t)|^2 = 1$ for all t.

[4]The Schrödinger equation (2.8) applies, provided no other physical systems couple to the register, apart from the operations specified by $H(t)$. The more general case of a register in contact with other systems or of not completely controlled operations leads to errors, as mentioned in Box 2.6. A generalized formalism is discussed in Chap. 4 on "Imperfect Quantum Operations".

The matrix $U_{xy}(t)$ must be unitary to preserve the normalization of the probability amplitudes and hence to be compatible with the probabilistic interpretation of the state vector.

2.2.2 Basics of atoms and molecules

The majority of quantum computer proposals utilize particular quantum mechanical properties, such as photon emission and absorption, electronic and nuclear spins, energy level transitions, etc., in atomic, molecular and optical systems. The physical law governing such phenomena is the Schrödinger equation[5]

$$i\hbar \frac{\partial}{\partial t}\psi = H\psi, \tag{2.11}$$

where ψ, the *wave function*, depends on both time $t \in \mathbb{R}$ and space $r = (x_1, x_2, \ldots, x_N) \in \mathbb{R}^N$; ψ is a complex-valued function, H in (2.11) is an operator, called the *Hamiltonian*, and $\hbar$ is the Planck constant. H consists of two parts,

$$H = K + V, \tag{2.12}$$

where

$$K = \text{the kinetic energy of the system,}$$
$$V = \text{the potential energy of the system.}$$

Both K and V are functions of (r, t) *classically.* As heuristics, we consider the motion of a harmonic oscillator which, classically, describes a 1-dimensional vibrating spring with an attached mass m and spring constant k:

$$K = \frac{1}{2m}p^2, \quad p = \text{ the linear momentum} = mv = m\dot{x},$$
$$V = \frac{1}{2}kx^2,$$

where x denotes the displacement of the mass from the equilibrium, and $\dot{x} = dx/dt$. The Hamiltonian principle gives us the equations of motion

$$\dot{x} = \partial H/\partial p = p/m, \quad \dot{p} = -\partial H/\partial x = -kx.$$

Thus,

$$\ddot{x} = \dot{p}/m = -kx/m,$$
$$m\ddot{x} + kx = 0,$$

---

[5]More precisely, this is the Schrödinger equation for the *spatial* wave functions, it is one of the many Schrödinger equations that are equivalent to each other. Another important component of the wave function of an elementary particle with *spin* is the wave function for the spin, which is not included in the model of (2.11) and (2.12) but may interact with the spatial wave function. See (2.29) later in this section. In Chap. 7 (especially Subsection 7.3.4), one can see examples of such a Schrödinger equation for an electron spin.

which is the equation of motion of the harmonic oscillator in Newtonian mechanics.

In quantum mechanics, K and V and, thus, H, are regarded as operators (for the spatial wave function) in a proper Hilbert space[6] through *quantization*:

$$p \mapsto \frac{\hbar}{i}\frac{d}{dx}; \quad x \mapsto x \cdot \quad \text{(i.e., } x \cdot (\psi) = x\psi, \text{ the multiplication operator)},$$

$$(2.13)$$

$$K = \frac{1}{2m}p^2 \longrightarrow -\frac{\hbar^2}{2m}\frac{d^2}{dx^2},$$
$$V = \frac{1}{2}kx^2 \longrightarrow \frac{1}{2}kx^2 \cdot,$$

and, thus, (2.11) becomes a time-dependent partial differential equation (PDE)

$$i\hbar\frac{\partial}{\partial t}\psi(x,t) = -\frac{\hbar^2}{2m}\frac{\partial^2}{\partial x^2}\psi(x,t) + \frac{1}{2k}x^2\psi(x,t); \quad x,t \in \mathbb{R}. \quad (2.14)$$

The proper abstract functional space setting for a rigorous mathematical investigation of the equation (2.14) is

$$L^2(\mathbb{R}) = \left\{ f \colon \mathbb{R} \to \mathbb{C} \mid \int_{-\infty}^{\infty} |f(x)|^2 dx < \infty \right\},^7 \quad (2.15)$$

called the space of square Lebesgue-integrable functions on $\mathbb{R}$. A function $f \in L^2(\mathbb{R})$, after normalization to norm 1, i.e.,

$$\hat{f} = f \Big/ \left[\int_{-\infty}^{\infty} |f(x)|^2 dx\right]^{1/2}, \quad \|\hat{f}\| \equiv \left[\int_{-\infty}^{\infty} |\hat{f}(x)|^2 dx\right]^{1/2} = 1$$

becomes a so-called *bound (pure) state* in quantum mechanics. We will return to the quantum-mechanical harmonic oscillator Eq. (2.14) in Section 3.2 of Chap. 3.

We may now proceed to consider a concrete system involving a few particles of electrons and nuclei in atomic and molecular physics. The Schrödinger equation modeling such a system provides satisfactory explanations of their chemical, electromagnetic and spectroscopic properties. Assume that the system under consideration is non-relativistic, and has N_1 nuclei with masses M_K and charges e^{Z_K}, e being the electron charge, for $K =$

---

[6]We defer until Def. 2.3.1 later in this section to introduce what a Hilbert space is and the relevant notation.

[7]The integral in (2.15) is the *Lebesgue* integral. But for all practical purposes, the reader may view it as the usual Riemann integral in calculus. Both the Riemann and the Lebesgue integrals obey the same linearity rules. They differ only in sequential convergence properties of "completeness".

$1, 2, \ldots, N_1$, and $N_2 = \sum_{K=1}^{N_1} Z_K$ is the number of electrons. The position vector of the Kth nucleus will be denoted as $\boldsymbol{R}_K$, while that of the kth electron will be $\boldsymbol{r}_k$, for $k = 1, 2, \ldots, N_2$. Let m be the mass of the electron. The Schrödinger equation for the overall system is given by

$$
i\hbar \frac{\partial}{\partial t} \Psi(\boldsymbol{R}, \boldsymbol{r}, t) = H\Psi(\boldsymbol{R}, \boldsymbol{r}, t) = \left[-\sum_{K=1}^{N_1} \frac{\hbar^2}{2M_K} \nabla_K^2 - \sum_{k=1}^{N_2} \frac{\hbar^2}{2m} \nabla_k^2 \right.
$$

$$
-\sum_{K=1}^{N_1} \sum_{k=1}^{N_2} \frac{Z_K e^2}{|\boldsymbol{R}_K - \boldsymbol{r}_k|} + \frac{1}{2} \sum_{\substack{k \neq k' \\ k, k' = 1}}^{N_2} \frac{e^2}{|\boldsymbol{r}_k - \boldsymbol{r}_{k'}|}
$$

$$
\left. + \frac{1}{2} \sum_{\substack{K \neq K' \\ K, K' = 1}}^{N_1} \frac{Z_K Z_{K'} e^2}{|\boldsymbol{R}_K - \boldsymbol{R}_{K'}|} \right] \Psi(\boldsymbol{R}, \boldsymbol{r}, t), \tag{2.16}
$$

where H is the Hamiltonian partial differential operator (for the spatial wave function) and

$$
R = (\boldsymbol{R}_1, \boldsymbol{R}_2, \ldots, \boldsymbol{R}_{N_1}), \quad r = (\boldsymbol{r}_1, \boldsymbol{r}_2, \ldots, \boldsymbol{r}_{N_2}), \quad \boldsymbol{r}_j = (x_j, y_j, z_j) \in \mathbb{R}^3.
$$

The above PDE has a total of $3(N_1 + N_2) + 1$ dependent variables and is too complex for practical purposes of studying atomic or molecular problems. Born and Oppenheimer [7] provide a reduced order model by approximation, permitting a particularly accurate decoupling of the motions of the electrons and the nuclei. The main idea is to assume that Ψ in (2.16) takes the separable form of a product

$$
\Psi(\boldsymbol{R}, \boldsymbol{r}) = G(\boldsymbol{R}) F(\boldsymbol{R}, \boldsymbol{r}),
$$

which, after assuming that the electronic motions have much faster time scale than those of nuclear motions, leads to

$$
i\hbar \frac{\partial}{\partial t} F(\boldsymbol{R}, \boldsymbol{r}, t) = \left(-\frac{\hbar^2}{2m} \sum_{k=1}^{N_2} \nabla_k^2 + \frac{1}{2} \sum_{\substack{k \neq k' \\ k, k' = 1}}^{N_2} \frac{e^2}{|\boldsymbol{r}_k - \boldsymbol{r}_{k'}|} - \sum_{K=1}^{N_1} \sum_{k=1}^{N_2} \frac{Z_K e^2}{|\boldsymbol{R}_K - \boldsymbol{r}_k|} \right.
$$

$$
\left. + \frac{1}{2} \sum_{\substack{K \neq K' \\ K, K' = 1}}^{N_1} \frac{Z_K Z_{K'} e^2}{|\boldsymbol{R}_K - \boldsymbol{R}_{K'}|} \right) F(\boldsymbol{R}, \boldsymbol{r}, t). \tag{2.17}
$$

In "typical" molecules, the time scale for the valence electrons to orbit about the nuclei is about once every 10^{-15} second (and that of the inner-shell electrons is even smaller, about attoseconds, i.e., 10^{-18} second), that of the bonds vibration is about once every 10^{-14} second, and that of the molecule rotation is every 10^{-12} second. This difference of time scale makes the Born–Oppenheimer approximation valid, as the electrons move so fast that they can instantaneously adjust their motions with respect to the vibration and rotation movements of the slower and much heavier nuclei.

The Born–Oppenheimer approximation breaks down in several cases, chief among them when the nuclear motion is strongly coupled to electronic motions, e.g., when the Jahn–Teller effects [28, 29] are present. It also requires corrections for loosely held electrons such as Rydberg atoms.

It will be convenient here for us to use the *atomic units*, where the unit of length is the radius of "the first Bohr orbit" $a = \hbar^2(me^2)$, and the unit of time is $t = \hbar^3/(me^4)$, which is the time for the electron to travel one radian in the first Bohr orbit. (Here what we mean by the first Bohr orbit is the quantum state corresponding to the principal quantum number $n = 1$ in Example 2.2.1 to follow.) Considering only steady-states for (2.17):

$$F(\boldsymbol{R}, \boldsymbol{r}, t) = F(\boldsymbol{r}, t) = e^{-iEt/\hbar}\psi(\boldsymbol{r}), \quad \boldsymbol{r} = (\boldsymbol{r}_1, \boldsymbol{r}_2, \ldots, \boldsymbol{r}_{N_2}),$$
$$\boldsymbol{r}_j = (x_j, y_j, z_j), \quad 1 \le j \le N_2,$$

we obtain the time-independent form of the Born–Oppenheimer approximation

$$H\psi = E\psi, \quad H = -\frac{1}{2}\sum_{j=1}^{N_2}\nabla_j^2 + \sum_{1 \le j < k \le N_e}\frac{1}{r_{jk}} - \sum_{j=1}^{N_2}\sum_{k=1}^{N_1}\frac{Z_k}{|\boldsymbol{r}_j - \boldsymbol{R}_k|}$$
$$+ \sum_{1 \le j < k \le N_1}\frac{Z_j Z_k}{|\boldsymbol{R}_j - \boldsymbol{R}_k|}, \tag{2.18}$$

where

$$\psi = \psi(\boldsymbol{r}_1, \boldsymbol{r}_2, \ldots, \boldsymbol{r}_{N_2}), \quad \nabla_j^2 \equiv \frac{\partial^2}{\partial x_j^2} + \frac{\partial^2}{\partial y_j^2} + \frac{\partial^2}{\partial z_j^2},$$
$$r_{jk} \equiv |\boldsymbol{r}_j - \boldsymbol{r}_k|, \quad j, k = 1, 2, \ldots, N_2.$$

Example 2.2.1. We use the simplest atom, the hydrogen H, as an example to illustrate a quantum mechanical model and the idea of *quantum numbers*. In the Born–Oppenheimer model (2.18), by letting, $N_2 = N_1 = 1$, with $Z_1 = 1$ and $\boldsymbol{R}_1 = 0$, Eq. (2.18) becomes

$$\left(-\frac{1}{2}\nabla^2 - \frac{1}{r}\right)\psi(\boldsymbol{r}) = E\psi(\boldsymbol{r}), \quad \boldsymbol{r} \in \mathbb{R}^3, \tag{2.19}$$

the Born–Oppenheimer separation of the hydrogen atom.

We write (2.19) in spherical coordinates in view of the symmetry involved:

$$\frac{1}{r}\frac{\partial^2}{\partial r^2}(r\psi) + \frac{1}{r^2}\Lambda^2\psi + \frac{2}{r}\psi - 2E\psi = 0, \tag{2.20}$$

where

$$\Lambda^2 \equiv \frac{1}{\sin^2\theta}\frac{\partial^2}{\partial\phi^2} + \frac{1}{\sin\theta}\frac{\partial}{\partial\theta}\left(\sin\theta\frac{\partial}{\partial\theta}\right),$$

(the angular momentum operator, or, the Legendrian); (2.21)

r = the radial variable, $0 < r < \infty$;

θ = the colatitude or zenith, or the polar angle, $0 \le \theta \le \pi$;

ϕ = the azimuth or longitudinal angle, $0 \le \phi \le 2\pi$.

Eq. (2.19) has separable solutions

$$\psi(\boldsymbol{r}) = \psi(r,\theta,\phi) = R(r)Y(\theta,\phi). \tag{2.22}$$

The angular variables are quantized first as we know that angular functions are the spherical harmonics

$$Y(\theta,\phi) = Y_{\ell m}(\theta,\phi) = \Theta_{\ell m}(\theta)\Phi_m(\phi), \qquad \ell = 0,1,2,\ldots,\ m = \ell, \ell-1,\ldots,-\ell, \tag{2.23}$$

on the unit sphere $\mathcal{S}_2 \equiv \{\boldsymbol{r} \in \mathbb{R}^3 \mid |\boldsymbol{r}| = 1\}$, satisfying

$$\Lambda^2 Y_{\ell m}(\theta,\phi) = -\ell(\ell+1)Y_{\ell m}(\theta,\phi), \tag{2.24}$$

where in (2.23),

$$\Phi_m(\phi) = (2\pi)^{-\frac{1}{2}}e^{im\phi}, \tag{2.25}$$

$$\Theta_{\ell m}(\theta) = \left\{\frac{(2\ell+1)(\ell-|m|)!}{2(\ell+|m|)!}\right\}^{\frac{1}{2}} P_\ell^{|m|}(\cos\theta), \tag{2.26}$$

($P_\ell^{|m|}$ is the associated Legendre function).

Using (2.22)–(2.24) in (2.19), we obtain the equation for the radial function

$$\frac{1}{r}(rR)'' - \frac{\ell(\ell+1)}{r^2}R + \left(\frac{2}{r} - 2E\right)R = 0. \tag{2.27}$$

Solutions to the eigenvalue problem (2.27) that are square integrable over $0 < r < \infty$ are known to be

$$R(r) = R_{n\ell}(r) = (-2)\left\{\frac{(n-\ell-1)!}{2n[(n+\ell)!]^3}\right\}(2r)^\ell L_{n+\ell}^{2\ell+1}(2r)e^{-r};$$

$$\ell = 0,1,2,\ldots,n-1, \tag{2.28}$$

$$E_n = -\frac{1}{2}\frac{1}{n^2}, \quad n = 1,2,\ldots, \text{ independent of } \ell,$$

where $L_{n+\ell}^{(2\ell+1)}$ are the associated Laguerre polynomials such that for $m,n = 0,1,2,\ldots$

$$xL_{m+n}^{(m)}{}'' + (m+1-x)L_{m+n}^{(m)}{}' + (m+n)L_{m+n}^{(m)} = 0,$$

$$L_{m+n}^{(m)}(x) = \frac{e^x x^{-(m+n)}}{(m+n)!}\frac{d^{m+n}}{dx^{m+n}}(e^{-x}x^{2m+n}),$$

(when $m = 0$, $L_n^{(0)}(x)$ is simply denoted as $L_n(x)$).

The electron state of the hydrogen can thus be characterized by three "quantum numbers" n, ℓ, m:

$$\left\{ \begin{array}{l} |n\ell m\rangle: \ R_{n\ell}(r)\Theta_{\ell m}(\theta)\Phi_m(\phi) \text{ is its spatial wave function, with a} \\ \text{proper normalization factor, } n = 1, 2, \ldots; \ell = 0, 1, 2, \ldots, n-1; \\ m = -\ell, -\ell+1, \ldots, 0, 1, \ldots, \ell, \text{ with energy level } E = E_n = -\dfrac{1}{2}\dfrac{1}{n^2}, \\ n = 1, 2, \ldots, \text{ independent of } \ell \text{ and } m. \end{array} \right.$$

The first quantum number n is called the *principal quantum number*. This n sometimes is referred to as the *shell number* such that all states with the same value of n make up the nth Bohr shell. For each given $n = 1, 2, \ldots$, there can be at most $2n^2$ electron "orbitals." The principal quantum number represents physically the radial "range" that the region extends from the center of the atom.

The second quantum number ℓ is called the *subsidiary* (or angular-momentum) *quantum number*. This value $\ell = 0, 1, 2, \ldots, n-1$, is often associated with the "shape" of the orbital. For example,

(i) $\ell = 0$ implies spherically shaped, denoted also as an "s" (sharp) orbital. There can be a maximum of 2 electrons with the assignment of s.

(ii) $\ell = 1$ implies double loped, denoted also as a "p" (principal) orbital. There can be a maximum of 6 electrons with the assignment of p.

(iii) $\ell = 2$ implies quadra-lobed, denoted also as a "d" (diffuse) orbital. There can be a maximum of 10 electrons with the assignment of d.

(iv) $\ell = 3$ implies octa-lobed, denoted also as an "f" (fundamental) orbital. There can be a maximum of 14 electrons with the assignment of f. The rest of the orbital assignments corresponding to $\ell = 4, 5, \ldots$, goes on by

$$g, h, i, j, k, \ldots .$$

The shapes of s, p and d orbitals are shown in Fig. 2.4.

The third quantum number m is called the *magnetic quantum number*. There can be a maximum of $2\ell + 1$ electrons with the same quantum numbers ℓ and n, with a spatial dependence like $e^{im\phi}$. This quantum number is associated with the *Zeeman effect*: when a magnetic field is applied, spectral lines from degenerate states of electrons with the same n, ℓ but different m split into several lines.

The *spin quantum number* is a fourth quantum number. The spin quantum number for an electron is either $+1/2$ (spin up) or $-1/2$ (spin down). The spin wave function can be coupled to the spatial wave function to constitute a multi-component (here, 2-component) system satisfying a Schrödinger equation whose Hamiltonian operator is a matrix differential operator. An introduction of spin is already given in Box. 1.1 of Chap. 1. □

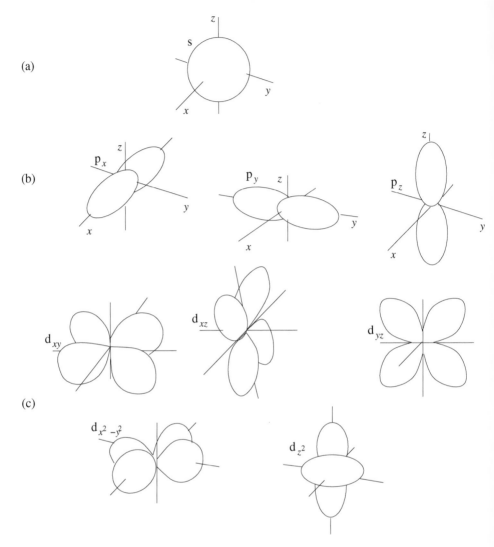

Figure 2.4: (Please see color insert following page 78.) Shapes of s, p and d orbitals; cf. (2, p. 76).
(a) s-orbital, spherical in shape;
(b) p-orbitals, with 2 lobes;
(c) d-orbitals, with 4 lobes.
Each of the orbitals displayed in (b) and (c) is obtained from superpositions with wave functions involving the azimuth phases $e^{+im\phi}$ and $e^{-im\phi}$. The s- and p-orbitals in (a)–(c) above are further shown sequentially in color in the first two color plates of the book. The d-orbitals in (c) as shown in the color plate there are in corresponding sequential order, except that for the last two d-orbitals $d_{x^2-y^2}$ and d_{z^2}, they are linear combinations of these above, here.
The color graphics are plotted using the Gaussian software (19, 21) with basis set 6-311++G(3df,2pd).

We now examine some further fundamentals of the atomic physics that are essential in any quantitative study of the manipulation of the quantum behavior of atoms. A complete description of the Hamiltonian (i.e., energy) of an atom contains 9 terms as follow [18]:

$$H = H_{el} + H_{CF} + H_{LS} + H_{SS} + H_{Ze} + H_{HF} + H_{Zn} + H_{II} + H_Q. \quad (2.29)$$

The first three terms have the highest order, called the atomic Hamiltonian. They are the *electronic Hamiltonian term*, *crystal field term* and the *spin-orbit interaction term*, respectively. The electronic Hamiltonian consists of kinetic energy of all electrons, $p_i^2/2m$, and two Coulomb terms: the potential energy of electrons relative to the nuclei, $-Z_n e^2/r_{ni}$, and the inter-electronic repulsion energy, e^2/r_{ij}:

$$H_{el} = \sum_i \frac{p_i^2}{2m} - \sum_{i>j} \frac{e^2}{r_{ij}} + \sum_{i,n} \frac{Z_n e^2}{r_{ni}}.$$

These three summations agree, respectively in sequential order, to the first three summations in (2.18).

The term H_{CF} is called the crystal field term and occurs when the atom is part of the arrangement of atoms, called a crystal. It comes from the interaction between the electron and the electronically charged ions forming the crystal, and is essentially a type of electrical potential energy:

$$V = -\sum_{i,j} \frac{Q_j e}{r_{ij}},$$

where Q_j is the ionic charge and r_{ij} is the distance from the electron to the ion. Normally, only those ions nearest to the electron are considered.

The third in the atomic Hamiltonian is the interaction between spin and orbit

$$H_{LS} = \lambda \vec{L} \cdot \vec{S}, \quad (2.30)$$

where $\vec{L}$ and $\vec{S}$ are the angular momenta of the orbit and spin, respectively, and λ is the coupling constant. In this section, we will use $\vec{S}$ for the electron spin and $\vec{I}$ for the nuclear spin.

The remaining six terms are called *spin Hamiltonians*. Terms H_{Ze} and H_{Zn} are two resulted from the application of an external magnetic field:

$$H_{Ze} = \beta \vec{B} \cdot (\vec{L} + \vec{S}),$$
$$H_{Zn} = -\sum_i g_{ni} \beta_n \vec{B} \cdot \vec{I}_i,$$

where $\vec{B}$ is the magnetic field strength. These two terms are called *Zeeman terms*, and they play big roles in NMR (nuclear magnetic resonance) and ESR (electron spin resonance).

The nuclear spin-spin interaction term H_{II} is also important in quantum computation,

$$H_{II} = \sum_{i>j} \vec{I}_i \cdot J_{ij} \cdot \vec{I}_j, \tag{2.31}$$

because it provides a mechanism for the interaction between qubits. Hyperfine interaction arises from the interaction between the nuclear magnetic moments and the electron,

$$H_{HF} = \vec{S} \cdot \sum_i A_i \cdot \vec{I}_i.$$

In (2.29), by letting the z-axis as the privileged direction of spin measurement, then the *spin-spin interaction term H_{SS}* is expressed as

$$H_{SS} = D \left[S_z^2 - \frac{1}{3} S(S+1) \right] + E(S_x^2 - S_y^2),$$

and the very last term in (2.29), called the *quadrupolar energy*, is given by

$$H_Q = \frac{e^2 Q}{4I(2I-1)} \left(\frac{\partial^2 V}{\partial Z^2} \right) (3I_z^2 - I(I+1) + \eta(I_x^2 - I_y^2)), \tag{2.32}$$

where V represents the electrostatic potential set up by the surrounding electron cloud and Z is the overall charge distribution.

For a specific system, only the Hamiltonian playing major roles is needed in the final model. For example, in the study of ESR, only three terms are retained and the Hamiltonian is written as

$$H = H_{Ze} + H_{HF} + H_{SS},$$

while in the NMR case

$$H = H_{Zn} + H_{II}.$$

2.3   Hilbert spaces

A rigorous mathematical study of linear evolution partial differential equations such as (2.12) was developed during the 1950s based on functional analysis and operator theory [10, 27, 60]; see (2.35) and (2.36). The partial differential operators "live" on a Hilbert space which is a space of functions.

Definition 2.3.1. Let $\mathcal{H}$ be a vector space distributed over $\mathbb{C}$. We say that $\mathcal{H}$ is a Hilbert space if $\mathcal{H}$ is equipped with an inner product $\langle \cdot | \cdot \rangle$ such that for each pair of vectors $x, y \in \mathcal{H}$, there is an associated complex number

$\langle x|y \rangle$ [8] satisfying

$$\langle y|x \rangle = \langle x|y \rangle^*, \text{ where "$*$" means complex conjugate,}$$
$$\langle x+y|z \rangle = \langle x|z \rangle + \langle y|z \rangle,$$
$$\langle \alpha x|y \rangle = \alpha^* \langle x|y \rangle,$$
$$\langle x|x \rangle \geq 0,$$
$$\langle x|x \rangle = 0 \quad \text{only if} \quad x = 0,$$

and if we define the norm $\|x\| \equiv [\langle x|x \rangle]^{1/2}$ of any $x \in \mathcal{H}$, the space $\mathcal{H}$ is complete with respect to the norm, i.e., for any (Cauchy) sequence $\{x_n | n = 1, 2, \ldots\}$ satisfying

$$\lim_{n,m \to \infty} \|x_n - x_m\| = 0,$$

this sequence has a limit $x_0 \in \mathcal{H}$,

$$\lim_{n \to \infty} x_n = x_0, \quad \text{i.e.,} \quad \lim_{n \to \infty} \|x_n - x_0\| = 0. \qquad \square$$

The space $L^2(\mathbb{R})$ we mentioned earlier is a Hilbert space, with the inner product defined by

$$\langle f|g \rangle = \int_{-\infty}^{\infty} f(x)^* g(x) dx;$$

while $C(\mathbb{R}, \mathbb{C})$ the space of all continuous functions mapping from $\mathbb{R}$ to $\mathbb{C}$ is not a Hilbert space because it cannot be equipped with an inner product that is complete.

Every Hilbert space arising from physical applications has a countable orthonormal basis (such Hilbert spaces are called *separable*), i.e., there exists a set $\{\phi_i | i = 1, 2, 3, \ldots\} \subseteq \mathcal{H}$ such that

$$\langle i|j \rangle = \delta_{ij} = \begin{cases} 1 & \text{if } i = j, \\ 0 & \text{if } i \neq j; \end{cases}$$

and we have an orthonormal expansion

$$| \ \rangle = \sum_{j=1}^{\infty} |j \rangle \langle j| \ \rangle.$$

---

[8]The notation for an inner product favored by most mathematicians is $\langle x, y \rangle$ instead of $\langle x|y \rangle$, the bra-ket notation due to Dirac. We may mention the following notational differences (and, yes, confusion!) between mathematicians and physicists. For two functions $\psi_\alpha(x)$ and $\psi_\beta(x)$, the L^2-inner product

$$\int \psi_\alpha^*(x) \psi_\beta(x) \, dx$$

is written by mathematicians as $\langle \psi_\alpha, \psi_\beta \rangle$, but by physicists as $\langle \alpha|\beta \rangle$. And physicists use $\psi_\alpha(x)$ to denote $\psi_\alpha(x) = \langle x|\alpha \rangle$ also. From now on, our notation will largely follow that of the physicist's.

Let $\mathcal{H}_1, \mathcal{H}_2$ be two Hilbert spaces and $A\colon \mathcal{H}_1 \to \mathcal{H}_2$ be a bounded linear operator, i.e., A satisfies

$$\|Af\|_2 \leq C\|f\|_1$$

for some constant $C \geq 0$, where $\|\ \|_1$ and $\|\ \|_2$ are, respectively, the norms on $\mathcal{H}_1$ and $\mathcal{H}_2$. In particular, when $\mathcal{H}_2 = \mathbb{C}$, A is said to be a *linear functional*. The vector space of all linear functionals endowed with the natural addition and scalar multiplication is a vector space denoted as $\mathcal{H}^*$, which is algebraically isomorphic to $\mathcal{H}$. (This is called the *Riesz Representation Theorem* in functional analysis.) Thus $\mathcal{H}^*$ is itself a Hilbert space, called the *dual space* of $\mathcal{H}$.

Let $\mathcal{H}$ be a *complex* Hilbert space with inner product $\langle\cdot|\cdot\rangle$. For any subset $S \subseteq \mathcal{H}$, define

$$\text{span } S = \text{closure of } \left\{ \sum_{j=1}^{n} \alpha_j \psi_j \mid \alpha_j \in \mathbb{C}, \psi_j \in S, j = 1, 2, \ldots, n, \text{ for all } \psi_j \in S \right\}$$

in $\mathcal{H}$. $\qquad\qquad$ (2.33)

In quantum mechanics, the Dirac bra-ket notation writes the vector ψ_k as $|k\rangle$, pronounced *ket* k, where k labels the vector. The role of ψ_j in the inner product $\langle j|k\rangle$ is that ψ_j is identified as a linear functional on $\mathcal{H}$ by

$$\psi_j^*\colon \mathcal{H} \longrightarrow \mathbb{C}; \quad \psi_j^*(\psi_k) \equiv \langle j|k\rangle.$$

The Dirac bra-ket notation writes ψ_j^* as $\langle j|$, pronounced *bra* j. The inner product is now written as

$$\langle \psi_j, \psi_k \rangle = \langle j|k\rangle.$$

Let $A\colon \mathcal{H} \to \mathcal{H}$ be a linear operator. Then in Dirac's notation we write $\langle \psi_j, A\psi_k \rangle$ as $\langle j|A|k\rangle$.

The set of all kets ($|k\rangle$) generates the linear space $\mathcal{H}$, and the set of all bras ($\langle j|$) also generates the dual linear space $\mathcal{H}^*$. These two linear spaces are called, respectively, the ket space and the bra space, which are dual to each other. The natural isomorphism from H onto H^*, or from H^* onto H, is called "taking the adjoint" by physicists. An expression like $\langle j|A|k\rangle$ has an intended symmetry: one can think of A acting on $|k\rangle$ on the right as a matrix times a column vector, or A acting on $\langle j|$ on the left as a row vector times a matrix.

The solution of the Schrödinger equation

$$i\hbar\frac{\partial}{\partial t}\psi(t) = H\psi(t), \quad \psi(0) = \psi_0 \in \mathcal{H} \text{ (initial condition)}, \qquad (2.34)$$

can now be written as

$$\psi(t) = e^{-iHt/\hbar}\psi(0), \qquad -\infty < t < \infty,$$

provided that H is independent of t. Note that $\{e^{-iHt/\hbar}| - \infty < t < \infty\}$ forms a *1-parameter group* of evolution operators on $\mathcal{H}$.

A linear operator A on $\mathcal{H}$ has a Hermitian conjugate (h.c.) $A^\dagger$ defined by the equation

$$\langle x|A|y\rangle = \langle y|A^\dagger|x\rangle^*, \text{ for all } x, y \in \mathcal{H};$$

(superscript $*$ means complex conjugate).

Operators A satisfying

$$A^\dagger A = AA^\dagger$$

are called *normal* operators. They represent physical properties. Their eigenvalues are the possible outcomes when the physical property is measured. Normal operators of particular significance are the Hermitian operators and the unitary operators. A linear operator A is said to be *Hermitian* if

$$A = A^\dagger$$

and *unitary* if

$$AA^\dagger = A^\dagger A = I, \text{ the identity operator on } H.$$

(We sometimes use $1_{\mathcal{H}}$ or just 1 to denote the identity operator I.) A Hermitian operator has real eigenvalues, while a unitary operator has unit eigenvalues which are phase factors. Unitary operators turn all kets and bras into equivalent ones and, therefore, they can describe geometrical operations such as rotations. Also, the description (of the quantum system) at any instant is structurally equivalent to that at any other time instant, so a unitary operator (or rather, a 1-parameter family of unitary operators) can describe the *temporal evolution* of the system.

The Hamiltonian H of a quantum system (2.34) is a Hermitian operator: $H = H^\dagger$. The real eigenvalues of H are the possible values for the energy of the quantum system. Assume that the Hamiltonian H in (2.34) is independent of time. Then as $H = H^\dagger$, for any $t \in \mathbb{R}$,

$$e^{-iHt/\hbar}(e^{-iHt/\hbar})^\dagger = e^{-iHt/\hbar}e^{iH^\dagger t/\hbar} = e^{-iHt/\hbar}e^{iHt/\hbar} = I,$$

consequently, $e^{-iHt/\hbar}$ is unitary for any $t \in \mathbb{R}$.

If H in (2.34) is time-dependent: $H = H(t)$, then for an *infinitesimal* time step, from t to $t + dt$, we have

$$I - \frac{i}{\hbar}H(t)dt$$

for the unitary evolution operator. For continuous evolution with respect to time t, the solution to (2.34) cannot be represented in terms of a 1-parameter evolutionary group. Rather, a 2-parameter group $\{U(t, s)| - \infty < t, s < \infty\}$ is needed:

$$\psi(t) = U(t, 0)\psi_0, \qquad -\infty < t < \infty, \qquad (2.35)$$

where

$$\left.\begin{array}{l} \dfrac{\partial}{\partial t}U(t,s) = H(t)U(t,s) \\[2ex] \dfrac{\partial}{\partial s}U(t,s) = -U(t,s)H(s) \end{array}\right\}, \qquad -\infty < t, s < \infty, \qquad (2.36)$$

and $U(t,s')U(s',s) = U(t,s)$ for $-\infty < t, s, s' < \infty$. Nevertheless, $U(t,s)$ remains unitary for any s, t. Thus, *all quantum operations are unitary*.

2.4   Complex finite dimensional Hilbert spaces and tensor products

Complex finite dimensional Hilbert spaces are particularly useful in quantum computing due to the special features of quantum computer bit representation and quantum device operations. We begin by considering a 2-dimensional complex Hilbert space $\mathcal{H}$ which is isomorphic to $\mathbb{C}^2$. The standard basis of $\mathbb{C}^2$ is

$$e_1 = \begin{bmatrix} 1 \\ 0 \end{bmatrix}, \quad e_2 = \begin{bmatrix} 0 \\ 1 \end{bmatrix}.$$

In quantum mechanics, the preferred notation for basis is, respectively, $|0\rangle$ and $|1\rangle$, where $|0\rangle$ and $|1\rangle$ generally refer to the "spin" state of a certain quantum system under discussion. "Spin up" and "spin down" form a binary quantum alternative and, therefore, a quantum bit or, a *qubit*. These spin states may then further be identified with e_1 and e_2 through, e.g.,

$$|0\rangle = \begin{bmatrix} 1 \\ 0 \end{bmatrix}, \quad |1\rangle = \begin{bmatrix} 0 \\ 1 \end{bmatrix}. \qquad (2.37)$$

For a column vector, say e_1, we use superscript T to denote its transpose, such as $e_1^T = [1 \ 0]$, for example. The spin states $|0\rangle$ and $|1\rangle$ form a standard orthonormal basis for the (single) qubit's ket space. The tensor product of a set of n vectors $|z_j\rangle \in \mathbb{C}_j^2$, specified by the quantum numbers z_j for $j = 1, 2, \ldots, n$, is written interchangeably as

$$|z_1\rangle \otimes |z_2\rangle \otimes \cdots \otimes |z_n\rangle = |z_1\rangle|z_2\rangle \cdots |z_n\rangle = |z_1 z_2 \cdots z_n\rangle. \qquad (2.38)$$

The tensor product space of n copies of $\mathbb{C}_2$ is defined to be

$$(\mathbb{C}_2)^{\otimes n} = \text{span}\{|j_1 j_2 \cdots j_n\rangle \mid j_k \in \{0,1\}, k = 1, 2, \ldots, n\}. \qquad (2.39)$$

It is a 2^n-dimensional complex Hilbert space with the induced inner product from $\mathbb{C}^2$. A *quantum state* is a vector in $(\mathbb{C}^2)^{\otimes n}$ (or in any Hilbert space $\mathcal{H}$) with the unit norm. (This type of quantum state is actually a more restrictive type, called a *pure state*. A pure state, say $|k\rangle$, could be identified with the projection operator $|k\rangle\langle k|$. Then a *general state* is a convex sum of such projectors, i.e., $\sum_k c_k |k\rangle\langle k|$ with $c_k \geq 0$ and $\sum_k c_k = 1$.) A quantum state is said to be *entangled* if it is *not* a tensor product of the form (2.19). A *quantum operation* is a *unitary linear transformation* on $\mathcal{H}$. The *unitary group* $U(\mathcal{H})$ on $\mathcal{H}$ consists of all unitary linear transformations on $\mathcal{H}$ with composition as the (noncommutative) natural multiplication operation of the group.

It is often useful to write the basis vectors $|j_1 j_2 \cdots j_n\rangle \in (\mathbb{C}^2)^{\otimes n}$, $j_k \in \{0,1\}$, $k = 1, 2, \ldots, n$, as column vectors according to the lexicographic ordering:

$$|00\cdots00\rangle = \begin{bmatrix} 1 \\ 0 \\ 0 \\ \vdots \\ 0 \\ 0 \\ 0 \end{bmatrix}, |00\cdots01\rangle = \begin{bmatrix} 0 \\ 1 \\ 0 \\ \vdots \\ 0 \\ 0 \\ 0 \end{bmatrix}, |00\cdots10\rangle = \begin{bmatrix} 0 \\ 0 \\ 1 \\ \vdots \\ 0 \\ 0 \\ 0 \end{bmatrix}, \cdots$$

$$|11\cdots10\rangle = \begin{bmatrix} 0 \\ 0 \\ 0 \\ \vdots \\ 0 \\ 1 \\ 0 \end{bmatrix}, |11\cdots11\rangle = \begin{bmatrix} 0 \\ 0 \\ 0 \\ \vdots \\ 0 \\ 0 \\ 1 \end{bmatrix}; \tag{2.40}$$

cf. (2.37).

2.5 Quantum Turing machines

In comparison with the definition of a classical Turing machine in Definition 2.1.1, we can now formulate a quantum Turing machine similar to a classical deterministic Turing machine. The main difference is that the transitions now allow for a unitary evolution of the state of the machine. A minor complication arises because it is desirable to represent the transition probabilities by computable numbers, as in the case of probabilistic Turing machines. Therefore, we define $\widetilde{C}$ to be the set of all complex numbers α such that there exists an algorithm that computes the real and imaginary parts of α to within 2^{-n} in time polynomial in n. The next definition is taken from [6].

Definition 2.5.1. A quantum Turing machine (QTM) M is given by a triplet (Σ, Q, δ), where Σ is a finite alphabet that contains a blank symbol B; Q is a finite set of states that contains an initial state q_0 and a final state $q_f \neq q_0$; and a function

$$\delta \colon Q \times \Sigma \longrightarrow \widetilde{C}^{\Sigma \times Q \times \{L, R\}}$$

that determines the transitions from one state of the Turing machine to the next. The TM has a 2-way infinite tape of cells indexed by the set of integers and a single read/write tape head that moves along the tape. The configurations, initial configurations, and final configurations are defined as in the case of deterministic Turing machines. Let S be a separable Hilbert space with an orthonormal basis labelled by the configurations of the Turing machine M.

The QTM M defines a linear operator $U_M \colon S \to S$ as follows. Suppose that M starts in the configuration c with current state p and scanned symbol σ; after one step, M will be in a superposition of configurations $\sum_i \alpha_i c_i$, where each nonzero α_i corresponds to a transition $\delta(p, \sigma)(\tau, q, d)$ and c_i is the new configuration that results from applying this transition to c. The operator U_M is then defined by linear extension. The QTM is said to be well-formed if the operator U_M is unitary. If a QTM M is measured, then the configuration c_i is observed with probability $|\alpha_i|^2$.

A final configuration of a QTM is any configuration in state q_f. If a QTM M is run with input x and at time T the superposition contains only final configurations, and at time less than T is contains no final configurations, then M is said to halt within running time T on input of x. $\qquad \square$

Programming a quantum Turing machines is a nontrivial matter, since the halting condition is quite difficult to guarantee. The paper by Bernstein and Vazirani [6] contains some examples and gives a few programming primitives. As such examples are quite lengthy, we omit them and refer any interested readers to the original reference [6]. Our subsequent discussions will instead be based on the quantum circuit model, which is equivalent to the quantum Turing machine model [66].

2.6   Universality of elementary quantum gates; quantum circuits

In a classical computer, a gate is a hardware device that executes a defined operation on a bit. In a quantum computer, a gate is a *quantum mechanical operation* on the qubit wave function specified by a Hamiltonian function. As we know by now, the Hamiltonian describes the energy of the quantum system as a function of coordinates such as position, momentum, angular momentum, and sometimes time. The application of defined external fields, which are applied according to a Hamiltonian for a specified time to the system, execute a particular computational operation on the qubit wave function. Since the fields are external to the qubit system, the energy of this system can be changed. Any 2×2 unitary matrix is an admissible quantum operation on one qubit. We call it a 1-bit gate. Similarly, we call a $2^k \times 2^k$ unitary transformation a k-bit gate. Let a 1-bit gate be

$$U = \begin{bmatrix} u_{00} & u_{01} \\ u_{10} & u_{11} \end{bmatrix};$$

we define the operator $\Lambda_m(U)$ [4] on $(m+1)$-qubits (with $m = 0, 1, 2, \ldots$) through its action on the basis by

$$\Lambda_m(U)(|x_1 x_2 \cdots x_m y\rangle) = \begin{cases} |x_1 x_2 \cdots x_m y\rangle & \text{if } \wedge_{k=1}^m x_k = 0, \\ u_{0y}|x_1 x_2 \cdots x_m 0\rangle + u_{1y}|x_1 x_2 \cdots x_m 1\rangle & \text{if } \wedge_{k=1}^m x_k = 1, \end{cases}$$

where "$\wedge$" denotes the Boolean operator AND. The matrix representation for $\Lambda_m(U)$ according to the ordered basis (2.40) is

$$\Lambda_m(U) = \begin{bmatrix} 1 & & & & \\ & \ddots & & \bigcirc & \\ & & 1 & & \\ & \bigcirc & & \begin{bmatrix} u_{00} & u_{01} \\ u_{10} & u_{11} \end{bmatrix} \end{bmatrix}_{2^{m+1} \times 2^{m+1}} . \tag{2.41}$$

This unitary operator is called the m-bit-controlled U operation.

An important logic operation on one qubit is the NOT-gate

$$\sigma_x = \begin{bmatrix} 0 & 1 \\ 1 & 0 \end{bmatrix} ; \tag{2.42}$$

which is one of the Pauli spin matrices. (The two other important Pauli matrices are

$$\sigma_y = \begin{bmatrix} 0 & -i \\ i & 0 \end{bmatrix}, \quad \sigma_z = \begin{bmatrix} 1 & 0 \\ 0 & -1 \end{bmatrix} .) \tag{2.43}$$

These Pauli matrices σ_x, σ_y and σ_z are often also denoted as X, Y and Z, respectively, as preferred by many researchers, in the rest of the book without further mention. From σ_x, define the important 2-bit operation $\Lambda_1(\sigma_x)$, which is the controlled-not gate, henceforth acronymed the CNOT gate. Its matrix representation is

$$\text{CNOT} = \begin{bmatrix} 1 & 0 & 0 & 0 \\ 0 & 1 & 0 & 0 \\ 0 & 0 & 0 & 1 \\ 0 & 0 & 1 & 0 \end{bmatrix} . \tag{2.44}$$

For any 1-bit gate A, denote by $A(j)$ the operation (defined through its action on the basis) on the slot j:

$$A(j)|x_1 x_2 \cdots x_j \cdots x_n\rangle = |x_1\rangle \otimes |x_2\rangle \otimes \cdots \otimes |x_{j-1}\rangle \otimes [A|x_j\rangle]$$
$$\otimes |x_{j+1}\rangle \otimes \cdots \otimes |x_n\rangle.$$

Similarly, for a 2-bit gate B, we define $B(j, k)$ to be the operation on the two qubit slots j and k [8]. A simple 2-bit gate is the swapping gate

$$U_{\text{sw}}|x_1 x_2\rangle = |x_2 x_1\rangle, \text{ for } x_1, x_2 \in \{0, 1\}. \tag{2.45}$$

Then $U_{\text{sw}}(j, k)$ swaps the qubits between the j-th and the k-th slots.

The tensor product of two 1-bit quantum gates S and T is defined through

$$(S \otimes T)(|x\rangle \otimes |y\rangle) = (S|x\rangle) \otimes (T|y\rangle), \text{ for } x, y \in \{0, 1\}.$$

A 2-bit quantum gate V is said to be *primitive* [8, Theorem 4.2] if

$$V = S \otimes T \quad \text{or} \quad V = (S \otimes T)U_{\text{sw}}$$

for some 1-bit gates S and T. Otherwise, V is said to be *imprimitive*.

It is possible to factor an n-bit quantum gate U as the composition of k-bit quantum gates for $k \leq n$. The earliest result of this *universality* study was given by Deutsch [11] in 1989 for $k = 3$. Then in 1995, Barenco [3], Deutsch, Barenco and Ekert [12], DiVincenzo [16] and Lloyd [37] gave the result for $k = 2$. Here, we quote the elegant mathematical treatment and results by J.-L. Brylinski and R. K. Brylinski in [8].

Definition 2.6.1 ([8, Def. 4.2]). A collection of 1-bit gates A_i and 2-bit gates B_j is called (exactly) *universal* if, for each $n \geq 2$, any n-bit gate can be obtained by composition of the gates $A_i(\ell)$ and $B_j(\ell, m)$, for $1 \leq \ell, m \leq n$. $\square$

Theorem 2.6.1 ([8, Theorem 4.1]). *Let V be a given 2-bit gate. Then the following are equivalent:*

(i) *The collection of all 1-bit gates A together with V is universal.*

(ii) *V is imprimitive.*

$\square$

Rather than following [8], which requires some knowledge of the Lie algebra theory, we will include an elementary proof of the universality of 2-bit gates at the end of this section.

A common type of 1-bit gate from AMO (atomic, molecular and optical) devices is the unitary rotation gate

$$U_{\theta,\phi} \equiv \begin{bmatrix} \cos\theta & -ie^{-i\phi}\sin\theta \\ -ie^{i\phi}\sin\theta & \cos\theta \end{bmatrix}, \qquad 0 \leq \theta, \phi \leq 2\pi. \qquad (2.46)$$

We have (the determinant) $\det U_{\theta,\phi} = 1$ for any θ and ϕ and, thus, the collection of all such $U_{\theta,\phi}$ is not dense in $U(\mathbb{C}^2)$. Any unitary matrix with determinant equal to 1 is said to be *special unitary* and the collection of all special unitary matrices on $(\mathbb{C}^2)^{\otimes n}$, $SU((\mathbb{C}^2)^{\otimes n})$, is a proper subgroup of $U((\mathbb{C}^2)^{\otimes n})$. It is known that the gates $U_{\theta,\phi}$ generate $SU(\mathbb{C}^2)$.

Another common type of 2-bit gate we shall encounter in the next two sections is the quantum phase gate (QPG)

$$Q_\eta = \begin{bmatrix} 1 & 0 & 0 & 0 \\ 0 & 1 & 0 & 0 \\ 0 & 0 & 1 & 0 \\ 0 & 0 & 0 & e^{i\eta} \end{bmatrix}, \qquad 0 \leq \eta \leq 2\pi. \qquad (2.47)$$

Theorem 2.6.2 ([8, Theorem 4.4])**.** *The collection of all the 1-bit gates $U_{\theta,\phi}$, $0 \le \theta, \phi,\; \phi \le 2\pi$, together with any 2-bit gate Q_η where $\eta \not\equiv 0$ (mod 2π), is universal.*   □

Note that the CNOT gate $\Lambda_1(\sigma_x)$ can be written as

$$\Lambda_1(\sigma_x) = U_{\pi/4,\pi/2}(2)Q_\pi U_{\pi/4,-\pi/2}(2); \tag{2.48}$$

cf. [15, (2.2)]. Thus, we also have the following.

Corollary 2.6.3. *The collection of all the 1-bit gates $U_{\theta,\phi}$, $0 \le \theta, \phi \le 2\pi$, together with the CNOT gate $\Lambda_1(\sigma_x)$, is universal.*   □

We now introduce the standard notation for several of the basic types of quantum gates as follows. (When several wires are present, the top wire always correspond to the leading qubit.)

NOT-gate σ_x (cf. (2.42))

unitary rotation gate $U_{\theta,\phi}$ (cf. (2.46))

The Walsh–Hadamard gate H:

$$\left. \begin{array}{l} H|0\rangle = \dfrac{1}{\sqrt{2}}(|0\rangle + |1\rangle) \\[2mm] H|1\rangle = \dfrac{1}{\sqrt{2}}(|0\rangle - |1\rangle) \end{array} \right\} \qquad H = \frac{1}{\sqrt{2}} \begin{bmatrix} 1 & 1 \\ 1 & -1 \end{bmatrix} \tag{2.49}$$

The CNOT-gate $\Lambda_1(\sigma_x)$

The 2-bit controlled-U gate $\Lambda_1(U)$ (cf. (2.41) with $m = 1$)

$$\Lambda_1(U) = \begin{bmatrix} 1 & 0 & 0 & 0 \\ 0 & 1 & 0 & 0 \\ & & \cdots\cdots & \\ 0 & 0 & \vdots & U \\ 0 & 0 & \vdots & \end{bmatrix}$$

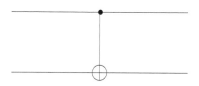

The QPG (cf. (2.47))

The tensor product $S \otimes T$ gate (where S and T are 1-bit gates)

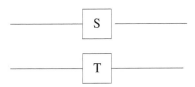

Also, the Toffoli quantum gate is the 3-bit controlled-not gate as shown in Fig. 2.5.

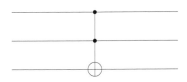

Figure 2.5: The Toffoli quantum gate $\Lambda_2(\sigma_x)$. (Compare with the classical Toffoli gate in Fig. 2.3.)

According to the above notation, then diagrammatically, the quantum circuit for (2.48) is given in Fig. 2.6, where the gates are applied from left to right.

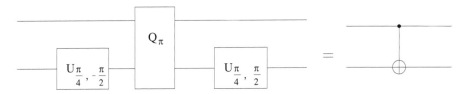

Figure 2.6: The quantum circuits for equation (2.48). The top wire represents the control qubit (remaining unchanged), while the second wire represents the second qubit that is changed conditioned on the value of the control qubit.

Let f be a Boolean function, where

$$f: \{0,1\} \times \{0,1\} \times \cdots \times \{0,1\} = \{0,1\}^n \to \{0,1\}. \qquad (2.50)$$

For $x \in \{0,1\}^n$ and $y \in \{0,1\}$, define the Boolean function evaluation (unitary) operator

$$U_f: \{0,1\}^{n+1} \times \{0,1\}^{n+1}$$
$$U_f |x\rangle |y\rangle = |x\rangle |y \otimes f(x)\rangle \qquad (y \oplus f(x) \equiv y + f(x) \bmod 2).$$

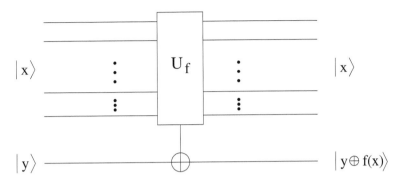

Figure 2.7: Quantum circuit for the Boolean evaluation U_f. In the above, $\oplus$ denotes the modulo 2 arithmetic.

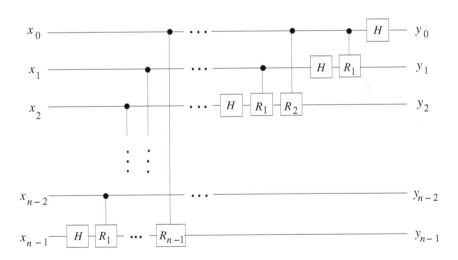

Figure 2.8: Quantum circuit for the quantum Fourier transform, where

$$R_d = \begin{bmatrix} 1 & 0 \\ 0 & e^{i\pi/2^d} \end{bmatrix}, d = 1, 2, \ldots, n-1$$

$$|x\rangle = |x_0 x_1 x_2 \ldots x_{n-1}\rangle, \quad |y\rangle = |y_0 y_1 y_2 \ldots y_{n-1}\rangle.$$

This operator is represented as in Fig. 2.7:

A very efficient operation in quantum computing is the n-bit quantum Fourier transform (QFT). QFT is a unitary operation

$$\textbf{QFT}: \{0,1\}^n \rightarrow \{0,1\}^n$$

$$\textbf{QFT} \, |x\rangle = \frac{1}{2^{n-1}} \sum_{y=0}^{2^n-1} e^{2\pi i x y/2^n} |y\rangle,$$

where in the exponents on the right hand side above, x and y are taken to be numbers in $\{0,1,2,\ldots,2^n-1\}$ corresponding to their binary representation. It is straightforward to verify that the circuit diagram for QFT can be given as in Fig. 2.8

We now address the property of universality of 2-bit quantum gates. We first give the following theorem, which is not exactly self-contained but contains some major ideas of a useful matrix factorization. In what follows, $U(2^n)$ denotes the group of all $2^n \times 2^n$ unitary matrices, while $SO(2^n)$ is the special orthogonal group of $2^n \times 2^n$ matrices, i.e., $M \in U(2^n)$ satisfies $M \in SO(2^n)$ if and only if the determinant of M is 1.

Theorem 2.6.4. *Let $V \in U(2^n)$. Then*

$$V = \prod_{i=1}^{2^n-1} \prod_{j=0}^{i-1} V_{ij}. \tag{2.51}$$

for a collection of matrices $V_{ij} \in U(2^n)$ such that

$$\left\{ \begin{array}{l} V_{ij}\colon \mathcal{S}_{ij} \longrightarrow \mathcal{S}_{ij} \text{ is the identity transformation,} \\ \mathcal{S}_{ij} \equiv span\{|m\rangle \mid m \in \{0,1,\ldots,2^n-1\}, m \neq i, m \neq j\}, \\ 0 \leq j < i \leq 2^n-1. \end{array} \right\} \tag{2.52}$$

In other words, each $V \in U(2^n)$ is a product of (generalized) controlled-$U(2)$ unitary matrices V_{ij}, which acts nontrivially only on $\mathcal{S}_{ij}^\perp = span\{|i\rangle, |j\rangle\}$.

Proof. We first quote the following fact [42, 50]: For any $V \in U(2^n)$, there exists a collection of unitary matrices $T_{i,j}$, $0 \leq j < i \leq 2^n-1$, and a D such that

$$V = \left(\prod_{i=1}^{2^n-1} \prod_{j=0}^{i-1} T_{i,j} \right) D, \tag{2.53}$$

where $T_{i,j} \in SO(2^n) \subseteq U(2^n)$ is a rotation involving $|i\rangle$ and $|j\rangle$ and satisfying (2.52), and D is a $2^n \times 2^n$ diagonal matrix whose diagonal entries are complex numbers of unit magnitude.

Now we can break up D into

$$D = \begin{pmatrix} d_0 & & & \\ & d_1 & & \\ & & \ddots & \\ & & & d_{2^n-1} \end{pmatrix} = D_1 D_2 \ldots D_{2^n-1} \tag{2.54}$$

where

$$
D_1 = \begin{pmatrix} d_0 & 0 & & & \\ 0 & d_1 & & & \\ & & 1 & & \\ & & & \ddots & \\ & & & & 1 \end{pmatrix}
\tag{2.55}
$$

and

$$
D_i = \begin{pmatrix} 1 & & & & \\ & \ddots & & & \\ & & d_i & & \\ & & & \ddots & \\ & & & & 1 \end{pmatrix}
\tag{2.56}
$$

for $i = 2, 3, \ldots, 2^n - 1$. It is easy to see that D_1 acts trivially except on $|0\rangle$ and $|1\rangle$, and the other D_i's act non-trivially only on $|i\rangle$. In addition, D_i's commute with each other, and each D_i commutes with $T_{k,l}$, $\forall 0 \le l < k < i$ as well. Thus,

$$
V = T_{2^n-1,2^n-2} \ldots T_{2^n-1,0} T_{2^n-2,2^n-3} \ldots T_{2^n-2,0} \ldots T_{2,1} T_{2,0} T_{1,0} D_1 D_2 \ldots D_{2^n-1}
$$

$$
\left.
\begin{array}{l}
= T_{2^n-1,2^n-2} T_{2^n-2,2^n-3} \ldots T_{2^n-1,0} D_{2^n-1} \\
\ \ T_{2^n-2,2^n-3} \ldots T_{2^n-2,0} D_{2^n-2} \\
\ \ \ldots \ldots \\
\ \ T_{2,1} T_{2,0} D_2 \\
\ \ T_{1,0} D_1
\end{array}
\right\} \ 2^n - 1 \text{ strings of products.}
\tag{2.57}
$$

For $0 \le j < i \le 2^n - 1$, define

$$
V_{ij} = \begin{cases} T_{i,j} & \text{if } j \ne 0, \\ T_{i,j} D_i = T_{i,0} D_i & \text{if } j = 0. \end{cases}
$$

Therefore we have reached

$$
V = \prod_{i=1}^{2^n-1} \prod_{j=0}^{i-1} V_{ij}
$$

where each V_{ij} is a unitary matrix which acts nontrivially only on the states $|i\rangle$ and $|j\rangle$ satisfying (2.52). $\qquad \square$

Each factor V_{ij} in (2.51) satisfies (2.52) and thus V_{ij} acts nontrivially only on the states $|i\rangle$ and $|j\rangle$. Denote the restriction of V_{ij} to the 2-dimensional subspace $\mathcal{S}_{ij}^{\perp} = \text{span}\{|i\rangle, |j\rangle\}$ by $\widehat{V}_{ij}$. Then $\widehat{V}_{ij} \in U(2)$. As pointed out in [4, p. 3465], each V_{ij} is not a standard $\Lambda_{n-1}(\widehat{V}_{ij})$ (in the notation of [4, p. 3458]) gate in the sense that the *controls are states rather than bits.*

Nevertheless, using Proposition 2.6.5 below, Barenco et al. [4, §VIII] point out how to rearrange basis states with a "gray code connecting state $|i\rangle$ to state $|j\rangle$" such that V_{ij} becomes unitarily equivalent to $\Lambda_{n-1}(\widehat{V}_{ij})$. In this sense, V_{ij} are *generalized* controlled-$\widehat{V}_{ij}$ gates.

Proposition 2.6.5. *The symmetric group S_{2^n} of permutations on the symbols $0, 1, 2, \ldots, 2^n - 1$ is generated by the 2-cycle $(2^n - 2, 2^n - 1)$ and the 2^n-cycle $(0, 1, 2, \ldots, 2^n - 1)$.*

Proof. This is a basic fact which can be found in most basic algebra or group theory books. $\square$

Incidentally, we note that the 2-cycle $(2^n - 2, 2^n - 1)$ is a permutation between the states $|\underbrace{1\,1\ldots 1\,0}_{n \text{ bits}}\rangle$ and $|\underbrace{1\,1\ldots 1}_{n \text{ bits}}\rangle$ and thus can be realized by the controlled-NOT gate with the nth qubit as the *target bit* and the first $(n-1)$ bits as the *control bits* as shown in Fig. 2.9.

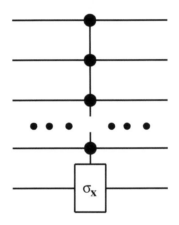

Figure 2.9: The n-bit controlled-NOT gate $\Lambda_{n-1}(\sigma_x)$, where σ_x is given by (2.42). This gate implements the two cycle $(2^n - 2, 2^n - 1)$ in Proposition 2.6.5.

On the other hand, the 2^n-cycle $(0, 1, 2, \ldots, 2^n - 1)$ makes the rotation of the states $|0\rangle \to |1\rangle \to \cdots \to |2^n - 2\rangle \to |2^n - 1\rangle \to |0\rangle$, i.e., the $|x\rangle \to |x + 1 \bmod 2^n\rangle$ operation. This can be implemented by the circuit as shown in Fig. 2.10.

Therefore, any permutation of the basis states $|x\rangle$, $x = 0, 1, 2, \ldots, 2^n - 1$, can be realized by finitely many controlled-NOT operations consisting of circuits as shown in Figs 2.9 and 2.10.

Thus, each factor V_{ij} in (2.51) can be realized by the circuit as shown in Fig. 2.11.

By concatenating together all the blocks V_{ij} as shown in Fig. 2.11 according to the factorization (2.51), we have constructed all $V \in U(2^n)$ with controlled-$\widehat{V}_{ij}$ gates according to (2.51).

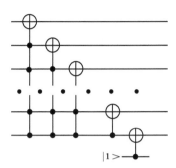

Figure 2.10: This circuit implements the operation $|x\rangle \to |x+1 \bmod 2^n\rangle$ or, equivalently, the 2^n-cycle $(0, 1, 2, \ldots, 2^n - 1)$ in Proposition 2.6.5. Note that the bit $|1\rangle$ at the bottom of the figure is the "scratch bit" which is sometimes omitted in circuit drawing. All the gates in this circuit are controlled-NOT gates.

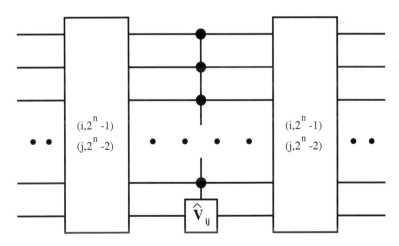

Figure 2.11: The unitary matrix V_{ij} in (2.51) as a controlled-$\widehat{V}_{ij}$ gate where $\widehat{V}_{ij} \in U(2)$. The operations $(i, 2^n - 1)$ and $(j, 2^n - 2)$ in the two boxes are cyclic permutations (which can be realized by concatenations of circuits in Figs. 2.9 and 2.10).

2.7   Quantum algorithms

In this section, we describe, in some brevity, several quantum algorithms and their circuit designs. Two of them, Grover's quantum search and Shor's factorization algorithms, have received the most recognition so far as their background problems are more "pragmatic," as they are related to *data mining* and *decryption*. But algorithmic development is an active area in quantum computing and we expect to see more and more new algorithms with speedup targeting a variety of real-world applications. Our description in this section is largely based on the work of Diao [14].

2.7.1   The Deutsch–Jozsa problem [13]

Consider a Boolean function $f: \{0,1\}^n \to \{0,1\}$ (cf. (2.50)). We say that a Boolean function is *constant* if

$$f(x) = c \quad \text{for some} \quad c \in \{0,1\}, \text{ for all } x \in \{0,1\}^n.$$

On the other hand, we say that f is *balanced* if

$$\text{card}(f^{-1}(\{0\})) = \text{card}(f^{-1}(\{1\})),$$

where card denotes the cardinality of a set. Then *Deutsch–Jozsa problem* is:

"Given a Boolean function $f: \{0,1\}^n \to \{0,1\}, f$ known to be either constant or balanced, determine whether f is constant or balanced."

With the classical computer, in general we need $2^{n-1} + 1$ evaluations of f to determine with certainty whether f is constant or balanced. But with the quantum computer, we need only one evaluation of f to make this determination with certainty. The Deutsch–Jozsa algorithm goes as shown in Box 2.2.

---

1. Prepare an $(n + 1)$-bit quantum register and initialize it to the state $(|0\rangle)^n |1\rangle$.
2. Apply the Walsh–Hadamard transform to all the qubits.
3. Apply function evaluation unitary transform U_f

$$U_f : |x\rangle|y\rangle \to |x\rangle|y \oplus f(x)\rangle$$

4. Apply the Walsh–Hadamard transform to the first n qubits (of x).
5. Measure the first n qubit. If the outcome is $|00\ldots0\rangle$, then f is constant. If the outcome is not $|00\ldots0\rangle$, then f is balanced.

Box 2.2 Steps for the Deutsch–Jozsa algorithm

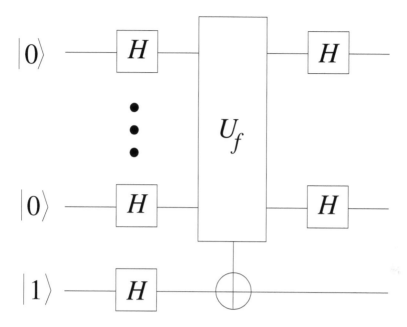

Figure 2.12: Circuit diagram for the Deutsch–Jozsa algorithm.

The circuit diagram is given in Fig. 2.12.

The first step of the algorithm, the application of the Walsh-Hadamard gates, puts the first n bits of the quantum register in a superposition state of all possible states. This state, denoted as $|s\rangle$, is the average of all basis states:

$$|00\ldots0\rangle \longmapsto$$
$$\frac{|0\rangle + |1\rangle}{\sqrt{2}}\frac{|0\rangle + |1\rangle}{\sqrt{2}}\cdots\frac{|0\rangle + |1\rangle}{\sqrt{2}} = \frac{1}{2^{n/2}}\left(|00\ldots0\rangle + |10\ldots0\rangle + \ldots |11\ldots1\rangle\right)$$
$$\equiv |s\rangle \tag{2.58}$$

The subsequent algorithmic operations are now performed with all possible register values in "parallel" which would be impossible on a classical computer.

Example 2.7.1. We provide a simple example for $n = 2$ to illustrate how the Deutsch–Jozsa algorithm works.

(a) Let $f(00) = f(01) = f(10) = f(11) = 1$. The following sequence of quantum states shows the flow of computation utilizing the Deutsch-

Jozsa algorithm. Start from the initial state $|001\rangle$. Then

$$|00\rangle|1\rangle$$
$$\rightarrow \frac{|0\rangle + |1\rangle}{\sqrt{2}} \frac{|0\rangle + |1\rangle}{\sqrt{2}} \frac{|0\rangle - |1\rangle}{\sqrt{2}} = \frac{|00\rangle + |01\rangle + |10\rangle + |11\rangle}{2} \frac{|0\rangle - |1\rangle}{\sqrt{2}}$$
$$\rightarrow \frac{|00\rangle + |01\rangle + |10\rangle + |11\rangle}{2} \frac{|1\rangle - |0\rangle}{\sqrt{2}} = -\frac{|0\rangle + |1\rangle}{\sqrt{2}} \frac{|0\rangle + |1\rangle}{\sqrt{2}} \frac{|0\rangle - |1\rangle}{\sqrt{2}}$$
$$\rightarrow -|00\rangle \frac{|0\rangle - |1\rangle}{\sqrt{2}}.$$

The first two qubits are in the basis state $|00\rangle$, and a measurement gives the value 00 with certainty. Hence, f is constant.

(b) Let $f(00) = f(01) = 0$, and $f(10) = f(11) = 1$. The following sequence of quantum states shows the result of computation utilizing Deutsch-Jozsa's algorithm. We start from the initial state $|001\rangle$. Then,

$$|00\rangle|1\rangle$$
$$\rightarrow \frac{|0\rangle + |1\rangle}{\sqrt{2}} \frac{|0\rangle + |1\rangle}{\sqrt{2}} \frac{|0\rangle - |1\rangle}{\sqrt{2}} = \frac{|00\rangle + |01\rangle + |10\rangle + |11\rangle}{2} \frac{|0\rangle - |1\rangle}{\sqrt{2}}$$

$$\rightarrow \frac{|00\rangle + |01\rangle - |10\rangle - |11\rangle}{2} \frac{|0\rangle - |1\rangle}{\sqrt{2}} = \frac{|0\rangle - |1\rangle}{\sqrt{2}} \frac{|0\rangle + |1\rangle}{\sqrt{2}} \frac{|0\rangle - |1\rangle}{\sqrt{2}}$$
$$\rightarrow |10\rangle \frac{|0\rangle - |1\rangle}{\sqrt{2}}.$$

The outcome of measuring the first 2 qubit is 10, hence f is balanced.□

2.7.2 The Bernstein–Vazirani problem [6]

Let $x = x_1 x_2 \ldots x_n$ and $y = y_1 y_2 \ldots y_n$ be two n-bit strings in $\{0, 1\}^n$. We define the **dot product** $x \cdot y$ by

$$x \cdot y = x_1 y_1 \oplus x_2 y_2 \oplus \ldots \oplus x_n y_n.$$

The *Bernstein–Vazirani problem* is

"Given a Boolean function $f_a : \{0, 1\}^n \rightarrow \{0, 1\}$ defined by $f_a(x) = a \cdot x$, where a is an unknown n-bit string in $\{0, 1\}^n$, determine the value of a."

As a contains n bits, with the classical computer, we need n evaluations of f_a to determine the value of a. But with the quantum computer, we need only one evaluation of f_a to determine a.

The Bernstein–Vazirani algorithm proceeds as shown in Box 2.3. It is virtually identical to the Deutsch–Jozsa algorithm.

1. Prepare an $(n + 1)$-bit quantum register and initialize it to the state $(|0\rangle)^n |1\rangle$.
2. Apply the Walsh–Hadamard transform to all the qubits.
3. Apply the function evaluation unitary transform U_{f_a}

$$U_{f_a} : |x\rangle|y\rangle \rightarrow |x\rangle|y \oplus f_a(x)\rangle.$$

4. Apply the Walsh–Hadamard transform to the first n qubits.
5. Measure the first n qubits. The outcome is $|a\rangle$.

Box 2.3 Steps for the Bernstein–Vazirani algorithm

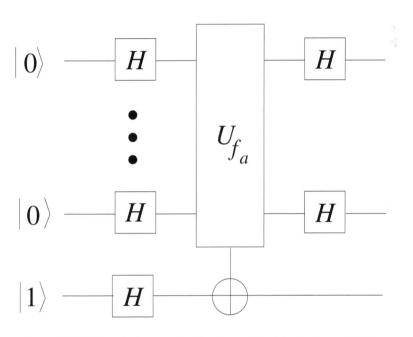

Figure 2.13: Circuit diagram for the Bernstein–Vazirani algorithm.

Example 2.7.2. We use $n = 2$ to illustrate some details as how the algorithm works. Let $a = 11$. The following sequence of quantum states shows the flow of computation utilizing the Bernstein-Vazirani algorithm. Start from the initial state $|001\rangle$. Then

$$|00\rangle|1\rangle$$

$$\rightarrow \frac{|0\rangle + |1\rangle}{\sqrt{2}} \frac{|0\rangle + |1\rangle}{\sqrt{2}} \frac{|0\rangle - |1\rangle}{\sqrt{2}} = \frac{|00\rangle + |01\rangle + |10\rangle + |11\rangle}{2} \frac{|0\rangle - |1\rangle}{\sqrt{2}}$$

$$\rightarrow \frac{|00\rangle - |01\rangle - |10\rangle + |11\rangle}{2} \frac{|0\rangle - |1\rangle}{\sqrt{2}}$$

$$\rightarrow |11\rangle \frac{|0\rangle - |1\rangle}{\sqrt{2}}.$$

The outcome of measuring the first 2 qubits is 11, hence $a = 11$. □

2.7.3 Simon's problem [57]

We say that a function $f : \{0,1\}^n \rightarrow \{0,1\}^n$ is **2–1** if for each $z \in Range(f)$, there are exactly two distinct n-bit strings x and y such that $f(x) = f(y) = z$.

A function $f : \{0,1\}^n \rightarrow \{0,1\}^n$ has a **period** a if $f(x) = f(x \oplus a)$, for all $x \in \{0,1\}^n$.

Simon's problem is stated as follows.

"Given a function $f : \{0,1\}^n \rightarrow \{0,1\}^n$ which is 2-1 and has period a, determine the period a."

It is known that with the classical computer, we need exponentially many (with respect to n) evaluations of f to determine the period a. But with the quantum computer, we need only $O(n)$ evaluations (expected) of f to determine the period a.

At this point, one first encounters the idea of "repeat sufficiently often" to get the desired result with a given success probability. This needs to be explained. In particular, in Box 2.3, the number of runs (n) is only "sufficient in order of magnitude" and depends on some threshold value for the probability of failure.

Simon's algorithm can be stated as shown in Box 2.4.

Example 2.7.3. As an illustration, let $n = 2$. Let $f(00) = 01$, $f(01) = 11$, $f(10) = 01$, and $f(11) = 11$. The following sequence of quantum states shows the flow and result of computation utilizing Simon's algorithm. We start from the initial state $|0000\rangle$. Then,

$$|0000\rangle$$

$$\rightarrow \frac{|0\rangle + |1\rangle}{\sqrt{2}} \frac{|0\rangle + |1\rangle}{\sqrt{2}} |00\rangle = \frac{1}{2}(|00\rangle + |01\rangle + |10\rangle + |11\rangle)|00\rangle$$

$$\rightarrow \frac{1}{2}(|0001\rangle + |0111\rangle + |1001\rangle + |1111\rangle) = \frac{1}{2}((|0\rangle + |1\rangle)|0\rangle|01\rangle$$

$$+ (|0\rangle + |1\rangle)|1\rangle|11\rangle)$$

$$\rightarrow \frac{1}{2}(|0\rangle(|0\rangle + |1\rangle)|01\rangle + |0\rangle(|0\rangle - |1\rangle)|11\rangle) = \frac{1}{2}(|0001\rangle$$

$$+ |0101\rangle + |0011\rangle - |0111\rangle).$$

The outcome of measuring the first 2 qubits yields either $|00\rangle$ or $|01\rangle$. Suppose that we have run the above computation twice and obtained $|00\rangle$ and $|01\rangle$, respectively. We now have a system of linear equations

$$\begin{cases} 00 \cdot a = 0 \\ 01 \cdot a = 0 \end{cases}$$

The solution is $a = 10$, the period of this function. □

---

1. Repeat the following procedure for n times.
   (a) Prepare a $2n$-bit quantum register and initialize it to the state
      $(|0\rangle)^n (|0\rangle)^n$.
   (b) Apply the Walsh–Hadamard transform to the first n qubits.
   (c) Apply function evaluation unitary transform U_{f_a}

   $$U_{f_a} : |x\rangle|y\rangle \rightarrow |x\rangle|y \oplus f_a(x)\rangle.$$

   (d) Apply the Walsh–Hadamard transform to the first n qubits.
   (e) Measure the first n qubits. Record the outcome $|x\rangle$.
2. With the n outcomes $x_1, x_2, \ldots, x_n$, solve the following system of linear equations:

   $$\begin{cases} x_1 \cdot a = 0 \\ x_2 \cdot a = 0 \\ \quad \vdots \\ x_n \cdot a = 0. \end{cases}$$

   The solution a is the period of f.

Box 2.4 Steps for Simon's algorithm

2.7.4 Grover's quantum search algorithm [25]

Suppose we are given an unsorted database with N items. We want to find a target item w. This problem can be formulated using an *oracle function* $f : \{0, 1, \ldots, N - 1\} \rightarrow \{0, 1\}$, where $w \in \{0, 1, \ldots, N - 1\}$ is given, and

$$f(x) = \begin{cases} 0, & \text{if } x \neq w, \\ 1, & \text{if } x = w. \end{cases} \tag{2.59}$$

The (single-item) search problem can now be stated as follows:

"Given an oracle function as in (2.59), find the w such that $f(w) = 1$."

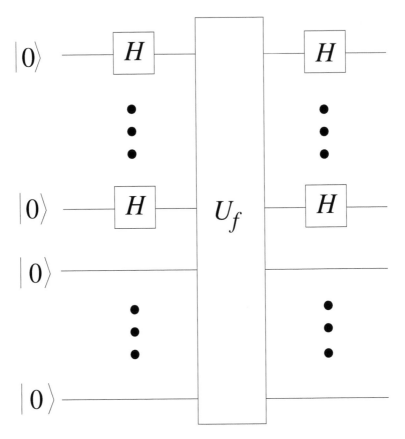

Figure 2.14: Circuit diagram for Simon's algorithm. Operation and measurement by this circuit need to be repeated $O(n)$ times.

This is an example of "data-mining."

It can be proved that with the classical computer, in average $\frac{N+1}{2}$ oracle calls are needed in order to find the search target. But with the quantum computer, on average, we need $O(\sqrt{N})$ oracle calls to find the search target. So there is a quadratic speedup. As a special case when $N = 4$, by using Grover's algorithm, exactly one oracle call suffices in order to find the search target with certainty.

L.K. Grover's quantum search algorithm proceeds as shown in Box 2.5.

Without loss of generality, let $N = 2^n$.

1. Prepare an $(n + 1)$-bit quantum register and initialize it to the state
   $(|0\rangle)^n|1\rangle$.
2. Apply the Walsh–Hadamard transforms individually and simultaneously to all the $n + 1$ qubits.

3. Repeat the following procedure for about $\frac{\pi}{4}\sqrt{N}$ times. See Fig. 2.15.

   (a) Apply function evaluation unitary transform U_f (selective sign flipping operator)

   $$U_f : |x\rangle|y\rangle \rightarrow |x\rangle|y \oplus f(x)\rangle.$$

   This is equivalent to the unitary operator $\mathcal{I}_w = \mathbb{I} - 2|w\rangle\langle w|$ on the first n qubits.

   (b) Apply unitary operator (inversion about the average operator) $\mathcal{I}_s = 2|s\rangle\langle s| - \mathbb{I}$ on the first n qubits, where $|s\rangle$ is defined by (2.58). See Fig. 2.16.

4. Measure the first n qubits. We obtain the search target with a high probability.

Box 2.5 Steps for Grover's quantum search algorithm

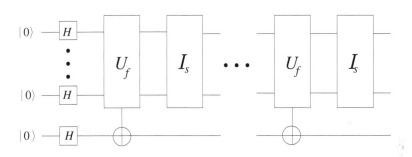

Figure 2.15: Block diagram for Grover's algorithm.

Example 2.7.4. Let $N = 2^2 = 4$, and Item 3 be the search target. The qubit representation of 3 is $|11\rangle$. The following sequence of quantum states shows how the computation works utilizing Grover's algorithm. Start from the initial state $|001\rangle$. Then

$$|00\rangle|1\rangle$$

$$\rightarrow \frac{|0\rangle + |1\rangle}{\sqrt{2}} \frac{|0\rangle + |1\rangle}{\sqrt{2}} \frac{|0\rangle - |1\rangle}{\sqrt{2}} = \frac{|00\rangle + |01\rangle + |10\rangle + |11\rangle}{2} \frac{|0\rangle - |1\rangle}{\sqrt{2}}$$

$$\rightarrow \frac{|00\rangle + |01\rangle + |10\rangle - |11\rangle}{2} \frac{|0\rangle - |1\rangle}{\sqrt{2}}$$

$$\rightarrow |11\rangle \frac{|0\rangle - |1\rangle}{\sqrt{2}}.$$

The outcome of measuring the first 2 qubits yields 11, which is the search target $w = 3$. □

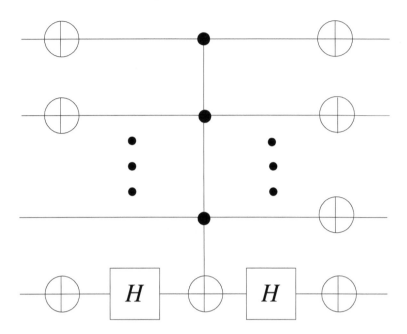

Figure 2.16: Circuit diagram for the inversion about the average operator $\mathcal{I}_s$.

Reviewing Grover's algorithm, we observe that the key to the improved efficiency in the quantum case comes about from the fact that although the coherent evolution of the superposition state is governed by the amplitude of the wave vector, the probability of reaching the target state is governed by the *square* of the amplitude [53].

One of the issues surrounding the implementation of the quantum search has been *whether or not quantum entanglement is needed in principle for its execution* [38, 41]. Indeed, the first implementations of the algorithm were carried out using optical and atomic setups where entanglement was absent [34, 1]. In another formulation of the quantum search [53], a series of $\sqrt{N}$ pulses were sufficient to excite a target atom in an array of atoms. Here again, it is quantum coherence between the atomic levels that enables the search rather than entanglement. On the other hand, it can be argued on very general terms that for any quantum algorithm based on a binary decomposition of the Hilbert space in terms of qubits, entanglement is essential for an efficiently scalable implementation [30].

As a pedagogical note here, we focus on a particularly simple case of the quantum search where the algorithm provides a *deterministic* solution, namely Example 2.7.4, a 4-element search. Usually, in an N-state search, the search algorithm yields the desired state with high probability (arbitrarily close to 1) but not with certainty. This is characteristic of quantum algorithms in general. However in the case of $N = 4$, implemented on two qubits, it can be shown [54] that a *single* run through the search algorithm

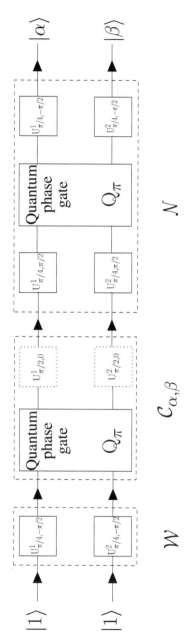

Figure 2.17: Level diagram for the implementation of Grover's algorithm.

produces the target state with 100 percent probability. The key transform needed is a 2-qubit phase gate, which is known to be the simplest realization of entanglement in unitary evolution.

In the three steps in each run of the search algorithm in Box 2.5, a coherent build up of amplitude in the target state is made. Step 3 (a) and (b) require an entangling operation between qubits in the system. In what follows, we express the logic gates in terms of single-qubit Pauli matrices (regarding each qubit as a spin-1/2 particle). Then, the entangling operations in Step 3 (a) and (b) become 2-spin transformations. Moreover, the sign flip and inversion about mean operations are shown to be simply related to each other via a joint spatial rotation of the spins.

Let us follow the discussions in [43]. First, note that a convenient representation of $U_{\theta,\phi}$ (cf. (2.46)) in terms of Pauli spin matrices is given by

$$U_{\theta,\phi} = \cos\theta \mathbf{1} - i\cos\phi\sin\theta\sigma_x - i\sin\phi\sin\theta\sigma_y. \tag{2.60}$$

The 2-bit quantum phase gate Q_η in (2.47) introduces a phase η only when both the qubits in the input state are 1. In the following we shall need the quantum phase gate only with $\eta = \pi$ for which we have

$$Q_\pi = |0,0><0,0| + |0,1><0,1| + |1,0><1,0| - |1,1><1,1|, \tag{2.61}$$

and since $|0><0| = (1+\sigma_z)/2$ and $|1><1| = (1-\sigma_z)/2$,

$$Q_\pi = \frac{1}{2}(\mathbf{1}_1\mathbf{1}_2 + \mathbf{1}_1\sigma_{z2} + \sigma_{z1}\mathbf{1}_2 - \sigma_{z1}\sigma_{z2}). \tag{2.62}$$

We now apply the above formalism to the 4-element search problem, as shown in Fig. 2.17. The Hilbert space of the two qubits consists of the four states $|0,0\rangle$, $|0,1\rangle$, $|1,0\rangle$, and $|1,1\rangle$. We assume that initially, the system starts out in the state

$$|\Psi_{\text{in}}\rangle = |1,1\rangle. \tag{2.63}$$

In the first step, we apply the Walsh-Hadamard transformation

$$W = \frac{(\mathbf{1}_1 + i\sigma_{y1})}{\sqrt{2}}\frac{(\mathbf{1}_2 + i\sigma_{y2})}{\sqrt{2}} = U^1_{\pi/4,-\pi/2}U^2_{\pi/4,-\pi/2} \tag{2.64}$$

which rotates each qubit from $|0>$ to $(|0> -|1>)/\sqrt{2}$ and $|1>$ to $(|0> +|1>)/\sqrt{2}$. The resulting state is

$$|\Psi_{\pi/2}\rangle = \frac{1}{2}(|0,0> +|0,1> +|1,0> +|1,1>). \tag{2.65}$$

In the second step the unitary operator $C_{\alpha,\beta}$ flips the sign of the desired target state $|\alpha,\beta>$ ($\alpha = 0$ or 1 and $\beta = 0$ or 1). In the original formulation of the algorithm [26], this is accomplished through the "sign-flip" operator $(1 - 2|\alpha,\beta><\alpha,\beta|)$. Here we follow a different approach. We first flip the sign of the state $|1,1>$ via a quantum phase gate Q_π followed by single-bit unitary operators that either retain the state of the qubit or change the state of the qubit from 0 to 1 or 1 to 0. The sign flip operators for the four possible states in this approach are given by

$$\mathcal{C}_{0,0} = -\sigma_{x1}\sigma_{x2}Q_{\pi} = U^1_{\pi/2,0}U^2_{\pi/2,0}Q_{\pi}, \tag{2.66}$$

$$\mathcal{C}_{0,1} = -i\sigma_{x1}1_2 Q_{\pi} = U^1_{\pi/2,0}Q_{\pi}, \tag{2.67}$$

$$\mathcal{C}_{1,0} = -i1_1\sigma_{x2}Q_{\pi} = U^2_{\pi/2,0}Q_{\pi}, \tag{2.68}$$

$$\mathcal{C}_{1,1} = Q_{\pi}. \tag{2.69}$$

Here the resulting state may have an unimportant overall phase factor. After application of one of the above four transformations, the state of the system is given by one of the following:

$$|\Psi_{\alpha\beta}\rangle = \begin{cases} [|0,0\rangle + |0,1\rangle + |1,0\rangle - |1,1\rangle]/2; & \alpha,\beta = 1,1 \\ [|0,0\rangle + |0,1\rangle - |1,0\rangle + |1,1\rangle]/2; & \alpha,\beta = 1,0 \\ [|0,0\rangle - |0,1\rangle + |1,0\rangle + |1,1\rangle]/2; & \alpha,\beta = 0,1 \\ [-|0,0\rangle + |0,1\rangle + |1,0\rangle + |1,1\rangle]/2; & \alpha,\beta = 0,0 \end{cases}. \tag{2.70}$$

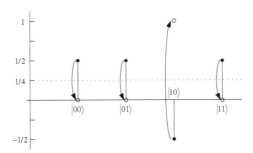

Figure 2.18: Level diagram for the implementation of Grover's algorithm.

Now we apply the most crucial, step in the quantum search process, namely, the inversion-about-mean operation. This is usually accomplished by the application of the 'diffusion' operator $\mathcal{N} = 1 - 2|\Psi_{\pi/2}\rangle\langle\Psi_{\pi/2}|$. We can see the effect that this has on our 4-element Hilbert space, as shown in Fig. 2.18. Each state has an amplitude of $1/2$ but the marked state alone has a negative sign. Thus, the mean of the four amplitudes is $(1/2 + 1/2 - 1/2 + 1/2)/4 = 1/4$. Inverting the amplitudes about the mean results in all other states becoming zero in amplitude, and the marked state gains unit amplitude. Thus, for this specific example of $N = 4$, we have a 100 percent probability of exciting the target state in a single run through the algorithm.

In order to find a circuit representation for $\mathcal{N}$ in terms of the operators $U_{\theta,\phi}$ and Q_{π} we first note from Eq. (2.64) that $\mathcal{W} = U^1_{\pi/4,-\pi/2}U^2_{\pi/4,-\pi/2}$ takes

$|1,1\rangle$ to the uniform superposition state $|\Psi_{\pi/2}\rangle$. Hence,

$$
\begin{aligned}
\mathcal{N} &= 1 - 2|\Psi_{\pi/2}\rangle\langle\Psi_{\pi/2}| \\
&= U^1_{\pi/4,-\pi/2}U^2_{\pi/4,-\pi/2}\left(1 - 2|1,1\rangle\langle1,1|\right)U^{1\dagger}_{\pi/4,-\pi/2}U^{2\dagger}_{\pi/4,-\pi/2} \\
&= U^1_{\pi/4,-\pi/2}U^2_{\pi/4,-\pi/2}\,Q_\pi\,U^{1\dagger}_{\pi/4,-\pi/2}U^{2\dagger}_{\pi/4,-\pi/2}.
\end{aligned}
\tag{2.71}
$$

That is, the inversion about mean is given by a rotated 2-qubit phase gate. The physical significance of this rotation becomes clear when we express the above relation in terms of Pauli matrices. First note that since

$$
|\Psi_{\pi/2}\rangle \rightarrow \frac{1}{\sqrt{2}}\begin{pmatrix}1\\1\end{pmatrix}_1 \otimes \frac{1}{\sqrt{2}}\begin{pmatrix}1\\1\end{pmatrix}_2,
\tag{2.72}
$$

we can write $|\Psi_{\pi/2}\rangle\langle\Psi_{\pi/2}|$ as follows:

$$
|\Psi_{\pi/2}\rangle\langle\Psi_{\pi/2}| \rightarrow \frac{1}{4}\begin{pmatrix}1&1\\1&1\end{pmatrix}_1 \otimes \begin{pmatrix}1&1\\1&1\end{pmatrix}_2
\tag{2.73}
$$

$$
\rightarrow \frac{1}{4}(1_1 + \sigma_{x1})(1_2 + \sigma_{x2}).
\tag{2.74}
$$

Thus we immediately have the following Pauli representation for $\mathcal{N}$:

$$
\begin{aligned}
\mathcal{N} &= 1_1 1_2 - \frac{1}{2}(1_1 + \sigma_{x1})(1_2 + \sigma_{x2}) \\
&= \frac{1}{2}(1_1 1_2 - 1_1\sigma_{x2} - \sigma_{x1}1_2 - \sigma_{x1}\sigma_{x2}).
\end{aligned}
\tag{2.75}
$$

To make contact with (2.71), we use the fact that rotating σ_z about the y axis by 90 degrees yields σ_x, that is

$$
\frac{(1 + i\sigma_y)}{\sqrt{2}}\sigma_z\frac{(1 - i\sigma_y)}{\sqrt{2}} = -\sigma_x,
\tag{2.76}
$$

Thus, we have the remarkable fact that the operator $\mathcal{N}$ as given by (2.75) is the spatially rotated quantum phase operator Q_π of Eq. (2.67), i.e.,

$$
\mathcal{N} = \exp\left(i\frac{\pi}{2}\sigma_{y1}\right)\exp\left(i\frac{\pi}{2}\sigma_{y2}\right)Q_\pi\exp\left(-i\frac{\pi}{2}\sigma_{y1}\right)\exp\left(-i\frac{\pi}{2}\sigma_{y2}\right).
\tag{2.77}
$$

The above representation holds more generally for more than two qubits, and gives some insight into how to implement the inversion about mean in a physical system. We point out to the reader that a scheme for implementing the search operations on atomic qubits interacting successively with classical Ramsey fields and quantized high-Q cavity fields is presented in [54].

2.7.5 Shor's factorization algorithm [56]

RSA cryptosystem, due to R. Rivest, A. Shamir and L.M. Adleman, is the most widely used public key cryptosystem presently. Decoding of RSA-encrypted messages depends on the factorization of a composite integer that is a product of two large prime numbers. It is known that the integer factorization problem is classically *intractable*. Given an integer N, the most efficient classical algorithm to date, *number field sieve*, has a time complexity of

$$O(\exp[(\log N)^{1/3}(\log \log N)^{2/3}]).$$

Shor's quantum factorization algorithm has a time complexity of

$$O((\log N)^2(\log \log N)(\log \log \log N)).$$

Hence it is a polynomial time algorithm. We consider the *integer factorization problem*:

"Given a composite positive integer N, factorize it into the product of its prime factors."

Shor's algorithm utilizes several fundamental theorems in number theory. We refer the reader to a quite accessible account by Lomonaco [39], so that we can omit the number-theoretic prerequisite here and go ahead to describe how the Shor algorithm works, as shown in Box 2.6.

---

1. Choose a random integer $a < N$, such that a and N are co-prime. This can be done by using a random number generator and the Euclidean algorithm on a classical computer.

2. Find the period T of the function $f_{a,N}(x) = a^x \bmod N$. This step can be expounded as follows:

   (a) Prepare two L-bit quantum registers in initial state

   $$\left(\frac{1}{\sqrt{2^L}} \sum_{x=0}^{2^L-1} |x\rangle\right)|0\rangle,$$

   where L is chosen such that $N^2 \leq 2^L < 2N^2$.

   (b) Apply $U_f : |x\rangle|0\rangle \rightarrow |x\rangle|f_{a,N}(x)\rangle$:

   $$\frac{1}{\sqrt{2^L}} \sum_{x=0}^{2^L-1} |x\rangle|0\rangle \rightarrow \frac{1}{\sqrt{2^L}} \sum_{x=0}^{2^L-1} |x\rangle|f_{a,N}(x)\rangle$$

   (c) Apply QFT to the first register:

   $$\frac{1}{\sqrt{2^L}} \sum_{x=0}^{2^L-1} |x\rangle|f_{a,N}(x)\rangle \rightarrow \frac{1}{2^L} \sum_{y=0}^{2^L-1} \left(\sum_{x=0}^{2^L-1} e^{2\pi i x y/2^L}|y\rangle\right)|f(x)\rangle$$

(d) Make a measurement on the first register, obtaining y.

(e) Find T from y via continued fractions for $\frac{y}{2^L}$. This step might fail- in that case, repeat (2a).

3. If T is odd, then repeat Step 1. If T is even and $N | a^{T/2} + 1$, repeat Step 1. If T is even and $N \nmid a^{T/2} + 1$, compute $d = gcd(a^{T/2} - 1, N)$, which is a nontrivial factor of N.

Box 2.6 Steps for Shor's factorization algorithm

Example 2.7.5. Let $N = 15$. Choose $L = 8$ such that $N^2 = 225 < 2^L < 450 = 2N^2$. Choose a random integer $a = 2$, which is coprime with 15. Thus $f_{a,N}(x) = 2^x \bmod 15$. The following sequence of quantum states shows the flow of computation utilizing Shor's algorithm:

$$|0\rangle|0\rangle \longrightarrow \frac{1}{2^4} \sum_{x=0}^{2^8-1} |x\rangle|0\rangle$$

$$\longrightarrow \frac{1}{2^4} \sum_{x=0}^{2^8-1} |x\rangle|f_{a,N}(x)\rangle$$

$$= \frac{1}{2^4}(|0\rangle|1\rangle + |1\rangle|2\rangle + |2\rangle|4\rangle + |3\rangle|8\rangle$$

$$+ |4\rangle|1\rangle + |5\rangle|2\rangle + |6\rangle|4\rangle + |7\rangle|8\rangle + \ldots$$

$$+ |2^8 - 2\rangle|4\rangle + |2^8 - 1\rangle|8\rangle)$$

$$\longrightarrow \frac{1}{2^4} \sum_{x=0}^{2^8-1} \frac{1}{2^4} \sum_{y=0}^{2^8-1} \omega^{xy}|y\rangle|2^x \bmod 15\rangle$$

$$\longrightarrow \frac{1}{2^8} \sum_{y=0}^{2^8-1} |y\rangle \sum_{x=0}^{2^8-1} \omega^{xy}|2^x \bmod 15\rangle,$$

where $\omega = e^{2\pi i/2^8}$. Suppose that the outcome of measuring the first n qubits is $|56\rangle$. We can compute the continued fraction of $\frac{56}{256}$ to be $[0, 4, 1, 1, 3]$, i.e.,

$$\frac{56}{256} = 0 + \cfrac{1}{4 + \cfrac{1}{1 + \cfrac{1}{1 + \cfrac{1}{3}}}}.$$

The second number 4 satisfies $2^4 = 1 \bmod 15$. So 4 is the period of $f_{a,N}$; $2^{4/2} - 1 = 3$ yields a factor of 15 and $15 = 3 \times 5$. □

An NMR implementation of Shor's algorithm based on this section will be given in Section 10.4, Chap. 10.

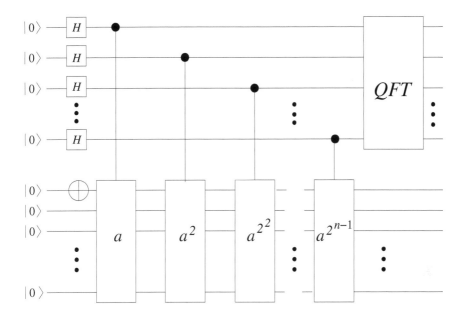

Figure 2.19: Quantum circuit for the Shor algorithm.

2.8 Quantum adder and multiplier

To show some basic quantum computer design concepts, such as how the quantum computer performs elementary arithmetic operations of addition and multiplication, in this section we illustrate the quantum networks of adder and multiplier design, based on Vedral, Barenco and Ekert [62]; cf. also [47, §2.4].

First, consider the quantum adder. We need several basic circuits, as indicated in Figs. 2.20, 2.21 and 2.22. They are S (sum), C (carry) and C^{-1} (inverse carry).

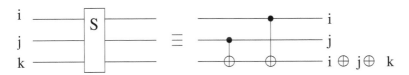

$$S: |i, j, k\rangle \rightarrow |i, j, i \oplus j \oplus k\rangle, \text{ for } i, j, k \in \{0,1\}$$

Figure 2.20: The quantum circuit for the S (sum) subroutine.

Using S, C and C^{-1} as described, then Fig. 2.23 provides the network for an n-bit quantum adder. The inputs are $a_n a_{n-1} \cdots a_1 a_0$ and $0 b_n b_{n-1} \cdots b_1 b_0$, and the outputs are $a_n a_{n-1} \cdots a_1 a_0$ and $a + b$. Note that the wire for b_{n+1} denotes a possible highest order carry qubit.

Remark 2.8.1. The designs as given here [62] utilize only the CNOT and Toffoli gates. They have the same complexity of the classical circuits and, thus, do not show any special advantage of quantum computing, when the adding operations involve only two integers.

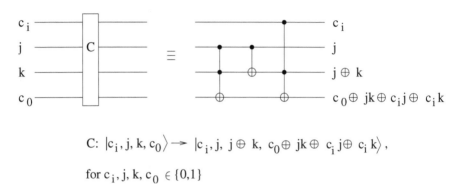

$$C: \; |c_i, j, k, c_0\rangle \rightarrow |c_i, j, \; j \oplus k, \; c_0 \oplus jk \oplus c_i j \oplus c_i k\rangle,$$

$$\text{for } c_i, j, k, c_0 \in \{0,1\}$$

Figure 2.21: The quantum circuit for the C (carry) subroutine. Here c_0 denotes the input carry bit (inherited from the preceding, lower qubits), while c_0, originally 0, at the output will be the next carry bit. The reader can easily check that the circuit works as intended.

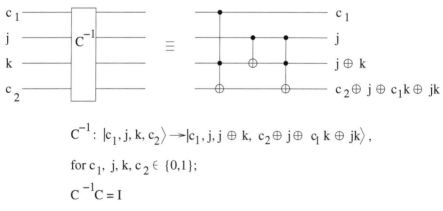

$$C^{-1}: \; |c_1, j, k, c_2\rangle \rightarrow |c_1, j, j \oplus k, \; c_2 \oplus j \oplus c_1 k \oplus jk\rangle,$$

$$\text{for } c_1, j, k, c_2 \in \{0,1\};$$

$$C^{-1} C = I$$

Figure 2.22: The quantum circuit for the C^{-1} (reverse carry) subroutine. It corresponds to the inverse operation of C.

Quantum computing's main advantages come from QC's capability to utilize *entanglement* and then by achieving *massive parallelism*. For example, we illustrate how to add 2 to 0, 1, 3 and 5 in one adder operation.

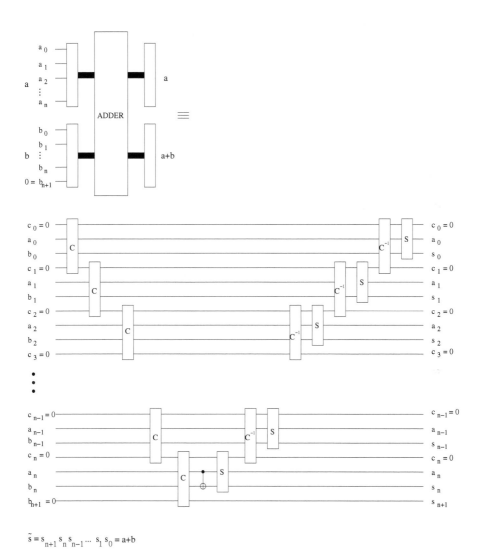

$$\tilde{s} = s_{n+1}\, s_n\, s_{n-1} \cdots s_1\, s_0 = a+b$$

Figure 2.23: Quantum adder network (given by (62)), where $s = a + b$ is the sum of two binary numbers a and b.

Write

$$|a\rangle = |0\ 1\ 0\rangle \quad (a = 2 \text{ in decimal representation}),$$
$$|b\rangle = \alpha_1|0\ 0\ 0\rangle + \alpha_2|0\ 0\ 1\rangle + \alpha_3|0\ 1\ 1\rangle + \alpha_4|1\ 0\ 1\rangle,$$

where $|b\rangle$ is a superposition of quantum states corresponding to 0, 1, 3 and 5. By inputing $|b\rangle$ into the quantum adder, we will obtain the outcome

$$|\psi\rangle = \alpha_1|0\ 1\ 0\rangle + \alpha_2|0\ 1\ 1\rangle + \alpha_3|1\ 0\ 1\rangle + \alpha_4|1\ 1\ 1\rangle.$$

This shows some ideas why entanglement plays a crucial role in the speedup by QC.   □

Once we have understood the circuit design for the quantum ADDER, we can easily follow [62] to give the design of the quantum circuit for the (controlled) MULTIPLIER. The main idea of the MULTIPLIER design for performing the multiplication of $x \cdot a$, where $x = x_n x_{n-1} \cdots x_1 x_0$ and $a = a_n a_{n-1} \cdots a_1 a_0$, is based on addition:

$$x \cdot a = x_0 2^0 \cdot a + x_1 2^1 \cdot a + x_2 2^2 \cdot a + \cdots + x_n 2^n \cdot a.$$

In Fig. 2.24, the top register entering ADDER j is exactly the number $x_{j-1} 2^{j-1} \cdot a$, for $j = 1, 2, \ldots, n+1$. Therefore, we obtain the desired outcome.

In Figs. 2.24 and 2.25, by following [62] an extra control bit c is added, which can either activate ($c = 1$) or deactivate ($c = 0$) the entire circuit. Thus, this circuit is called the controlled multiplier.

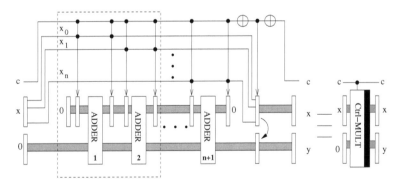

Figure 2.24: Quantum circuit of a controlled multiplier design (adapted from (62)) for performing the multiplication $x \cdot a$, where x and a have, respectively, binary representation $x_n x_{n-1} \cdots x_1 x_0$ and $a_n a_{n-1} \cdots a_1 a_0$. Details of the area framed within the dotted lines are given in Fig. 2.25.

Also, note that the number a is *not explicitly* shown in the circuit Fig. 2.24. The matter of fact is that $2^j a$ is generated by a set of Toffoli gates. This is illustrated in Fig. 2.25 in some detail.

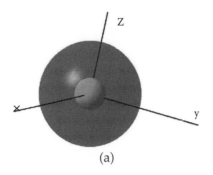

(a)

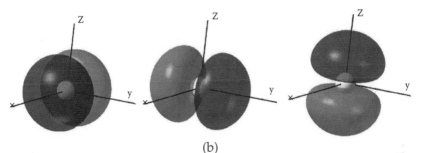

(b)

Figure 2.4: Shapes of s, p and d orbitals; cf. Atkins [2, p. 76 in the references of Chapter 2].
(a) s-orbital, spherical in shape;
(b) p-orbitals, with 2 lobes;
(c) (see following page) d-orbitals, with 4 lobes.
Each of the orbitals displayed in (b) and (c) is obtained from superpositions with wave functions involving the azimuth phases $e^{+im\phi}$ and $e^{-im\phi}$.
The color graphics are plotted using the Gaussian software.

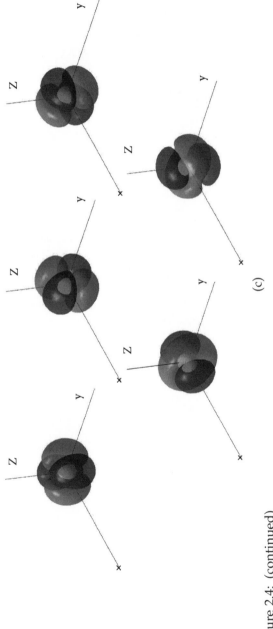

Figure 2.4: (continued)

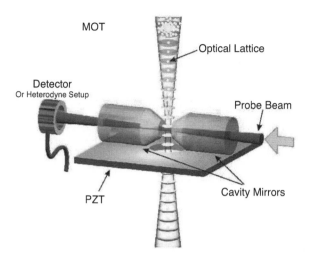

Figure 3.7: Design scheme of a cavity setup using a laser beam to shuttle atoms in and out of the cavity. Courtesy of Michael Chapman, Georgia Institute of Technology, Atlanta, Georgia, U.S.A. Two counterpropatating laser beams focused through the cavity in the vertical direction produce an optical lattice. Translating the lattice transports atoms collected in the magneto-optic trap (MOT) into the cavity mode below.

Figure 3.9: Photograph illustrating experiments with individual atoms in an optical cavity: the atoms are first trapped and cooled with laser light (bright spot left). They are then transferred by a laser beam (blue horizontal line) into the cavity (center). They are held inside the cavity using an 'optical tweezer' (green line). The distance between the cavity mirrors (viewed from the side, conical forms above and below the green line) is 0.5 mm, the atoms are transported along a distance of 14 mm. Courtesy of G. Rempe, Max-Planck-Institut für Quantenoptik, Garching, Germany.

Figure 3.10: Photograph of a partially mounted experimental setup with a high-quality cavity for the microwave range. In the central circular hole, one cavity mirror can be seen as a highly reflecting surface. Courtesy of J.-M. Raimond, Laboratoire Kastler-Brossel, Ecole Normale Supérieure, Paris.

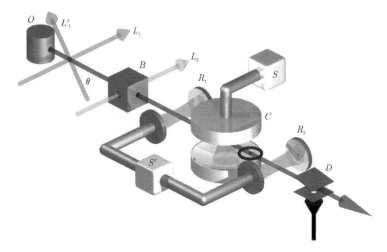

Figure 3.14: Design scheme of an experiment where an atomic beam is crossing a microwave cavity. The atoms are prepared in an oven (O), are prepared in a Rydberg state (see Fig. 3.4) with laser beams L_1, L'_1, L_2. In the zones R_1, R_2, single-qubit operations are performed with microwaves from a common source S'. The source S feeds photons into the cavity C. The atoms are detected in D. The dark circle at the cavity exit illustrates the orientation of the electron orbit of the involved atomic levels. Courtesy of J.-M. Raimond, Laboratoire Kastler-Brossel, Ecole Normale Supérieure, Paris.

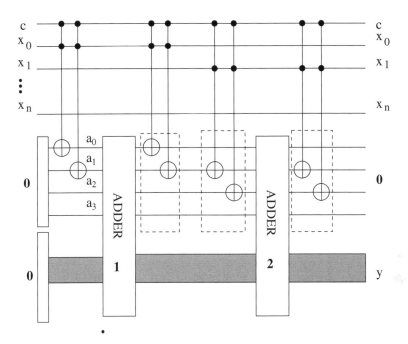

Figure 2.25: This circuit explains part of Fig. 2.24 with some details. We use $a = |0011\rangle$ as an example. The leftmost two Toffoli gates prepare a for ADDER1. Between ADDER1 and ADDER2, a set of four Toffoli gates prepares $2a$ for the top register if $x_1 = 1$. Otherwise if $x_1 = 0$, then what is prepared for the top register is just 0. To the right of ADDER2, the rightmost two Toffoli gates recover the top register to 0 and set it up for ADDER3.

One can further design the exponentiation $a^x \bmod N$ quantum operation using the ADDER and MULTIPLIER circuits. Thus this and all the other quantum arithmetic operations required for the Shor algorithm can be carried out successively by the corresponding circuits.

2.9 Quantum error correction codes

In the implementation of long quantum computations, there are several major sources of errors. See Box 2.7. Therefore, *error correction* is indispensable in quantum computing, even more so than in classical computing. In general, errors in quantum computing systems are more complicated than their classical counterparts and, therefore, correction schemes are also more sophisticated, as well as device-dependent. For example, for NMR systems, because of its long decoherence time (in the order of 10^4 sec) and gate operation time (order 10^{-3} sec) and the ensemble nature of NMR, errors tend to average out and the system is more fault-tolerant than others. But the opposite also tends to be true, i.e., if the component quantum system is made up of just a few single photons, electrons and atoms, then the component system will not be as fault-tolerant in general.

---

(1) Any quantum system in the real world interacts with the ambience, causing *decoherence* that tends to cause phase shifts or destroy the *relative phase information* (i.e., dephasing) in superposition quantum states;

(2) *Dissipation* due to the spontaneous emission of photons or phonons happens often uncontrollably, causing qubit flips and loss of energy;

(3) *Discrete, depolarization errors of spin and phase flips* may happen due to noise in communication channels;

(4) Laser noise;

(5) Unmodeled dynamics;

(6) *Measurement errors*;

(7) Not accurately controlled quantum gates lead to inaccuracies in quantum operations which build up after lengthy cascades of the computational process, rendering such computations unreliable.

Box 2.7 Sources of quantum computing errors

Recall that in classical computing, a basic idea for error correction is by *majority voting*. This can be simply explained as in Example 2.9.1.

Example 2.9.1 (Majority voting). The binary 0 and 1 bits are replicated several times and coded, e.g., as follows

$$0 \mapsto \bar{0} \equiv 0\,0\,0\,0\,0, \quad 1 \mapsto \bar{1} \equiv 1\,1\,1\,1\,1,$$

where $\bar{0}$ and $\bar{1}$ are, respectively, the coded words for 0 and 1. Thus, for instance,

$$13 = 1 \cdot 2^3 + 1 \cdot 2^2 + 0 \cdot 2^1 + 1 \cdot 2^0 = 1\,1\,0\,1 \mapsto \bar{1}\,\bar{1}\,\bar{0}\,\bar{1}$$

$$= 1\,1\,1\,1\,1 \vdots 1\,1\,1\,1\,1 \vdots 0\,0\,0\,0\,0 \vdots 1\,1\,1\,1\,1.$$

If we have a coded word $1\,1\,1\,0\,1 \vdots 1\,1\,1\,1\,1 \vdots 0\,0\,0\,1\,0 \vdots 1\,1\,1\,1\,1$, then we know that two errors have happened at, respectively, the fourth and the fourteenth bits and they can be simply corrected. The number of repeated qubits representing the very same qubit (here, five) is determined by the error rates of the quantum gates and quantum channels. If the error rates are low, then we can use fewer number of repeated qubits for majority voting. Conversely, if the error rates are high, then we must use more qubits.

If the quantum computer's error rates are "way too high", it can be expected that no amazing error correction schemes can help. That is, error corrections have limitations as what they can do. Error-correcting schemes are dependent on what types of errors and how prone they may happen in the machines and networks.

The fidelity or error rates of quantum processes will be studied in Chap. 4. □

One could immediately think of generalizing the idea of majority voting to quantum computing, say by encoding a quantum state $a|0\rangle + b|1\rangle \mapsto a|0\,0\,0\,0\rangle + b|1\,1\,1\,1\rangle$ in an analogous way as in Example 2.7.1. However, there is a severe hurdle as replicating or copying is absolutely forbidden in quantum mechanics. This is the famous *No-Cloning Theorem* that says cloning is a nonlinear operation and cannot be realized by any unitary operator.

Theorem 2.9.1 (The No-Cloning Theorem). *Let $\mathcal{H}$ be a complex Hilbert space. Then there does not exist a unitary transformation $U\colon \mathcal{H}\otimes\mathcal{H} \to \mathcal{H}\otimes\mathcal{H}$ such that there exists a $|s\rangle \in \mathcal{H}$ satisfying*

$$U(|\psi\rangle|s\rangle) = |\psi\rangle|\psi\rangle, \quad \text{for all} \quad |\psi\rangle \in \mathcal{H}.$$

(The state $|s\rangle$ above reserves the "slot for copying".)

Proof. If a unitary U exists, then

$$U(|\psi\rangle|s\rangle) = |\psi\rangle|\psi\rangle, \quad U(|\phi\rangle|s\rangle) = |\phi\rangle|\phi\rangle.$$

By unitarity, we have

$$\langle\psi, s|U^+U|s, \phi\rangle = \langle\psi|\phi\rangle\langle s|s\rangle = \langle\psi|\phi\rangle = \langle\psi, \psi|\phi, \phi\rangle = (\langle\psi|\phi\rangle)^2.$$

Thus, by letting $x = \langle\psi|\phi\rangle$, we have $x = x^2$. Hence $x = 0$ or $x = 1$. By simply choosing $|\psi\rangle$ and $|\phi\rangle$ in $\mathcal{H}$ such that $\langle\psi|\phi\rangle \neq 0$ or 1, we obtain a contradiction. □

Remark 2.9.1. From the No-Cloning Theorem, we now see why a classical fan-out gate (2.1) is not allowable in quantum computing because it contains the COPY operation(s) in Table 2.1. A similar operation of (2.3) of setting an input to an arbitrarily given quantum state is also an implicit COPY operation and, thus, is disallowed because of the No-Cloning Theorem. The classical fan-in gate (2.2) would require that all the inputs be identical (and thus are copies of each other) and it is a reverse operation of fan-out. Since fan-out operation is disallowed by the No-Cloning Theorem and since quantum computing is reversible, fan-in is also disallowed by reversibility and the No-Cloning Theorem. □

Due to the No-Cloning Theorem, for a good while many people thought that quantum error correction was impossible. But P. Shor's work [55] changed all that. The 9-qubit quantum error correction code (QECC) published by Shor [55] is the earliest and also the simplest type QECC. It constitutes an excellent example for illustrating how quantum error correction codes work.

Example 2.9.2 (Shor's 9-qubit code). In Box 2.7 item (3), we have mentioned depolarization errors. They can be specified as follows:

(i) *bit-flip error:* $\sigma_x \begin{bmatrix} a \\ b \end{bmatrix} = \begin{bmatrix} b \\ a \end{bmatrix}$,

(ii) *phase-flip error:* $\sigma_z \begin{bmatrix} a \\ b \end{bmatrix} = \begin{bmatrix} a \\ -b \end{bmatrix}$,

(iii) *phase- and bit-flip error* $\sigma_y \begin{bmatrix} a \\ b \end{bmatrix} = -i \begin{bmatrix} b \\ -a \end{bmatrix}$.

Shor's 9-qubit encodes as follows: We encode

$$|0\rangle \rightarrow |\bar{0}\rangle \equiv |+ + +\rangle, \quad |1\rangle \rightarrow |\bar{1}\rangle \equiv |- - -\rangle,$$

where $|+\rangle \equiv \frac{1}{\sqrt{2}}(|0\ 0\ 0\rangle + |1\ 1\ 1\rangle)$, $|-\rangle \equiv \frac{1}{\sqrt{2}}(|0\ 0\ 0\rangle - |1\ 1\ 1\rangle)$. Thus, the *codewords* for $|0\rangle$ and $|1\rangle$ are

$$\left. \begin{array}{l} |\bar{0}\rangle = 2^{-3/2}[(|0\ 0\ 0\rangle + |1\ 1\ 1\rangle)(|0\ 0\ 0\rangle + |1\ 1\ 1\rangle)(|0\ 0\ 0\rangle + |1\ 1\ 1\rangle)], \\ |\bar{1}\rangle = 2^{-3/2}[(|0\ 0\ 0\rangle - |1\ 1\ 1\rangle)(|0\ 0\ 0\rangle - |1\ 1\ 1\rangle)(|0\ 0\ 0\rangle - |1\ 1\ 1\rangle)]. \end{array} \right\} \quad (2.78)$$

The encoding circuit is shown in Fig. 2.26

The decoding circuit is given in Fig. 2.27. Let us explain how it works. We represent any 9-qubit term in the tensor products on the right-hand side of (2.78) as

$$|u_1 u_2 u_3\rangle |u_4 u_5 u_6\rangle |u_7 u_8 u_9\rangle, \quad u_j \in \{0, 1\}, j = 1, 2, \ldots, 9.$$

Suppose a bit-flip error happens to one of the u's in $|u_1 u_2 u_3\rangle$, say, u_2 is erroneously flipped:

$$\left. \begin{array}{l} |\bar{0}\rangle \rightarrow 2^{-3/2}[(|0\ 1\ 0\rangle + |1\ 0\ 1\rangle)(|0\ 0\ 0\rangle + |1\ 1\ 1\rangle)(|0\ 0\ 0\rangle + |1\ 1\ 1\rangle)], \\ |\bar{1}\rangle \rightarrow 2^{-3/2}[(|0\ 1\ 0\rangle - |1\ 0\ 1\rangle)(|0\ 0\ 0\rangle + |1\ 1\ 1\rangle)(|0\ 0\ 0\rangle + |1\ 1\ 1\rangle)]. \end{array} \right\} \quad (2.79)$$

We want to detect this $\sigma_x^{(2)}$ error while preserving superpositions. The special trick is to add two ancilla qubits (a_1 and b_1 in Fig. 2.27) and perform two CNOT operations on $|u_1 u_2 u_3\rangle$. From Fig. 2.27, we see that

$$A_1 = a_1 + u_1 + u_2 = u_1 + u_2, \quad \text{if we set} \quad a_1 = 0,$$
$$B_1 = b_1 + u_1 + u_3 = u_1 + u_3, \quad \text{if we set} \quad b_1 = 0.$$

Thus (2.79) gives

$$A_1 = 1, \quad B_2 = 2 = 0 \quad (\text{mod}2), \quad \text{after measurements.}$$

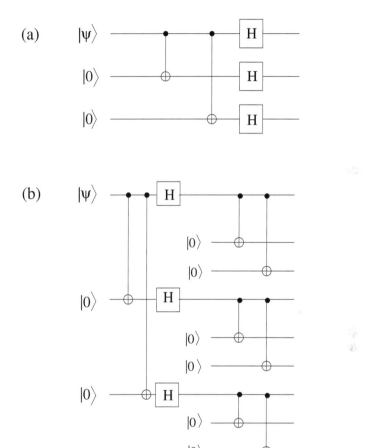

Figure 2.26: The encoding circuit for the Shor 9-qubit code, where H is the Hadamard gate given in (2.49). Part (a) is the circuit that constructs the first 3 qubits in (2.78). Part (b) is the concatenation of part (a) that gives the whole 9 qubits in (2.78).

We see that $A_1 = 1$ implies that one of the u's in $u_1 u_2$ is bit-flipped (because its "parity" is odd as A_1 is an odd integer), while $B_1 = 2$ implies that u_1 and u_3 are correct as $u_1 u_3$ has the normal "even parity" as B_1 is an even integer. Therefore, we pinpoint that u_2 is the qubit where bit-flip has occurred and we can just apply the Pauli matrix $\sigma_x^{(2)}$ to u_2 and recover the correct codewords.

Note that this procedure works if at most one qubit was flipped erroneously. But for a low enough error rate, the probability of two flips is much smaller, and error correction repairs the large majority of errors.

The measurements of $(A_1, B_1), (A_2, B_2)$ and (A_3, B_3) in the circuit of Fig. 2.27 provides the *error syndromes* and the corrective unitary operations. See Table 2.2.

Next, we address phase-flip errors. We first note the following identities:

$$\left. \begin{array}{l} (H \otimes H \otimes H)(|0\,0\,0\rangle + |1\,1\,1\rangle) = |0\,0\,0\rangle + |0\,1\,1\rangle + |1\,0\,1\rangle + |1\,1\,0\rangle, \\ (H \otimes H \otimes H)(|0\,0\,0\rangle - |1\,1\,1\rangle) = |0\,0\,1\rangle + |0\,1\,0\rangle + |1\,0\,0\rangle + |1\,1\,1\rangle. \end{array} \right\} \tag{2.80}$$

Note that this operator $H \otimes H \otimes H$ is applied to $|u_1 u_2 u_3\rangle$, $|u_4 u_5 u_6\rangle$ and $|u_7 u_8 u_9\rangle$ separately for a total of 6 items. Note that

$$\left. \begin{array}{l} C_1 = c_1 + (u_1 + u_2 + u_3) + (u_4 + u_5 + u_6) = (u_1 + u_2 + u_3) \\ \qquad + (u_4 + u_5 + u_6), \text{ if } c_1 = 0, \\ C_2 = c_2 + (u_1 + u_2 + u_3) + (u_7 + u_8 + u_9) = (u_1 + u_2 + u_3) \\ \qquad + (u_7 + u_8 + u_9), \text{ if } c_2 = 0; \end{array} \right\} \tag{2.81}$$

cf. C_1 and C_2 in Fig. 2.27. When a phase flip occurs, say at u_2, then we have the erroneous codewords

$$\left. \begin{array}{l} |\bar{0}\rangle \rightarrow (|0\,0\,0\rangle - |1\,1\,1\rangle)(|0\,0\,0\rangle + |1\,1\,1\rangle)(|0\,0\,0\rangle + |1\,1\,1\rangle) \\ |\bar{1}\rangle \rightarrow (|0\,0\,0\rangle + |1\,1\,1\rangle)(|0\,0\,0\rangle - |1\,1\,1\rangle)(|0\,0\,0\rangle - |1\,1\,1\rangle). \end{array} \right\} \tag{2.82}$$

Utilizing (2.82), (2.80) and (2.81) in sequential order, we see that for (the wrongly encoded) $|\bar{0}\rangle$,

$$C_1 = (u_1 + u_2 + u_3) + (u_4 + u_5 + u_6) = \text{odd} + \text{even} = 1(\text{mod}2),$$
$$C_2 = (u_1 + u_2 + u_3) + (u_7 + u_8 + u_9) = \text{odd} + \text{even} = 1(\text{mod}2).$$

error syndrome (A_j, B_j) $j = 1, 2, 3$	corrective unitary operation
(0,0)	1
(0,1)	$\sigma_x^{(3j)}$
(1,0)	$\sigma_x^{(3j-1)}$
(1,1)	$\sigma_x^{(3j-2)}$

Table 2.2: Bit-flip error syndromes and corresponding corrective operations for Shor's 9-bit QEEC

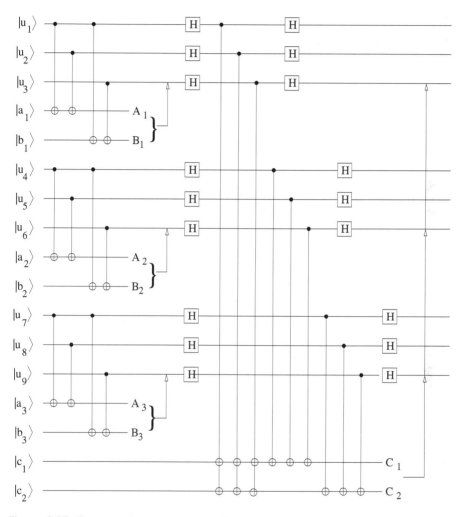

Figure 2.27: The quantum error-correction circuit for Shor's 9-qubit QECC. Measurements are performed at (A_j, B_j) for $j = 1, 2, 3$ for determining the bit-flip error syndrome, and at (C_1, C_2) to determine the phase-flip error syndrome. Corrective unitary operations (as Pauli matrices) will then be applied at the arrows depending on the detected parities. (This diagram is adapted from Pittenger (47, p. 90).)

This contradicts the correct parity for $|\bar{0}\rangle$ as C_1 and C_2 should both be even for $|\bar{0}\rangle$. (Similarly, for the correctly encoded $|\bar{1}\rangle$, the parity for C_1 and C_2 should be both odd.) Therefore, we just apply $\sigma_z^{(j)}$ at the top rightmost arrow place in Fig. 2.27, by taking $j = 1$. (It is also OK to choose $j = 2$ or 3. The outcome is the same.) The error syndromes and corrections are given in Table 2.3.

Since phase-and-bit flip error σ_y is a composition of σ_x and σ_z: $\sigma_y = -i\sigma_z\sigma_x$, we see that Shor's 9-bit QECC works also for the σ_y-type error.

Therefore, Shor's 9-bit QECC works for any *single qubit depolarization* errors.   □

There are more effective QECCs than Shor's 9-bit QECC. For example, the *7-qubit CSS QECC* due to Calderbank and Shor [9] and Steane [58, 59], which encodes by

$$|0\rangle \rightarrow |\bar{0}\rangle \equiv \frac{1}{\sqrt{8}}[|0\,0\,0\,0\,0\,0\,0\rangle + |0\,0\,0\,1\,1\,1\,1\rangle + |0\,1\,1\,0\,0\,1\,1\rangle + |0\,1\,1\,1\,1\,0\,0\rangle$$
$$+ |1\,1\,0\,1\,0\,0\,1\rangle + |1\,0\,1\,0\,1\,0\,1\rangle + |1\,1\,0\,0\,1\,1\,0\rangle + |1\,0\,1\,1\,0\,1\,0\rangle],$$

$$|1\rangle \rightarrow |\bar{1}\rangle \equiv \frac{1}{\sqrt{8}}[|1\,1\,1\,1\,1\,1\,1\rangle + |1\,1\,1\,0\,0\,0\,0\rangle + |1\,0\,0\,1\,1\,0\,0\rangle + |1\,0\,0\,0\,0\,1\,1\rangle$$
$$+ |0\,0\,1\,0\,1\,1\,0\rangle + |0\,1\,0\,1\,0\,1\,0\rangle + |0\,0\,1\,1\,0\,0\,1\rangle + |0\,1\,0\,0\,1\,0\,1\rangle].$$

$$(2.83)$$

The above encoding is adapted from the *classical [7, 4, 3] Hamming code* (see, e.g., the book [40]), with a "generator matrix" G and a "parity check" matrix H:

$$G = \begin{bmatrix} 0 & 0 & 0 & 1 \\ 0 & 0 & 1 & 0 \\ 1 & 1 & 0 & 0 \\ 1 & 0 & 0 & 0 \\ 0 & 1 & 1 & 0 \\ 0 & 1 & 0 & 1 \\ 1 & 0 & 1 & 1 \end{bmatrix}, \quad H = \begin{bmatrix} 0 & 0 & 0 & 1 & 1 & 1 & 1 \\ 0 & 1 & 1 & 0 & 0 & 1 & 1 \\ 1 & 0 & 1 & 0 & 1 & 0 & 1 \end{bmatrix},$$

error syndrome (C_1, C_2)	corrective unitary operation
(0,0)	1
(1,0)	$\sigma_z^{(4)}$ (or $\sigma_z^{(5)}, \sigma_z^{(6)}$)
(0,1)	$\sigma_z^{(7)}$ (or $\sigma_z^{(8)}, \sigma_z^{(9)}$)
(1,1)	$\sigma_z^{(1)}$ (or $\sigma_z^{(2)}, \sigma_z^{(3)}$)

Table 2.3: Phase-flip error syndromes and corresponding corrective operations for Shor's 9-bit QECC.

where $HG = 0$. One can use H for error syndrome analysis and make corrective operations. The CSS QECC can correct any single-qubit depolarization (Pauli-type) errors.

The QECC using the fewest qubits is due to Laflamme, Miguel, Paz and Zurek [35] with 5 qubits. Its encoding is done by

$$|0\rangle \to |\bar{0}\rangle \equiv \frac{1}{\sqrt{8}}[|0\,0\,0\,0\,0\rangle - |0\,1\,1\,1\,1\rangle - |1\,0\,0\,1\,1\rangle + |1\,1\,1\,0\,0\rangle$$
$$+ |0\,0\,1\,1\,0\rangle + |0\,1\,0\,0\,1\rangle + |1\,0\,1\,0\,1\rangle + |1\,1\,0\,1\,0\rangle],$$

$$|1\rangle \to |\bar{1}\rangle \equiv \frac{1}{\sqrt{8}}[|1\,1\,1\,1\,1\rangle - |1\,0\,0\,0\,0\rangle - |0\,1\,1\,0\,0\rangle + |0\,0\,0\,1\,1\rangle$$
$$+ |1\,1\,0\,0\,1\rangle + |1\,0\,1\,1\,0\rangle + |0\,1\,0\,1\,0\rangle + |0\,0\,1\,0\,1\rangle]. \tag{2.84}$$

It can also correct any single-qubit depolarization errors. Details of the constructions that lead to (2.83) and (2.84) will take considerable space, and we refer the reader to the specific literature. For a general account of the theory of QECCs, see, e.g., [24, 33].

To close this section, we include as an example containing the idea of *decoherence-free subspace* which may handle certain decoherence effects of physical qubits.

Example 2.9.3. Assume that the physical qubit of quantum system suffers from the following *collective dephasing* effect:

$$|0\rangle \to |0\rangle, \quad |1\rangle \to e^{i\theta}|1\rangle.$$

Thus, relative phase information becomes incorrect in any superposition. However, if we encode as follows:

$$|0\rangle \to |\bar{0}\rangle = \frac{1}{\sqrt{2}}(|0\rangle|1\rangle - i|1\rangle|0\rangle),$$
$$|1\rangle \to |\bar{1}\rangle = \frac{1}{\sqrt{2}}(|0\rangle|1\rangle + i|1\rangle|0\rangle),$$

then we see that

$$\alpha|0\rangle + \beta|1\rangle \to e^{i\theta}(\alpha|\bar{0}\rangle + \beta|\bar{1}\rangle),$$

for an arbitrary superposition. Hence coherence is preserved, as the global phase factor $e^{i\theta}$ causes no harm at all. This encoding is said to map qubits into a *decoherence free subspace* of the relevant Hilbert space.

Experimental work done by Kielpinski et al. [32] showed that for an ion trap device (cf. Chap. 5) such an encoding increased the ion storage time by up to an order of magnitude. See more discussions in Sections 5.7 and 5.10. □

As mentioned in Box 2.7, there are many sources of quantum computing errors one needs to give heed to. More recently, proposals of *topological* and *anyonic quantum computing* have been made (see an introduction in [49], e.g.), which offers a great promise for quantum computing with a greatly reduced error rate.

2.10   Lasers: a heuristic introduction

The laser is one of the greatest inventions of the 20th Century. LASER is the acronym of Light Amplification by Stimulated Emission of Radiation. Its foundation was first laid by Charles H. Townes with his graduate students James P. Gordon and Herbert J. Zeiger in the form of *pulsed microwave* radiation called a *maser* in a paper by Gordon, Zeiger and Townes [22] in 1954. The Soviet scientists Nikolay Basov and Aleksandr Prokhorov worked independently on the quantum oscillator and solved the problem of *continuous* output systems by using more than two energy levels. (C. Townes, N. Basov and A. Prokhorov shared the Nobel prize in 1964 with the citation "for fundamental work in the field of quantum electronics, which has led to the construction of oscillators and amplifiers based on the maser-laser principle".)

Most of the quantum computing devices to be discussed in this book are laser-driven. Thus, it is appropriate for us to give it a brief, heuristic introduction here for the benefit of the readers who are not physicists by training. We will follow the useful on-line material from Wikipedia [63, 64, 65] and Gormally [23]. For an advanced account, see the books [51, 52].

A laser is composed of several components:

(1) an external energy *pump source*. Examples of pump sources include flash lamps, arc lamps, light from another laser, electrical discharges, chemical reactions and even explosive devices, depending on the gain medium;

(2) an active laser amplifying or *gain medium*. It is a material of controlled purity, size, and shape, based on the quantum mechanical effect of *stimulated emission* to amplify the beam. There are four basic types: liquids, gases, solids and semiconductors. It is the major factor that determines the power and general wavelength of operation among other properties of the laser.

(3) a *resonant optical cavity*, which determines the wavelength and optical mode operation. It usually consists of two reflecting mirrors in an arrangement as shown in Fig. 2.28.

The laser operation begins by pumping energy from the source to the gain medium. The pumped energy is absorbed by the gain medium to generate excited quantized states in the medium. When the number of particles in one excited state exceeds the number of particles in some lower state, *population inversion* is achieved. In this condition, an optical beam passing through the medium at the frequency separation of the two levels produces more stimulated emission than stimulated absorption so the beam is amplified. The function of the medium here is thus that of energy storage and resonant light amplification. The resonant optical cavity extends the effective length of the interaction of light with the medium by repeated reflections.

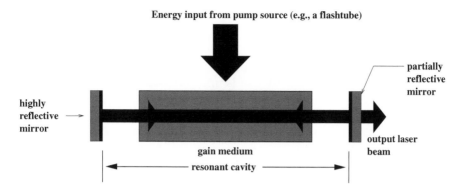

Figure 2.28: Schematic of a typical laser.

The light generated by stimulated emission is very similar to the input signal in its wavelength, phase, and polarization. This gives laser light its characteristic coherence, and allows it to maintain the uniform polarization and monochromaticity established by the optical cavity design. In quantum optics, mathematically, the laser's monochromaticity, action and propagation are quantized and modelled as a *simple harmonic oscillator*. See Chap. 3, Section 3.2.

Thus, the resonant cavity contains a coherent light beam propagating back and forth between reflective surfaces so that each photon passes through the gain medium multiple times before being emitted from the output aperture or lost to diffraction or absorption. The cavity ensures that the divergence of the beam is small. Only light that travels in a direction closely parallel to the axis of the cavity can undergo multiple reflections at the mirrors and make multiple passes through the amplifying medium. Those rays whose paths are not closely parallel to the axis execute a zig-zag way within the cavity, diverge and escape or get attenuated without getting amplified. Thus, laser cavity "purifies" the laser beam, which constitutes what is called a *cavity mode*.

As light or photons bounce repeatedly inside the cavity, passing through the gain medium, if the gain (the amplification factor) in the medium is stronger than the loss, the power of the propagating light can increase exponentially. However, each stimulated emission event decreases the population inversion, reducing the capacity of the gain medium for further amplification unless further pump energy is transferred to the medium. When this effect becomes strong, the gain is said to be *saturated*. The three competing and/or compensating factors: pump power, gain saturation and cavity loss finally reach an equilibrium value of the intracavity laser power which determines the operating point of the laser. If the pump power is chosen too small (below the "laser threshold"), the gain is not sufficient to overcome the resonator losses, and lasing will not occur.

Laser mirrors, made with specially coated surfaces, usually do not reflect all wavelengths or colors (in the visible range) of light equally; their reflectivity is matched to the wavelength or color at which the laser operates. In appearance, they do not look like ordinary mirrors and are transparent at some wavelengths. One of the mirrors reflects almost all of the light at the laser wavelength that falls upon it (the highly reflective mirror in Fig. 2.28). The other mirror reflects between 20% and 98% of the incident light depending upon the type of laser, the light that is not reflected being transmitted through the mirror. This transmitted portion constitutes the output beam of the laser.

Lasers may be classified into the following types [65]:

(1) Gas lasers (Helium-neon gas laser, Argon ion gas laser, Krypton ion gas laser, Xenon ion gas laser, Nitrogen gas laser, Hydrogen fluoride chemical laser, Chemical Oxygen-Iodine Laser (COIL), Carbon dioxide ($CO2$) gas laser, Carbon monoxide (CO) gas laser, Excimer chemical lasers);

(2) Dye lasers (made of liquid dye);

(3) Metal-vapor lasers (Helium-cadmium (HeCd) laser, Helium-mercury (HeHg) laser, Helium-Selenium (HeSe) laser, Copper vapor laser, Gold vapor laser);

(4) Solid-state lasers (Ruby laser, Neodymium YAG (Nd:YAG) laser, Neodymium YLF (Nd:YLF) laser, Neodymium doped YVO4 (Nd:YVO) laser, Neodymium Glass (Nd:Glass) laser, Titanium sapphire (Ti:sapphire) laser, Thulium YAG (Tm:YAG) laser, Ytterbium YAG (Yb:YAG) laser, Ytterbium doped glass laser (rod, plate/chip, and fiber), Holmium YAG (Ho:YAG) laser, Cerium doped lithium strontium(or calcium) aluminum fluoride, Promethium 147 doped phosphate glass (147Pm+3:Glass) laser, Chromium doped Chrysoberyl (Alexandrite) laser, Erbium doped and Erbium-Ytterbium codoped glass lasers, Trivalent Uranium doped calcium fluoride (U:CaF2) laser, Divalent Samarium doped calcium fluoride (U:CaF2) laser);

(5) Semiconductor lasers (Semiconductor laser diode);

(6) Other types of lasers (Free electron laser, "Nickel-like" Samarium laser; Raman laser, uses inelastic Stimulated Raman Scattering in a nonlinear media, mostly fiber, for amplification).

Cavity QED, one of the important quantum computing devices to be addressed in Chap. 3, works on the basic design principle which is the same as that of the light amplifier of a laser, except that the gain medium is just a single atom or perhaps an ion trap embedded in a resonant cavity. The mirror enclosing the resonant cavity, though, is totally reflective, made of multilayer dielectrics with a high reflectivity of 99.9999%.

The most common types of lasers used for the control of quantum computing devices are the continuous wave (CW) (versus pulsed) lasers. Dye lasers have broad wavelengths of operation, but limited ranges with each dye. For example, rhodamine 6G lasers usually only in the $570 - 640 \ nm$ range, but other dyes can be used between wavelengths from inside the ultra-violet into the infra-red. Titanium sapphire (Ti:sapphire) solid-state laser has operation wavelengths of $650 - 1100 \ nm$ and has lately replaced dye laser to a great extent as it has the advantage of being highly tunable. Semiconductor laser diodes are available for a broad range of operation wavelengths as well, depending on the device material, e.g., $0.4 \ \mu m$ (GaN) or 0.63-1.55 μm (AlGaAs) or 3-20 μm (lead salt). They have the advantage of being inexpensive and compact. Nevertheless, they are tunable only over a very narrow bandwidth.

2.11 Quantum computing devices and requirements

At present, several types of elementary quantum computing devices have been developed, based on AMO (atomic, molecular and optical) or semiconductor physics and technologies. We may roughly classify them into the following types as shown in Box 2.8.

---

atomic — ion and atom traps, cavity QED;
molecular — NMR;
optical — linear optics;
semiconductor — coupled quantum dots, silicon (Kane) (31);
crystal structure — nitrogen-vacancy (NV) diamond;
superconductivity — SQUID.

Box 2.8 Quantum device types

The above classification is not totally rigorous as new types of devices, such as quantum dots, or ion traps embedded in cavity-QED, have emerged which are of the *hybrid* nature. Also, laser pulse control, which is of the optical nature, seems to be omnipresent. In [5], a total of 12 types of quantum computing proposals have been listed.[9] But hundreds of proposals can be found in the literature.

---

[9]The additional proposals not listed above but given in [5] are quantum Hall qubits, electrons in liquid helium, and spin spectroscopies. The liquid helium proposal uses a potential qubit with a technology dubbed "electrons on helium" [48]. It would work by making a thin film of liquid helium to cover a Si (silicone) surface, where electrons are bound by an atomic-like microscopy image force above the helium surface to form an electronic 2-level system. Qubits are laterally defined by electrostatic gate patterns in the Si substrate, where circuits are also designed for qubit control, coupling and readout.

The proposal with NV-diamond in Box 2.8 will be described in Section 10.3 of Chap. 10, where the design of a solid-state NMR quantum computer is discussed. That design builds on

DiVincenzo has summarized five essential scientific requirements that any quantum computer based on a particular approach must have [17].

---

(1) *The system must have well-defined qubits, and be scalable.*

(2) *The system must have long coherence times,* to permit the quantum mechanical phase of the system wavefunction to be preserved through the many steps of computation so that the error rate must be sufficiently low.

(3) *Universal quantum gates must be possible to perform all elementary logical operations.*

(4) *There must be an efficient means to measure the states of the qubits.* At the end of a calculation, the state of each qubit state must be measured.

(5) *The system must be initialized in a reliable manner.* Initialization sets the quantum system to a reproducible reference state.

Box 2.9 D. DiVincenzo's five criteria for a working quantum computer

---

Scalability [item (1) in Box 2.9] means that the cost of increasing the number of bits n in the quantum register does not grow prohibitively fast, at most polynomially in n.

Due to the universality theorem (Section 2.6), it is sufficient to implement qubit gates acting on an arbitrary, single qubit and, for any pair of qubits, one of the entangling 2-qubit gates such as the quantum phase gate or the CNOT gate.

DiVincenzo's criterion item (2) can be quantified as follows. Two numbers T_1 and T_2, as characterizations of decoherence, are defined by

T_1: the *relaxation time*, is the average time the system takes for it to decay for $|1\rangle$ to $|0\rangle$;

T_2: the *phase coherence time*, is the average time over which the qubit energy-level difference does not vary.

We also define the operation time, τ_{op}, to be the time required to execute one quantum gate. The quantum computer must be able to execute thousands of gating operations within time duration T_2, i.e., $T_2/\tau_{op} \approx 10^3 \sim 10^4$. In Table 2.4, we list some reference data for some of the device systems treated in the book.

Our main task in the subsequent chapters is to address the items in Box 2.9, especially item (3), along with the relevant issues therein for the construction of a working QC.

Kane's original design of a solid-state quantum computer [31] (1998) (which may be regarded as a hybrid between the quantum dot and NMR quantum computer) but with a few further advantages.

Chapter	System		τ_{op}	T_2	T_2/τ_{op}
3	Cavity-QED	optical	10^{-14}	10^{-5}	10^9
		microwave	10^{-4}	10^0	10^4
5	Ion Traps (In)		10^{-14}	10^{-1}	10^{13}
7	Quantum dots	electron charge (GraAs)	10^{-13}	10^{-10}	10^3
		spintronics	10^{-7}	10^{-3}	10^4
9	SQUID	charge	10^{-11}	10^{-6}	10^5
		charge-flux	10^{-11}	10^{-6}	10^5
		flux	10^{-11}	10^{-6}	10^5
		phase	10^{-11}	10^{-6}	10^5
10	NMR		$10^{-3} \sim 10^{-6}$	$10^{-2} \sim 10^8$	$10^5 \sim 10^{14}$

Table 2.4: The rightmost column of this table gives the *coherence quality factor*, i.e., the number of 1-qubit operations achievable before decoherence takes over, of five types of devices treated in the book. The unit of τ_{op} and T_2 is in seconds. The data is excerpted from [17, 45, 67].

References

[1] J. Ahn, T.C. Weinacht, and P.H. Bucksbaum, *Science* **287** (2000), 463; see also P.L. Knight, *Science* **287** (2000), 441.

[2] P.W. Atkins, *Molecular Quantum Mechanics*, Second Edition, Oxford University Press, Oxford, UK, 1983.

[3] A. Barenco, A universal two bit gate for quantum computation, *Proc. Royal Soc. London A* **449** (1995), 679–693.

[4] A. Barenco, C.H. Bennett, R. Cleve, D.P. DiVincenzo, N. Margolus, P. Shor, T. Sleator, J.A. Smolin and H. Weinfurter, Elementary gates for quantum computation, *Phys. Rev. A* **52** (1995), 3457–3467.

[5] S. Bartlett, http://www.physics.usyd.edu.au/ bartlett/ISIT2005/.

[6] E. Bernstein and U. Vazirani, Quantum complexity theory, *SIAM J. Comput.* **26** (1997), 1411–1473.

[7] M. Born and J.R. Oppenheimer, *Ann. Physik* **84** (1927), 457.

[8] J-L Brylinski and R. Brylinski, Universal Quantum Gates, in *Mathematics of Quantum Computation*, R. Brylinski and G. Chen, (eds.), Chapman & Hall/CRC, Boca Raton, Florida, 2002, 101–116.

[9] A.R. Calderbank and P. Shor, Good quantum error correcting codes exist, *Phys. Rev. A* **54** (1996), 1098–1105.

[10] G. Chen and J. Zhou, *Vibration and Damping in Distributed Parameter Systems*, CRC Press, Boca Raton, FL, 1993.

[11] D. Deutsch, Quantum computational networks, *Proc. Royal Soc. London* **A425** (1989), 73.

[12] D. Deutsch, A. Barenco and A. Ekert, Universality in quantum computation, *Proc. Royal Soc. London A* **449** (1995), 669–677.

[13] D. Deutsch and R. Jozsa, Rapid solution of problems by Quantum Computation, *Proc. Royal Soc. London* **A439** (1992), 553–558.

[14] Z. Diao, Quantum computing, in *"Handbook of Linear Algebra"*, L. Hogben (ed.), Chapman & Hall/CRC, Boca Raton, Florida, 2006, to appear.

[15] Z. Diao, M.S. Zubairy and G. Chen, A quantum circuit design for Grover's algorithm, *Zeit. für Naturforsch.* **56a** (2001), 879–888.

[16] D.P. DiVincenzo, Two-bit quantum gates are universal for quantum computation, *Phys. Rev. A* **51** (1995), 1015–1022.

[17] D.P. Divincenzo, The physical implementation of quantum computation, in *Scalable Quantum Computers*, S.L. Braunstein and H.K. Lo, (eds.), Wiley-VCH, Berlin, 2001, pp. 1–11.

[18] J.W. Emsley, J. Feeney, and L.H. Sutcliffe, *High Resolution Nuclear Magnetic Resonance Spectroscopy*, Pergamon Press, Oxford, 1965.

[19] J.B. Foresman, and Æeen Frisch, *Exploring Chemistry with Electronic Structure Methods*, Gaussian, Inc., Pittsburgh, PA, 1996.

[20] E. Fredkin and T. Toffoli, Conservative logic, *Int. J. Theoretical Phys.* **21** (1982), 219–253.

[21] Æeen Frisch, M.J. Frisch, and G.W. Trucks, *Gaussian 03 User's Reference*, Gaussian, Inc., Carnegie, PA, 2003.

[22] J.P. Gordon, H.J. Zeiger, and C.H. Townes, Molecular microwave oscillator and new hyperfine structure in the microwave spectrum of NH3, *Phys. Rev.* **95** (1954), 282–284.

[23] J. Gormally, http://members.aol.com/WSRNet/tut/ut5.htm.

[24] M. Grassl, Algorithmic aspects of quantum error-correcting codes, in *Mathematics of Quantum Computation*, Chapter 9, pp. 223–252, R. Brylinski and G. Chen, (eds.), Chapman and Hall/CRC, Boca Raton, Florida, 2002.

[25] L. Grover, A fast quantum algorithm for database search, *Proc. 28th Annual ACM Symposium on Theory of Computing*, ACM, New York, 1996, 212–219.

[26] L.G. Grover, Quantum mechanics helps in searching for a needle in a haystack, *Phys. Rev. Lett.* **79** (1997), 325.

[27] E. Hille and R.S. Phillips, *Functional Analysis and Semigroups*, Amer. Math. Soc., Providence, R.I., 1957.

[28] H.J. Jahn and E. Teller, *Proc. Roy. Soc.* **A161** (1937), 220.

[29] H.J. Jahn and E. Teller, *Proc. Roy. Soc.* **A164** (1938), 117.

[30] R. Jozsa in *The Geometric Universe: Science, Geometry and the Work of Roger Penrose*, S. Huggett, L. Mason, K.P. Tod, S.T. Tsou, and N.M.J. Woodhouse (eds.) (Oxford Univ. Press, 1997), p. 369.

[31] B.E. Kane, A silicon-based nuclear spin quantum computer, *Nature* **393** (1998), 133–137.

[32] D. Kielpinski, V. Meyer, M.A. Rowe, C.A. Sackett, W.M. Itano, C. Monroe and D. J. Wineland, A decoherence-free quantum memory using trapped ions, *Science* **29** (2001), Feb. 2001.

[33] A. Klappenecker, *Quantum Error Correction Codes*, book in preparation, Cambridge Univ. Press, Cambridge, U.K., in preparation.

[34] P.G. Kwiat, J.R. Mitchell, P.D.D. Schwindt, and A.G. White, *J. Mod. Optics* **47** (1999), 257.

[35] R. Laflamme, C. Miquel, J.-P. Paz, and W.H. Zurek, A perfect quantum error-correcting code, *Phys. Rev. Lett.* **7** (1996), 198–201.

[36] R. Landauer, Irreversibility and heat generation in the computing process, *IBM J. Res. Dev.* **5** (1961), 183.

[37] S. Lloyd, Almost any quantum logic gate is universal, *Phys. Rev. Lett.* **75** (1995), 346–349.

[38] S. Lloyd, *Phys. Rev.* **A 61** (2000), 0103011.

[39] S.J. Lomonaco, Jr, Shor's quantum factoring algorithm, quant-ph/0010034, in *Proc. of Symposia in Appl. Math., Amer. Math. Soc.*, Vol. 58, Providence, R.I., 2002, 161–179.

[40] F.J. MacWilliams and N.J.A. Sloane, *The Theory of Error Correcting Codes*, North Holland, Amsterdam, 1977.

[41] D.A. Meyer, Sophisticated quantum search without entanglement, *Phys. Rev. Lett.* **85** (2000), 2014.

[42] F.D. Murnaghan, *The Unitary and Rotation Groups*, Spartan, Washington, D.C., 1962.

[43] A. Muthukrishnan, M. Jones, M.O. Scully, and M.S. Zubairy, *J. Mod. Opt.* **16** (2004), 2351.

[44] M.H.A. Newman, Alan Mathison Turing, *Biographical Memoirs of Fellows of the Royal Society of London* **1** (1955), 253–263.

[45] M.A. Nielsen and I.L. Chuang, *Quantum Computation and Quantum Information*, Cambridge University Press, Cambridge, U.K., 2000.

[46] J.J. O'Connor and E.F. Robertson, article about the biography of Alan Turing, in http://www-groups.dcs.st-and.ac.uk/~history/Mathematicians/Turing.htm.

[47] A.O. Pittenger, *An Introduction to Quantum Computing Algorithms*, Birkhäuser, Boston, 2000.

[48] P.M. Platzman and M.L. Dykman, Quantum Computing with electrons floating on liquid helium, *Science* **284** (1999), 1967.

[49] J. Preskill, Lecture Notes for Physics 229, Quantum Information and Computation, http://www.theory.caltech.edu/people/preskill/ph229.

[50] M. Reck, A. Zeilinger, H.J. Bernstein, and P. Bertani, Experimental realization of any discrete unitary operator, *Physical Review Letters* **73** (1994), 58–61.

[51] M. Sargent III, M.O. Scully, and W.E. Lamb, Jr., *Laser Physics*, Westview Press, 6th printing, 1993.

[52] M.O. Scully and M.S. Zubairy, *Quantum Optics*, Cambridge University Press, Cambridge, UK, 1997.

[53] M.O. Scully and M.S. Zubairy, *Phys. Rev.* **A 64** (2001), 022304.

[54] M.O. Scully and M.S. Zubairy, *Proc. Nat. Acad Sci.* **98** (2001), 9490.

[55] P. Shor, Scheme for reducing decoherence in quantum computer memory, *Phys. Rev. A* **52** (1995), 2493–2496.

[56] P. Shor, Polynomial-time algorithms for prime factorization and discrete logarithms on a quantum computer, *Siam J. Computing* **26** (1997), 1484–1509.

[57] D. Simon, On the power of quantum computation, *Proc. 35th Annual Symposium on Foundations of Computer Science*, IEEE Computer Society Press, Los Alamitos, CA, 1994, 116–123.

[58] A.M. Steane, Error correcting codes in quantum theory, *Phys. Rev. Lett.* **77** (1996), 793–797.

[59] A.M. Steane, Multi-particle interference and quantum error correction, *Proc. Roy. Soc. London A* **452** (1996), 2551–2577.

[60] H. Tanabe, *Equations of Evolution*, Pitman, London, 1979.

[61] A.M. Turing, On computable numbers, with an application to the Entscheidungs problem, *Proc. London Math. Soc.* **2** (1936), 42:230.

[62] V. Vedral, A. Barenco, and A. Ekert, Quantum networks for elementary arithmetic operations, quant-ph/9511018, *Phys. Rev. A* **54** (1996), 147.

[63] Wikipedia, http://en.wikipedia.org/wiki/Laser.

[64] Wikipedia, http://en.wikipedia.org/wiki/Laser_construction.

[65] Wikipedia, http://en.wikipedia.org/wiki/List_of_laser_types.

[66] A. Yao, Quantum circuit complexity, *Proc. the 34th IEEE Symposium on Foundations of Computer Science*, IEEE Computer Society Press, Los Alamitos, CA, 1993, 352–360.

[67] J.Q. You and F. Nori, Superconducting circuits and quantum information, *Physics Today* (Nov. 2005), 42–47.

Chapter 3

Two-Level Atoms and Cavity QED

- Two-level atoms and the Rabi oscillation

- The harmonic oscillator

- Cavity-QED hardware and its application to quantum entanglement and gating

- Quantum erasure

The main objective of this chapter is to show that the 1-bit unitary gates (2.46) and the 2-bit QPG (2.47) can be implemented using the methods of, respectively, 2-level atoms and cavity QED. For clarity, we divide the discussion into four sections: background and reference material from atomic physics (Section 3.1), the quantized light field (Section 3.2), photons in cavities (Section 3.3), and the interaction between an atom and zero or one photon in a cavity (Section 3.4). Quantum erasers, deferred from Section 1.3 of Chap. 1, will be studied in Sections 3.5 and 3.6.

3.1 Two-level atoms

The atomic energy levels are very susceptible to excitation by electromagnetic radiation. The structure of electronic eigenstates and the interaction between electrons and photons can be quite complicated. Here we give a quick introduction of atomic transitions and photon emission and absorption in Box 3.1 but we will utilize them on a regular basis henceforth in this and the subsequent chapters throughout the book, as they are of fundamental importance in quantum device designs.

Though an atom has infinitely many energy levels, under certain assumptions what we have in effect is a 2-level atom. These assumptions are as follows:

(i) the difference in energy levels of the electrons approximately matches the energy of the incident photon;

(ii) the symmetries of the atomic structure and the resulting "selection rules" allow the transition of the electrons between the two levels; and

(iii) all the other levels are sufficiently "detuned" in frequency separation with respect to the frequency of the incident field such that there is no transition to those levels.

Then the model of a 2-level atom provides a good approximation.

The emission and absorption of photons by atoms is intimately associated with changes in the *quantum numbers* (cf. Example 2.2.1) of the electrons and the nucleus of the atoms, leading to *selection rules*, and with the interaction of atomic moments with electric and magnetic fields. We follow the convention that the external magnetic field points along the z-axis. Experimentally an atom has no intrinsic electric dipole moment, but an external electric field will induce a dipole moment by changing the equilibrium position of the electron charge relative to the nuclear charge. Permanent magnetic dipole moments, which are proportional to the z-component of angular momentum and spin, do exist in atoms. These moments are involved in electric and magnetic dipole transitions. Higher order electric and magnetic moments are associated with the longer-lived decays (sometimes called "forbidden" decays) of metastable atomic levels. *Metastability* occurs when angular momentum selection rules *prevent* electric or magnetic dipole decays. Dipole transition rates are typically proportional to ω_0^3, where ω_0 is the frequency separation of the quantized energy levels involved in the photon emission. Atomic levels separated by relatively small frequency differences, often observed with magnetic dipole transitions, may also be very metastable, although the transitions are dipole-allowed.

Atomic transition rates are proportional to the square of matrix elements with the general form $\langle a|\mathbf{p}e^{i\mathbf{k}\cdot\mathbf{r}}|b\rangle$, where $|a\rangle$ and $|b\rangle$ represent the atomic wave functions of the levels involved in the transition, $|\mathbf{k}| = 2\pi/\lambda$, $\mathbf{r}$ is the position vector of the electron relative to the nucleus, and $\mathbf{p}(= m\mathbf{v})$ is the momentum of the electron involved in the transition. Since for optical or lower frequencies, $\lambda \gg r, e^{i\mathbf{k}\cdot\mathbf{r}} \approx 1$, so for the active electron the matrix element of a dipole transition becomes $m\langle a|\mathbf{v}|b\rangle \cong -i\omega m\langle a|\mathbf{r}|b\rangle$, assuming harmonic motion of the oscillating electron. When the driving frequency ω satisfies $\omega \approx \omega_0$, the transition probability becomes appreciable.

Electronic orbital angular momentum is proportional to the magnetic moment of the orbit; its quantum numbers are denoted ℓ and m_ℓ, with the magnitude of the total orbital angular momentum $|\mathbf{L}| = (\ell(\ell+1))^{1/2}\hbar$ and z-component $L_z = m_\ell\hbar$. The quantum number ℓ

takes the values $\ell = 0, 1, 2, 3, \ldots (n-1)$, where n is the principal quantum number of the level, and the first 4 values of ℓ are denoted by the letters s, p, d, f. The quantum number m_ℓ has the values $0, \pm 1, \ldots \pm 1$. Electrons, protons and neutrons are "spin-1/2" particles, with quantum numbers $m_s = \pm 1/2$, corresponding to $s = 1/2$ for electrons. The nuclear quantum number associated with the combined spins of the protons and neutrons is denoted I, with z-component m_I. Spin is a quantum mechanical property associated with the magnetic dipole moment of the particle, which is treated analogously to angular momentum. The photon has angular momentum quantized with a single integer unit of $\hbar$, with z-components $\pm \hbar$ only. Selection rules associated with photon emission are a consequence of the conservation of angular momentum.

Orbital $\boldsymbol{L}$ and spin $\boldsymbol{S}$ angular momentum add according to the rules of quantum mechanics to form the total electronic angular momentum $\boldsymbol{J} = \boldsymbol{L} + \boldsymbol{S}$. $\boldsymbol{J}$ couples to the nuclear spin angular momentum $\boldsymbol{I}$ to form the total atomic angular momentum which is often denoted $\boldsymbol{F}$. For electric dipole transitions, the selection rules for the changes in angular momentum are $\Delta \ell = \pm 1$ only, and $\Delta F = 0, \pm 1, \Delta m_F = 0, \pm 1$. If the nuclear spin $I = 0$, then the rules for F become the rules for J. For magnetic dipole transitions, $\Delta \ell = 0$ is allowed.

Box 3.1 Photon emission and absorption by atoms and ions

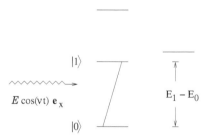

Figure 3.1: An atom with some energy levels (horizontal lines), two of which ($|0\rangle$ and $|1\rangle$) are connected by an electric dipole transition (slant line, resonance frequency $(E_1 - E_0)/\hbar$) that is driven by a near-resonant laser field (wavy arrow, frequency ν, electric field vector $\mathcal{E}\,\mathbf{e}_x$).

3.1.1 Atom–light interaction

We focus here on simple atoms like hydrogen and the alkali atoms where the interaction with visible light involves a single electron. This electron is bound to the atomic core and coupled to an external electric field $\boldsymbol{E}(\boldsymbol{r}, t)$ by its electric charge e, see Fig. 3.1. Its wave function $\psi(\boldsymbol{r}, t)$ determines the probability density $|\psi(\mathbf{r}, t)|^2$ of finding that electron at position $\mathbf{r} = (x, y, z)$ at time t, and obeys the Schrödinger equation

$$i\hbar \frac{\partial}{\partial t}\psi(\mathbf{r}, t) = H\psi(\mathbf{r}, t) = (H_0 + H_1)\psi(\mathbf{r}, t), \tag{3.1}$$

where

$$H_0 \equiv -\frac{\hbar^2}{2m}\mathbf{\nabla}^2 + V(\mathbf{r}) \tag{3.2}$$

is the differential operator for the unperturbed Hamiltonian of the electron, with the gradient operator $\mathbf{\nabla} = (\partial/\partial x, \partial/\partial y, \partial/\partial z)$ and the electrostatic potential $V(\mathbf{r})$ between the electron and the nucleus, and

$$H_1 \equiv -e\mathbf{r} \cdot \mathbf{E}(\mathbf{r}_0, t) \tag{3.3}$$

is the interaction Hamiltonian between the field and the electron of the atom.

Remark 3.1.1. The wavelength of visible light, typical for atomic transitions, is about a few thousand times the diameter of an atom. Therefore, there is no significant spatial variation of the electric field across an atom and so we can replace $\mathbf{E}(\mathbf{r}, t)$ by $\mathbf{E}(\mathbf{r}_0, t)$, the field at a reference point inside the atom such as the position of the nucleus or the center of mass. Consistent with this long-wavelength approximation (known as *dipole approximation* in the physics literature) is that the magnetic field $\mathbf{B}$ satisfies $\mathbf{B}(\mathbf{r}, t) \cong 0$, so that the Lorentz force of the radiation field on the electron is $e\mathbf{E}(\mathbf{r}_0, t)$, the negative gradient of $-e\mathbf{r} \cdot \mathbf{E}(\mathbf{r}_0, t)$, which adds to the potential energy $V(\mathbf{r})$ in the Hamiltonian operator. $\qquad\square$

For the unperturbed atom system, i.e., without the external electromagnetic field, let the eigenstates of the electron be $\psi_n = |n\rangle$ with energy eigenvalues $\hbar\omega_n$ where the index $n = 0, 1, \ldots$ enumerates the eigenstates starting from the lowest energy eigenvalue. The eigenstates are solutions of

$$H_0\psi_n = \hbar\omega_n\psi_n, \quad \text{or} \quad H_0|n\rangle = \hbar\omega_n|n\rangle; \qquad \langle n|n'\rangle = \delta_{nn'}. \tag{3.4}$$

To the state $|n\rangle$ corresponds the wave function $\psi_n(\mathbf{r}) \equiv \langle \mathbf{r}|n\rangle$, and the scalar product used here is the natural one on the Hilbert space $L^2(\mathbb{R}^3)$,

$$\langle n|n'\rangle = \int \overline{\psi_n(\mathbf{r})}\psi_{n'}(\mathbf{r})\, d^3r \tag{3.5}$$

where the overbar denotes complex conjugation and $d^3r = dx\,dy\,dz$ is the Riemann–Lebesgue measure on $\mathbb{R}^3$. The eigenstates are orthogonal because the Hamiltonian H_0 (3.2) is hermitian with respect to the scalar product (3.5).

Let the external electromagnetic field be a monochromatic plane-wave field linearly polarized along the x-axis, which interacts with the atom placed at $\mathbf{r}_0 = 0$. The electric field takes the form

$$\mathbf{E}(0, t) = \mathcal{E}\cos(\nu t)\mathbf{e}_x; \quad \mathbf{e}_x \equiv (1, 0, 0)^T. \tag{3.6}$$

To solve the Schrödinger equation (3.1), we expand the electron wave function in the basis of the unperturbed eigenstates,

$$\psi(\boldsymbol{r}, t) = \sum_{n=0}^{\infty} C_n(t)\psi_n(\boldsymbol{r}), \quad \text{or} \quad |\psi(t)\rangle = \sum_{n=0}^{\infty} C_n(t)|n\rangle. \tag{3.7}$$

We get a system of linear ordinary differential equations for the time-dependent coefficients by projecting the Schrödinger equation (3.1) onto one of the eigenstates and using the orthogonality condition (3.4):

$$\frac{d}{dt}C_n = -i\omega_n C_n - \frac{i}{\hbar}\mathcal{E}\cos(\nu t)\sum_{n'=0}^{\infty} P_{nn'}C_{n'}. \tag{3.8}$$

In (3.8), the $P_{n'n}$ are the matrix elements of the electric dipole operator, projected onto the electric field polarization:

$$P_{nn'} = \langle n|e\boldsymbol{r}\cdot\boldsymbol{e}_x|n'\rangle \tag{3.9}$$

$$= \langle n|ex|n'\rangle = \int \overline{\psi_n(\boldsymbol{r})}\, ex\, \psi_{n'}(\boldsymbol{r})d^3r$$

Typically, the Hamilton operator H_0 is invariant under space reflection $\boldsymbol{r} \mapsto -\boldsymbol{r}$ so that the eigenfunctions can be chosen to satisfy $\psi_n(-\boldsymbol{r}) = \pm\psi_n(\boldsymbol{r})$. (The sign occurring here fixes the quantum number "parity" for the eigenstate $|n\rangle$.) Hence, the diagonal matrix elements of the dipole operator vanish,

$$\langle n|ex|n\rangle = e\int x\,|\psi_n(\boldsymbol{r})|^2 d^3r = 0, \tag{3.10}$$

because x is an odd function while $|\psi_n(\boldsymbol{r})|^2$ is even. A generalization of this argument shows that in (3.9), $P_{nn'} = 0$ whenever the states $|n\rangle$, $|n'\rangle$ have the same parity. Additional "selection rules" (related to angular momentum conservation) apply when H_0 is rotationally symmetric, see, e.g., Messiah's textbook [40, Chap. XIII]. □

3.1.2 Reduction to a 2-level atom

The linear system of differential equations (3.8) has time-dependent coefficients and exact solutions of such equations are not always easy to come by. We now show that the complexity can be reduced to a system of two equations with time-independent coefficients if the external electromagnetic field is nearly resonant with a particular atomic transition. To this end, we apply perturbation theory and work out the coefficients

$$c_n(t) = C_n(t)e^{i\omega_n t} \tag{3.11}$$

in a power series in the electric dipole interaction H_1. In zeroth order, $c_n^{(0)}(t) = c_n(0)$. Putting this into the right hand side of (3.8) and integrating, we get to first order

$$c_n^{(1)}(t) = -\frac{\mathcal{E}}{2\hbar} \sum_{n'} P_{n'n} \frac{e^{i(\omega_n - \omega_{n'} + \nu)t} - 1}{\omega_n - \omega_{n'} + \nu} c_{n'}(0)$$

$$-\frac{\mathcal{E}}{2\hbar} \sum_{n'} P_{n'n} \frac{e^{i(\omega_n - \omega_{n'} - \nu)t} - 1}{\omega_n - \omega_{n'} - \nu} c_{n'}(0). \qquad (3.12)$$

In this expression, the smallest denominators are those satisfying the "resonance condition" $|\omega_n - \omega_{n'}| \approx \nu$. Typical atomic energy spectra have non-equidistant energy levels: the energy eigenvalues are such that the resonance condition is only fulfilled with a particular pair of eigenstates, say $n = 0$, $n' = 1$ or vice versa. Even if rotational symmetry implies that the same energy eigenvalue corresponds to more than one eigenfunction (a "degenerate energy level"), the selection rules of angular momentum conservation restrict the couplings to essentially two states. Experimentally, additional static electric or magnetic fields can be applied to split the transition frequencies, pushing all states except two out of resonance.

With such an understanding, we can identify the *2-level approximation* with the asymptotic limit

$$\frac{|\omega_n - \omega_{n'}| - \nu}{|\omega_1 - \omega_0| - \nu} \to \infty \quad \text{for} \quad n, n' \notin \{0, 1\}, n \neq n', \qquad (3.13)$$

$$\frac{\omega_1 - \omega_0}{\nu} = \text{const.} \qquad (3.14)$$

In the following, we denote ω the *atomic transition frequency*:

$$\omega \equiv \omega_1 - \omega_0 > 0. \qquad (3.15)$$

The perturbative solution (3.12) becomes

$$c_0^{(1)}(t) = -\frac{\mathcal{E}}{2\hbar} P_{10} \frac{e^{i(-\omega+\nu)t} - 1}{-\omega + \nu} c_1(0)$$

$$-\frac{\mathcal{E}}{2\hbar} P_{10} \frac{e^{i(-\omega-\nu)t} - 1}{-\omega - \nu} c_1(0), \qquad (3.16)$$

$$c_1^{(1)}(t) = -\frac{\mathcal{E}}{2\hbar} P_{01} \frac{e^{i(\omega+\nu)t} - 1}{\omega + \nu} c_0(0)$$

$$-\frac{\mathcal{E}}{2\hbar} P_{01} \frac{e^{i(\omega-\nu)t} - 1}{\omega - \nu} c_0(0). \qquad (3.17)$$

with all other coefficients vanishing to first order, hence a dramatic reduction of dimensionality.

It is clear that the solution (3.16), (3.17) can be generated by applying perturbation theory to the interaction Hamiltonian restricted to the 2-dimensional subspace

$$\mathcal{S} = \text{span}\{|0\rangle, |1\rangle\}. \qquad (3.18)$$

Since $|0\rangle\langle 0| + |1\rangle\langle 1|$ is the projector onto S, the restricted interaction Hamiltonian is of the form

$$
\begin{aligned}
[|0\rangle\langle 0| &+ |1\rangle\langle 1|] H_1 [|0\rangle\langle 0| + |1\rangle\langle 1|] \\
&= -e\mathcal{E}\{[\langle 0|x|0\rangle]|0\rangle\langle 0| + [\langle 0|x|1\rangle]|0\rangle\langle 1| \\
&\quad + [\langle 1|x|0\rangle]|1\rangle\langle 0| + [\langle 1|x|1\rangle]|1\rangle\langle 1|\} \cos(\nu t) \\
&= -\mathcal{E}\{P_{01}|0\rangle\langle 1| + P_{10}|1\rangle\langle 0|\} \cos(\nu t),
\end{aligned}
\tag{3.19}
$$

where we have used the parity selection rule (3.10) in the last line. This operator is hermitian as it should since by construction $P_{01} \equiv e\langle 0|x|1\rangle$, $P_{10} \equiv e\langle 1|x|0\rangle$, and therefore $P_{10} = \overline{P}_{01}$.

In the subspace S, the Schrödinger equation (3.1) is now equivalent to the system

$$
\frac{d}{dt}\begin{bmatrix} c_0(t) \\ c_1(t) \end{bmatrix} = \begin{bmatrix} 0 & i\Omega_R e^{-i\phi - i\omega t}\cos\nu t \\ i\Omega_R e^{i\phi + i\omega t}\cos\nu t & 0 \end{bmatrix}\begin{bmatrix} c_0(t) \\ c_1(t) \end{bmatrix},
\tag{3.20}
$$

where

$$
\Omega_R \equiv \frac{|P_{10}|\mathcal{E}}{\hbar} = \text{the Rabi frequency,}
\tag{3.21}
$$

and

$$
\phi = \text{phase of } P_{10}: \quad P_{10} = |P_{10}|e^{i\phi}.
\tag{3.22}
$$

As a final step, we make the so-called *rotating wave or dipole approximation* in (3.20) and drop the rapidly oscillating terms $\propto e^{\pm i(\omega + \nu)t}$ that generate in perturbation theory, after integration, small contributions involving the non-resonant denominator $\omega + \nu$. This approximation corresponds to the asymptotic limit

$$
\frac{\omega + \nu}{\omega - \nu} = \frac{2\omega}{\Delta} - 1 \to \infty
\tag{3.23}
$$

where the *detuning* is defined as

$$
\Delta = \omega - \nu.
\tag{3.24}
$$

The *rotating wave approximation* has the equivalent effect of replacing the interaction Hamiltonian (3.19) to the following:

$$
H_1 \mapsto -\frac{1}{2}\mathcal{E}\left\{P_{01}|0\rangle\langle 1|e^{i\nu t} + P_{10}|1\rangle\langle 0|e^{-i\nu t}\right\}
\tag{3.25}
$$

Remark 3.1.2. In the regime (3.23), the atomic transition frequency ω actually disappears from the theory and only frequency differences occur. The rotating wave approximation has a similar status as the 2-level approximation: typically, $\omega + \nu$ is comparable to the frequencies $\omega_n - \omega_{n'}$ of the non-resonant transitions in (3.13), and both contributions must be neglected simultaneously for consistency. From an experimental point of view, terms like $e^{\pm i(\omega+\nu)t}$ represent high frequency oscillations which cannot be observed with laboratory instrumentation. $\square$

3.1.3   Single atom qubit rotation

Within the rotating wave approximation, we obtain from (3.20), using the detuning (3.24),

$$
\begin{cases}
\dot{c}_0(t) = \dfrac{i\Omega_R}{2} e^{-i\phi} e^{i\Delta t} c_1(t), \\[2mm]
\dot{c}_1(t) = \dfrac{i\Omega_R}{2} e^{i\phi} e^{-i\Delta t} c_0(t),
\end{cases}
\tag{3.26}
$$

which in turn leads to a second order single ODE with constant coefficients

$$
\ddot{c}_0 - i\Delta \dot{c}_0 + \frac{\Omega_R^2}{4} c_0 = 0.
\tag{3.27}
$$

The atomic transition frequency no longer appears here, except via the detuning. Solving for c_0 and c_1 in (3.26) and (3.27), we obtain the explicit solution in terms of the initial condition $(c_0(0), c_1(0))$:

$$
\begin{bmatrix} c_0(t) \\ c_1(t) \end{bmatrix} =
\begin{bmatrix}
e^{i\frac{\Delta}{2}t} \left[\cos\left(\frac{\Omega}{2}t\right) - i\frac{\Delta}{\Omega} \sin\left(\frac{\Omega}{2}t\right) \right] & e^{i\frac{\Delta}{2}t} \cdot i \cdot \frac{\Omega_R}{\Omega} e^{-i\phi} \sin\left(\frac{\Omega}{2}t\right) \\[2mm]
e^{-i\frac{\Delta}{2}t} \cdot i\frac{\Omega_R}{\Omega} e^{i\phi} \sin\left(\frac{\Omega}{2}t\right) & e^{-i\frac{\Delta}{2}t} \left[\cos\left(\frac{\Omega}{2}t\right) + i\frac{\Delta}{\Omega} \sin\left(\frac{\Omega}{2}t\right) \right]
\end{bmatrix}
\begin{bmatrix} c_0(0) \\ c_1(0) \end{bmatrix},
\tag{3.28}
$$

where the generalized Rabi frequency is $\Omega = \sqrt{\Omega_R^2 + \Delta^2}$. This is illustrated in Fig. 3.2. One checks that the perturbative result (3.16) is recovered by performing the expansion to lowest order in Ω_R/Δ. The advantage of the 2-level and rotating-wave approximations is that one can obtain analytically the non-perturbative result (3.28).

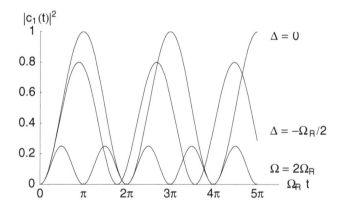

Figure 3.2: Probability $|c_1(t)|^2$ (3.28) of finding the atom in the excited state, after illuminating it for a time t with a near-resonant laser field. The atom is initially in the ground state: $(c_0(0), c_1(0)) = (1, 0)$. Time is plotted in units of the inverse Rabi frequency Ω_R. The selected detunings are $\Delta = 0, -\Omega_R/2, \sqrt{3}\,\Omega_R$ (from top to bottom).

Consider as an example the case of a laser at resonance,

$$\Delta = \omega - \nu = 0, \tag{3.29}$$

and, thus

$$\frac{\Delta}{\Omega} = 0, \quad \Omega_R = \Omega. \tag{3.30}$$

The above yields

$$\begin{bmatrix} c_0(t) \\ c_1(t) \end{bmatrix} = \begin{bmatrix} \cos\left(\frac{\Omega}{2}t\right) & ie^{-i\phi}\sin\left(\frac{\Omega}{2}t\right) \\ ie^{i\phi}\sin\left(\frac{\Omega}{2}t\right) & \cos\left(\frac{\Omega}{2}t\right) \end{bmatrix} \begin{bmatrix} c_0(0) \\ c_1(0) \end{bmatrix}. \tag{3.31}$$

(Note that if, instead, we relate $C_j(t)$ to $C_j(0)$ for $j = 0, 1$, then because of (3.11), the corresponding unitary matrix will contain some additional phase factor(s), unlike the (special) unitary matrix in (3.31) whose determinant is 1.) Write

$$\theta = \frac{\Omega}{2}t \tag{3.32}$$

$$\phi' = -\phi + \pi,$$

and re-name ϕ' as ϕ. Then the above matrix becomes

$$U_{\theta,\phi} \equiv \begin{bmatrix} \cos\theta & -ie^{i\phi}\sin\theta \\ -ie^{-i\phi}\sin\theta & \cos\theta \end{bmatrix}. \tag{3.33}$$

Note that a shift of the initial phase of the external electric field, $\nu t \mapsto \nu t + \varphi$ is equivalent to a shift of ϕ. Note also that θ in (3.33) depends on t. We have just proven the following.

Theorem 3.1.1. *An electric field near-resonant with an atomic transition, when applied for a finite time, realizes an arbitrary* 1-bit *(Rabi) rotation unitary gate (3.33), with parameters controlled by the electric field strength, its phase, and its detuning relative to the atomic resonance.* ☐

Remark 3.1.3. If we write the interaction Hamiltonian as [57, (5.2.44), p. 157]

$$H_1 = -\frac{\hbar\Omega}{2}[e^{-i\phi}|1\rangle\langle 0| + e^{i\phi}|0\rangle\langle 1|], \tag{3.34}$$

then we obtain

$$e^{-\frac{i}{\hbar}H_1 t} = \begin{bmatrix} \cos\left(\frac{\Omega}{2}t\right) & ie^{-i\phi}\sin\left(\frac{\Omega}{2}t\right) \\ ie^{i\phi}\sin\left(\frac{\Omega}{2}t\right) & \cos\left(\frac{\Omega}{2}t\right) \end{bmatrix}, \tag{3.35}$$

which is the same as the matrix in (3.31). The Hamiltonian (3.34) is thus called the *effective (interaction) Hamiltonian* for the 2-level atom. It represents the essence of the original Hamiltonian in (3.1) after simplifying assumptions. From now on, for the many physical systems under discussion, we will simply use the effective Hamiltonians for such systems as supported by theory and/or experiments, rather than to derive the Hamiltonians from scratch. ☐

3.1.4   Two-level atom hardware

In this section, we review in a simplified manner some of the atomic physics background of atom-light interactions. More information can be found in any textbook on atomic physics, for example [21].

Figs. 3.3 and 3.4 illustrate the above discussion with transitions in the Rubidium (Rb) atom. Atomic levels are represented by thick horizontal lines, and electric (magnetic) dipole transition by thin (dashed) lines connecting the levels, respectively. The symbols at the levels give some of the quantum numbers that characterize them in atomic spectroscopy. Energy increases from bottom to top.

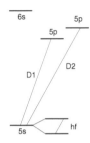

Figure 3.3: Ground state and lowest electronically excited states in the Rubidium 87 atom. D1, D2: strong electric dipole transitions, hf: splitting between the hyperfine components of the ground state.

Rb 87 atom	
D1 transition frequency	$\omega_{D1}/2\pi = 377\,\text{THz}$
D2 transition frequency	$\omega_{D1}/2\pi = 384\,\text{THz}$
D1 decay rate	$\Gamma_{D1}/2\pi = 5.75\,\text{MHz}$
D2 decay rate	$\Gamma_{D2}/2\pi = 6.07\,\text{MHz}$
D1 saturation intensity	$I_{D1} = 1.49\,\text{mW/cm}^2$
D2 saturation intensity	$I_{D2} = 1.67\,\text{mW/cm}^2$
hf splitting frequency	$\omega_{hf}/2\pi = 3.04\,\text{GHz}$

Table 3.1: Characteristic parameters in the Rubidium 87 atom.

Fig. 3.3 shows the ground state of the Rb atom, with the two main transitions that give rise to the D1 and D2 lines in the emission spectrum. Their frequencies are given in Table 3.1. The letters s and p indicate the "angular momentum quantum number" cf. Section 2.2; a selection rule (cf. Box 3.1) states that these must differ for an electric dipole transition. The excited 5p state is split in two levels with different energies due to spin-orbit coupling (cf. (2.30)), this is called the "fine structure". The ground state shows a similar splitting (electron-nucleus spin-spin coupling, "hyperfine structure"), but on a much smaller energy scale.

The excited states of the D1 and D2 transitions are unstable and decay to the ground state by spontaneously emitting a photon with the corresponding energy $\hbar\omega_{D1,D2}$. The corresponding decay rates are given in Table 3.1, a formula can be found in (4.58). When one of the qubit states coincides with an electronically excited state, the gate operation has to be performed fast compared to the inverse decay rates, in order to avoid spontaneous emission that changes the qubit state in an uncontrolled way. An alternative qubit implementation uses the hyperfine components of the ground state that are both stable. This qubit can be manipulated using either magnetic radiation near-resonant with the hyperfine splitting frequency ("microwave transition", see Table 3.1), or two optical fields that are near-resonant with either the D1 or D2 line and whose frequency difference is close to the hyperfine splitting (a "Raman transition").

In Fig. 3.4, we display some so-called Rydberg levels of the Rb 87 atom. These correspond to stationary states with a large principal quantum number (denoted n) and close to the dissociation edge. These states have similarities to Kepler orbits in the gravitational field of a star, with the planet playing the role of the electron. Electric dipole transitions between these levels are in the microwave range (see Table 3.2), and the rates for spontaneous decay are significantly smaller than that for Fig. 3.3. Another advantage is their large electric dipole moment, typically thousand times larger than that for the D1 lines of the alkali atoms. This allows to achieve with very weak fields a strong coupling (Rabi frequency larger than the spontaneous decay rate).

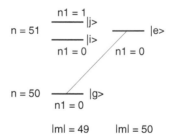

Figure 3.4: Selected Rydberg levels in the Rubidium 87 atom. The quantum numbers are n, n_1, and m, with values as shown. The states $|m| < n - 1$ acquire slightly different energies in a static electric field, see Table 3.2. A qubit is encoded in the subspace spanned by the states $|g\rangle$ and $|e\rangle$ that are connected by an electric dipole transition. The states $|i\rangle$ and $|j\rangle$ are used as auxiliary states for preparation and manipulation.

From the Tables given here, one can check the accuracy of the 2-level and rotating-wave approximations. For the Rb D1 line, a typical "far detuned" field might lie right in between the D1 and D2 transitions. In this situation, the ratio between resonant and non-resonant frequency denomi-

Rb 87 atom Rydberg levels	
ge transition frequency	$\omega_{ge}/2\pi = 51\,099\,\text{MHz}$
ei frequency shift	$\omega_{\beta\gamma}/2\pi = 0.97\,\text{MHz}/(\text{V/m})$
ij frequency shift	$\omega_{\beta\gamma}/2\pi = 1.94\,\text{MHz}/(\text{V/m})$
e decay rate	$\Gamma_e/2\pi = 5.3\,\text{Hz}$
g decay rate	$\Gamma_g/2\pi = 5.3\,\text{Hz}$

Table 3.2: Transition frequencies between selected Rydberg levels in the Rubidium 87 atom.

nators (see (3.23)) is of the order of

$$\frac{\omega_{D1} + \omega_{D2}}{\omega_{D1} - \omega_{D2}} \approx 100, \tag{3.36}$$

and even larger values are obtained closer to resonance.

As mentioned before, an accurate manipulation of qubit states requires that the relevant operation time is shorter than the lifetime $1/\Gamma$ of the involved atomic levels. This requires a sufficiently large Rabi frequency. From the parameter "saturation intensity" given in Table 3.1, one can compute

$$\frac{\Omega_{D1}}{\Gamma_{D1}} = \sqrt{\frac{I}{2I_{D1}}}, \tag{3.37}$$

where $I = \frac{1}{2}\epsilon_0 c |\mathcal{E}|^2$ is the laser intensity (for the plane wave of (3.6)). A glance at Table 3.1 also shows that one typically has $\Omega \ll \omega_{D1}$.

3.2 Quantization of the electromagnetic field

The quantization of the electromagnetic radiation field as simple harmonic oscillators is important in quantum optics. This fundamental contribution is due to Dirac. Here we provide a motivation following the approach of [57, Chap. 1, pp. 3–4].

3.2.1 Normal mode expansion

We begin with the classical description of the field based on Maxwell's equations. These equations relate the electric and magnetic field vectors E and B, respectively. Maxwell's equations lead to the following wave equation for the electric field:

$$\nabla^2 E - \frac{1}{c^2}\frac{\partial^2 E}{\partial t^2} = 0, \tag{3.38}$$

along with a corresponding wave equation for the magnetic field. The electric field has the spatial dependence appropriate for a cavity resonator of length L, as illustrated in Fig. 3.5. An example is the laser resonant optical

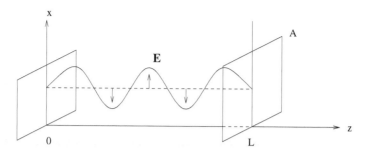

Figure 3.5: Illustration of a normal mode for the electric field (3.39) in a cavity (length L and cross-section A). The curve illustrates the variation of the electric field vector along the cavity axis; see (3.39).

cavity in Section 2.10. We take the electric field to be linearly polarized in the x-direction and expand in the normal modes (so-called in the sense that they constitute orthogonal coordinates for oscillations) of the cavity

$$E_x(z,t) = \sum_j A_j q_j(t) \sin(k_j z), \qquad (3.39)$$

where q_j is the normal mode amplitude with the dimension of a length, $k_j = j\pi/L$, with $j = 1, 2, 3, \ldots$, and

$$A_j = \left(\frac{2\nu_j^2 m_j}{V\epsilon_0} \right)^{1/2}, \qquad (3.40)$$

with $\nu_j = j\pi c/L$ being the cavity eigenfrequency; $V = LA$ (A is the transverse area of the optical resonator) is the volume of the resonator and m_j is a constant with the dimension of mass. The constant m_j has been included only to establish the analogy between the dynamical problem of a single mode of the electromagnetic field and that of the simple harmonic oscillator. The equivalent mechanical oscillator will have a mass m_j, and a Cartesian coordinate q_j. The nonvanishing component of the magnetic field B_y in the cavity is obtained from (3.39) as

$$B_y = \sum_j A_j \left(\frac{\dot{q}_j \epsilon_0}{k_j} \right) \cos(k_j z). \qquad (3.41)$$

The classical Hamiltonian for the field is

$$H_{cl} = \frac{1}{2} \int_V d^3r (\epsilon_0 E_x^2 + \frac{1}{\mu_0} B_y^2), \qquad (3.42)$$

where the integration is over the volume of the cavity and $\epsilon_0 \mu_0 c^2 = 1$. It follows, on substituting from (3.39) and (3.41) for E_x and B_y, respectively, in (3.42), that

$$H_{cl} = \frac{1}{2} \sum_j (m_j \nu_j^2 q_j^2 + m_j \dot{q}_j^2)$$

$$= \frac{1}{2} \sum_j \left(m_j \nu_j^2 q_j^2 + \frac{p_j^2}{m_j} \right), \tag{3.43}$$

where $p_j = m_j \dot{q}_j$ is the canonical momentum of the jth mode. (3.43) expresses the Hamiltonian of the radiation field as a sum of independent oscillator energies. This suggests correctly that each mode of the field is dynamically equivalent to a mechanical harmonic oscillator.

3.2.2   Field mode quantization; the harmonic oscillator

The present dynamical problem can be quantized by replacing q_j and p_j with operators which obey the commutation relations

the commutator of q_j with $p_{j'} = [q_j, p_{j'}] = i\hbar \delta_{jj'}$, \qquad (3.44)

$$[q_j, q_{j'}] = [p_j, p_{j'}] = 0. \tag{3.45}$$

In the following we shall restrict ourselves to a single mode of the radiation field modeled by a simple harmonic oscillator. The corresponding Hamiltonian is therefore given by

$$H = \frac{1}{2m} p^2 + \frac{1}{2} m \nu^2 x^2, \tag{3.46}$$

where

$p = $ the momentum operator,

$m = $ a constant with the dimensions of mass,

$x = $ the position operator (corresponding to the q variable above),

$\nu = $ the natural (circular) frequency of the oscillator

(a parameter related to the potential depth). \qquad (3.47)

The choice

$$p = \frac{\hbar}{i} \frac{d}{dx}, \tag{3.48}$$

for the momentum operator attains the commutation relations (3.44) and (3.45) because for any sufficiently smooth function ϕ on $\mathbb{R}$, one has

$$x \frac{d}{dx} \phi - \frac{d}{dx}(x\phi) = \phi. \tag{3.49}$$

This implementation of the commutation relations is called the "position representation". The Hamiltonian (3.46) becomes a linear differential operator on a vector space of function (called "wave functions"), so that the

Schrödinger equation is represented by a partial differential equation. Of particular interest are the eigenfunctions ψ of the Hamiltonian operator that satisfy the "stationary Schrödinger equation"

$$H\psi = -\frac{\hbar^2}{2m}\frac{d^2\psi}{dx^2} + \frac{1}{2}m\nu^2 x^2\psi = E\psi. \tag{3.50}$$

This differential equation is the *harmonic oscillator*, of fundamental importance in quantum mechanics. It can be re-scaled using the variables

$$y = \sqrt{\frac{m\nu}{h}}\, x, \quad \lambda = \frac{E}{\hbar\nu}, \tag{3.51}$$

thus (3.50) becomes

$$\frac{1}{2}\left[\frac{d^2\psi}{dy^2} - y^2\psi\right] = -\lambda\psi. \tag{3.52}$$

We now define two operators

$$a = \frac{1}{\sqrt{2}}\left(\frac{d}{dy} + y\right), \quad a^\dagger = -\frac{1}{\sqrt{2}}\left(\frac{d}{dy} - y\right). \tag{3.53}$$

Note that $a^\dagger$ is the Hermitian adjoint operator of a with respect to the $L^2(\mathbb{R})$ inner product. Then it is easy to check that for any sufficiently smooth function ϕ on $\mathbb{R}$,

$$(aa^\dagger - a^\dagger a)\phi = \phi, \tag{3.54}$$

i.e.,

$$\text{the commutator of } a \text{ with } a^\dagger = [a, a^\dagger] = 1, \tag{3.55}$$

where 1 denotes the identity operator but is often simply written as (the scalar) 1 in the physics literature.

3.2.3 Energy spectrum and stationary states

It is now possible to verify the following:
(i) Let $\widetilde{H} \equiv -\frac{1}{2}\left(\frac{d^2}{dy^2} - y^2\right)$. Then

$$\widetilde{H} = a^\dagger a + \frac{1}{2}1 = aa^\dagger - \frac{1}{2}1. \tag{3.56}$$

(ii) If ψ_λ is an eigenstate of $\widetilde{H}$ satisfying

$$\widetilde{H}\psi_\lambda = \lambda\psi_\lambda \quad \text{(i.e., (3.52))}, \tag{3.57}$$

then so are $a\psi_\lambda$ and $a^\dagger\psi_\lambda$:

$$\widetilde{H}(a\psi_\lambda) = (\lambda - 1)\,a\psi_\lambda,$$
$$\widetilde{H}(a^\dagger\psi_\lambda) = (\lambda + 1)\,a^\dagger\psi_k.$$

(iii) If λ_0 is the lowest eigenvalue (or energy level) of $\widetilde{H}$, then

$$\lambda_0 = \frac{1}{2}. \tag{3.58}$$

(iv) All of the eigenvalues of $\widetilde{H}$ are given by

$$\lambda_n = \left(n + \frac{1}{2}\right), \qquad n = 0, 1, 2, \dots . \tag{3.59}$$

(v) Denote the n-th eigenstate ψ_n by $|n\rangle$, for $n = 0, 1, 2, \dots$. Then

$$|n\rangle = \frac{(a^\dagger)^n}{(n!)^{1/2}} |0\rangle. \tag{3.60}$$

Furthermore,

$$\begin{aligned}
a^\dagger a |n\rangle &= n|n\rangle, & n &= 0, 1, 2, \dots; \\
a^\dagger |n\rangle &= \sqrt{n+1}\, |n+1\rangle, & n &= 0, 1, 2, \dots; \\
a|n\rangle &= \sqrt{n}\, |n-1\rangle, & n &= 1, 2, \dots; \quad a|0\rangle = 0.
\end{aligned} \tag{3.61}$$

(vi) The completeness relation is

$$\mathbf{1} = \sum_{n=0}^{\infty} |n\rangle\langle n|. \tag{3.62}$$

(vii) The wave functions are

$$\psi_j(y) = N_j H_j(y) e^{-y^2/2}, \qquad y = \left(\frac{m\nu}{\hbar}\right)^{1/2} x, \tag{3.63}$$

$j = 0, 1, 2, \dots$, where $H_j(y)$ are the Hermite polynomials of degree j, and N_j is a normalization factor. See Fig. 3.6 for an illustration.

Remark 3.2.1.   (i) The energy levels $\left(n + \frac{1}{2}\right) \hbar\nu$ of H (after converting back to the x-coordinate from the y-coordinate in (3.59)) may be interpreted as the presence of n quanta or photons of energy $\hbar\nu$. The eigenstates $|n\rangle$ are called the Fock states or the photon number states.

   (ii) The energy level $E = \frac{1}{2}\hbar\nu$ (from (3.51) and (3.58)) is called the ground state energy.

   (iii) Because of (3.61), we call a and $a^\dagger$, respectively, the annihilation and creation operators. $\square$

Remark 3.2.2. By applying the outlined quantization procedure to all cavity modes, we get the following expansion for the electric field operator

$$E_x(z, t) = \sum_j \mathcal{E}_j \left(a_j(t) \sin(k_j z) + \text{h.c.}\right) \tag{3.64}$$

$$\mathcal{E}_j = \left(\frac{\hbar\nu_j}{LA\epsilon_0}\right)^{1/2} \tag{3.65}$$

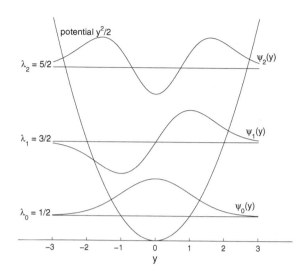

Figure 3.6: Illustration of the lowest stationary wave functions for a single cavity mode. The coordinate y is proportional to the electric field strength, and $|\psi(y)|^2$ gives the probability density of finding the corresponding value. For the "particle in a potential" analogy (see the Hamiltonian (3.46)), the parabola gives the "potential" $\frac{1}{2}y^2$ appearing in (3.52). The horizontal lines indicate the energy eigenvalues $\frac{1}{2}, \frac{3}{2}, \frac{5}{2}$. These lines also serve as (shifted) horizontal axes to plot the wave functions (3.63).

Here, the constant $\mathcal{E}_j$ has the dimension "electric field" and gives the order of magnitude of the field due to a single photon. The annihilation operator has the time-dependence

$$a_j(t) = a_j e^{-i\nu_j t}, \tag{3.66}$$

that follows from the operator equivalent of the Schrödinger equation (the "Heisenberg equation"),

$$\frac{da_j}{dt} = -\frac{i}{\hbar}\left[H, a_j\right]. \tag{3.67}$$

We use the initial condition $a_j(0) = a_j$ and the Hamiltonian $H = \sum_j \hbar\nu_j \left(a_j^\dagger a_j + \frac{1}{2}\right)$. $\qquad\square$

3.3 Cavity QED

Cavity QED (quantum electrodynamics) is a system that enables the coupling of single atoms to only a few photons in a resonant cavity. It is realized by applying a large laser electric field in a narrow band of frequencies within a small Fabry–Perot cavity consisting of highly reflective mirrors.

See, for example, the schematic setup of Fig. 3.7. Inside the cavity, the mode structure of the radiation field is drastically changed. In particular, there is a single privileged cavity mode that is resonant with the atomic transition and dominates the atom-field dynamics so completely that the influence of all other modes is negligible. This section provides a historical overview (Subsection 3.3.1) and illustrative device designs to facilitate the understanding of the technical analysis to be given in the next section.

3.3.1   A brief historical account of cavity QED

Historically, in the development of cavity QED, the research interests evolved through several topics. The early theoretical predictions were concerned with the modifications of atomic energy shifts and spontaneous decay rates by reflecting boundaries (the Casimir–Polder force [11], the Purcell effect [30, 45]), which have become paradigms of "perturbative cavity QED" [2, 24]. The groups led by E. Hinds (Yale Univ., New Haven, Connecticut, now in London), S. Haroche (Ecole Normal Supérieure in Paris, France), and D. Heinzen and M. Feld (Cambridge, Massachusetts, Heinzen now in Austin, Texas) have performed pioneering experiments in this connection that demonstrated cavity-modified spontaneous decay [18, 62], shifted transition frequencies [22], and vacuum-induced forces [52, 58]. A review written by Hinds is available [25]. The manipulation of few-photon fields by atoms sent through a cavity started in the 1980s by the group of H. Walther (Max Planck Institute for Quantum Optics, Garching near Munich in Germany) [39], and has led to the development of the micromaser [65]. The groups of Haroche (Ecole Normal Supérieure in Paris, France) and of H. Kimble (Caltech, Pasadena, California) have demonstrated that many atoms that couple collectively to the same field mode show phenomena such as superradiance [47] and collective Rabi oscillations [7, 28]. More recently, the controlled entanglement of individual atoms and photons has opened the possibility for realizing quantum bits of information and implementing their swap between different physical carriers [19, 46].

We focus in the following on the physics explored with single atoms in high-quality cavities. Two routes have been taken to achieve highly reflecting mirrors and long photon storage times. At microwave frequencies and low temperatures, superconducting mirrors have been used since the 1980s by the groups of Haroche and of Walther, with photons now bouncing at least 10^{10} times back and forth. In the visible range, highly reflecting multilayer dielectrics (99.9999%) have become available at the beginning of the 1990s (group of Kimble, and that of G. Rempe first in Konstanz, Germany, and later in Garching, Germany). The two frequency bands require different atomic transitions and experimental schemes. In particular, microwave photons are difficult to detect, so that information about the cavity field is mostly extracted from atoms that crossed it, as in the case of the micromaser [39]. In the visible range, in contrast, the field transmitted through the cavity mirrors is directly measurable. Those data can be used to detect the presence of as few as a single atom in the cavity [51] and to monitor its dynamics [33, 41]. For a review, see Pinkse and Rempe [44].

The control over the atoms injected between the cavity mirrors has improved over the years. The first experiments employed atomic beams with a sufficiently low flux so that the cavity volume contained at most one atom at any time. The interaction time can either be post-selected via a velocity measurement, or set to a desired value by chopping the beam with rotating shutters. The groups led by Walther and Haroche sent a beam of the so-called excited Rydberg atoms through a microwave cavity to replenish it with photons. This has led to the development of a laser working in the microwave band with the smallest possible active medium, the "micromaser" [10, 39]. The generation of photon number states is another feat of this technique [6, 63, 66]. These experiments have opened the doorway to the so-called "strong coupling" regime (cf. (2.104)). Significant effects such as the Rabi oscillations occur in this regime already on the level of individual atoms and photons.

A fundamental experiment in this setting was performed by Rempe and Walther [50]: they observed how the frequency of Rabi oscillations for a single atom in a microwave cavity depends on the photon number. For a cavity state with a broad quantum-mechanical uncertainty in the photon number (a so-called "coherent state"), the Rabi oscillations get "out of step" and their envelope follows a "collapse" and "revival" scenario. The Haroche group reported an improved experiment where up to five Rabi oscillations could be resolved [8]. Another key achievement is the demonstration by the Kimble group that a single atom can change the resonance frequency of a cavity in the visible band: a frequency splitting is observed when the atomic transition is on resonance with the cavity mode, even when there are no photons in the cavity ("vacuum Rabi splitting") [60].

With the advent of laser cooling and trapping (Nobel prize 1997 awarded to S. Chu, C. Cohen-Tannoudji, and W. Phillips), atoms with a much lower velocity became available, allowing far longer interaction times. For example, with a so-called "atomic fountain", cold atoms are injected into the cavity by launching them upwards from a spatially separated laser trap. Their passage through a cavity can be detected by a change in the transmitted light, as demonstrated by the Kimble group [26, 33]. When the atoms reach the highest point of their trajectory, the cavity resonance is shifted and the atoms can be "caught" by increasing the strength of the cavity field with the proper timing [27, 43, 67]. It took some time to construct an independent laser trap within the cavity because the laser beams required for the usual magneto-optical traps (MOT) are blocked by the mirrors (sometimes only a fraction of a millimeter apart). A "trap within the cavity" was achieved by the Walther group, who surround the cavity with electrodes that create a confining potential for ions [20]. Recently, the so-called "far detuned optical dipole traps" have been created by shining an additional laser beam through the cavity mirrors, as reported by the groups of Kimble and Rempe [36, 67]. This has permitted the construction of a single-atom laser emitting in the visible band [38] and the possibility to re-examine cavity mode splittings due to individual atoms [4, 35]. It has also become possible to use a focused laser beam to deposit the atoms in the cavity ("optical tweezer" or "shuttle") [53], as demonstrated by the group led by M. Chapman (Georgia Tech, Atlanta, Georgia). Finally, we mention that atoms can also be cooled directly with the cavity field, by dissipating the excess energy through the nonzero transmission of the mirrors. This has been suggested theoretically by the groups of H. Ritsch (Innsbruck, Austria) and of S. Chu (Stanford, California) [16, 64, 12] and confirmed experimentally by the Rempe group [34].

The potential of cavity QED for quantum information processing was recognized early [61, 13], one of the original motivations being the realization of simple models (a single 2-state system coupled to a single field mode); see the review by Walther [65]. Important steps in this context are the exploration of the quantum measurement itself [9], and the nondestructive detection of the cavity photon number [42] by the Haroche group. More details on the quantum entanglement between atom and cavity mode can be found in the review papers [19, 32, 46]. A particularly promising feature of cavity QED is that it may provide an "interface" between photonic qubits (useful for long-distance communication) and atomic qubits (useful for long-time storage). Efficient sources that produce "single photons on demand" have already been implemented with atoms stored in cavities, by the Rempe and Kimble groups [23, 31, 37].

3.3.2 Cavity hardware

We focus on two examples: cavities for the optical range (near resonant with the atomic transitions shown in Fig. 3.3) and for the microwave range (atomic levels of Fig. 3.4).

Fig. 3.7 shows schematically an experimental setup where atoms are trapped in a laser beam and can be "shuttled" into and out of a cavity. The cavity mirrors are located at the conical tips; a larger mirror area is not required since the cavity mode can be chosen with a transverse, Gaussian confinement (see (3.69) below).

Fig. 3.8 shows in more detail the design of a cavity that is optimized for the optical frequency range. In Fig. 3.9, an actual photograph of the cavity is shown. Superimposed is an image of a cloud of cold trapped atoms and a laser beam that is used for the transfer into the cavity.

The mirrors of the cavity should have a very high reflectivity R to store the photons as long as possible. A typical number to characterize this is the decay rate κ for the field inside the cavity. It can be written in the form

$$\kappa = \frac{c}{2L}\left(1 - R\right) \tag{3.68}$$

where $2L/c$ is the time for bouncing back and forth between the mirrors and $1 - R$ the mirror transmission. (Actually, R is the average reflectivity

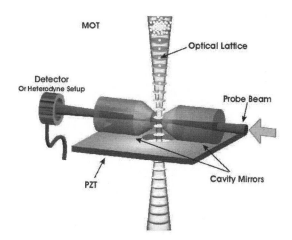

Figure 3.7: (Please see color insert following page 78.) Design scheme of a cavity setup using a laser beam to shuttle atoms in and out of the cavity. Courtesy of M. Chapman, Georgia Tech, Atlanta. Two counterpropagating laser beams focused through the cavity in the vertical direction produce an optical lattice, translating the lattice transports atoms collected in the magneto-optic trap (MOT) into the cavity mode below.

of both mirrors, and $1 - R$ also includes absorption losses.) High-quality cavities used in cavity QED experiments are based on dielectric (multilayer) mirrors and have a photon decay rate of the order of 1 MHz at frequencies in the visible ($\nu/2\pi \sim 500\,\mathrm{THz}$). With $1 - R \sim 10^{-6}$, the photons are reflected $1/(1 - R) =$ a million times before being transmitted or absorbed. This figure is also known as the cavity "finesse".

Typical dimensions for optical cavities are a spacing $L = 0.1\,mm$ between the mirrors and a mirror diameter of a few mm. The cavity modes can be classified by "quantum numbers" according to the number of nodes transversal to the cavity axis and along this axis. One example is the integer j in the expression $k_j = j\pi/L$ for the axial mode wavevector; see (3.39). In the transverse directions, the simplest mode is the one with quantum numbers "TEM$_{00}$" that has a Gaussian shape:

$$|\boldsymbol{E}_{00}(x,y,z)|^2 = (N/w(z)^2)\exp\left[-(x^2+y^2)/w(z)^2\right]\sin(k_j z)^2, \qquad (3.69)$$

$$w(z) = w\sqrt{1 + z^2/(k_j^2 w^4)}, \qquad (3.70)$$

where N is a normalization constant. For identical mirrors, the transverse width of the mode is smallest at the cavity center (located here at $z = 0$) where it is equal to the "waist" w; the mode becomes wider towards the mirrors, similar to a rotation hyperboloid. (3.69) is valid in the "paraxial" limit where the transverse behavior of the field is slow on the scale of the wavelength $2\pi/k_j$. This requires $k_j w \gg 1$. The transverse area A occurring

Figure 3.8: Design scheme of a high-quality cavity. The mirrors are circular and curved to confine the mode function around the cavity axis. The iso-field surface between the mirrors illustrates the antinodes of the mode. Courtesy of P. Pinkse, Max-Planck-Institut für Quantenoptik, Garching, Germany.

in the quantized field (see (3.64)) is of the order of the cross-section πw^2 of the Gaussian (3.69). The frequency spacing for axial modes is $c(k_{j+1} - k_j)/(2\pi) = c/2L \sim 1\,\mathrm{THz}$, and once a cavity mode is tuned near an atomic resonance, the other modes are far detuned and can be neglected. Due to the narrow cavity resonance, active stabilization of the cavity length is required to maintain the resonance conditions with an atomic transition. Further details can be found in [32].

Cavity QED based on Rydberg atoms requires cavities for microwave frequencies where metallic mirrors provide good reflectivity, in particular if they are superconducting. Even if the microwave wavelength is of the order of millimeters, the mirror surface has to be polished to better than $10\,nm$ roughness to avoid scattering loss. Photon lifetimes $1/\kappa \sim 1\,ms$ have been achieved. A photograph of an experimental setup is shown in Fig. 3.10. Typical dimensions are in the cm range for cavity length and diameter, the waist of the lowest mode being $w \approx 6\,mm$. Experiments with atoms in microwave cavities are reviewed in [46].

Figure 3.9: (Please see color insert.) Photograph illustrating experiments with individual atoms in an optical cavity: the atoms are first trapped and cooled with laser light (bright spot left). They are then transferred by a laser beam (blue horizontal line) into the cavity (center). They are held inside the cavity using an "optical tweezer" (green line). The distance between the cavity mirrors (viewed from the side, conical forms above and below the green line) is $0.5\ mm$, the atoms are transported along a distance of $14\ mm$. Courtesy of G. Rempe, Max-Planck-Institut für Quantenoptik, Garching, Germany.

3.4 Cavity QED for the quantum phase gate

In the following, we construct a model for a 2-qubit gate that can be realized with an atom placed inside the cavity.

3.4.1 Atom-cavity Hamiltonian

A 3-level atom is injected into the cavity. An electron in the atom has three levels, $|\alpha\rangle$, $|\beta\rangle$ and $|\gamma\rangle$, as shown in Fig. 3.11. Actually, the state $|\alpha\rangle$ will be used only as an *auxiliary* level because we choose to implement the first qubit, $|1\rangle$ and $|0\rangle$, in the states $|\beta\rangle$ and $|\gamma\rangle$, respectively. Once the atom enters the cavity, the strong electromagnetic field of the privileged cavity mode causes transitions of the electron between $|\alpha\rangle$ and $|\beta\rangle$, and a photon or photons are released or absorbed in this process. The cavity photons are far detuned with respect to the transition $|\alpha\rangle \leftrightarrow |\gamma\rangle$ so that the state $|\gamma\rangle$ remains a "spectator" or "detached". Of the photon states $|n\rangle = |0\rangle$, $|1\rangle$, $|2\rangle, \ldots$ inside the cavity only $|0\rangle$ and $|1\rangle$ will be important (which physically mean 0 or 1 photons inside the cavity), and they define the second qubit.

Figure 3.10: (Please see color insert.) Photograph of an partially mounted experimental setup with a high-quality cavity for the microwave range. In the central circular hole, one cavity mirror can be seen as a highly reflecting surface. Courtesy of J.-M. Raimond, Laboratoire Kastler-Brossel, Ecole Normale Supérieure, Paris.

The Hamiltonian for the atom-cavity field interaction is given by

$$H = H_0 + H_1 + H_2, \tag{3.71}$$

where

$H_0 = \dfrac{\hbar \omega_{\alpha\beta}}{2} (|\alpha\rangle\langle\alpha| - |\beta\rangle\langle\beta|) = $ the atom's Hamiltonian,

$H_1 = \hbar\nu a^\dagger a = $ the Hamiltonian of the laser electric field of the cavity,

$$\tag{3.72}$$

and

$H_2 = \hbar g(|\alpha\rangle\langle\beta|a + |\beta\rangle\langle\alpha|a^\dagger) = $ the interaction Hamiltonian of $\qquad$ (3.73) the laser field with the atom; with $g > 0$.

The coupling constant (with dimension frequency) is given by $g = |P_{\alpha\beta}|\mathcal{E}_c/\hbar = |P_{\alpha\beta}|(\nu/\hbar V\epsilon_0)^{1/2}$ where $\mathcal{E}_c$ is the electric field amplitude for a single photon in the cavity and $P_{\alpha\beta}$ the electric dipole transition moment, as defined in (3.9). See the derivation of (3.71) in [48]. The creation and annihilation operators $a^\dagger$ and a act on the cavity states $|n\rangle$ for $n = 0, 1, 2, \ldots, \infty$, according to (3.61). Note that they operate only on the second bit, while the rest of the operators operate only on the first bit (the atomic state). Also, the operator $|\alpha\rangle\langle\alpha| - |\beta\rangle\langle\beta|$ can be written as σ_z (see (2.43)), because its matrix representation with respect to the ordered basis $\{|\alpha\rangle, |\beta\rangle\}$ is exactly σ_z.

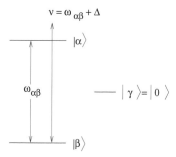

Figure 3.11: Atomic energy levels used in cavity QED. The cavity photons excite the electric dipole transition between $|\beta\rangle$ and $|\alpha\rangle$, leading to a phase shift for the qubit state $|\beta\rangle = |1\rangle$ under suitable conditions. The second qubit state $|\gamma\rangle = |0\rangle$ is unaffected by the cavity photons.

Since the state $|\gamma\rangle$ is detached from the others, its energy can be set to zero without loss of generality, leading to

$$H|\gamma\rangle = 0. \tag{3.74}$$

Lemma 3.4.1. *Let the underlying Hilbert space be*

$$\mathcal{H} = span\{|\alpha, n\rangle, |\beta, n\rangle, |\gamma, n\rangle | n = 0, 1, 2, \ldots\}, \tag{3.75}$$

where span S of a set S is the vector subspace formed (i.e., spanned) by finite linear combinations of all vectors in the set S. Then the Hamiltonian operator (3.71) has a family of 2-dimensional invariant subspaces

$$V_n = span\{|\alpha, n - 1\rangle, |\beta, n\rangle\}, \qquad n = 1, 2, \ldots . \tag{3.76}$$

Indeed, with respect to the ordered basis in (3.76), the matrix representation of H on V_n is given by

$$H|_{V_n} = \hbar \begin{bmatrix} \frac{1}{2}\omega_{\alpha\beta} + \nu(n-1) & g\sqrt{n} \\ g\sqrt{n} & -\frac{1}{2}\omega_{\alpha\beta} + \nu n \end{bmatrix}, \qquad n = 1, 2, 3, \ldots . \tag{3.77}$$

Proof. Even though the verification is straightforward, let us provide some details for the ease of future reference. We have, from (3.61), and (3.71)–(3.73),

$$H|\alpha, n - 1\rangle = H_0|\alpha, n - 1\rangle + H_1|\alpha, n - 1\rangle + H_2|\alpha, n - 1\rangle$$

$$= \left[\frac{\hbar\omega_{\alpha\beta}}{2}|\alpha, n - 1\rangle\right] + [\hbar\nu(n-1)|\alpha, n - 1\rangle] + [\hbar g\sqrt{n}|\beta, n\rangle]$$

$$= \hbar\left[\frac{\omega_{\alpha\beta}}{2} + \nu(n-1)\right]|\alpha, n - 1\rangle + \hbar[g\sqrt{n}]|\beta, n\rangle. \tag{3.78}$$

Similarly,

$$H|\beta, n\rangle = \hbar\left[-\frac{\omega_{\alpha\beta}}{2} + \nu n\right]|\beta, n\rangle + \hbar[g\sqrt{n}]|\alpha, n - 1\rangle. \tag{3.79}$$

Therefore, we obtain (3.77). $\qquad\qquad\qquad\qquad\qquad\qquad\qquad\square$

Lemma 3.4.2. *On the invariant subspace V_n in (3.76), the Hamiltonian operator H has two eigenstates*

$$\begin{cases} |+\rangle_n \equiv \cos\theta_n |\alpha, n-1\rangle - \sin\theta_n |\beta, n\rangle, \\ |-\rangle_n \equiv \sin\theta_n |\alpha, n-1\rangle + \cos\theta_n |\beta, n\rangle, \end{cases} \qquad (3.80)$$

where

$$\sin\theta_n \equiv \frac{\Omega_n - \Delta}{D}, \quad \cos\theta_n \equiv \frac{2g\sqrt{n}}{D},$$
$$D \equiv [(\Omega_n - \Delta)^2 + 4g^2 n]^{1/2},$$
$$\Omega_n \equiv (\Delta^2 + 4g^2 n)^{1/2},$$
$$\Delta \equiv \nu - \omega_{\alpha\beta} = \textit{the detuning frequency}, \qquad (3.81)$$

with eigenvalues (i.e., energy levels)

$$E_{\pm(n)} = \hbar \left[n\nu + \frac{1}{2}(-\nu \mp \Omega_n) \right], \qquad (3.82)$$

such that

$$H|+\rangle_n = E_{+(n)}|+\rangle_n, \quad H|-\rangle_n = E_{-(n)}|-\rangle_n. \qquad (3.83)$$

Proof. These eigenvalues and eigenvectors can be computed in a straightforward way, using the 2×2 matrix representation of H on V_n, (3.77), with respect to the ordered basis of V_n chosen as in (3.76). $\qquad \square$

Remark 3.4.1. The states $|+\rangle_n$ and $|-\rangle_n$ in (3.80) are called the *dressed states* in the sense that atoms are dressed by electromagnetic fields. These two states are related to the splitting of the spectral lines due to the electric field. An illustration can be found in Figure 3.12. $\qquad \square$

3.4.2   Large detuning limit

Now, let us assume *large detuning*:

$$|\Delta| \gg 2g\sqrt{n}. \qquad (3.84)$$

Also, $|\Delta|$ remains small in comparison to the transition frequency $\omega_{\alpha\beta}$ so that approximate resonance is maintained. Accordingly, this is satisfied for $n \ll (\omega_{\alpha\beta}/g)^2$. This is not a stringent condition since the upper limit is typically of the order of 10^{12}, depending on the physical implementation.

If the detuning is large,

$$\Omega_n = (\Delta^2 + 4g^2 n)^{1/2} = \Delta \left(1 + \frac{4g^2 n}{\Delta^2} \right)^{1/2} \approx \Delta + \frac{2g^2 n}{\Delta};$$
$$\sin\theta_n = (\Omega_n - \Delta)/D \approx 0; \quad \cos\theta_n = 2g\sqrt{n}/D \approx 1. \qquad (3.85)$$

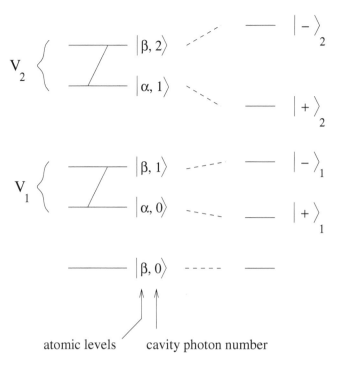

atomic levels   cavity photon number

Figure 3.12: Left: Energy levels of the composite system "2-level atom plus cavity mode". The invariant subspaces (3.76) are spanned by the levels connected by the slant dashed lines. Right: Dressed energy levels (3.82).

Thus, from (3.80),

$$|+\rangle_n \approx |\alpha, n-1\rangle, \quad |-\rangle_n \approx |\beta, n\rangle. \tag{3.86}$$

Remark 3.4.2. By the assumption (3.83) of large detuning, from (3.82) using (3.81) and (3.84), we now have

$$
\begin{aligned}
E_{+(n)} &= \hbar \left[n\nu + \frac{1}{2}(-\nu - \Omega_n) \right] \\
&\approx \hbar \left[\left(n - \frac{1}{2} \right)\nu - \frac{1}{2}\left(\Delta + \frac{2g^2 n}{\Delta} \right) \right] \\
\\
&= \hbar \left[\left(n - \frac{1}{2} \right)\nu - \frac{1}{2}(\nu - \omega_{\alpha\beta}) - \frac{g^2 n}{\Delta} \right] \\
&= \hbar \left[\frac{\omega_{\alpha\beta}}{2} + \nu(n-1) \right] - \frac{\hbar g^2 n}{\Delta}.
\end{aligned} \tag{3.87}
$$

Similarly, we have

$$E_{-(n)} \approx \hbar \left[-\frac{\omega_{\alpha\beta}}{2} + \nu n \right] + \frac{\hbar g^2 n}{\Delta}, \qquad (3.88)$$

such that

$$H|+\rangle_n \approx H|\alpha, n - 1\rangle = E_{+(n)}|\alpha, n - 1\rangle,$$
$$H|-\rangle_n \approx H|\beta, n\rangle = E_{-(n)}|\beta, n\rangle, \qquad (3.89)$$

where $E_{\pm(n)}$ are now given by (3.87) and (3.88).

Thus, we see that the assumption (3.83) of large detuning causes the off-diagonal terms in the matrix (3.77) (which are contributed by the *interaction Hamiltonian* H_2 in (3.72)) to disappear. The de facto or, *effective interaction Hamiltonian*, has now become

$$\widetilde{H}_2 = -\frac{\hbar g^2}{\Delta} (aa^\dagger |\alpha\rangle\langle\alpha| - a^\dagger a|\beta\rangle\langle\beta|), \qquad (3.90)$$

such that $\widetilde{H} \equiv H_0 + H_1 + \widetilde{H}_2$ now admits a diagonal matrix representation

$$\widetilde{H}|_{V_n} = \begin{bmatrix} E_{+(n)} & 0 \\ 0 & E_{-(n)} \end{bmatrix} \qquad (3.91)$$

with respect to the ordered basis $\{|\alpha, n - 1\rangle, |\beta, n\rangle\}$ of V_n, with $E_{\pm(n)}$ here given by (3.87) and (3.88). $\qquad \square$

3.4.3   Two-qubit operation

We now define the first qubit by

$$|1\rangle = |\beta\rangle, \quad |0\rangle = |\gamma\rangle, \qquad (3.92)$$

where $|\gamma\rangle$ is the state in Fig. 3.11 that is *detached* (or *off-resonance*) from the Hamiltonian H.

For the second qubit, consider the Fock states $|0\rangle$, $|1\rangle$ of the cavity field with zero or one photon. The cavity field is near-resonant with the $|\alpha\rangle \leftrightarrow |\beta\rangle$ transition: $\nu = \omega_{\alpha\beta} + \Delta$ (see Fig. 3.11), while being too far detuned to couple the state $|\gamma\rangle$ to any of the $|\alpha\rangle$, $|\beta\rangle$.

Theorem 3.4.3. *With respect to the effective Hamiltonian $\widetilde{H}_2$ (3.90),*

$$\widetilde{V} \equiv span\{|0,0\rangle, |0,1\rangle, |1,0\rangle, |1,1\rangle\} \qquad (3.93)$$

is an invariant subspace in the underlying Hilbert space (3.75), such that

$$\begin{cases} \widetilde{H}_2|0,0\rangle = 0, \quad \widetilde{H}_2|0,1\rangle = 0, \quad \widetilde{H}_2|1,0\rangle = 0, \\ \widetilde{H}_2|1,1\rangle = E_{1,1}|1,1\rangle, \ where \ E_{1,1} = \frac{\hbar g^2}{\Delta}. \end{cases} \qquad (3.94)$$

Consequently, with respect to the ordered basis of $\widetilde{V}$ in (3.93), $\widetilde{H}_2$ admits a diagonal matrix representation

$$\widetilde{H}_2|_{\widetilde{V}} = \begin{bmatrix} 0 & & & \bigcirc \\ & 0 & & \\ & & 0 & \\ \bigcirc & & & E_{1,1} \end{bmatrix}, \tag{3.95}$$

with the evolution operator

$$e^{-i\widetilde{H}_2 t/\hbar}|_{\widetilde{V}} = \begin{bmatrix} 1 & & & \bigcirc \\ & 1 & & \\ & & 1 & \\ \bigcirc & & & \exp\left(-i\frac{E_{1,1}t}{\hbar}\right) \end{bmatrix}. \tag{3.96}$$

Proof. We have

$$\widetilde{H}_2|0,0\rangle = \widetilde{H}_2|\gamma,0\rangle = 0 \quad \text{and} \quad \widetilde{H}_2|0,1\rangle = \widetilde{H}_2|\gamma,1\rangle = 0, \tag{3.97}$$

because $\langle\alpha|\gamma\rangle = \langle\beta|\gamma\rangle = 0$. Also,

$$\widetilde{H}_2|1,0\rangle = \widetilde{H}_2|\beta,0\rangle = 0, \tag{3.98}$$

because $a^\dagger a|n\rangle = 0$ for the zero-photon state $n = 0$. But

$$\widetilde{H}_2|1,1\rangle = \widetilde{H}_2|\beta,1\rangle = E_{1,1}|1,1\rangle. \tag{3.99}$$

So both (3.95) and (3.96) follow. $\qquad\square$

The unitary operator in (3.96) gives us the QPG (quantum phase gate, cf. (2.47))

$$Q_\eta = \begin{bmatrix} 1 & & & \\ & 1 & & \bigcirc \\ & & 1 & \\ \bigcirc & & & e^{i\eta} \end{bmatrix} \tag{3.100}$$

with

$$\eta \equiv -\frac{E_{1,1}}{\hbar}t. \tag{3.101}$$

This operation does not factorize into two single qubit operations and therefore entangles the two qubits. This is the key difference with respect to the time evolution under the free Hamiltonian

$$\widetilde{H} = H_0 + H_1 = \hbar\left[\frac{\omega_{\alpha\beta}}{2}\left(|\alpha\rangle\langle\alpha| - |\beta\rangle\langle\beta|\right) + \nu a^\dagger a\right] \tag{3.102}$$

that generates the unitary operation

$$e^{-i\widetilde{H}t/\hbar}|_{\widetilde{V}} = \begin{bmatrix} 1 & \bigcirc \\ \bigcirc & e^{i\omega_{\alpha\beta}t/2} \end{bmatrix} \otimes \begin{bmatrix} 1 & \bigcirc \\ \bigcirc & e^{-i\nu t} \end{bmatrix}$$

$$= \begin{bmatrix} 1 & & & \bigcirc \\ & e^{-i\nu t} & & \\ & & e^{i\omega_{\alpha\beta}t/2} & \\ \bigcirc & & & e^{i\omega_{\alpha\beta}t/2 - i\nu t} \end{bmatrix} \tag{3.103}$$

The tensor product is expanded here with respect to the ordered basis (3.93).

We can now invoke Theorem 2.6.2 in Section 2.6 to conclude the following.

Theorem 3.4.4. *The collection of 1-bit gates $U_{\theta,\phi}$ in (3.33) from the 2-level atoms and the 2-bit gates Q_η in (3.100) from cavity QED is universal for quantum computation.* $\qquad\qquad\square$

3.4.4   Cavity QED based quantum phase gate: another variant

In the preceding subsections, to implement quantum logic gates, the two qubits are represented by two separate systems: the internal states of an atom represent one qubit and the quantum state of the field inside the cavity represent the other. However, experimental realization of typical quantum algorithms may require the *two qubits to be treated on an equal footing*. We discuss a quantum phase gate based on cavity QED in which the two qubits are represented by two different modes of the radiation field inside the cavity and the gate is implemented by passing a 3-level atom through the cavity. Cavity QED with long lived Rydberg states and high Q cavities thus provides certain advantages. In this subsection, we follow the discussions in [71].

The transformation for a 2-bit quantum phase gate is given by $Q_\eta|\alpha_1,\beta_2\rangle = \exp(i\eta\delta_{\alpha_1,1}\delta_{\beta_2,1})|\alpha_1,\beta_2\rangle$, where $|\alpha_1\rangle$ and $|\beta_2\rangle$ stand for the basis states $|0\rangle$ or $|1\rangle$ of the qubits 1 and 2, respectively. Thus the quantum phase gate introduces a phase η only when both the qubits in the input states are 1. A representation of the quantum phase gate is given by the operator

$$Q_\eta = |0_1, 0_2\rangle\langle 0_1, 0_2| + |0_1, 1_2\rangle\langle 0_1, 1_2|$$
$$+ |1_1, 0_2\rangle\langle 1_1, 0_2| + e^{i\eta}|1_1, 1_2\rangle\langle 1_1, 1_2|, \tag{3.104}$$

and since $|0\rangle\langle 0| = (1 + \sigma_z)/2$ and $|1\rangle\langle 1| = (1 - \sigma_z)/2$, Eq. (3.104) has the matrix representation

$$Q_\eta = 1_1 1_2 - \frac{1}{4}(1 - e^{i\eta})(1_1 1_2 - 1_1\sigma_{z2} - \sigma_{z1}1_2 + \sigma_{z1}\sigma_{z2}). \tag{3.105}$$

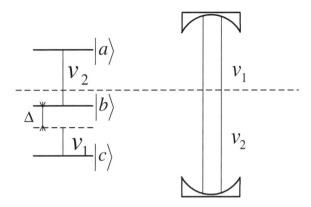

Figure 3.13: The schematics of the quantum phase gate. The cavity can hold two modes of frequencies ν_1 and ν_2. The atomic levels are such that $\omega_{ab} = \nu_2$ and $\omega_{bc} = \nu_1 + \Delta$.

Here we discuss a quantum phase gate with $\eta = \pi$.

The schematics of the quantum phase gate is shown in Fig. 3.13. The cavity has resonant frequencies at ν_1 and ν_2, and the photon number states $|0\rangle$ and $|1\rangle$ represent logic 0 and 1, respectively. Thus the possible cavity field states are $|0_1, 0_2\rangle$, $|0_1, 1_2\rangle$, $|1_1, 0_2\rangle$, and $|1_1, 1_2\rangle$. We consider a 3-level atom in cascade configuration such that the upper two levels are resonant with the cavity mode 2, i.e., $\omega_{ab} = \nu_2$, and the lower levels are detuned by an amount Δ from the cavity mode 1, i.e., $\omega_{bc} = \nu_1 + \Delta$. A quantum phase gate with a π phase shift is implemented if the atom in its ground state $|c\rangle$ passes through the cavity such that: (1) The detuning Δ is equal to g_2, and (2) the interaction time τ of the atom with the cavity is such that $g_1\tau = \sqrt{2}\pi$. Here g_i ($i = 1, 2$) are the vacuum Rabi frequencies associated with the interaction of the cavity modes with the respective atomic states. Under these conditions, the atomic state is decoupled from the photon states both before and after the interaction and it remains in the ground state $|c\rangle$. The cavity states also remain unaffected for the initial states $|0_1, 0_2\rangle$, $|0_1, 1_2\rangle$, and $|1_1, 0_2\rangle$, and acquire a π phase shift only for the state $|1_1, 1_2\rangle$.

In the following we discuss this phase gate in a dressed state picture that brings out the essential physics rather clearly.

First, we note that the atom (which is initially in the ground state $|c\rangle$) will remain completely decoupled with the cavity field if there is no photon in the mode 1, i.e., when the cavity field states are $|0_1, 0_2\rangle$ and $|0_1, 1_2\rangle$. Also in the case when there is one photon in mode 1 and no photon in mode 2 ($|1_1, 0_2\rangle$), the atom will again remain almost decoupled from the cavity field states if the detuning Δ is sufficiently large. The interesting situation is, however, when there is one photon each in the two modes of the cavity, i.e., the initial cavity field state is $|1_1, 1_2\rangle$. We analyze this case in the following.

The effective Hamiltonian for the interaction, in the dipole and rotating-wave approximations, is

$$H = \overline{H}_0 + \overline{H}_1, \tag{3.106}$$

where

$$
\begin{aligned}
\overline{H}_0 &= \hbar\nu_1 a_1^\dagger a_1 + \hbar\nu_2 a_2^\dagger a_2 + \hbar\omega_{bc}\,|b\rangle\,\langle b| + \hbar\omega_{ac}\,|a\rangle\,\langle a|\,, \tag{3.107} \\
\overline{H}_1 &= \hbar g_1(a_1\,|b\rangle\,\langle c| + a_1^\dagger\,|c\rangle\,\langle b|) \\
&\quad + \hbar g_2(a_2\,|a\rangle\,\langle b| + a_2^\dagger\,|b\rangle\,\langle a|), \tag{3.108}
\end{aligned}
$$

where a_i $(a_i^\dagger)$ are the photon annihilation and creation operators for the two modes. The effective Hamiltonian $\mathcal{H}_I$ in interaction picture is

$$
\begin{aligned}
\overline{H}_I &\equiv e^{-i\overline{H}_0 t/\hbar}\,\overline{H}_1\,e^{i\overline{H}_0 t/\hbar} \\
&= H_1 + H_2,
\end{aligned}
$$

where

$$
\begin{aligned}
H_1 &= \hbar g_1\left(a_1\,|b\rangle\,\langle c|\,e^{-i\Delta t} + a_1^\dagger\,|c\rangle\,\langle b|\,e^{i\Delta t}\right) \\
&= \hbar g_1\left(|b,0,1\rangle\,\langle c,1,1|\,e^{-i\Delta t} + |c,1,1\rangle\,\langle b,0,1|\,e^{i\Delta t}\right). \tag{3.109} \\
H_2 &= \hbar g_2\left(a_2\,|a\rangle\,\langle b| + a_2^\dagger\,|b\rangle\,\langle a|\right) \\
&= \hbar g_2\left(|a,0,0\rangle\,\langle b,0,1| + |b,0,1\rangle\,\langle a,0,0|\right). \tag{3.110}
\end{aligned}
$$

Here we have used the fact that, for an initial $|1_1, 1_2\rangle$ state for the cavity and the state $|c\rangle$ for the atom, the only allowed states for the atom-field system are $|a,0,0\rangle$, $|b,0,1\rangle$, and $|c,1,1\rangle$.

At this point we resort to a dressed state picture and define the symmetric and anti-symmetric states with respect to the field mode 2,

$$
\begin{aligned}
|+\rangle &\equiv \frac{1}{\sqrt{2}}\left(|a,0,0\rangle + |b,0,1\rangle\right), \\
|-\rangle &\equiv \frac{1}{\sqrt{2}}\left(|a,0,0\rangle - |b,0,1\rangle\right).
\end{aligned}
$$

In terms of these states H_2 can be rewritten as

$$H_2 = \hbar g_2\left\{|+\rangle\,\langle+| - |-\rangle\,\langle-|\right\} \tag{3.111}$$

with eigenvalues $\hbar g_2$ and $-\hbar g_2$. Thus the net effect of the field at frequency ν_2 is the dynamic Stark splitting. The Hamiltonian H_1 in the interaction picture of H_2 is given by

$$
\begin{aligned}
H_{1I} &= \frac{\hbar g_1}{\sqrt{2}}\left\{|+\rangle\,\langle c,1,1|\,e^{-i(\Delta+g_2)t} - |-\rangle\,\langle c,1,1|\,e^{-i(\Delta-g_2)t}\right. \\
&\quad \left. + |c,1,1\rangle\,\langle+|\,e^{i(\Delta+g_2)t} - |c,1,1\rangle\,\langle-|\,e^{i(\Delta-g_2)t}\right\}. \tag{3.112}
\end{aligned}
$$

When $g_2 = \Delta$, the interaction Hamiltonian simplifies and we obtain

$$
\begin{aligned}
H_{1I} = \quad & \frac{\hbar g_1}{\sqrt{2}} \left\{ |+\rangle \langle c,1,1| \, e^{-2i\Delta t} \right. \\
& + |c,1,1\rangle \langle +| \, e^{2i\Delta t} \\
& \left. - |-\rangle \langle c,1,1| - |c,1,1\rangle \langle -| \right\}.
\end{aligned}
\tag{3.113}
$$

For sufficiently large detuning we can ignore the oscillating terms $\exp(\pm 2i\Delta t)$, resulting in

$$
H_{1I} = -\frac{\hbar g_1}{\sqrt{2}} \left\{ |-\rangle \langle c,1,1| + |c,1,1\rangle \langle -| \right\}.
\tag{3.114}
$$

The effective Rabi frequency between the levels $|-\rangle$ and $|c,1,1\rangle$ is $g_1/\sqrt{2}$. Thus for an interaction time between the atom and the cavity field τ such that $g_1\tau = \sqrt{2}\pi$ we have $|c,1,1\rangle \to -|c,1,1\rangle$. This completes the description of the quantum phase gate Q_π.

3.4.5 Atom-cavity hardware

We summarize here again some of the experimental details and requirements to perform the quantum gate experiment described above. More information can be found, e.g., in [19, 32].

In a general perspective, the quantum gate can work perfectly if its operation time, say τ, is short compared to the time scales for (i) spontaneous decay of the atom and (ii) photon loss from the cavity. We therefore require $\Gamma\tau \ll 1$ and $\kappa\tau \ll 1$ in terms of the spontaneous decay rate Γ given in Tables 3.1 and 3.2, and the cavity linewidth κ defined in (3.68). The operation time τ is connected to the phase η of the gate by (3.101). Using a typical value $\eta \sim \pi$ and the large detuning limit (3.84), we obtain the requirements of the

$$\text{strong coupling regime:} \quad \Gamma, \kappa \ll g. \tag{3.115}$$

In physical terms, this means that a single photon in the cavity leads to atomic Rabi oscillations that are faster than the loss processes of atom or cavity. Current experiments are able to reach this regime, see, e.g., [32, 48, 53].

Two strategies can be identified to control the interaction time τ of the atom with the cavity:

1. Atoms are emitted from a pulsed source, cross the cavity in free flight, and are detected after leaving the cavity. This allows to post-select those experimental runs where atoms had a given velocity v and therefore an interaction time $\tau \sim w/v$, where w is the transverse cavity mode size (see Subsection 3.3.2). This approach is used in current experiments with Rydberg transitions and cavities for microwave frequencies; see Fig. 3.14 and [48]. A variant of this approach uses laser-cooled atoms that are released from a trap at a given time and fall through the cavity; see [32].

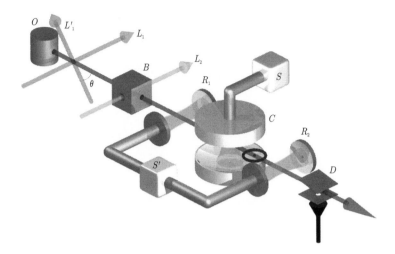

Figure 3.14: (Please see color insert.) Design scheme of an experiment where an atomic beam is crossing a microwave cavity. The atoms are prepared in an oven (O), are prepared in a Rydberg state (see Fig. 3.4) with laser beams L_1, L_1', L_2. In the zones R_1, R_2, single-qubit operations are performed with microwaves from a common source S'. The source S feeds photons into the cavity C. The atoms are detected in D. The dark circle at the cavity exit illustrates the orientation of the electron orbit of the involved atomic levels. Courtesy of J.-M. Raimond, Laboratoire Kastler–Brossel, Ecole Normale Supérieure, Paris.

2. Laser-cooled atoms move sufficiently slowly that it is possible to "catch" them when they have entered the cavity field. The presence of the atom changes the energies of the cavity modes and this can be measured by transmitting a weak probe beam through the cavity mirrors. Once the atom is detected, one shifts the cavity frequency (by adjusting the cavity length) or increases the probe beam intensity. This increases the mechanical potential provided by the position-dependent energy levels of the atom-cavity system, and can lead to a bound, oscillatory motion of the atomic center of mass. This approach is followed with optical cavities and strong transitions from the atomic ground state [32]. It is even possible to reduce the kinetic energy of the atomic motion ("cavity cooling").

A more efficient control is possible when the atom is cooled and trapped in a laser beam, as shown schematically in Figs. 3.7 and 3.9. In this case, the atom can be spatially localized within one half wavelength along the laser beam axis, due to an effective mechanical potential. This potential can be shifted by changing parameters of the laser system so that the atom can be moved in and out of the cavity [53].

These approaches can be modeled in the following way. We have to re-store in the previous discussion the dependence of the cavity field on the atomic center of mass position, $\mathcal{E}(\mathbf{r}_A, t)$. One can describe an atom moving along a prescribed trajectory $\mathbf{r}_A(t)$ by evaluating $\mathcal{E}(\mathbf{r}_A(t), t)$. This gives an additional time dependence to the Rabi frequency Ω_R ((3.21)): as the atom crosses the cavity mode, the Rabi frequency changes smoothly from zero to some maximum value and to zero again. This method of incorporating the atomic motion is appropriate for high-velocity atomic beams whose motion is essentially unaffected by the light field.

For laser-cooled atoms, the back action of the light field on the atomic motion becomes a key issue. In that case, the atomic center of mass position must also be quantized, similar to the quantization of the cavity mode variables in Subsection 3.2.2. (See Subsection 5.2.2, and Fig. 5.10, e.g., for a similar procedure with ions.) For the 2-dimensional subspace V_1 (3.76) spanned by the states $|\alpha, 0\rangle$ and $|\beta, 1\rangle$, one introduces a spinor-valued wave function $\Psi(\mathbf{r}_A, t) = [\psi_{\alpha,0}(\mathbf{r}_A, t), \psi_{\beta,1}(\mathbf{r}_A, t)]^T$ that depends on the center of mass coordinate $\mathbf{r}_A$ and satisfies the Schrödinger equation

$$i \frac{d}{dt} \begin{bmatrix} \psi_{\alpha,0}(\mathbf{r}_A, t) \\ \psi_{\beta,1}(\mathbf{r}_A, t) \end{bmatrix} = \frac{-\hbar}{2 m_A} \nabla_A^2 \begin{bmatrix} \psi_{\alpha,0}(\mathbf{r}_A, t) \\ \psi_{\beta,1}(\mathbf{r}_A, t) \end{bmatrix}$$
$$+ \begin{bmatrix} \frac{1}{2}\Delta & -g(\mathbf{r}_A)e^{-i\phi(\mathbf{r}_A)} \\ -g(\mathbf{r}_A)e^{i\phi(\mathbf{r}_A)} & -\frac{1}{2}\Delta \end{bmatrix} \begin{bmatrix} \psi_{\alpha,0}(\mathbf{r}_A, t) \\ \psi_{\beta,1}(\mathbf{r}_A, t) \end{bmatrix}, \quad (3.116)$$

where the ∇_A is the gradient with respect to the center of mass position $\mathbf{r}_A$, and $\Delta = \omega - \nu$ is the detuning (3.24) of the atomic transition with respect to the laser frequency. The phase $\phi(\mathbf{r}_A)$ is included for generality and permits to cover also the case of an electric field in plane wave form.

The coupling matrix in (3.116) can be diagonalized which leads to two scalar Schrödinger equations. This decoupling is not perfect, however, because the transformation matrix does not commute with the differentiation operation ∇_A. In the large detuning limit, this gives only a small correction, and ignoring it, one gets two mechanical potentials given by the position-dependent generalization of (3.87) and (3.88):

$$\tilde{V}_{\alpha,0}(\mathbf{r}_A) = \frac{1}{2}\Delta + \frac{g^2(\mathbf{r}_A)}{\Delta}, \qquad \tilde{V}_{\beta,1}(\mathbf{r}_A) = -\tilde{V}_0(\mathbf{r}_A). \quad (3.117)$$

The potential minimum for the wave function $\psi_{\beta,1}(\mathbf{r}_A)$ is located at the maximum Rabi frequency (if $\Delta > 0$), i.e., the ground state atom can be trapped in the antinode of the cavity field, as illustrated in Fig. 3.15. Close to the minimum, the potential is approximately quadratic, and the atomic eigenstates separate into products of the same harmonic oscillator states found in Section 3.2.2 for the quantized cavity mode, (3.63).

The quantized description in terms of a wave function $\psi_{\beta,1}(\mathbf{r}_A)$ is not very convenient under typical experimental conditions because one would have to keep track of many eigenfunctions of $\tilde{V}_{\beta,1}(\mathbf{r}_A)$ (it is difficult to cool the atom to the ground state), and because cavity damping and spontaneous

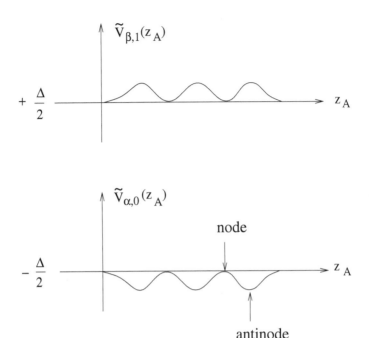

Figure 3.15: Mechanical potentials obtained from a local diagonalization of the potential matrix in (3.116), in the large detuning limit. Along the cavity axis, the equilibrium positions in the potential $V_{\beta,1}(z_A)$ are the antinodes of the field (for $\Delta > 0$).

emission make it impossible to ascribe a single quantum state to the atomic motion at any given time. A statistical description is required that goes beyond Schrödinger's equation for a closed system. Mathematical methods for that context are outlined in Chap. 4.

3.5   Quantum eraser

In Section 1.3, Chap. 1, we have promised to address the important notion of quantum erasure. We are now ready to do so in the present and the subsequent sections. Our main reference for this section is [1].

Complementarity lies at the heart of quantum mechanics [5]. A classic example is the Young's double-slit experiment in which a "particle" exhibits *wave-like* behavior by displaying interference if it goes through both slits. The interference disappears and a *particle-like* behavior is exhibited if the which-path information becomes available [15]. The question then is: What would happen if we *erase* the which-path information after the particle has passed through the slits? Would such a *quantum eraser* process restore the interference fringes? The answer is yes [54, 55] and this has been verified experimentally [29, 69]. This erasure and fringe retrieval can be achieved even *after* the atoms hit the screen [56].

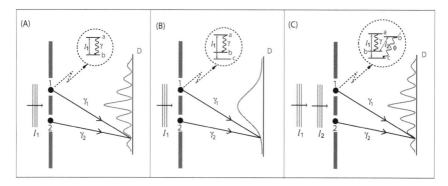

Figure 3.16: Here we consider three possible configurations of atoms that are placed at sites 1 and 2. In (a) we consider a 2-level atom initially in the state b. The incident pulse l_1 excites one of the two atoms to state a from where it decays to state a emitting a γ photon. In (b) the atom initially in the ground state c is excited by the pulse l_1 to state a from where it decays to state b. In (c) a fourth level is added. A pulse l_2 excites the atom to state b' after the atom has decayed to state b. The atom in the state b' emits a ϕ photon and ends up in state c.

To illustrate this point, an excellent example is the delayed-choice quantum eraser, proposed by Scully and Drühl [54] (1982). They analyzed a photon correlation experiment designed to probe the extent to which information accessible to an observer and its erasure affects measured results. Here we present a simple description of the quantum eraser that brings out the counterintuitive aspects related to time in the quantum mechanical domain. We consider the scattering of light from two atoms located at sites 1 and 2 on the screen D (Fig. 3.16) and analyze three different cases:

(a) Resonant light impinges from left on 2-level atoms (Fig. 3.16(a)) located at sites 1 and 2. An atom excited to level a emits a γ photon. There are two possibilities for the atom, either it remains in the ground state b or it can get excited to the state a by the incident light and emit a γ photon. We look at the interference of these photons at the screen. Since both atoms are finally in the state b after the emission of photons, it is not possible to determine which atom contributed the γ photon. A large number of such experiments are carried out, i.e., any one photon will yield one count on the screen and it takes many such photon events to build up a pattern. The resulting distribution of the detected photons exhibits an interference pattern as shown in Fig. 3.16(a). This is an analogue of the usual Young's double slit experiment. Instead of the usual light beams through two pin holes, we have considered scattered light from two atoms. The key to the appearance of the interference is the lack of the which-path information for the photons.

(b) In a case when the atoms have three levels (Fig. 3.16(b)), the drive field excites the atoms from the ground state c to the excited state a. The atom in state a can then emit a γ photon and end up in state b. Here the photon detected on the screen leaves behind the which-path information, i.e., the atom responsible for contributing the γ photon is in level b whereas the other atom remains in level c. Thus a measurement of the internal states of the atoms provides us the which-path information and no interference is observed. That is, the state of the atom acts as an observer state. The precise mathematical description of photons γ_1 and γ_2 is the same in cases a and b. It is only the presence of the passive observer state that kills the interference. However the loss of coherence in the present scheme does not invoke the uncertainty principle.

The question, however, is now can we erase the which-path information stored in the atom(s) and thus regain interference? If the loss of interference were due to some kind of noise or uncertainty due to quantum fluctuations the answer would be no. But this is not the case and the interference can be recovered. The question then is whether it is possible to wipe out the which-path information and recover the interference.

(c) As shown in Fig. 3.16(c), this can possibly be done by driving the atom by another field that takes the atom from level b to b' and after an emission of a ϕ photon at $b' - c$ transition ends up in level c. Now the final state of both the atoms is c and a measurement of internal states can not provide us the which-path information. It would therefore seem that the interference fringes will be restored. But a careful analysis indicates that the which-path information is still available through the ϕ photon. A measurement on ϕ photon can tell us which atom contributed the γ photon. Can we erase the which-path information contained in the ϕ photon and recover the interference fringes? Scully and Drühl considered an ingenious device based on an electro-optic shutter that can absorb the ϕ photon in such a way that the which-path information is erased [54]. A slightly modified version of such an eraser using a parametric process involving nonlinear crystal (instead of single atoms) has been experimentally realized by Shih, et al. [29] (2000).

3.6   Quantum disentanglement eraser

The notions of quantum eraser and entanglement are interrelated. For example, in any set-up for quantum eraser, the which-path information, and therefore the disappearance of the fringes, is achieved by entangling the state of the particle with another controlling *qubit* that contains the which-path information. As shown above, we can however restore the fringes by erasing the which-path information contained in the controlling qubit.

Our discussion in this section is based on the reference [70]. We consider an interesting new class of quantum erasers, called *disentanglement erasers* [17]. These consist of at least three subsystems A, B, and T. The AB

subsystem is prepared in the entangled Bell state

$$|\psi_{AB}\rangle = \frac{1}{\sqrt{2}}(|0_A, 1_B\rangle + |1_A, 0_B\rangle) \tag{3.118}$$

If the two parts $|0_A, 1_B\rangle$ and $|1_A, 0_B\rangle$ of this entangled state are tagged with the third qubit labelled T such that the state of the whole system is

$$|\psi_{ABT}\rangle = (|0_A, 1_B\rangle|0_T\rangle + |1_A, 0_B\rangle|1_T\rangle)/\sqrt{2}, \tag{3.119}$$

then the purity of the entanglement of the subsystem AB is lost. The state of the AB subsystem is described by the statistical mixture

$$\rho_{AB} = (|0_A, 1_B\rangle\langle 0_A, 1_B| + |1_A, 0_B\rangle\langle 1_A, 0_B|)/2. \tag{3.120}$$

However, if the tagged information can somehow be erased, then the entanglement of the AB subsystem will be restored. In order to see this we define the superposition states of the tagged state $|\pm_T\rangle = (|0_T\rangle \pm |1_T\rangle)/\sqrt{2}$. The state of the combined ABT system (3.119) can then be rewritten as

$$|\psi_{ABT}\rangle = \frac{1}{\sqrt{2}}\left[|+_T\rangle[\frac{1}{\sqrt{2}}(|0_A, 1_B\rangle + |1_A, 0_B\rangle] \right.$$
$$\left. +|-_T\rangle[\frac{1}{\sqrt{2}}(|0_A, 1_B\rangle - |1_A, 0_B\rangle)\right]. \tag{3.121}$$

An outcome $|+_T\rangle$ for the tagged state therefore restores the original state (1) for the AB subsystem whereas the outcome $|-_T\rangle$ yields $(|0_A, 1_B\rangle - |1_A, 0_B\rangle)/\sqrt{2}$. A phase shift then restores the original state. Thus a measurement of the tagging qubit in one of these superposition states restores the entangled state. This *disentanglement eraser* of Garisto and Hardy shows that "the entanglement of any two particles that do not interact (directly or indirectly), never disappears but rather is encoded in the ancilla of the system". An implementation of such eraser has been demonstrated in NMR systems [59].

Here we consider the usage of two high quality cavities to realize this new type of quantum eraser as proposed by Garisto and Hardy [17]. We use an auxiliary atom's passage through the first cavity to control the degree of entanglement between the two cavities. We further discuss how these changes can be studied by using a probe atom and post measurement on the auxiliary atom.

We consider a system of two high Q microwave cavities A and B initially in the vacuum states. The cavities can be prepared in the entangled state (3.118) by first passing an atom 1 in excited state $|a\rangle$ through the two cavities (Fig. 3.16). The interaction times between the atom and the cavities are chosen such that the interaction time with cavity A corresponds to a $\pi/2$ pulse, with the resulting state:

$$|\psi_1\rangle = \frac{1}{\sqrt{2}}(|0_A\rangle|a_1\rangle + |1_A\rangle|b_1\rangle)|0_B\rangle, \tag{3.122}$$

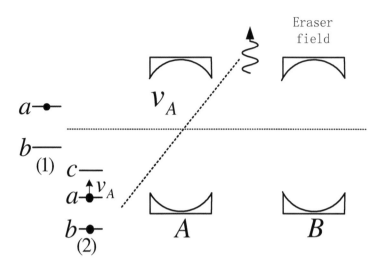

Figure 3.17: Figure depicting the preparation of entangled states. Cavities A and B, each resonant with the $|a\rangle - |b\rangle$ transition are prepared in an entangled state (3.123) by passing the 2-level atom 1 initially in the excited state $|a\rangle$ with appropriate passage times through the two cavities. The entangled state given by (3.125) is prepared via the passage of the 3-level atom 2 through the cavity A.

and the interaction time with cavity B corresponds to a π pulse, yielding the final state

$$|\psi_2\rangle = \frac{1}{\sqrt{2}}(|0_A\rangle|1_B\rangle + |1_A\rangle|0_B\rangle)|b_1\rangle. \qquad (3.123)$$

Thus the photons in the cavities A and B are entangled and the atom 1 is decoupled from the AB system.

In the second step we *tag* a qubit with the entangled state (3.123). The tagging qubit in our system is a 3-level atom (atom 2) that passes through cavity A only (Fig. 3.16). The level structure of atom 2 is such that the field inside the cavity A is nonresonant with the $|a\rangle \rightarrow |b\rangle$ transition but is dispersively coupled with the $|c\rangle \rightarrow |a\rangle$ transition with $\omega_{ca} = \nu_A + \Delta$. Here the atomic levels $|a\rangle$ and $|b\rangle$ represent the states $|0\rangle$ and $|1\rangle$ for the tagging qubit, respectively. The effective Hamiltonian for the atom and the cavity field is given by $H_{eff} = -(\hbar g^2/\Delta)\left(aa^\dagger|c\rangle\langle c| - a^\dagger a|a\rangle\langle a|\right)$ (Eq. (3.90)), where g is a coupling coefficient, a and $a^\dagger$ are destruction and creation operators for the field state inside the cavity [57]. The atom 2 is initially prepared in a superposition state $(|a\rangle + |b\rangle)/\sqrt{2}$. After passage through the cavity A, a quantum phase gate is made with a phase shift $\eta = g^2\tau/\Delta$. Here τ is the interaction time between the atom and the cavity. Such a quantum phase gate has been discussed and experimentally implemented in [49].

The resulting state after the passage of the atom through the cavity is

$$|\psi_3\rangle = e^{-iH_{eff}\tau}|\psi_2\rangle$$

$$= \frac{1}{\sqrt{2}}\left[|b_2\rangle[\frac{1}{\sqrt{2}}(|0_A,1_B\rangle + |1_A,0_B\rangle)]\right.$$

$$\left. + |a_2\rangle[\frac{1}{\sqrt{2}}(|0_A,1_B\rangle + e^{i\eta}|1_A,0_B\rangle)]\right]. \quad (3.124)$$

We chose the interaction time τ such that $\eta = \pi$. Then $|\psi_3\rangle$ is of the form (3.121). If atom 2 then interacts with a classical Ramsey field such that $(|a_2\rangle + |b_2\rangle)/\sqrt{2} \to |a_2\rangle$ and $(-|a_2\rangle + |b_2\rangle)/\sqrt{2} \to |b_2\rangle$, we obtain

$$|\psi_4\rangle = \frac{1}{\sqrt{2}}[|0_A\rangle|1_B\rangle|a_2\rangle + |1_A\rangle|0_B\rangle|b_2\rangle]|b_1\rangle. \quad (3.125)$$

We have thus obtained an entangled state of the form (3.119) and the entanglement between photons inside the cavities A and B is controlled by atom 2.

We now consider an example where the entanglement-based phenomena can be controlled by the tagging qubit. We consider an atomic double cavity Ramsey interferometry where the appearance or disappearance of interference fringes in the final atomic state probability depends on the presence or absence of entanglement between the photons of the two cavities.

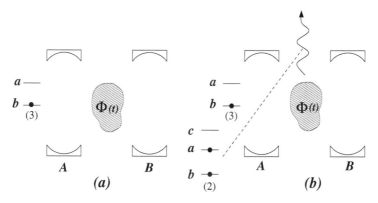

Figure 3.18: (a) A 2-level atom initially in its ground state $|b\rangle$ passes through the entangled cavities A and B. The atom in state $|a\rangle$ acquires a phase shift while passing between the two cavities. The probability of finding the atom in the excited or ground state exhibits interference fringes. (b) The system is the same as in (a) but a tagging qubit can partially or completely erase the interference fringes via dispersive coupling of a 3-level atom with cavity A.

We proceed to analyze the passage of a 2-level atom 3 initially in its ground state $|b\rangle$ through two cavities that are initially in entangled state

(3.123). First we discuss the the system in the absence of tagging qubit (Fig. 3.17). A theoretical study and an experimental demonstration of a similar set-up are discussed in [3, 14, 68]. The initial atom-field state is

$$|\psi_1\rangle = \frac{1}{\sqrt{2}}(|0_A, 1_B\rangle + |1_A, 0_B\rangle)|b_3\rangle. \tag{3.126}$$

After the passage through cavity A with a passage time corresponding to a π pulse, we have

$$|\psi_2\rangle = \frac{1}{\sqrt{2}}(|b_3, 1_B\rangle + |a_3, 0_B\rangle)|0_A\rangle. \tag{3.127}$$

If the atom interacts with a classical field dispersively such that a phase $\Phi(t)$ is acquired when the atom is in the excited state $|a\rangle$ and no phase when it is in ground state $|b\rangle$, then the resulting state is

$$|\psi_3\rangle = \frac{1}{\sqrt{2}}(|b_3, 1_B\rangle + e^{i\Phi(t)}|a_3, 0_B\rangle)|0_A\rangle. \tag{3.128}$$

Finally the atomic passage through cavity B with interaction time corresponding to a $\pi/2$ pulse yields

$$|\psi_4\rangle = \frac{1}{2}(|b_3, 1_B\rangle(1 + e^{i\Phi(t)}) + |a_3, 0_B\rangle(1 - e^{i\Phi(t)}))|0_A\rangle. \tag{3.129}$$

The probability of finding the atom 3 in the excited and ground states exhibit antisymmetric interference fringes, i.e.,

$$P_{a_3} = 1 - P_{b_3} = \frac{1}{2}(1 - \cos\Phi(t)). \tag{3.130}$$

The appearance of the fringes (say in P_{a_3}) is due to the interference between two paths: The atom absorbs a photon in cavity A or in cavity B.

The system is however more complicated when the tagging atom 2 passes through cavity A (Fig. 3.17). In this case, the initial state is

$$|\psi_1'\rangle = \frac{1}{\sqrt{2}}[|0_A\rangle|1_B\rangle|a_2\rangle + |1_A\rangle|0_B\rangle|b_2\rangle]|b_3\rangle. \tag{3.131}$$

The atom 3 now passes through the cavities A and B with a phase shift $\Phi(t)$ acquired by the level $|a\rangle$ while the atom passes in between the two cavities. The final state after the passage of atom 3 can be determined as above and is given by

$$
\begin{aligned}
|\psi_4'\rangle &= \frac{1}{2}[|b_3, 1_B\rangle(|a_2\rangle + |b_2\rangle e^{i\Phi(t)}) \\
&\quad + |a_3, 0_B\rangle(|a_2\rangle - |b_2\rangle e^{i\Phi(t)})]|0_A\rangle.
\end{aligned} \tag{3.132}
$$

The probabilities for atom 3 to be in the excited and ground states are

$$P_{a_3} = P_{b_3} = \frac{1}{2} \tag{3.133}$$

and the fringes disappear. The disappearance of the fringes is due to the availability of which-path information. The state of the tagged atom controls the which-path information. We assume that the atom is found in the ground state. Thus the atom must have followed the path absorbing a photon in cavity A and emitting the photon in cavity B if the tag atom 2 is in state $|b_2\rangle$, Similarly if it followed the path with no interaction with either cavity A or B, the tag atom 2 must be in state $|a_2\rangle$.

How do we restore the interference fringes? In a "delayed choice" quantum eraser we might like to do so even after the atom 3 has been detected. A quantum eraser would require erasing the which-path information.

Let the tag atom 2 interact with a classical Ramsey field, such that

$$
\begin{aligned}
|a_2\rangle &\rightarrow \cos\theta|a_2\rangle + \sin\theta|b_2\rangle \\
|b_2\rangle &\rightarrow -\sin\theta|a_2\rangle + \cos\theta|b_2\rangle
\end{aligned}
\tag{3.134}
$$

Here the angle θ depend on the Rabi frequency of the field and the interaction time.

A subsequent measurement of tag atom in state $|a_2\rangle$ yields the following expressions for the probabilities

$$
P_{a_3} = 1 - P_{b_3} = \frac{1}{2}[1 + \sin2\theta\cos(\Phi(t))]
\tag{3.135}
$$

Similar expressions are obtained if the tag atom is detected in the ground state $|b_2\rangle$.

The visibility of the fringes is given by $\mathcal{V} = \sin2\theta$. It is clear that the interference fringes are fully restored (with unit visibility) for $\theta = \pi/4$. We then obtain

$$
P_{a_3} = \frac{1}{2}[1 + \cos(\Phi(t))]
\tag{3.136}
$$

if the atom 2 is found in state $|a_2\rangle$ and

$$
P_{a_3} = \frac{1}{2}[1 - \cos(\Phi(t))]
\tag{3.137}
$$

if the atom 2 is found in state $|b_2\rangle$.

References

[1] Y. Aharonov and M.S. Zubairy, *Science* **307** (2005), 875.

[2] G. Barton, in *Cavity quantum electrodynamics, Adv. At. Mol. Phys.*, edited by P.R. Berman (Academic Press, New York, 1994), suppl. 2.

[3] P. Bertet, S. Osnaghi, A. Rauschenbeutel, G. Nogues, A. Auffeves, M. Brune, J.-M. Raimond, and S. Haroche, *Nature (London)* **411** (2001), 166.

[4] A. Boca, R. Miller, K.M. Birnbaum, A.D. Boozer, J. McKeever, and H.J. Kimble, *Phys. Rev. Lett.* **93** (2004), 233603.

[5] N. Bohr, *Naturwissenschaften* **16** (1928), 245.

[6] S. Brattke, B.T.H. Varcoe, and H. Walther, *Phys. Rev. Lett.* **86** (2001), 3534.

[7] R. Brecha, L. Orozco, M. Raizen, M. Xiao, and H. Kimble, *J. Opt. Soc. Am. B* **12** (1995), 2329.

[8] M. Brune *et al.*, *Phys. Rev. Lett.* **76** (1996), 1800.

[9] M. Brune *et al.*, *Phys. Rev. Lett.* **77** (1996), 4887.

[10] M. Brune, J.-M. Raimond, P. Goy, L. Davidovich, and S. Haroche, *Phys. Rev. Lett.* **59** (1987), 1899.

[11] H.B. Casimir and D. Polder, *Phys. Rev.* **73** (1948), 360.

[12] P. Domokos and H. Ritsch, *Phys. Rev. Lett.* **89** (2002), 253003.

[13] P. Domokos, J.-M. Raimond, M. Brune, and S. Haroche, *Phys. Rev. A* **52** (1995), 3554.

[14] B.-G. Englert, H. Walther, and M.O. Scully, *Appl. Phys. B* **54** (1992), 366.

[15] R.P. Feynman, R. Leighton, and M. Sands, *The Feynman Lectures on Physics*, Vol. III, (Addison Wesley, Reading, MA 1965).

[16] M. Gangl and H. Ritsch, *Eur. Phys. J. D* **8** (2000), 29.

[17] R. Garisto and L. Hardy, *Phys. Rev* **A 60** (1999), 827.

[18] P. Goy, J.-M. Raimond, M. Gross, and S. Haroche, *Phys. Rev. Lett.* **50** (1983), 1903.

[19] P. Grangier, G. Reymond, and N. Schlosser, Implementations of quantum computing using cavity quantum electrodynamics schemes,

[20] G.R. Guthöhrlein, M. Keller, K. Hayasaka, W. Lange, and H. Walther, *Nature* **414** (2001), 49.

[21] H. Haken and H.C. Wolf, *The Physics of Atoms and Quanta*, Advanced Texts in Physics, 6th ed. (Springer, Heidelberg Berlin, 2000).

[22] D.J. Heinzen and M.S. Feld, *Phys. Rev. Lett.* **59** (1987), 2623.

[23] M. Hennrich, T. Legero, A. Kuhn, and G. Rempe, *Phys. Rev. Lett.* **85** (2000), 4872.

[24] E.A. Hinds, in *Cavity Quantum Electrodynamics*, Adv. At. Mol. Opt. Phys., edited by P. R. Berman (Academic, New York, 1994), suppl. 2.

[25] E.A. Hinds, K.S. Lai, and M. Schnell, *Phil. Trans. R. Soc. London A* **355** (1997), 2353.

[26] C. Hood, M. Chapman, T. Lynn, and H. Kimble, *Phys. Rev. Lett.* **80** (1998), 4157.

[27] C. Hood, T. Lynn, A. Doherty, A. Parkins, and H. Kimble, *Science* **287** (2000), 1447.

[28] Y. Kaluzny, P. Goy, M. Gross, J.-M. Raimond, and S. Haroche, *Phys. Rev. Lett.* **51** (1983), 1175.

[29] Y.-H. Kim, R. Yu, S.P. Kulik, Y. Shih, and M.O. Scully, *Phys. Rev. Lett.* **84** (2000), 1.

[30] D. Kleppner, *Phys. Rev. Lett.* **47** (1981), 233.

[31] A. Kuhn, M. Hennrich, and G. Rempe, *Phys. Rev. Lett.* **89** (2002), 067901.

[32] A. Kuhn, M. Hennrich, and G. Rempe, Strongly coupled atom-cavity systems, in *Quantum Information Processing*, edited by G. Leuchs and T. Beth (Wiley-VCH, Weinheim, 2003), pp. 182–95.

[33] H. Mabuchi, Q. Turchette, M. Chapman, and H. Kimble, *Opt. Lett.* **21** (1996), 1393.

[34] P. Maunz, T. Puppe, I. Schuster, N. Syassen, and P.W.H. Pinkse and G. Rempe, *Nature* **428** (2004), 50.

[35] P. Maunz, T. Puppe, I. Schuster, N. Syassen, P.W.H. Pinkse, and G. Rempe, *Phys. Rev. Lett.* **94** (2005), 033002.

[36] J. McKeever *et al.*, *Phys. Rev. Lett.* **90** (2003), 133602.

[37] J. McKeever *et al.*, *Science* **303** (2004), 1992.

[38] J. McKeever, A. Boca, A.D. Boozer, J.R. Buck, and H.J. Kimble, *Nature* **425** (2003), 268.

[39] D. Meschede, H. Walther, and G. Müller, *Phys. Rev. Lett.* **54** (1985), 55154.

[40] A. Messiah, *Mécanique quantique*, Vol. II, Dunod, Paris, 1995, new edition.

[41] P. Münstermann, T. Fischer, P. Maunz, P. Pinkse, and G. Rempe, *Phys. Rev. Lett.* **82** (1999), 3791.

[42] G. Nogues, A. Rauschenbeutel, S. Osnaghi, M. Brune, J.-M. Raimond, and S. Haroche, *Nature* **400** (1999), 239.

[43] P.W.H. Pinkse, T. Fischer, P. Maunz, and G. Rempe, *Nature* **404** (2000), 365.

[44] P. Pinkse and G. Rempe, in *Cavity-enhanced spectroscopies*, Vol. 40 of *Experimental Methods in the Physical Sciences*, edited by R. van Zee and P. Looney (Academic Press, San Diego, 2003).

[45] E.M. Purcell, *Phys. Rev.* **69** (1946), 681. **82** (1999), 3795.

[46] J.-M. Raimond, M. Brune, and S. Haroche, Manipulating quantum entanglement with atoms and photons in a cavity, *Rev. Mod. Phys.* **73** (2001), 565.

[47] J.-M. Raimond, P. Goy, M. Gross, C. Fabre, and S. Haroche, *Phys. Rev. Lett.* **49** (1982), 1924.

[48] A. Rauschenbeutel, G. Nognes, S. Osnaghi, P. Bertet, M. Brune, J.-M. Raimond and S. Haroche, Coherent operation of a tunable quantum phase gate in cavity QED, *Phys. Rev. Lett.* **83** (1999), 5155.

[49] A. Rauschenbeutel, G. Nogues, S. Osnaghi, P. Bertet, M. Brune, J.-M. Raimond, and S. Haroche, *Phys. Rev. Lett.* **83** (1999), 5166.

[50] G. Rempe, H. Walther, and N. Klein, *Phys. Rev. Lett.* **58** (1987), 353.

[51] G. Rempe, R. Thompson, H. Kimble, and R. Lalezari, *Opt. Lett.* **17** (1992), 363.

[52] V. Sandoghdar, C.I. Sukenik, E.A. Hinds, and S. Haroche, *Phys. Rev. Lett.* **68** (1992), 3432.

[53] J.A. Sauer, K.M. Fortier, M.S. Chang, C.D. Hamley, and M.S. Chapman, Cavity QED with optically transported atoms, *Phys. Rev. A*, **69** (2004), 051804.

[54] M.O. Scully and K. Drühl, *Phys. Rev. A* **25** (1982), 2208.

[55] M.O. Scully, B.-G. Englert, and H. Walther, *Nature (London)* **351** (1991), 111.

[56] M.O. Scully and H. Walther, *Found. Phys.* **28** (1998), 399; U. Mohrhoff, *Am. J. Phys.* **64** (1996), 1468; B.-G. Englert, M.O. Scully, and H. Walther, *Am. J. Phys.* **67** (1999), 325; U. Mohrhoff, *Am. J. Phys.* **67** (1999), 330.

[57] M.O. Scully and M.S. Zubairy, *Quantum Optics*, Cambridge Univ. Press, Cambridge, U.K., 1997.

[58] C.I. Sukenik, M.G. Boshier, D. Cho, V. Sandoghdar, and E.A. Hinds, *Phys. Rev. Lett.* **70** (1993), 560.

[59] G. Teklemariam, E.M. Fortunato, M.A. Pravia, T.F. Havel, and D.G. Cory, *Phys. Rev. Lett.* **86** (2001), 5845; G. Teklemariam, E.M. Fortunato, M.A. Pravia, Y. Sharf, T.F. Havel, D.G. Cory, A. Bhattaharyya, and J. Hou, *Phys. Rev. A* **66** (2002), 012309.

[60] R.J. Thompson, G. Rempe, and H.J. Kimble, *Phys. Rev. Lett.* **68** (1992), 1132.

[61] Q. Turchette, C. Hood, W. Lange, H. Mabuchi, and H. Kimble, *Phys. Rev. Lett.* **75** (1995), 4710.

[62] A.G. Vaidyanathan, W.P. Spencer, and D. Kleppner, *Phys. Rev. Lett.* **47** (1981), 1592.

[63] B.T.H. Varcoe, S. Brattke, M. Weidinger, and H. Walther, *Nature* **403** (2000), 743.

[64] V. Vuletić and S. Chu, *Phys. Rev. Lett.* **84** (2000), 3787

[65] H. Walther, *Physica Scripta* **1** (1998), 138.

[66] M. Weidinger, B.T.H. Varcoe, R. Heerlein, and H. Walther, *Phys. Rev. Lett.* **82** (1999), 3795.

[67] J. Ye, D.W. Vernooy, and H. J. Kimble, *Phys. Rev. Lett.* **83** (1999), 4987.

[68] S.-B. Zheng, *Opt. Commun.* **173** (2000), 265.

[69] X.Y. Zou, L.J. Wang, and L. Mandel, *Phys. Rev. Lett.* **67** (1991), 318; see also P.G. Kwait, A.M. Steinberg, and R.Y. Chiao, *Phys. Rev. A* **47** (1992), 7729.

[70] M.S. Zubairy, G.S. Agarwal, and M.S. Zubairy, *Phys. Rev. A* **70** (2004), 012316.

[71] M.S. Zubairy, M. Kim, and M.O. Scully, *Phys. Rev. A* **68** (2003), 033820.

Chapter 4

Imperfect Quantum Operations

- Density operators and completely positive maps

- The Kraus representation theorem

- von Neumann equation and master equation

- Calculation of fidelity

4.1 Fidelity

In this chapter, we introduce the basic concepts in modeling the effects of imperfections in qubit dynamics that occur for neutral atoms and optical cavities. Recall that such imperfections are linked to the design in quantum error correction in Section 2.9 of Chap. 2, as they cause the loss of fidelity and generate errors. The most significant impact is that it is no longer possible to describe the quantum system in terms of a "pure state" (a wave function or a state vector). A more general class of Hilbert space operators is needed that corresponds—for the finite-dimensional Hilbert spaces we restrict ourselves here to—to hermitian, positive semi-definite matrices. These objects are commonly called "density matrices" and denoted by ρ. Their time evolution is no longer described by the Schrödinger equation, but is given by a so-called *completely positive map*, at least in many physically relevant cases. For quantum computation, the most significant concept is the "fidelity" F of a quantum operation. If $\rho_{|i\rangle}(t)$ is the density matrix of the system after an evolution of time duration t for an initial state $|i\rangle$ and $|f\rangle = U(t)|i\rangle$ is the "nominal" target state that the quantum gate would have produced under perfect conditions, then the fidelity can be defined as [2]

$$F = \min_{|i\rangle} \text{tr}\left(\mathbb{P}_{|f\rangle}\rho_{|i\rangle}(t)\right) = \min_{|i\rangle} \text{tr}\left(U(t)\mathbb{P}_{|i\rangle}U^{\dagger}(t)\rho_{|i\rangle}(t)\right). \tag{4.1}$$

We denote tr the trace of a matrix (see (4.5)), and use the definition

$$\mathbb{P}_{|f\rangle} = |f\rangle\langle f| \tag{4.2}$$

for the "projector" onto the state $|f\rangle$. The minimum in (4.1) is taken over all possible initial states. We give a more detailed introduction to density matrices in Section 4.2 below.

The definition (4.1) of the fidelity can be roughly rephrased in the following sentence: the probability that the imperfect gate produces the nominal output state is at least F, independent of the input state. The independence of the input state is essential for quantum computing because the quantum operation can only exploit the massively parallel character allowed by quantum mechanics if it works equally well with arbitrary superpositions of input states.

Different physical processes lead to imperfections and require a description in terms of density matrices:

- imperfect knowledge of the initial state, i.e., the initial state can only be specified by a set of states and the corresponding probabilities;

- the fields involved in the quantum operation are not completely known, and have to be described in terms of some stochastic processes. The corresponding unitary evolution has to be averaged with respect to the distribution function of the process;

- in the context of cavity QED implementations, one has to model the fact that atoms in electronically excited states are prone to "spontaneous decay" by emitting a photon into the continuum of field modes instead of into the cavity mode. This happens at unpredictable instants of time, and the state of the atom after the photon emission must be described by a statistical average;

- similarly, the photons in the cavity can leak out because the mirrors are (infinitesimally) partially transmitting. This leads to a growing lack of information about the precise photon number in the cavity.

4.2 Density matrices

We recall from Box 1.2 that if a system is described by the state vector $|\psi\rangle$ (one then speaks of a "pure state", see Def. 4.2.2 below), the expectation values $\langle A \rangle$ of relevant system observables A can be computed from

$$\langle A \rangle = \langle \psi | A | \psi \rangle \tag{4.3}$$

$$= \sum_n \langle \psi | n \rangle \langle n | A | \psi \rangle = \sum_n \langle n | A | \psi \rangle \langle \psi | n \rangle$$

$$= \operatorname{tr} \left(A | \psi \rangle \langle \psi | \right). \tag{4.4}$$

In the first step, we have introduced the identity operator in terms of projectors onto a complete, orthonormal Hilbert space basis $\{|n\rangle\}$. In (4.4), we have introduced the trace of an operator O

$$\operatorname{tr} O \equiv \sum_n \langle n | O | n \rangle \tag{4.5}$$

in terms of that basis. It is easy to see that the trace is independent of the actual choice of basis.

Remark 4.2.1. In (4.4), we may identify the projector

$$\mathbb{P}_{|\psi\rangle} \equiv |\psi\rangle\langle\psi| \tag{4.6}$$

onto the system state. These projectors onto "pure states" are actually system observables as well: they are hermitian by construction, and for a Hilbert space vector $|\chi\rangle$, the expectation value of $\mathbb{P}_{|\chi\rangle}$ is the probability of finding the system in the state $|\chi\rangle$. Indeed, according to quantum mechanics, this probability is given by $|\langle\chi|\psi\rangle|^2 = \operatorname{tr}\left(\mathbb{P}_{|\chi\rangle}\mathbb{P}_{|\psi\rangle}\right)$. □

Incomplete knowledge about the state of a quantum system can be represented in the following way: one specifies a set of states $\{|i\rangle\}$ and a set of positive real numbers $\{p_i\}$ with $\sum_i p_i = 1$ and constructs the hermitian operator

$$\rho = \sum_i p_i \mathbb{P}_{|i\rangle} \tag{4.7}$$

which is called the *density matrix*. Expectation values of system observables A can now be given by the definition

$$\langle A \rangle \equiv \operatorname{tr}\left(A\rho\right), \tag{4.8}$$

which generalizes (4.3). □

Remark 4.2.2. At first glance, the construction looks as if the system were probabilistically distributed over the states $|i\rangle$, with probabilities p_i. This interpretation is only true, however, if the states $|i\rangle$ are mutually orthogonal: according to Remark 4.2.1, the probability of finding the system in the state $|n\rangle \in \{|i\rangle\}$ is given by the expectation value of the projector

$$\langle \mathbb{P}_{|n\rangle} \rangle = \sum_i p_i \operatorname{tr}\left(\mathbb{P}_{|n\rangle}\mathbb{P}_{|i\rangle}\right) \tag{4.9}$$

$$= p_n + \sum_{i \neq n} p_i \, |\langle n | i \rangle|^2,$$

and the sum in the last line is only zero if the state $|n\rangle$ is orthogonal to all $|i\rangle$ for $i \neq n$.

Indeed, it is easy to see that the same density matrix can be obtained from different sets of states $\{|i\rangle\}$ if these states are not mutually orthogonal or if they appear with equal probabilities. All predictions of quantum mechanics are encoded, however, in the expectation values of system observables, and hence in the system density matrix. One therefore can not attribute physical reality to a particular expansion of the form (4.7) because other expansions can be found that lead to the same physics. (The labels $\{i\}$ are an example of "hidden variables". See Section 1.2, Chap. 1.) $\square$

Earlier in Section 1.2 of Chap. 1, we had some discussions of density matrices. The following is a rigorous definition.

Definition 4.2.1. A density matrix is a linear operator ρ on the system's Hilbert space $\mathcal{H}$ with the following properties:

1. ρ is hermitian;

2. ρ is positive, i.e., for all $|\psi\rangle \in \mathcal{H}$ we have $\mathrm{tr}\left(\mathbb{P}_{|\psi\rangle}\rho\right) \geq 0$;

3. $\mathrm{tr}\,\rho = 1$. $\square$

The first property makes ρ an observable, the second one is required by the probability interpretation of quantum mechanics, and the last one ensures the normalization of probabilities.

As a consequence of this definition, the eigenvectors of a density matrix are orthogonal, and its eigenvalues p_n are non-negative. A set of eigenvectors and eigenvalues provides a particular expansion (4.7) of the density matrix where the coefficients p_i can be interpreted as probabilities. This observation motivates the following characterization of a pure state.

Definition 4.2.2. A system is in a *pure state* if its density matrix has only a single eigenvector with eigenvalue 1, all other eigenvalues being zero. As a consequence, the density matrix is a projector, i.e., $\rho\rho = \rho$.

On the other hand, we say that a system is in a *mixed state* if its density matrix ρ satisfies

$$\rho\rho \neq \rho,$$

i.e., ρ is not a projector. $\square$

Note that the phase of a pure state vector in the Hilbert space is not fixed by this definition. (Its norm is, because ρ has unit trace.) Indeed, the predictions of quantum mechanics are invariant with respect to a change in the phases of Hilbert space vectors (which is called "global U(1) symmetry").

Example 4.2.1. A particularly important class of density matrices describes systems in thermal equilibrium. This is the generalization of Boltzmann statistics to quantum mechanics. Let the states $|i\rangle$ in the expansion (4.7) be eigenstates of the system Hamiltonian H with eigenvalues E_i, and be chosen mutually orthogonal. (This is an additional requirement only in the case of degeneracies.) Then the probabilities are given by

$$p_i = \frac{e^{-\beta E_i}}{Z}, \qquad Z = \sum_i e^{-\beta E_i}, \qquad (4.10)$$

where $\beta = 1/(k_B T)$ is the inverse absolute temperature of the system, and Z is called the partition function. The thermal equilibrium density matrix can also be written as the operator $\rho = \exp(-\beta H)/Z$. It is a well-defined operator if β is positive and if the Hamiltonian H is bounded from below.

Consider the special case of a single-mode cavity with the Hamiltonian $H = \hbar \nu a^\dagger a$ that we studied in Section 3.2. Its eigenstates are the photon number states $\{|n\rangle; n = 0, 1, \ldots\}$ with energies $n\hbar\nu$, see (3.61). At a given inverse temperature β, the average value of the annihilation operator is zero because

$$\langle a \rangle = \sum_{n=0}^{\infty} p_n \langle n|a|n \rangle \qquad (4.11)$$

$$= \sum_{n=1}^{\infty} p_n \sqrt{n} \langle n|n-1 \rangle = 0 \qquad \text{(see (3.61))}$$

A similar calculation gives the average photon number

$$\langle a^\dagger a \rangle = \sum_{n=0}^{\infty} \frac{e^{-\beta n \hbar \nu}}{Z} \langle n|a^\dagger a|n \rangle = \frac{1}{e^{\beta \hbar \nu} - 1}, \qquad (4.12)$$

a result known as the "Bose-Einstein distribution function". At room temperature, $\langle a^\dagger a \rangle$ is vanishingly small for photon frequencies $\nu \gg 10^{11} \text{s}^{-1}$ which includes the range of visible light. Microwave cavities ($\nu \sim 10^9 \text{s}^{-1}$) contain one photon on average only if the cavity is cooled to temperatures around a few degree K above absolute zero.   □

4.3   Time evolution of density matrices

4.3.1   The von Neumann equation

If the system is initially prepared in a pure state, isolated from its environment and evolves according to the Hamiltonian H, the time dependence of the density matrix ρ satisfies the von Neumann equation ($\hbar = 1$)

$$\frac{d}{dt}\rho = -i\,[H, \rho]. \qquad (4.13)$$

This can be easily seen by taking the time derivative of $\rho(t) = |\psi(t)\rangle\langle\psi(t)|$ and using the Schrödinger equation (3.1), $i(d/dt)|\psi\rangle = H|\psi\rangle$. For an initially prepared non-pure state described by a density matrix, the same equation applies, following from linearity. We now discuss how this equation is modified for the evolution of an imperfect quantum system.

4.3.2   Quantum operations

We start with an axiomatic approach by introducing the notion of a "quantum operation". An important result is the representation theorem (4.18), whose proof is given for a finite-dimensional Hilbert space.

Definition 4.3.1. A quantum operation is a mapping Φ from an initial density matrix ρ to a final density ρ_f:

$$\Phi : \rho \mapsto \rho_f = \Phi(\rho), \tag{4.14}$$

which satisfies the following requirements:

1. Φ is linear in ρ.

2. $\Phi(\rho)$ is a density matrix for any initial density matrix ρ. In particular it is positive. (A quantum operation with this property is called a "positive map".)

3. Φ is a "completely positive" map. This means the following: let $\mathcal{H} \otimes \tilde{\mathcal{H}}$ be an extension of the system Hilbert space and extend Φ to a linear mapping $\Phi' = \Phi \otimes \mathbb{1}$ on products of positive operators ρ and $\tilde{\rho}$ on $\mathcal{H}$ and $\tilde{\mathcal{H}}$, respectively, by the rule

$$\Phi'(\rho \otimes \tilde{\rho}) \equiv \Phi(\rho) \otimes \tilde{\rho}. \tag{4.15}$$

Then Φ is completely positive if Φ' is positive for any dimension of the added Hilbert space $\tilde{\mathcal{H}}$. □

These conditions ensure that the final density matrix ρ_f can be used to make predictions for system observables. The last one embodies the physically sensible fact that the dynamics of the system are not changed when it is augmented by adding another independent system that does not interact with it.

Remark 4.3.1. The domain of the quantum operation Φ can be extended to include also "skew operators" using linearity with complex coefficients. For vectors $|\psi\rangle, |\chi\rangle \in \mathcal{H}$, we call $|\psi\rangle\langle\chi|$ a skew operator and define

$$\Phi(|\psi\rangle\langle\chi|) = \frac{1}{4} \left[\Phi(\mathbb{P}_{+1}) - \Phi(\mathbb{P}_{-1}) \right.$$
$$\left. -i\Phi(\mathbb{P}_{+i}) + i\Phi(\mathbb{P}_{-i}) \right], \tag{4.16}$$
$$\mathbb{P}_{e^{i\phi}} \equiv \mathbb{P}_{(|\psi\rangle + e^{i\phi}|\chi\rangle)/\sqrt{2}}, \tag{4.17}$$

where Φ is applied to projectors onto superposition states with suitably chosen relative phase ϕ. The D^2-dimensional vector space spanned by skew operators is sometimes called a Liouville space. □

4.3.3 The Kraus representation theorem

Theorem 4.3.1. *For any completely positive map* Φ*, there exists a countable set of operators* $\{\Omega_k\}$ *such that the map admits the representation*

$$\Phi(\rho) = \sum_k \Omega_k \rho \Omega_k^\dagger. \tag{4.18}$$

The operators Ω_k *satisfy*

$$\sum_k \Omega_k^\dagger \Omega_k = 1. \tag{4.19}$$

Proof. It is easy to check that (4.18) defines a completely positive map. To prove the converse, we extend the Hilbert space to $\mathcal{H} \otimes \mathcal{H}$. Consider a vector $|\phi\rangle \in \mathcal{H} \otimes \mathcal{H}$ that has a nonzero overlap with both sectors and construct the operator

$$P = (\Phi \otimes 1)(\mathbb{P}_{|\phi\rangle}). \tag{4.20}$$

Since Φ is completely positive, P is the concatenation of two positive operators and hence positive itself. Since P is hermitian, its spectral representation exists and can be written in the form

$$P = \sum_k |\varphi_k\rangle\langle\varphi_k|, \qquad |\varphi_k\rangle \in \mathcal{H} \otimes \mathcal{H}. \tag{4.21}$$

In the above, we have lumped the non-negative eigenvalues into the non-normalized eigenvectors $|\varphi_k\rangle$.

We now construct for any given $|\psi\rangle \in \mathcal{H}$ the following linear map $\Omega_k|\psi\rangle$ on the system Hilbert space. Let $|\chi\rangle \in \mathcal{H}$ and set

$$\langle\chi|\Omega_k|\psi\rangle = ((\langle\chi| \otimes \langle\psi^\star|)|\varphi_k\rangle. \tag{4.22}$$

We use the notation $\langle\chi| \otimes \langle\psi^\star|$ for the tensor product between the linear forms $\langle\chi|$ and $\langle\psi^\star|$. The linear form $\langle\psi^\star|$ is defined with respect to a basis $\{|n\rangle\}$ of $\mathcal{H}$, by the equation

$$\langle\psi^\star|n\rangle \equiv \langle n|\psi\rangle \tag{4.23}$$

and its linear extension. We can now check that for arbitrary $|\chi\rangle, |\chi'\rangle, |\psi\rangle \in \mathcal{H}$, the following equalities hold:

$$\sum_k \langle\chi|\Omega_k|\psi\rangle\langle\psi|\Omega_k^\dagger|\chi'\rangle = \sum_k ((\langle\chi| \otimes \langle\psi^\star|)|\varphi_k\rangle\langle\varphi_k|(|\chi'\rangle \otimes |\psi\rangle))$$
$$= ((\langle\chi| \otimes \langle\psi^\star|)P(|\chi'\rangle \otimes |\psi\rangle)) \qquad \text{(by (4.21))}$$
$$= ((\langle\chi| \otimes \langle\psi^\star|)(\Phi \otimes 1)(\mathbb{P}_{|\phi\rangle})(|\chi\rangle \otimes |\psi\rangle)) \qquad \text{(by (4.20))}.$$
$$\tag{4.24}$$

We now specialize to the following form for the vector $|\phi\rangle \in \mathcal{H} \otimes \mathcal{H}$:

$$|\phi\rangle = \sum_n |n\rangle \otimes |n\rangle \tag{4.25}$$

(this vector is a so-called maximally entangled state on the product Hilbert space). Its projector admits the following expansion

$$\mathbb{P}_{|\phi\rangle} = \sum_{n,m} (|n\rangle \otimes |n\rangle)(\langle m| \otimes \langle m|)$$

$$= \sum_{n,m} (|n\rangle\langle m|) \otimes (|n\rangle\langle m|) \tag{4.26}$$

in terms of skew operators. It is a truly remarkable fact that the full knowledge about Φ can be obtained by applying its extension ((4.20)) to this single projector. Using the definition of the extended quantum operation $\Phi \otimes 1$ and making use of the generalized linearity noted in Remark 4.3.1, we obtain

$$(\Phi \otimes 1)(\mathbb{P}_{|\phi\rangle}) = \sum_{n,m} \Phi(|n\rangle\langle m|) \otimes (|n\rangle\langle m|).$$

Taking the matrix elements required in (4.24) and using the definition (4.23), we find

$$\sum_{n,m} ((\langle \chi| \otimes \langle \psi^\star|) \left[\Phi(|n\rangle\langle m|) \otimes (|n\rangle\langle m|) \right] (|\chi'\rangle \otimes |\psi^\star\rangle)$$

$$= \sum_{n,m} \langle \chi| \Phi(|n\rangle\langle m|)|\chi'\rangle \langle n|\psi\rangle \overline{\langle m|\psi\rangle}$$

$$= \langle \chi| \Phi(|\psi\rangle\langle \psi|)|\chi'\rangle. \tag{4.27}$$

In the last step, we have used the expansion of $|\psi\rangle$ with respect to the basis $\{|n\rangle\}$. Hence, we have proven the operator identity (4.18) for the special case of a pure state $\rho = \mathbb{P}_{|\psi\rangle}$. The proof is now extended to a mixed state by decomposing ρ into projectors $\mathbb{P}_{|\psi_i\rangle}$ onto eigenvectors with non-negative weights (eigenvalues) p_i, and using the linearity of Φ. $\qquad\square$

The proof above implies that for a Hilbert space with dimension D, a maximum number D^2 of operators Ω_k is needed to represent a quantum operation.

4.3.4   Quantum Markov processes

We now return to the time evolution of a system's density matrix. We focus on the case that the future evolution of the system for $t \geq 0$ is determined by the knowledge of its density matrix $\rho(0)$ alone. Situations where this is true are, for example: (i) a system that has not yet interacted with its environment, or (ii) system-environment correlations that have been generated for $t < 0$ decay so rapidly in time that they do not influence the evolution for $t \geq 0$ on the time scales relevant for the system dynamics.

Under these conditions, the system's density matrix $\rho(t)$ at a fixed later time t is the result of a quantum operation

$$\rho(t) = \Phi_t(\rho(0)). \tag{4.28}$$

This is because the linearity of quantum mechanics implies that $\rho(t)$ depends linearly on the initially prepared state $\rho(0)$, and because $\rho(t)$ must again be a density matrix. Such a quantum operation Φ_t is called a *dynamical map*.

In case (ii) discussed above, it can be assumed that the correlations that have built up between the system and its surroundings, during the time t, are completely taken into account by the dynamical map Φ_t. They decay sufficiently fast not to influence the future evolution for times $t' > t$. This also applies for many systems of the case (i) above. In this situation, the system evolves in a "Markovian" way: its future again depends only on $\rho(t)$ and is described by a quantum operation $\Phi_{t',t}$. Since the initial state can be chosen arbitrarily, we obtain

$$\Phi_{t'} = \Phi_{t',t}\Phi_t, \tag{4.29}$$

the composition property for the time evolution of a Markovian open quantum system. If the system dynamics are not explicitly time-dependent, the intermediate time t can be chosen arbitrarily,[1] so we have the so-called quantum semigroup property

$$0 \le t \le t' : \qquad \Phi_{t'} = \Phi_{t'-t}\Phi_t. \tag{4.30}$$

This functional equation is solved by an operator of the form $\Phi_t = \exp(\mathcal{L}t)$ where $\mathcal{L}$ is called the "Liouvillian" (super)operator. Under conditions of continuity that are typically satisfied, the Liouvillian operator can be used to formulate a time-local differential equation for the system density matrix,

$$\frac{d}{dt}\rho(t) = \mathcal{L}\rho(t). \tag{4.31}$$

This equation is called *the master equation* and generalizes the von Neumann equation (4.13) to a dynamical map. The following theorem gives the general form of Markovian master equations, using the Kraus representation theorem 4.3.1.

Theorem 4.3.2. *Let $\Phi_{\Delta t} = \exp(\mathcal{L}\Delta t)$ be a dynamical map that satisfies the continuity condition*

$$\lim_{\Delta t \to 0} \operatorname{tr}\left[A\Phi_{\Delta t}(\rho) - A\rho\right] = 0 \tag{4.32}$$

for all bounded system observables A and for which the derivative

$$\lim_{\Delta t \to 0} \frac{1}{\Delta t}\left[\Phi_{\Delta t}(\rho) - \rho\right] \tag{4.33}$$

exists for all density matrices ρ. Then the corresponding Liouvillian operator can be written in the form

$$\mathcal{L}\rho = -i\,[H, \rho] + \sum_k \left(A_k\rho A_k^\dagger - \frac{1}{2}\left\{A_k^\dagger A_k, \rho\right\}\right) \tag{4.34}$$

---

[1]The contrary would be the case for a driven system, for example. Note also that choosing t too small compared to the decay time of system-environment correlations leads to inconsistencies.

where H is a Hamiltonian operator and the $\{A_k\}$ is a countable set of operators. The notation above,

$$\{A,\, B\} \equiv AB + BA, \tag{4.35}$$

signifies the "anti-commutator" of two operators A and B. The formula (4.34) is called the Lindblad form of the master equation, the operators A_k are called Lindblad operators.

Proof. Let $\Delta t > 0$ and write $\rho = \rho(t)$ for simplicity. We use the representation theorem (4.18) for the family $\Phi_{\Delta t}$ of quantum operations and obtain

$$\rho(t + \Delta t) = \Phi_{\Delta t}(\rho) = \sum_k \Omega_k \rho \Omega_k^\dagger \tag{4.36}$$

The operators occurring in (4.36) can be written in the form

$$\Omega_k = \omega_k \mathbf{1} + V_k, \tag{4.37}$$

where the V_k are uniquely defined by the requirement that their trace be zero. Note that ω_k and V_k depend in general on Δt.

The change in the density matrix is computed to be

$$\Phi_{\Delta t}(\rho) - \rho = \left(\sum_k |\omega_k|^2 - 1 \right) \rho + \sum_k \left(\overline{\omega}_k V_k\, \rho + \rho \omega_k V_k^\dagger \right)$$
$$+ \sum_k V_k \rho V_k^\dagger. \tag{4.38}$$

Using the continuity condition (4.32) for all operators A and ρ, we find the limits

$$\lim_{\Delta t \to 0} \sum_k |\omega_k|^2 = 1, \tag{4.39}$$

$$\lim_{\Delta t \to 0} \sum_k \overline{\omega}_k V_k = 0, \tag{4.40}$$

$$\lim_{\Delta t \to 0} \sum_k V_k\, \rho V_k^\dagger = 0, \tag{4.41}$$

where the last line applies to any density matrix ρ. Since the derivative (4.33) exists, we can introduce the derivatives

$$g \equiv \lim_{\Delta t \to 0} \frac{\sum_k |\omega_k|^2 - 1}{\Delta t}, \tag{4.42}$$

$$\Gamma - iH \equiv \lim_{\Delta t \to 0} \frac{\sum_k \overline{\omega}_k V_k}{\Delta t}, \tag{4.43}$$

where Γ and H are both hermitian.

The trace of the derivative (4.33) must be zero because the the dynamical map preserves the trace of the density matrix. We thus find

$$0 = \lim_{\Delta t \to 0} \frac{\operatorname{tr}\left[\Phi_{\Delta t}(\rho) - \rho\right]}{\Delta t}$$

$$= \operatorname{tr}\left[g\rho + (\Gamma - iH)\rho + \rho(\Gamma - iH) + \lim_{\Delta t \to 0} \frac{1}{\Delta t} \sum_k V_k \rho V_k^\dagger\right]$$

$$= \operatorname{tr}\left[g\rho + 2\Gamma\rho + \lim_{\Delta t \to 0} \frac{1}{\Delta t} \sum_k V_k^\dagger V_k \rho\right] \quad \text{(cyclic permutation)}. \quad (4.44)$$

Since this must hold for any density matrix ρ, we find yet another derivative

$$\lim_{\Delta t \to 0} \frac{\sum_k V_k^\dagger V_k}{\Delta t} = -g - 2\Gamma \quad (4.45)$$

We can thus introduce the Lindblad operators A_k by the limiting procedure

$$A_k \equiv \lim_{\Delta t \to 0} \frac{V_k}{\sqrt{\Delta t}}. \quad (4.46)$$

Finally, the time derivative of the operator equation (4.38) assumes the form

$$\frac{d}{dt}\rho = -i\left[H, \rho\right] + \sum_k \left(A_k \rho A_k^\dagger - \frac{1}{2}\left\{A_k^\dagger A_k, \rho\right\}\right), \quad (4.47)$$

which is the standard diagonal form (4.34) of the Lindblad master equation. □

Remark 4.3.2. The Lindblad operators A_k are not unique: the master equation (4.34) is unchanged under the following operations

1. a unitary transformation $A_k \mapsto B_k = \sum_l u_{kl} A_l$; this transforms the master equation into a non-diagonal form;

2. a displacement by a complex number a_k: $A_k \mapsto A_k + a_k$, combined with the change in the Hamiltonian $H \mapsto H - (i/2) \sum_k \left(\bar{a}_k A_k - a_k A_k^\dagger\right) + b$, where b is a real constant. This implies that the Lindblad operators can be chosen to be traceless. □

4.3.5 Non-Markovian environments

We have encountered in Subsection 4.3.4 the concept of a quantum system (S) coupled to an environment or reservoir (R). Given a Hamiltonian for both system and environment, this setting provides a more general model for *open* quantum systems. As mentioned before, master equations are a good approximation when the correlations between system and environment are weak and decay rapidly.

We comment here on the opposite case of long-lived correlations ("non-Markovian case"), where the Lindblad form (4.34) for the master equation does not apply. There are even examples where the dynamical map is not completely positive [3]. In fact, for non-Markovian environments, the limit $\Delta t \to 0$ in the above proof conflicts with the requirement that Δt be larger than the system-environment correlation time.

In the system+environment setting, averages of observables pertaining to the system alone can be computed from the so-called "reduced density matrix" ρ_S. The elements of this matrix are defined by the "partial trace"

$$\langle n|\rho_S|m\rangle = \sum_\nu \langle n \otimes \nu|\rho_{SR}|m \otimes \nu\rangle \equiv \langle n|\mathrm{tr}_R\rho_{SR}|m\rangle \qquad (4.48)$$

where ρ_{SR} is the density matrix in the combined Hilbert space for system and environment, $|m\rangle$, $|n\rangle$ are states in the system Hilbert space, and $\{|\nu\rangle\}$ is a basis for the environment Hilbert space. The problem in the non-Markovian case is that the correlations between system and environment cannot be retrieved from the knowledge of ρ_S alone, but being long-lived, they still influence the future evolution.

Let $U(t)$ be the unitary evolution operator generated by the Hamiltonian in the combined Hilbert space for system and environment, and assume that $\rho_{SR}(0) = \rho_S(0) \otimes \rho_R(0)$ factors. One can then show that the mapping

$$\rho_S(0) \mapsto \rho_S(t) = \mathrm{tr}_R\rho_{SR}(t)$$
$$= \mathrm{tr}_R\left[U(t)\rho_{SR}(0)U^\dagger(t)\right] \qquad (4.49)$$

is a completely positive map. But since $\rho_{SR}(t)$ does not factor, an iteration of this mapping is not possible. Hence, the semigroup property (4.30) does not hold which has been used in the derivation of the Lindblad form.

Consider finally the case that even the initial state $\rho_{SR}(0)$ does not factor. This occurs, for example, for a system at a thermal equilibrium with its environment and strongly coupled to it. It is then possible that the mapping (4.49) generates a non-positive system operator [3]. This means that it must be restricted to a subspace of density matrices $\rho_{RS}(0)$.

4.4 Examples of master equations

We outline in this section some examples where the conditions for Markovian reduced system dynamics are satisfied. The key step will be the identification of the Lindblad operators in the system Hilbert space that appear in the master equation (4.34). We focus on examples relevant for cavity QED implementations of quantum computing.

4.4.1 Leaky cavity

The mirrors of a cavity are never perfectly reflecting in practice. The leakage of the field to the outside world also provides experimentalists with a way to monitor the evolution of the cavity field (see illustration in Fig. 4.1).

Let ρ be the density matrix for the cavity field and 2κ the rate (3.68) at which the average photon number decays. Then the Lindblad master equation is given in terms of the single Lindblad operator $A = \sqrt{2\kappa}a$ where a is the photon annihilation operator. Explicitly, we have the Liouvillian

$$\mathcal{L}_{\text{cav}}\rho = -i\,[H,\,\rho] + \kappa\left(2a\rho a^\dagger - \{a^\dagger a,\,\rho\}\right), \tag{4.50}$$

where $H = \nu a^\dagger a$ is the cavity field Hamiltonian. (We have set $\hbar = 1$ and removed the zero point energy.) A simple calculation gives the following equations of motion for the average field operator and average photon number:

$$\frac{d}{dt}\langle a\rangle = -i\nu\langle a\rangle - \kappa\langle a\rangle, \tag{4.51}$$

$$\frac{d}{dt}\langle a^\dagger a\rangle = -2\kappa\langle a^\dagger a\rangle. \tag{4.52}$$

They can be solved and show an exponential decay for the average cavity field at the rate κ. The average photon number vanishes at large times so that the cavity mode decays to its ground state (vacuum). This shows that the outside world acts like a zero-temperature bath for the cavity field.

If finite temperature effects play a role, one has to add a second Lindblad operator $A' = \sqrt{2\kappa'}a^\dagger$. It has the effect of increasing the cavity photon number at a rate $2\kappa'$ so that at large times, $\langle a^\dagger a\rangle \to \kappa'/\kappa$ (see Fig.4.1). By identifying this with the Bose-Einstein distribution (4.12), one obtains the ratio $\kappa'/\kappa = (e^{\hbar\nu/k_B T} - 1)^{-1}$.

The Liouvillian (4.50) can be derived from a Hamiltonian model for the cavity and its environment with an interaction Hamiltonian of the form $H_I = g\left(aB^\dagger + a^\dagger B\right)$, where g is a coupling constant and B contains an infinite sum over a continuum of photon modes. It represents the electric field operator outside the cavity. To estimate the timescale for correlations within the environment, one analyzes the 2-time correlation function $\langle B(t)B^\dagger(t')\rangle_R$ which is an expectation value with respect to the (equilibrium) state of the environment. This function depends only on $t - t'$ (for a

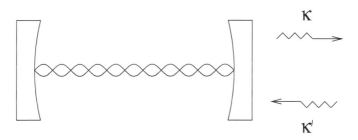

Figure 4.1: Illustration of photons leaking out of a cavity through the right mirror. The left mirror is supposed to be perfect. Here κ' is the rate of photons entering from the outside world into the cavity.

stationary environment) and decreases to zero for $|t - t'| \to \infty$. The corresponding time scale (i.e., the correlation time) is related to the behavior of $N(\omega)$, the density of continuum modes per unit frequency interval, by the well-known relations between Fourier transform pairs. A smooth function $N(\omega)$ thus leads to a very short correlation time—this is the situation where the Markov approximation applies. In the approximation of weak coupling to the continuum, one finds the following for the dissipation rates

$$2\kappa = g^2 \int_{-\infty}^{+\infty} d\tau \, \langle B(t + \tau) B^\dagger(t) \rangle_R e^{i\nu\tau}, \tag{4.53}$$

$$2\kappa' = g^2 \int_{-\infty}^{+\infty} d\tau \, \langle B^\dagger(t + \tau) B(t) \rangle_R e^{i\nu\tau}. \tag{4.54}$$

More details about the microscopic derivation of (4.50) can be found in the book by Gardiner [1].

4.4.2   Unstable 2-level system

Consider an atomic 2-level system with the ground and excited states $|g\rangle$, $|e\rangle$ and an energy splitting $\hbar\omega$, placed inside a cavity (see illustration in Fig. 4.2). As described in Section 3.1.1, the atom couples to the electromagnetic field, with an enhanced coupling to cavity modes. But a planar cavity with finite-area mirrors leaves the continuum of modes with wavevector perpendicular to the cavity axis ("*perpendicular modes*") essentially unaffected. The coupling to this mode continuum makes the atomic states unstable: $|e\rangle$ can decay to $|g\rangle$ by emitting a photon into a perpendicular mode, and the absorption from a perpendicular mode (if populated) gives rise to a transition $|g\rangle \to |e\rangle$. The fact that these modes form a continuum allows us to describe these transitions in terms of a Markovian master equation in Lindblad form.

The Lindblad operator for emission processes involves the 2-state lowering operator $\sigma = |g\rangle\langle e|$ and the Liouvillian becomes

$$\mathcal{L}_{\text{at}} \rho = -i \left[H, \, \rho \right] + \gamma \left(2\sigma\rho\sigma^\dagger - \{\sigma^\dagger\sigma, \, \rho\} \right). \tag{4.55}$$

Recall that the atomic Hamiltonian, with the rotating wave approximation, is given by (Eqs. (3.4) and (3.34))

$$H = \frac{\Delta}{2} \left(\mathbb{P}_{|e\rangle} - \mathbb{P}_{|g\rangle} \right) - \frac{\Omega}{2} \left(e^{i\phi}\sigma + e^{-i\phi}\sigma^\dagger \right), \tag{4.56}$$

with the second term describing the coupling with a coherent laser field at the frequency $\nu = \omega - \Delta$.

The probability p_e of finding the atom in the excited state is given by the diagonal ee entry of the density matrix. Its equation of motion is found as

$$(d/dt)p_e = \text{tr}(\mathbb{P}_{|e\rangle} \mathcal{L}_{\text{at}} \rho) = -i \, \text{tr} \left([\mathbb{P}_{|e\rangle}, H] \, \rho \right) - 2\gamma p_e. \tag{4.57}$$

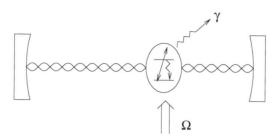

Figure 4.2: Illustration of a 2-level atom placed inside a cavity and driven by a laser field (Rabi frequency Ω). The rate γ characterizes the rate of the "spontaneous emission" of a photon in a mode other than the cavity; see (4.58). In this process, the atom makes a jump from the upper to the lower energy level, as indicated by the wavy arrow connecting the levels.

The first term on the rightmost part of the equation gives the Hamiltonian contribution to the excited state population. Coupling to the field continuum makes the excited state decay at a rate 2γ.

If the surrounding mode continuum is the same as in free space and if it is at zero temperature (or at $k_B T \ll \hbar\omega$), the decay rate γ is given by

$$\gamma = \frac{|P_{ge}|^2 \omega^3}{6\pi\epsilon_0 \hbar c^3}, \tag{4.58}$$

where P_{ge} is the matrix element of the electric dipole operator $-er$ between the states $|g\rangle$ and $|e\rangle$ and ϵ_0 is the vacuum permittivity (also called Coulomb constant). This result can be shown by introducing the quantized electromagnetic field as the environment for the atom, with a coupling given by the electric dipole interaction (3.3). One works out the dynamical map for the atomic system by using the mode expansion (3.64) for the electric field operator, taking into account the atom-field coupling to second order and performing the partial trace over the field Hilbert space. More details can be found in the book by Scully and Zubairy [4, Chap. 8]. In that calculation, the rate (4.58) turns out to be given by an expression similar to (4.53),

$$\gamma = \frac{|P_{ge}|^2}{2\hbar^2} \int\limits_{-\infty}^{+\infty} d\tau \, e^{i\omega\tau} \langle E_a(t+\tau) E_a(t)\rangle_R, \tag{4.59}$$

where the 2-time correlation function is taken in the vacuum state for the field, and the subscript a denotes the component of the field vector along the direction of the electric dipole moment connected to the 2-level atom. With a stationary state for the mode continuum, the rate (4.59) does not actually depend on t. The field correlation time is again related to the frequency scale on which the field density of modes $N(\omega)$ varies around the atomic transition frequency. In free space, $N(\omega)$ is smooth, the correlation time is very short, and the Markov approximation applies.

At finite temperature (when the assumption $k_B T \ll \hbar \omega$ is removed), the decay rate (4.58) becomes $\gamma(1 + \bar{n}(\omega))$ where $\bar{n}(\omega)$ is the average photon number (4.12) at the transition frequency ω according to the Bose-Einstein distribution. The Liouvillian (4.55) has to be modified accordingly, and a second Lindblad operator proportional to $\sigma^\dagger$ has to be taken into account. Indeed in this case, the ground state also is unstable because it can absorb a thermal photon. The absorption occurs at a rate $\gamma \bar{n}(\omega)$, so the ground state population evolves like

$$\frac{d}{dt} p_g = -i \operatorname{tr}\left(\left[\mathbb{P}_{|g\rangle}, H\right]\rho\right) - \gamma \bar{n}(\omega) p_g + \gamma(1 + \bar{n}(\omega)) p_e. \tag{4.60}$$

In the limit $t \to \infty$ and with the laser field switched off, one gets $p_e/p_g \to \bar{n}(\omega)/(1 + \bar{n}(\omega)) = e^{-\beta \hbar \omega}$, hence the Boltzmann statistics of thermal equilibrium.

4.4.3 Dephasing

The last example of master equations here is provided by fluctuations of the qubit transition frequency that randomize the relative phase between qubit superposition states ("*dephasing*"). With neutral atoms, this occurs when the laser field has frequency fluctuations, leading to a Hamiltonian (recall (4.56))

$$H_I = -\frac{1}{2} \sigma_3 \delta\nu(t) \tag{4.61}$$

with a random $\delta\nu(t)$ and

$$\sigma_3 = \mathbb{P}_{|e\rangle} - \mathbb{P}_{|g\rangle}. \tag{4.62}$$

Dephasing is also the main source of imperfections for qubits implemented in a solid; there it occurs due to the coupling to lattice vibrations or phonons. In the Markov limit (short-lived correlations of the phase fluctuations), this process can be represented by the Liouvillian (where only the dissipative part is written)

$$\mathcal{L}\rho = 2\gamma_d \left(\sigma_3 \rho \sigma_3 - \rho\right). \tag{4.63}$$

We have made use of $\sigma_3^2 = 1$ to simplify the terms involving the anticommutator. The rate γ_d characterizes the decay of qubit superposition states. In terms of the correlation function of $\delta\nu(t)$, it is given by

$$\gamma_d = \frac{1}{4} \int\limits_{-\infty}^{+\infty} d\tau \, \langle \delta\nu(t+\tau)\delta\nu(t)\rangle. \tag{4.64}$$

While the diagonal elements of the density matrix (the basis states $\mathbb{P}_{|g\rangle}$, $\mathbb{P}_{|e\rangle}$) are unaffected by the Liouvillian (4.63), off-diagonal elements decay exponentially according to

$$(d/dt)\rho_{eg} = \langle e|\mathcal{L}\rho|g\rangle = -2\gamma_d \rho_{eg}. \tag{4.65}$$

A superposition state with equal weights $|\psi\rangle = (1/\sqrt{2}, e^{i\alpha}/\sqrt{2})^T$ is thus characterized after a time t by the density matrix

$$\rho(t) = \frac{1}{2} \begin{pmatrix} 1 & e^{-i\alpha - 2\gamma_d t} \\ e^{i\alpha - 2\gamma_d t} & 1 \end{pmatrix}. \tag{4.66}$$

The probability of finding the initial (or "target") state is given by $F_{|\psi\rangle} = \mathrm{tr}\left(\mathbb{P}_{|\psi\rangle}\rho(t)\right) = \frac{1}{2}(1 + e^{-2\gamma_d t})$. It tends to 50% at large times. In this limit, the state (4.66) becomes proportional to the unit matrix (a classical, unbiased mixture of any pair of orthogonal states in the 2-dimensional Hilbert space). It is intuitively clear that the target state is found with 50% probability in this mixture.

4.5 Fidelity calculations

We discuss here two examples of 1-qubit gates with imperfections. The first one is based on statistically distributed parameters of the gate. A density matrix is introduced as a classical average of pure state projectors depending on random parameters. The second example illustrates the competition between a "coherent operation" (Rabi oscillation in a laser field) and an "incoherent" decay (spontaneous emission of the excited state).

4.5.1 Fluctuating gate parameters

An additional type of situation where a statistical treatment has to be used occurs if the parameters that determine single- or 2-qubit gates are not completely known. Denote $U(t, \Gamma)$ the unitary evolution operator for a fixed parameter Γ (this can be a multicomponent or infinite-dimensional object) and $p(\Gamma)$ the corresponding (classical) probability. The simplest case is that this distribution is known or can be computed with some statistical modeling. In that case, the only acceptable choice for the evolved system density matrix is given by the expectation value

$$\rho(t) = \mathrm{E}[U(t, \Gamma)\rho_i U^\dagger(t, \Gamma)] \tag{4.67}$$

$$= \sum_\Gamma p(\Gamma) U(t, \Gamma)\rho_i U^\dagger(t, \Gamma). \tag{4.68}$$

The second line above is of the form of the representation theorem (4.18), showing that this is a quantum operation. If complete knowledge about the distribution of the parameters Γ is not available, a best guess for $p(\Gamma)$ can be inferred from the knowledge of some expectation values using a maximum likelihood argument, for example.

Example 4.5.1. Consider the single qubit operation (3.33)

$$U_{\theta,\phi} \equiv \begin{bmatrix} \cos\theta & -ie^{i\varphi}\sin\theta \\ -ie^{-i\varphi}\sin\theta & \cos\theta \end{bmatrix}, \tag{4.69}$$

that describes a partial Rabi oscillation. Imagine the parameter φ (the phase of the applied laser field) to be a random variable. It is clear that the final, averaged density matrix involves expressions like $E[e^{i\varphi}]$. Assuming a Gaussian probability distribution (average $\bar{\varphi}$ and variance σ^2), we have

$$E[e^{i\varphi}] = e^{i\bar{\varphi} - \frac{1}{2}\sigma^2}, \tag{4.70}$$

which allows us to compute the final density matrix (4.67). We have to compare this density matrix with the "target state" $U_{\theta_t, \varphi_t} \rho_i U^\dagger_{\theta_t, \varphi_t}$ where the parameters θ_t, φ_t may differ from θ and $\bar{\varphi}$ above. To simplify the formulas, we focus on the special case that the initial state is the atomic ground state $|\psi\rangle = (1, 0)^T$. The probability of finding the target state after the randomized Rabi oscillation is then given by

$$F_{|g\rangle} = \cos^2(\theta_t - \theta) - \frac{\sin(2\theta_t)\sin(2\theta)}{2}\left(1 - \cos(\varphi_t - \bar{\varphi})e^{-\frac{1}{2}\sigma^2}\right), \tag{4.71}$$

as can be calculated as follows. The target state is

$$\rho_t = U_{\theta_t, \varphi_t} \rho_i U^+_{\theta_t, \varphi_t} = |\psi_t\rangle\langle\psi_t|, \quad \text{if} \quad \rho_i = |\psi_i\rangle\langle\psi_i|,$$

where

$$|\psi_t\rangle = U_{\theta_t, \varphi_t}|\psi_i\rangle.$$

The state after averaging is

$$\bar{\rho} = E[U_{\theta,\varphi}\rho_i U^+_{\theta,\varphi}] = E[|\psi\rangle\langle\psi|], \text{ where } |\psi\rangle = U_{\theta,\varphi}|\psi_i\rangle.$$

Then the probability of finding the "target state" is

$$\begin{aligned}
F &= tr(\rho_t \bar{\rho}) \\
&= tr(|\psi_t\rangle\langle\psi_t|E[|\psi\rangle\langle\psi|]) \\
&= E[tr(|\psi_t\rangle\langle\psi_t|\psi\rangle\langle\psi|]] \\
&= E[|\langle\psi_t|\psi\rangle|^2].
\end{aligned}$$

Now, compute ψ_t and ψ for initial state $|\psi_i\rangle = (1, 0)^T$:

$$|\psi_t\rangle = U_{\theta_t, \varphi_t}|\psi_i\rangle = (\cos\theta_t, -ie^{-i\varphi_t}\sin\theta_t)^T,$$
$$|\psi\rangle = U_{\theta,\varphi}|\psi_i\rangle = (\cos\theta, -ie^{-i\varphi}\sin\theta)^T;$$
$$\langle\psi_t|\psi\rangle = \cos\theta_t\cos\theta + ie^{i(\varphi_t - \varphi)}\sin\theta_t\sin\theta,$$
$$|\langle\psi_t|\psi\rangle|^2 = \cos^2\theta_t\cos^2\theta + \sin^2\theta_t\sin^2\theta + 2\cos\theta_t\cos\theta\sin\theta_t\sin\theta \cdot \text{Re}(e^{i(\varphi_t - \varphi)}).$$

But, by (4.70),

$$E[e^{i(\varphi_t - \varphi)}] = e^{i(\varphi_t - \bar{\varphi}) - \sigma^2/2},$$

and, thus,

$$E[\text{Re } e^{i(\varphi_t - \varphi)}] = \cos(\varphi_t - \bar{\varphi})e^{-\sigma^2/2}.$$

Hence

$$
\begin{aligned}
F &= E[|\langle\psi_t|\psi\rangle|^2] \\
&= \cos^2\theta_t \cos^2\theta + \sin^2\theta_t \sin^2\theta + 2\cos\theta_t \cos\theta \sin\theta_t \sin\theta \\
&\quad + 2\cos\theta_t \cos\theta \sin\theta_t \sin\theta[\cos(\varphi_t - \bar{\varphi})e^{-\sigma^2/2} - 1] \\
&= (\cos\theta_t \cos\theta + \sin\theta_t \sin\theta)^2 + \frac{1}{2}\sin(2\theta_t)\sin(2\theta)[\cos(\varphi_t - \bar{\varphi})e^{-\sigma^2/2} - 1],
\end{aligned}
$$

as given by (4.71). As expected, this probability is maximized if $\theta_t = \theta \,(\text{mod}\, 2\pi)$, and $\varphi_t = \bar{\varphi}\,(\text{mod}\, 2\pi)$. It is interesting that the probability (4.71) is immune against laser phase fluctuations if $\theta_t = 0\,(\text{mod}\,\pi)$ where the target state is either the ground or the excited state.

The computation of the fidelity (4.1) requires an optimization with respect to all initial (pure) states. For a given concrete example, it is often possible from symmetry considerations to guess the state(s) that is (are) most perturbed by the imperfections in the gate. Upper bounds to the fidelity are thus relatively easy to find. □

4.5.2 Spontaneous decay

Consider a 2-level atom subject to a coherent laser pulse and coupled to the electromagnetic continuum at finite temperature. The master equation for the density matrix then is of the form (4.55), with a term

$$
\gamma_g \left(2\sigma^\dagger\rho\sigma - \{\sigma\sigma^\dagger, \rho\}\right) \tag{4.72}
$$

added to describe absorption from the continuum. Taking matrix elements, it is easy to find the following equations of motion (the "optical Bloch equations")

$$
\dot{\rho}_{ee} = \frac{i}{2}\Omega\left(\rho_{eg} - \rho_{ge}\right) - 2\gamma\rho_{ee} + 2\gamma_g, \tag{4.73}
$$

$$
\dot{\rho}_{eg} = 2\Omega\left(\rho_{gg} - \rho_{ee}\right) - (i\Delta + \gamma)\rho_{eg}. \tag{4.74}
$$

We have put $\gamma = \gamma_e + \gamma_g$, where γ_e is the excited state decay rate. The equation for ρ_{gg} is redundant since $\text{tr}\rho = \rho_{gg} + \rho_{ee} = 1$ is conserved. The equation for ρ_{ge} is obtained by complex conjugation from the one given here. For simplicity, we have redefined the phase of ρ_{ge} such that the laser phase $\phi = 0$ in the Hamiltonian (4.56).

The linear system of differential equations (4.73), (4.74), can be straightforwardly solved, but the resulting expressions are cumbersome. They simplify at resonance, $\Delta = 0$, and for an atom in the ground state at $t = 0$:

$$\rho_{ee}(t) = \rho_{ee}(\infty) \left[1 - e^{-3\gamma t/2} \left(\cos(\lambda t) + \frac{3\gamma}{2\lambda} \sin(\lambda t) \right) \right], \tag{4.75}$$

$$\rho_{eg}(t) = \frac{2i\gamma_g}{\Omega} + \frac{2i\gamma}{\Omega} \rho_{ee}(\infty) \left[1 - e^{-3\gamma t/2} \left(\cos(\lambda t) - \frac{\Omega^2 - \gamma^2}{2\gamma\lambda} \sin(\lambda t) \right) \right], \tag{4.76}$$

$$\rho_{ee}(\infty) = \frac{\Omega^2/2 + 2\gamma\gamma_g}{\Omega^2 + 2\gamma^2}, \tag{4.77}$$

$$\lambda = \sqrt{\Omega^2 - \gamma^2/4}. \tag{4.78}$$

We see that for weak dissipation, the Rabi oscillations are damped. If the damping exceeds a critical value, $\gamma \geq 2\Omega$, they disappear completely.

This result is illustrated in Figure 4.3, where we plot the probability $F_{|g\rangle}$ of finding the "target state". The latter is defined by Eqs. (4.75), (4.76) with $\gamma \mapsto 0$, $\gamma_g \mapsto 0$, and $\Omega \mapsto \Omega_t$. The probability F is minimal if the target state has a large excited state component, as expected from the instability of this state. Only a slight improvement is found by adjusting $\Omega_t = \lambda$ (dashed line), taking into account the slowdown of the Rabi oscillations due to damping ((4.78)).

The computation of the fidelity (4.1) of the 1-qubit quantum gate requires a minimization with respect to the set of all initial states. This is generally more difficult to perform explicitly, and one has to resort to numerical techniques.

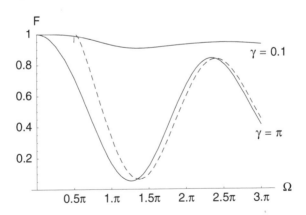

Figure 4.3: Overlap $F_{|g\rangle} = \mathrm{tr}(\rho_t \rho)$ between the target state ρ_t for un-damped Rabi oscillations and the density matrix ρ with elements (4.75), (4.76), for two different spontaneous decay rates $\gamma = 0.1$ and $\gamma = \pi$. Solid line: nominal Rabi frequency $\Omega_t = \Omega$, dashed line: $\Omega_t = (\Omega^2 - \gamma^2/4)^{1/2}$ (result plotted for $\Omega \geq \pi/2$ only). Other parameters: $t = 1$ and $\gamma_g = 0$.

References

[1] C.W. Gardiner, *Handbook of Stochastic Methods*, (Springer, Berlin, 1983).

[2] M.A. Nielsen and I.L. Chuang, *Quantum Computation and Quantum Information*, Cambridge University Press, Cambridge, U.K., 2001.

[3] P. Pechukas, Reduced dynamics need not be completely positive, *Phys. Rev. Lett.* **73** (1994), 1060. Comment R. Alicki, *Phys. Rev. Lett.* **75** (1995), 3020; reply: *id.* **75** (1995), 3021.

[4] M.O. Scully and M.S. Zubairy, *Quantum Optics*, Cambridge University Press, Cambridge, U.K., 1997.

Chapter 5

Quantum Computation Using Cold, Confined Atomic Ions

- Types of ion traps

- Qubit setups

- Various designs of universal quantum gates, due to Cirac–Zoller, Debye–Waller, and Sørensen–Mølmer

- Device architecture, scalability, and error correction

- Assessment

5.1 Introduction

A technology to physically implement quantum computation has many requirements; some conceptual and some technical. In 1995, Cirac and Zoller [13] proposed to implement quantum computation using recently developed experimental techniques [3] which used laser beams to cool atomic ions confined in vacuum inside an ion trap, until they formed a stable linear ion array under the joint forces due to their mutual repulsion and the confining potential gradient. Since each cold ion of the stable array was identifiable, two internal electronic levels of each ion could act as a computational quantum bit, or qubit. A quantum gate, or quantum computational unit, requires a correlated action on at least two qubits. Cold ion internal level qubits have no direct coupling, since the separation of the ions in the linear array is large compared to atomic interaction distances. However, since the minimum of the ion trapping potential is essentially harmonic, the ion array oscillates coherently in the trap at the center of mass (CM) angular frequency ω_{CM}, which remains the same, regardless of the number of ions. This and other quantized normal modes of oscillation of the ions in the trap can be cooled to their lowest energy state using laser beams, in order to act as a "motional qubit" to couple the ions on demand.

A laser beam of the proper frequency, focussed on a single ion of the array, could simultaneously interact with the internal electronic qubit, and also the motional qubit. The motional qubit consists of the lowest, or ground, level and first excited level of the CM motion, which couples all of the ions in the array at the characteristic CM frequency. If the coupling interaction between the internal and motional qubits depends on the state of the motional qubit, then its *conditional* character produces *quantum entanglement* of the system wave function, which is necessary to implement quantum information processing between coupled qubits. Quantum entanglement refers to an internal correlation of quantum coordinates, which cannot be associated with an individual qubit. A quantum gate consists of this entangling operation, plus electron Rabi rotations of single internal qubit states, all induced by focussed laser beams having well-defined frequencies. *Universal* quantum gates can be used to execute any desired numerical algorithm. In principle, a quantum computer could be constructed using an array of many identical cold ions in a linear trap, each acting as a qubit participating in universal 2-qubit gates with the CM motional qubit. Cirac and Zoller described a particular protocol to implement gates of this type [13].

This seminal proposal by Cirac and Zoller has led to many experimental and theoretical investigations of ways to build practical quantum computer gates using cold, confined ions, and alternative means to implement them. Current thinking, and the state of the experimental art, particularly as expressed and developed by a group at the National Institute of Standards and Technology (NIST) at Boulder, CO, are discussed in this chapter. This discussion here, as well as the others throughout the rest of the book, is not intended to be an exhaustive review, but rather it is hoped to summarize the basis of the technique, plus perceived strengths and weaknesses of this technology uncovered by recent experimental work, in a way that is accessible to non-physicists involved in the quantum computing field. Sufficient progress has been made to warrant an early assessment of progress toward a functioning quantum computer with many qubits. At this time, there is some controversy over whether laser-excited ions in cold confined ion arrays are competitive as a technical choice for large scale quantum computation. The internal and external interactions of this system occur primarily with well-defined electric and magnetic fields, and consequently are well described by theory. Since this theory is available, experimental departures from it can be quantified, and associated with imperfections of the system operation, or with coupling to the "environment", i.e., anything not described precisely by the theory. Even if these arrays should ultimately prove not to be the best choice for quantum computation, the fact that the interactions can be clearly described theoretically in most aspects, and quantitative calculations of details made, make this scheme a useful subject for study in evaluating any practical alternative.

Recall the five criteria of D. DiVincenzo for a working QC in Box 2.9 in Chap. 2. Box 5.1 offers a "preview" of this chapter as to how ion traps work apropos the five criteria.

---

(1) *Well-defined qubits and scalability:* Two of the internal levels of each isolated ion form a well-defined qubit, and if the ions form a regular linear array, these qubits can be individually identified and manipulated using focussed laser light. The qubits are coupled through the CM motion of the linear array, which can be conditionally coupled to the qubit state by optical excitation. There is no inherent limitation on the number of ions in the array, and the CM frequency of the motional qubit is independent of ion number, so the scale of the system is arbitrary.

(2) *Sufficiently long coherence times:* The isolated ions in vacuum, and the motional qubit, interacting with well-defined internal and external fields, potentially retain their coherence for very long times compared to the time to execute many quantum gates (but as usual, the devil is in the details; see more discussions in this chapter).

(3) *Universal quantum gates:* The universal quantum mechanical phase gate, and the related controlled NOT (CNOT) gate, were demonstrated in principle by a protocol in the initial proposal by Cirac and Zoller (13). Subsequently there have been experimental demonstrations of several alternative ways to implement related universal gates using confined ions, and additional theoretical protocols for simpler or alternative gates to be executed by other laser interactions with the ions. These quantum gates are deterministic, having no arbitrary or uncontrolled character, and are executed at chosen times.

(4) *Reliable readout:* By focusing a resonant laser beam on an individual ion, an electron in one of the qubit levels can be repetitively excited to a higher ion electronic level in a cycling transition, which brings the electron back to the original qubit level. The characteristic of this induced electric dipole cycling transition is that, following each excitation, the emission of a fluorescent photon signifies that there was an electron in the original qubit level. The fluorescence photons can be collected and recorded with reasonable efficiency, so that if fluorescence photons are recorded, the ion was in the qubit level addressed. If fluorescence photons are absent, the other qubit level was occupied. The efficiency and reliability of this binary measurement can be made very large.

(5) *Initializability:* It is feasible to reduce the confined ion temperature sufficiently using laser light to populate the lowest level of the motional qubit with high probability. Laser transitions are also used to pump each ion into a defined qubit level, by using appropriate laser beam frequencies and polarizations to cause

a repetitive cycle of excitation and fluorescence to end in that level. This completes the initialization process, which could be verified by subsequent measurements if desired.

Box 5.1 Ion traps in meeting D. DiVincenzo's five criteria

Satisfactory *experimental* demonstrations of these conceptual criteria have made cold, confined ions a viable alternative for a quantum computation technology. Several ion types have been proposed or studied to demonstrate quantum gates or investigate specific aspects of the problem. Ion parameters of interest include having one active electron outside of closed electron shells, the availability of long-lived metastable excited states, or hyperfine structure levels in the ground state, which can be used as qubits, and the rates of both allowed and "forbidden" transitions between potential qubit levels. Ions such as Be^+ and Mg^+ have no metastable excited states, so the internal qubit states will be hyperfine structure levels in the lowest electronic (ground) state. Heavier ions such as Ca^+, Cd^+, or Hg^+ additionally have metastable excited electronic levels, which can be reached by higher polarity transitions, such as magnetic dipole or electric quadrupole. These metastable levels have low rates of spontaneous decay, cf. Subsection 4.5.2, a random process which destroys the qubit coherence. They maintain sufficient coherence so that the ground and metastable levels might be used as the qubit. The limitations set by nature and available technology, and the parameters of quantum computation, will eventually determine the best ion and internal level scheme to choose.

The $^9Be^+$ ion is emphasized in the present discussion, because this ion has been used extensively in the experimental work on quantum computation by the group at NIST-Boulder [3], and has also been useful to the second author (D.A. Church) in his separate collaborative research on cold, confined highly-charged ions [23]. The NIST-Boulder group has also published extensive discussions on the experimental and theoretical aspects of the details of the method [52, 54]. Although the Be^+ ion is likely not the best ultimate choice for quantum computation, it provides a specific focus for a general discussion of the experimental technique. Each candidate ion has its partisans, and the weighting of all parameters necessary to make the optimum choice of ion has not yet reached universal agreement.

Although the basic concept of a cold, confined ion quantum computer as presented by Cirac and Zoller is very elegant, experimentally there are many details, which complicate or limit its implementation. In the organization of this chapter, an attempt is made to initially discuss dominant topics in each section, and then to discuss major experimental supporting results and limitations to these topics. Progress toward a practical quantum computer is described, and the potential for computation based on present developments is assessed. A summary of each major section is presented. Alternative theoretical schemes for implementation of gates or other topics, which build on the basis established by Cirac and Zoller, are discussed in detail only when experimental results have been obtained.

5.2 Ion confinement, cooling, and condensation

Ion confinement, cooling, and condensation of the ions into a stable array are the basic techniques which make quantum computation using cold, confined ions feasible. The chosen techniques are briefly discussed below, in a general way, but which is intended to make clear their strengths and limitations.

5.2.1 Confinement: several types of ion traps

Several basic ways have been developed to confine one or more low energy ions in space and time, each with variants. The four major types are:

(i) The **Kingdon ion trap** [12] is an electrostatic device (see Fig. 5.1(a)), and as such, has no minimum in potential in space away from an electrode. Ions are confined by the conservation of angular momentum, as they orbit a central wire in a radial plane, while oscillating axially between end plates. A trap which depends on angular momentum for confinement cannot confine cold ions.

(ii) A **magnetic bottle** [51] consists of an axially non-uniform magnetic field, with stronger field regions at the ends than in the middle, Fig. 5.1(b). An orbiting charged particle can be confined in the radial plane by the Lorentz force exerted by the magnetic field on the moving charge. The charge is confined axially through conditional conservation of an adiabatic invariant, which is a composite parameter called the "action" that remains constant when individual parameters like the magnetic field vary slowly in space or time. In quantum mechanics, adiabatic invariance implies no spontaneous or uncontrolled changes in the populations of quantized energy levels, such as electromagnetic transitions between energy levels, which increase entropy or disorder. A magnetic bottle cannot confine very cold ions, which would have the minimal orbit, and its spatially varying magnetic field would change internal ion energies in an undesirable way for quantum computation.

(iii) A **Penning ion trap** (Fig. 5.1(c)) [12] uses the Lorentz force exerted by a *uniform* magnetic field for ion confinement in the radial plane, together with an electric potential varying quadratically in space for axial confinement. The defocusing character of the radial part of the electric quadrupole potential (cf. (2.32)) is overcome by a sufficiently strong confining magnetic Lorentz force. Since the total ion confinement results from two interactions, electric and magnetic, charged particles will experience an effective potential minimum in space, and remain confined even when cooled to very

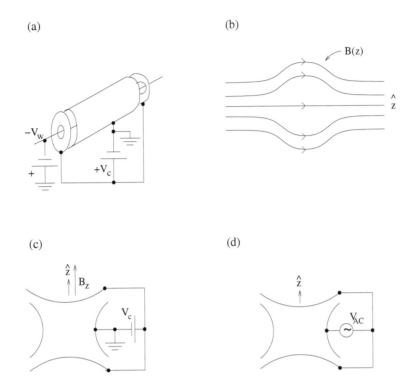

Figure 5.1: Schematics of four types of ion traps:

(a) Perspective view of an electrostatic Kingdon style ion trap, showing the outer cylinder and central wire used for radial confinement, and two end electrodes for axial confinement. Positive ions orbit the central wire, and are reflected axially by the end caps.

(b) Schematic of the field lines of a magnetic bottle. The axial motion of a charge rotating about the magnetic field lines can be reflected when the field becomes sufficiently strong, as axial kinetic energy is adiabatically converted into rotational kinetic energy.

(c) An axially symmetric Penning ion trap, based on a uniform axial magnetic field for radial ion confinement, and end caps with a dc potential for axial confinement of a positive charge.

(d) An axially symmetric Paul or radio-frequency ion trap. Ion confinement is based on the non-uniform high frequency electric field, which forces charges toward the minimum field at the center of symmetry.

low energy. The motion of the charges in the joint potential is basically harmonic, having well-defined frequencies, satisfying the Cirac–Zoller (and DiVincenzo) conditions (cf. Box 5.1) for a quantum computer. However, the uniform magnetic field must be strong for effective ion confinement, and must not change in space or time over the region in space where the ions are confined to preserve coherence. The stability of a magnetic field, which affects the energies of the magnetic sublevels of an ion electronic state by the Zeeman effect, may compromise wave function coherence unless carefully controlled. These conditions can be difficult to meet.

(iv) The **Paul, or radio frequency, ion trap** [12, 36] is chosen for quantum computation for several reasons. It provides a stable, 3-dimensional minimum in space of the effective electric potential using only radio frequency and dc electric fields, Fig. 5.1(d). A magnetic field may additionally be imposed, but is not required for confinement, and can be kept small. Since the electric fields vanish at the potential minimum, the effects on the energies of the ion electronic levels by the Stark effect[1] is very small, when the ions are cold. The motion of very cold ions in the trap is harmonic, i.e., the restoring force toward the equilibrium position is linear in space.

A linear Paul trap design [37] is usually selected, with the ions confined in a cylindrical space within four linear conducting electrodes oriented along the z-axis, each separated from the axis by a distance r_0. A time-varying potential $(V/2)\cos\Omega_{rf}t$ is applied to two opposite electrodes, and $-(V/2)\cos\Omega_{rf}t$ to the other opposite pair, see Fig. 5.2, or alternatively $V\cos\Omega_{rf}t$ is applied to one electrode pair and the other pair is grounded, an electrically simpler equivalent arrangement.

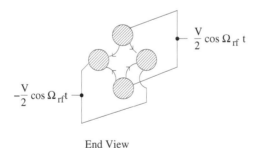

$$\frac{V}{2}\cos\Omega_{rf}t$$

$$-\frac{V}{2}\cos\Omega_{rf}t$$

End View

Figure 5.2: End view of a 4-rod linear quadrupole radio-frequency ion trap, showing the voltage biasing arrangement and the internal electric field line configuration at time t.

The time varying electric quadrupole potential within the volume de-

---

[1]The Stark effect, discovered in 1913 by J. Stark, is the splitting of a spectral line into several components in the presence of an electric field. This splitting is called a Stark Shift. The electric field may be externally applied, but in many cases it is an internal field caused by the neighboring ions or atoms. The effect is due to the interaction between the electric dipole moment of an electron with the electric field. See more discussions in Example 6.3.1 in Subsection 6.3.1.

fined by the rods produces a spatially non-uniform time varying electric field

$$E(x, y, t) = (V/r_0^2)(x - y) \cos \Omega_{rf} t \tag{5.1}$$

in the x and y dimensions of the linear trap. An ion with charge q and mass m will oscillate rapidly at frequency Ω_{rf} in this field, but due to the field gradient in space the ion will have a time-averaged displacement toward the region of weaker field [14], leading to a much slower secular motion which may be confined in both the x and y directions (Fig. 5.3).

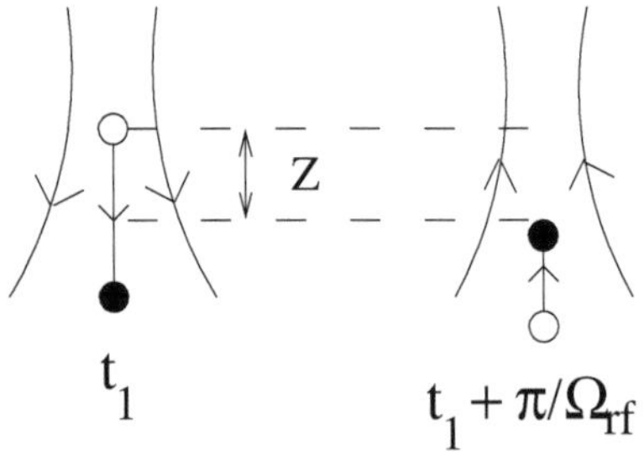

Figure 5.3: Illustration of the motion of a charge q in a spatially non-uniform time-varying electric field with frequency $\Omega_{rf}/2\pi$. The left diagram shows the first half of a cycle, and the right diagram shows the second half. The initial position of the charge is shown as an open circle, and the final position as a dark dot. Due to the spatial non-uniformity, the net force on the charge is toward the region of weaker electric field, resulting in a net time-averaged displacement Z.

Although there is an exact theory for a well-defined geometry, to understand the ion motion in general, a useful secular approximation [28] is used to average in time over the high frequency oscillation at Ω_{rf}. For each spatial dimension $\zeta = x, y$, or z, $\langle F(Z,t)\rangle/m = -\partial \langle U(Z,t)\rangle/\partial Z$, where F is the 1-dimensional time varying force, U is an effective potential, the time average is denoted $\langle\ \rangle$, and Z is the slowly varying part of the total motion described by coordinate ζ. The time averaged effective potential is

$$\langle U(Z,t)\rangle = q^2 E^2(Z)/2m\Omega_{rf}^2. \tag{5.2}$$

Upon taking the spatial derivative, the resulting equation is

$$d^2 Z/dt^2 + (2)^{-1}(qV/m\Omega_{rf}r_0^2)^2 Z = 0, \tag{5.3}$$

which describes a harmonic oscillation. The design of the trapping electrodes of a linear Paul trap results in confining harmonic motion in both the x and y directions, with frequency

$$\omega = qV/2^{1/2}m\Omega_{rf}r_0^2, \tag{5.4}$$

subject to the condition for the secular approximation $\omega \ll \Omega_{rf}$. Fig. 5.4 sketches an imaginary orbit and its time averaged form. In general, even if the electrodes are curved [11], or form a rounded corner, cold ions will be confined harmonically, with their motion described by the secular approximation (referring, e.g., to time average, $\langle\ \rangle$, in (5.2)), since the Taylor series expansion about the potential minimum is quadratic to first approximation. However, the binding frequency will in general not be the same as for a straight section. The approximate theory outlined above also applies to general confinement conditions, such as the 3-dimensional hyperbolic trap, Fig. 5.1(d), similar in electrode geometry to the Penning trap (Fig. 5.1(c)) which is axially symmetric about the symmetry (z) axis [19], or to traps which have more complex structures [16].

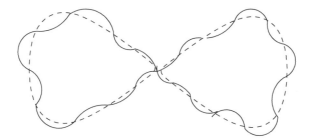

Figure 5.4: An illustration of an imagined closed 2-dimensional orbit of a confined ion in a Paul trap. The dashed line (- - -) shows the time averaged motion Z, while the solid line (—) shows the "actual" motion, which includes micromotion about the time average. The origin of coordinates is assumed to be at the crossing point. Actual orbits are generally not closed, nor do they generally pass through the origin.

The motion of the ions in the linear Paul trap along the z axis must also be confined. This is done by using a positive dc potential that increases as the positive ion moves in either direction along the z axis away from the linear trap center. This dc potential can be applied to additional electrodes located symmetrically about $z = 0$, or by splitting each rod at positions $\pm z_0$ so that the dc potential can be applied to each of the four rods for $z > |z_0|$. This is shown in Fig. 5.5. Because of the shielding of the dc potential by the conducting rods, it is not in general harmonic, but a Taylor series expansion of the potential $V_{dc}(z)$ about $z = 0$ will be harmonic in the small region of space near $z = 0$ occupied by very cold ions. Since the Laplace equation

$$\nabla^2 V_{dc}(x, y, z) = 0 \tag{5.5}$$

must also be satisfied, there is additionally a modification in the transverse frequency ω given by Eq. (5.4), due to the dc potential. The result is overall harmonic confinement of the ions in all three dimensions, but with different frequencies in different dimensions.

When cold ions are confined in this linear trap, they will lie in a string along the symmetry axis z as shown in Fig. 5.5, if $\omega \approx \omega_x \approx \omega_y \gg \omega_z$. If this condition is not met, more complex ordered arrays of ions are observed [54].

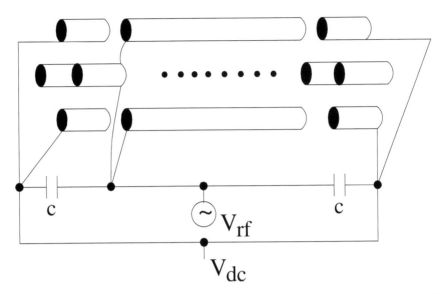

Figure 5.5: Illustration of axial confinement in a linear rf trap. Only an opposite pair of the four rods of Fig. 5.1 are shown. The gap in the rod structure permits a dc voltage on the ends of the rods, while capacitors C permit the rf voltage on all parts of the structure. A similar bias arrangement holds for the two rods not shown. A string of cold confined ions is shown as black dots.

An analytical estimate gives $\omega/\omega_z \cong 0.73N^{0.86}$, where N is the number of ions in the string, to preserve the linear array [44]. The ions in the linear array are separated axially by a balance between the Coulomb repulsion of the charges, and the force exerted by the confining axial electric field. The equilibrium ion spacing of the ions along the z axis tends to decrease near the trap center. Nevertheless, ion separations in the array are measured in micrometers, while zero point oscillation amplitudes of cold ions are measured in nanometers, and ion internal state wave function distributions having significant probability in space are still smaller. This means that each ion is physically isolated from the others in the array unless a deliberate coupling is externally initiated. It is useful to note that on average the electric field vanishes at the equilibrium site of each ion on the axis, and hence perturbations of the internal ion structure due to electric fields (the Stark effect) must be small.

Singly charged metal ions such as Be^+ are produced in the trap by thermally evaporating Be atoms from a coated filament, and ionizing them in the trap with a pulse of electrons from a nearby electron source [48], as diagrammed in Fig. 5.6.

The number of ions confined can be approximately controlled by the parameters of this process. The probability of ion loss is negligible for neutralizing electron capture collisions of the ions with neutral molecules, in part

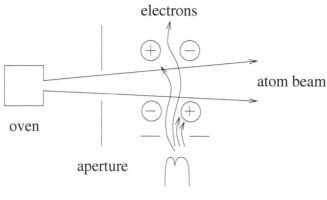

electrons

atom beam

oven

aperture

Filament

Figure 5.6: Diagram of a possible ionization apparatus to create metallic atomic ions inside the electrodes of a linear trap. An oven containing a metal is heated until evaporation occurs, resulting in a beam of atoms (straight arrows) which is defined by an aperture. A hot filament produces electrons, which are accelerated toward trap electrodes (curved arrows). Electrons striking beam atoms can result in ionization of a metal atom electron, leaving the metal ion confined in the trap.

because the vacuum is made very good to make collisions rare, but mainly because electron capture in low energy ion-molecule collisions is not energetically favorable for most metal ions in the ground state. The resulting ion confinement times in the trap can be days. There are still elastic ion-molecule collisions [48] which occur at average intervals given by

$$\tau_C = (nk_C)^{-1} \tag{5.6}$$

$\approx 3000s$ at neutral densities n corresponding to pressures near 10^{-11} Torr. (A Torr, named for Torricelli, is a pressure corresponding to 1/760 atmospheres, or a displacement of $1\,mm$ of mercury in a mercury manometer. 1 Torr corresponds to about 3.5×10^{16} molecules per cm^3 at room temperature). In the time constant equation, k_C denotes the elastic collision rate. Although elastic collisions do not destroy the ion, they can result in uncontrolled level-dependent phase changes in the ionic wave function, which result in decoherence, so excellent vacuum is essential.

5.2.2 Ion cooling and condensation

Ions produced in the trap as described above will generally have energies on the order of electron volts (eV), and will move as constituents of a charged gas. Since the charged ions are weakly coupled by long-range ion-ion Coulomb collisions, a thermal equilibrium described by a temperature will eventually evolve. By using a laser light beam, it is feasible to remove kinetic energy from these confined ions, to cool them to an ordered state [24, 25], where they lie in an array along the axis of the linear ion trap, as mentioned in the previous section.

Laser Doppler cooling [25] is used to cool confined ions from a relatively high temperature to temperatures of about 1 milliKelvin (mK). Let us explain how it works in the following paragraphs; cf., particularly, Fig. 5.8. Assume that an ion has two internal electronic energy levels separated by an energy $\hbar\omega_C$, where $\hbar$ is Planck's constant divided by 2π, with each level specified by quantum numbers. When the active electron of the ion is excited from the lower to the upper level by absorption of a photon from the laser beam at angular frequency $\omega_L \approx \omega_C$, the quantum numbers of the levels are such that the ion promptly returns to the same lower state by the random process of spontaneous emission of a photon. This is called a cycling transition. Assume that the photon resulting from spontaneous emission has a random direction in space, and an energy $\hbar\omega_C$ in the rest frame of the ion. The absorption profile of an ion near rest is described by a Lorentzian equation as a function of frequency, which has a large, sharp peak at $\omega = \omega_C$. The natural width of this absorption peak is given by the Heisenberg uncertainty principle as $\Gamma = \tau^{-1}$, where τ is the lifetime of the excited level, or the average time spent by the ion in that level after excitation, before a photon is spontaneously emitted as the electron drops to a lower state. Both the absorption and emission of the photon transfer momentum **p** to the ion, with $|p| = h/\lambda = \hbar\omega_C/c$, which results in small corrections to the resonance equations describing the interaction. At resonance, the cross section, or effective area for absorbing a photon from the laser beam, is $\sigma_C \cong \lambda^2/2$; a very large cross section, since the laser wavelength λ is much bigger than atomic dimensions. In accord with the Lorentzian lineshape, the cross section as a function of frequency is written (neglecting smaller details of ion recoil)

$$\sigma(\omega_L) \cong \sigma_C (\Gamma/2)^2 / \{(\omega_L - \omega_C)^2 + (\Gamma/2)^2\}. \tag{5.7}$$

This cross section drops to half its value at $\omega_L = \omega_C \pm \Gamma/2$, as sketched in Fig. 5.7. Farther from resonance, the interaction of a laser photon with the ion electronic state is weaker, although not zero, and the probability of excitation of the state is greatly reduced. The angular frequency difference between the laser frequency and the resonance frequency $(\omega_L - \omega_C)$ is called the *detuning* Δ.

When an ion moves relative to a laser beam, the frequency of the laser light at the position of the ion is shifted by the Doppler effect, by an amount $\Delta\omega_D \approx \omega_L v/c$ for small ion speed v compared to the speed of light c. As pictured in Fig. 5.8, for Doppler cooling, suppose that the frequency of the laser beam is tuned below the ion resonance, $\omega_L < \omega_C$. During the fraction of the oscillation cycle when the ion is moving toward the laser beam, the laser frequency is Doppler shifted upward in the ion frame of reference to be nearer the resonance of the absorption cross section, so photon absorption is more likely. Conversely, when the ion moves in the direction of the

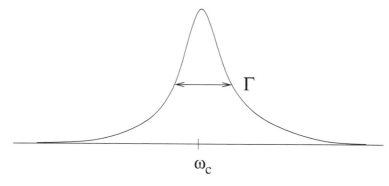

Figure 5.7: A sketch of an atomic resonance or cross section plot of interaction area vs. laser frequency ω_L. The resonance frequency ω_C and the full width at half maximum Γ are shown. If actually plotted according to Eq. (5.7), the peak would be much higher and narrower.

laser beam, the laser frequency is Doppler shifted farther away from resonance in the ion frame of reference, so photon absorption is even less likely. The net result is that absorption of laser photon energy with $\hbar\omega_L < \hbar\omega_C$ is favored. The extra energy needed to excite to the upper internal ion level must come from the ion motion, slowing the speed v. The spontaneous emission following absorption occurs at $\hbar\omega_C$ in the ion frame, so in an average cycle of this type, more energy is emitted than is absorbed from the laser photon, resulting in cooling of the ion motion. Alternatively, one may consider momentum conservation in terms of photon momentum $\hbar\omega/c$, and ion recoil momentum. Each laser photon absorbed with high probability from the laser beam while an ion moves toward the laser slows the ion, while on the average over many transitions, relatively little net momentum is transferred in spontaneous emission, due to the random direction of that emission.

Laser Doppler cooling is effective, because the cross section at resonance is large, the width of the resonance is narrow, and the transition cycle is short. For Be^+, the excited level lifetime is about 10 nanoseconds (ns), so the frequency half width $\Gamma/4\pi$ is about 8 MegaHertz (MHz). At resonance, about 5×10^7 photons per second could be absorbed and re-emitted. On the other hand, for ions near room temperature, the much larger Doppler half width $\Delta\omega_D/4\pi \approx 1500$ MHz. As the ions cool, and the Doppler width is reduced, the laser must be tuned closer to resonance to speed up the cooling process. The minimum temperature occurs when $\omega_L - \omega_C = \Gamma/2$, or $T_{\min} \approx \hbar\Gamma/k_B \approx 1$ mK. Near this temperature, the universal Coulomb coupling constant Γ_{CC}, which is the ratio of the repulsive Coulomb potential energy between two ions to their average kinetic energy [24], exceeds about 180. This means that the ions will enter an ordered phase [24, 25], and form a linear array along the ion trap axis, as depicted in Fig. 5.9.

It is not necessary to laser cool each ion, since while the ions cool, the

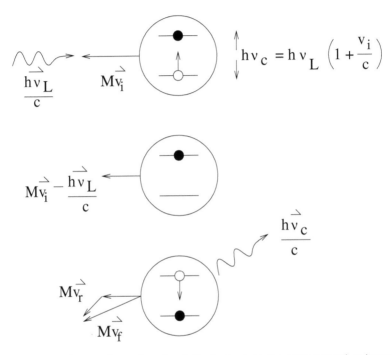

Figure 5.8: Doppler cooling of an ion is illustrated. In the upper part, a laser photon with momentum $h\nu_L/c$ and energy $h\nu_L$ excites an ion with momentum $-Mv$. When $\nu_L < \nu_C$, the excitation energy $h\nu_C$ comes partly
from the ion motion, through the Doppler effect. In the central part,
the excited ion has been slowed by the photon momentum. In the
lower part, the excited ion spontaneously decays, emitting a photon
with energy $h\nu_C$ in a random direction. Recoil momentum of the ion
contributes to this final momentum. The recoil momentum averages
to zero over many photon absorptions and random emissions, resulting in a net average slowing of the ion. Ions moving in the direction
of the photon have a smaller probability of photon absorption, due
to the Doppler shift and the resonant nature of the interaction.

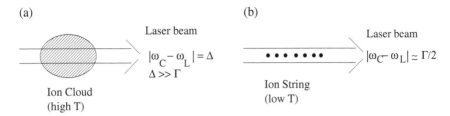

Figure 5.9: (a) A laser beam tuned below resonance interacts with a portion of a warm cloud of ions (cross-hatched oval). As the ions in the beam are Doppler-cooled, they interact via their electric fields with other ions outside the laser beam, cooling the whole cloud, which shrinks in size.
(b) Eventually, the ions cool to a regular array (dots) under the joint interaction of their mutual repulsion and the confining force of the trap electrodes.

coupling between them quantified by Γ_{CC} becomes stronger, as the elastic Coulomb collision rate increases as lower ion kinetic energies are reached. Although the collisions are elastic, and hence conserve kinetic energy, there can be energy and momentum transfer between the colliding ions. This results in a kinetic coupling within a cloud of ions, which are cooled together even when only some of them are laser cooled. This "sympathetic" cooling[2] also applies when a different type of ion is present, which will also be cooled, or which may be used to cool different qubit ions [54].

A milliKelvin temperature for the ions may seem very low, but it is insufficient for the Cirac-Zoller scheme to function. It is necessary to cool the ordered array of ions to a temperature at which the axial CM oscillations of the array in the trap are in the lowest vibrational quantum level. This occurs, according to the Boltzmann statistical distribution, when the population ratio of the two lowest levels of the quantized CM oscillation

$$N_1/N_0 = \exp(-\hbar\omega_{CM}/k_B T) \approx 0, \text{ so that } \hbar\omega_{CM} \gg k_B T. \qquad (5.8)$$

Taking the center of mass angular frequency $\omega_{CM} = \omega_z = 2\pi \times 10$ MHz, $T \leq 10^{-4} K$ in order to make $N_1/N_0 < 2\%$, so even lower temperatures than this are desired. To reach lower temperatures than 10^{-3} K, an entirely different laser cooling method, called *sideband cooling* [53], is required.

Doppler cooling reduces the Doppler width of the ion cooling transition to a value that is smaller than the natural width Γ, so that the combined widths approach Γ. If a confined ion in the array oscillates with small amplitude at frequency ω in the periodically varying electric field of a low-intensity near-resonant laser beam, as the laser frequency is swept, sideband frequencies at $\omega_C \pm n\omega$ (with n a positive integer) appear about the

---

[2]If one particle is cooled, then through collisional coupling with neighboring particles associated with the electric fields, those neighboring particles also get cooled indirectly (in sympathy).

central frequency of the internal transition at ω_C. These sideband frequencies appear at multiples of the ion oscillation frequencies, which include all the normal modes, or distinct ways that ions in an array can oscillate, that couple to the internal transition; see Fig. 5.10.

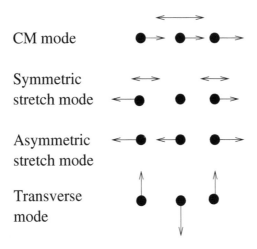

CM mode

Symmetric
stretch mode

Asymmetric
stretch mode

Transverse
mode

Figure 5.10: Examples of oscillation modes of three ions in a linear array are imagined. In the center of mass (CM) mode, the ions vibrate in phase together. In the symmetric stretch mode, the outer ions move in opposite directions, leaving the center ion at rest. In the asymmetric stretch mode, two ions move one way, while the third ion moves the other way with twice the amplitude. In a transverse mode, the ions move perpendicular to the string axis. Each distinct mode has a characteristic motional frequency.

The sideband frequencies are generated by a process similar to frequency modulation of the laser beam, as the ions move in the electric field of this beam. These sidebands are analogs of the Stokes or anti-Stokes transition frequencies observed in molecular spectroscopy, arising from molecular vibrations that are coupled to the electronic transition, resulting in possible energy levels and transitions shown in Fig. 5.11. The spectrum of relative excitation amplitudes of these motional sidebands, shown in Fig. 5.12, and of the central frequency ω_C, are determined by the Lamb–Dicke parameter [54]

$$\eta = kz_0, \text{ where } z_0 = (\hbar/2m\omega)^{1/2}. \tag{5.9}$$

The amplitude of the "zero point" oscillation (minimum amplitude uncertainty) of an ion in the lowest vibrational state of a mode with frequency ω is z_0, and $k = 2\pi/\lambda$ is the wave number of the laser beam with wavelength λ. The sidebands become important when $\eta \leq 1$. Note that η depends on the trapping voltages, through ω.

If the sidebands of the CM oscillation at frequencies $\omega_C \pm \omega_{CM}$ are separated by a frequency of about Γ from the central oscillation frequency ω_C,

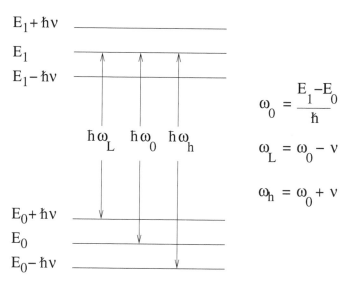

Figure 5.11: Internal energies and transitions of a harmonically confined ion coupled by laser light to its vibrational energy $\hbar v$ are displayed. The energies of the vibrational levels are greatly magnified in this diagram. The vibrational energies are all positive, but are referenced to the internal ion energy in the diagram. The lower (E_0) and upper (E_1) internal ion energies are coupled by the laser frequency ω_0. An ion having internal energy E_0 and relative vibrational energy $\hbar v$ can be coupled to E_1 by laser frequency ω_L, while laser frequency ω_H couples E_1 to $E_0 - \hbar v$. Similar transitions couple E_0 to $E_1 \pm \hbar v$. This coupling occurs only for cold ions, when the oscillation amplitude is comparable to or smaller than the wavelength of the laser light.

then they are *resolved*, and can be individually excited by a laser tuned to resonance at the sideband frequencies $\omega_C \pm \omega_{CM}$. Resonant excitation of the longer wavelength "red" sideband at $\omega_C - \omega_{CM}$ can be thought of as an induced internal electronic transition of the ion with energy $\hbar\omega_C$. The energy $\hbar(\omega_C - \omega_{CM})$ is supplied by the laser, and the energy $\hbar\omega_{CM}$ comes from a single quantum change of the CM normal mode energy. On average, excitation of this red sideband transition consequently cools the CM motion of the array of ions by one quantum, since the re-emitted optical decay photon has energy $\hbar\omega_C$. Clearly, an average cycle of excitation at the red sideband, followed by spontaneous emission, cools the normal mode; see Fig. 5.13. The lowest energy state of the normal mode can be reached by repeated application of this process. Other modes of oscillation can be similarly cooled, by slowly sweeping the laser frequency across mode frequencies below resonance. The lowest energy is the "zero point" energy of each vibration mode.

A limit to the effective quantum number for the lowest energy level of

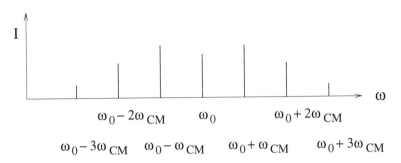

Figure 5.12: A schematic spectrum of an ion, in absorption or emission, showing the sidebands that appear around the resonant frequency ω_0 when the Lamb–Dicke parameter η is less than one. It is assumed that the ion moves harmonically in the trap with frequency ω_{CM}. The number and relative intensity I of the sidebands depends on the value of η. The frequencies are shown as lines, but are broadened by the rate of spontaneous photon decay during ion cooling, and by the Rabi frequency during stimulated Raman transitions.

the CM vibration is given by the relation [54]

$$\langle n_{\min}\rangle \approx (\Gamma/2\omega_{CM})^2 \ll 1 \tag{5.10}$$

For Be$^+$, $\Gamma \cong 2\pi \times 8$ MHz, and for $\omega_{CM} \approx 2\pi \times 12$ MHz the average minimum vibrational quantum number is $\langle n_{\min}\rangle \approx 0.1$, so with proper conditions on Γ and ω_{CM}, sideband cooling will suffice to implement the Cirac–Zoller proposal.

5.3   Ion qubits

Presented in the preceding sections were simplified discussions of the basic experimental methods used to confine ions, and to laser Doppler-cool the confined ions to a temperature at which they condense into a stable array in the trap. Laser Doppler cooling requires an ion with internal structure producing electron energy levels separated by convenient frequencies for laser work, and which permit cycling transitions to return the excited electron to its initial state. Laser sideband cooling of the ordered ions to populate the lowest vibrational level of the CM normal mode, part of the initialization process, was discussed. This cooling technique additionally requires resolved motional sidebands, which appear only when the condensed ions are very cold and tightly bound, as shown in Fig. 5.12. At this point the use of internal levels of ions as qubits is considered.

Recall that in a classical computer, a bit is a 2-state device holding a unit of information with a value of 0 or 1. A string of n bits can be interpreted as a single integer number expressed in binary form, with a single value between $(2^n - 1)$ and 0, as depicted in Fig. 5.15. Also recall from Chap. 2

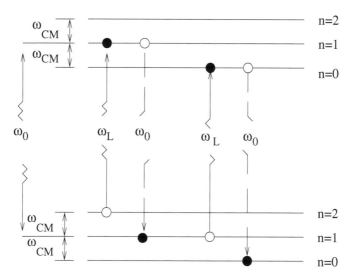

Figure 5.13: Sideband cooling of a harmonically bound ion is illustrated. The internal ion levels are separated by ω_0, and the lowest vibration levels, separated by ω_{CM}, are also shown. An ion in the lowest internal level, but in the second vibration level, is illustrated by an open circle. A laser is tuned to excite the system to the upper internal level and the first vibration level, as shown by a black dot. A subsequent spontaneous decay which does not change the vibrational quantum number n is shown (although decay to any n is possible). This cycle is repeated, with the ion ending in the lowest internal and vibrational levels, possibly after several cycles.

that in a quantum computer, a qubit is a 2-state quantum system such as a photon with two orthogonal polarization states, or two internal states of an atom or ion. Quantum states are waves, described mathematically by a wave function, but also as vectors in mathematical Hilbert space. The general quantum state can exist in a *linear superposition* of the two qubit basis states $|0\rangle$ and $|1\rangle$, i.e., a general state $|\psi\rangle = (a|0\rangle + b|1\rangle)$, where a and b are complex numbers (amplitudes) satisfying $|a|^2 + |b|^2 = 1$. Thus, a qubit can represent 0 (in state $|0\rangle$ if $a = 1$) or 1 (in state $|1\rangle$ if $b = 1$) or *both* numbers 0 and 1 *simultaneously*, if a and b have equal magnitudes. A string of n qubits, each in an *equal* superposition of the two basis states, can be interpreted as all of the integer numbers between $(2^n - 1)$ and 0, *existing simultaneously* as shown for $n = 4$ in Fig. 5.15. A *single* mathematical operation applied to this string of qubits is applied simultaneously to all of the numbers, resulting in massive quantum parallelism, and an enormous speed-up of computation.

The internal qubit levels of the ion may be chosen as a ground level and a long-lived excited level, called a metastable level, with energy separation $\hbar\omega_C$. During quantum computations, relatively fast transitions can

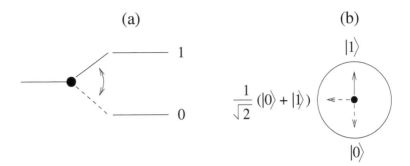

Figure 5.14: (a) A depiction of a computation bit, or binary digit, as a switch with possible settings of 0 and 1.
(b) A depiction of a qubit, with possible quantum basis states $|0\rangle$, $|1\rangle$, and with an equal superposition of basis states $0\rangle$ and $|1\rangle$ simultaneously.

be induced by laser photons between these two levels, without an appreciable probability of decay of the upper level, which would reduce coherence. Because of the long life of the upper level, the natural width Γ of the transition between the levels would be very narrow. Consequently, these levels are convenient for inducing sideband transitions, which are used for entangling operations as well as for sideband cooling. To maintain the best coherence during a gate operation, the laser linewidth $\Delta\omega_L$ should also be very narrow. The coherence and stability of present lasers may set limitations on the use of levels of this type for qubits, and only certain ions have very long-lived metastable states. Some additional comments and data may be gleaned from Section 6.8 and Tables 6.1 and 6.2 therein.

Alternatively, the qubit levels of the ion may be two different hyperfine structure levels within the ground state separated by $\hbar\omega_C$ in energy. Hyperfine structure describes the energy of interaction between the active electron and the nucleus of an ion or atom. These energy level separations may occur at radio-frequency or microwave transition frequencies. Because of the relatively low transition frequencies between the two levels, the life-

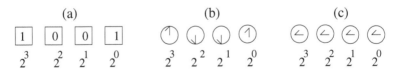

Figure 5.15: (a) A four bit register with the number 9 stored.
(b) A four qubit register with the number 9 stored. Arrows pointing upward refer here to qubits in the basis state $|1\rangle$, and arrows pointing downward refer to state $|0\rangle$.
(c) A four qubit register with all numbers from 0 to 15 simultaneously stored (see Fig. 5.14).

times are extremely long, and spontaneous transitions directly between the levels are certainly negligible. Radiation sources at these lower frequencies can be made very stable, with narrow widths. On the other hand, the transition wavelengths λ are so long that the Lamb–Dicke parameter η is essentially zero. This means that coupling to the motional sidebands is also negligible, so the entangling operations between the internal and motional qubits have low probability, and take long times, undesirable for computation.

A solution to the effective use of two ground state hyperfine levels as qubits is to employ two laser beams to induce *stimulated Raman transitions* between the levels [54]. This technique in principle permits the efficient adiabatic transfer of an electron directly from one hyperfine level to the other via a virtual intermediate state, which is not populated, preserving coherence. The transitions also have optical wavelengths, which permit coupling to the sideband frequencies with reasonable efficiency. A stimulated Raman transition is illustrated in Fig. 5.16. Since spontaneous transitions are assumed to be negligible, the linewidths are small and sideband resolution is high. We will see later that the use of these transitions will limit the choice of a candidate ion for quantum computing, however. For the Be ion, which has no long-lived excited metastable levels, the use of two hyperfine levels as qubits is required.

Initialization of the internal qubit levels can also be accomplished by laser beam-induced transitions in individual ions. A metastable qubit level is self initializing, since it eventually decays to the ground state. Better yet, it can be coupled by a faster laser transition to a higher lying level, which decays to the ground level promptly.

To initialize the ground state hyperfine qubit levels, the active electron of each qubit ion of the array must be placed in a specific ground state qubit level, say $|0\rangle$. The ion is "pumped" to this energy level from other ground energy states, by tuning the laser to induce single photon transitions to excited states, which decay by spontaneous emission of a photon bringing the electron back to several possible levels including the qubit level. The qubit level is itself not excited, or is excited only in a cycling transition. Over some number of cycles, the active electron of the ion will end up in the qubit level, initializing it, and leaving other ground levels unpopulated; see Fig. 5.17. This can be carried out for each individual ion of the array simultaneously, using an unfocussed laser beam, or individually, using a focussed laser beam.

The question whether the ion is indeed in a specific qubit level, and can be measured to be there, is accomplished by inducing a rapid cycling single photon transition from that qubit level to a higher ion excited level using a resonant laser beam, as discussed in Section 5.1. If the natural width of the transition is Γ, then $\approx \Gamma T$ fluorescent photons will be emitted in time T. By placing a lens or mirror close to the trap, the product of the fractional solid angle for photon collection with the efficiency for photon detection by a photomultiplier tube can be made $\approx 5 \times 10^{-3}$. For a transition

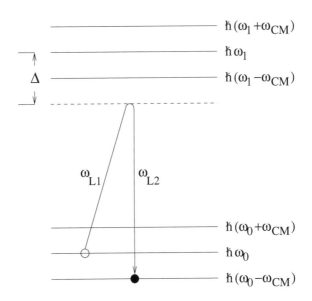

Figure 5.16: A stimulated Raman transition between an initial state with energy $\hbar\omega_0$ and a final state with energy $\hbar\omega_0 - \hbar\omega_{CM}$ is plotted for a single ion. The ion internal energies are $E_0 = \hbar\omega_0$ and $E_1 = \hbar\omega_1$. The vibrational energies are $\pm n\hbar\omega_{CM}$ (with $n = 1$). A "virtual level" detuned from $\hbar\omega_1$ by $\hbar\Delta$ is shown by a dashed line. This "level" is associated with weighted interactions with all internal ion levels (primarily E_1 here), and is set by the chosen laser frequencies. The stimulated Raman transition accomplished, using two laser frequencies, a coupling between the internal ion energy and its vibration, resulting in a new motional state of the ion. If several ions in a string have the center of mass frequency w_{CM}, then the CM motion of all the ions is changed by the laser interaction with a single ion.

rate $\Gamma \geq 10^7 s^{-1}$, ≥ 10 photons can be detected in $200\,\mu s$, if the level is populated; see Fig. 5.18. If the level is not populated, no fluorescent photons will be detected, although there may be a "dark count" randomly produced by the detector during the interval. If about 10 photons are detected in a fixed time interval, the probability of error in determining if that qubit is populated (rather than ten random dark counts occurring) is given by Poisson statistics as $\approx \exp(-10) \approx 5 \times 10^{-5}$. Slightly longer detection intervals would quickly provide even higher certainty [54].

5.4 Summary of ion preparation

Cold ions can be confined harmonically in fairly general configurations of standard electrode structures using a combination of radio-frequency and dc fields, since near the potential minimum the potential expansion is dominated by the quadratic term. In room temperature ultrahigh vacuum, un-

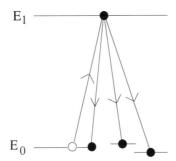

E_1

E_0

Figure 5.17: An illustration of optical pumping from a particular sublevel of level E_0 of an ion to other sublevels. Light absorption produces a transition of an electron to level E_1, followed by spontaneous emission of light in transitions to several possible sublevels of E_0. Each sublevel, is distinguished by a particular parameter (the angular momentum projection) which can change only in the transition. If an individual ion is returned to its original sublevel, it is excited again until it does not return. The sublevels need not have different energies, as shown in the diagram. The process ends with all ions in the same sublevel.

controlled ion-neutral collision intervals (Eq. (5.6)) can be made $\geq 10^3 s$. A confined array of cold ions is strongly coupled by elastic ion-ion Coulomb collisions, which result in normal modes of harmonic oscillation by the ions of the linear array. The CM mode of oscillation has a frequency independent of the number of ions in the array.

Confined ions have been cooled to a linear array using a laser beam exciting individual ions, first to milliKelvin temperatures by Doppler cooling, and subsequently to the lowest vibrational levels of the normal modes using sideband cooling. Similar techniques are used to initialize the internal ion qubits, which may consist of a ground and a metastable level, or of two ground state hyperfine structure levels of an ion. Qubit population is measured by cycling resonant laser transitions, with fluorescent photon detection. Efficient coupling of internal ground state ion levels to the CM motion requires 2-photon stimulated Raman transitions. The Lamb–Dicke parameter η quantifies the coupling between the internal qubit levels and the normal modes of the confined ion motion. The stimulated Raman transitions can excite only the internal ion qubit (the carrier transition), if the frequency difference between the co-propagating beams equals the internal ion transition frequency and the wave vectors are parallel. However, if the frequency difference equals the internal ion transition frequency plus or minus a motional ion frequency, and the wave vector difference lies along the direction of ion motion, then transitions of the internal qubit and the motional qubit can be simultaneously excited, coupling these qubits. This coupling is conditional, resulting in entangled states, in which the state of the internal qubit is correlated with the state of the motional qubit.

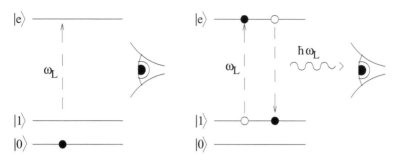

Figure 5.18: Detection of the electron in a qubit level is shown. In the left diagram, the qubit level $|0\rangle$ is assumed occupied, so a laser tuned to the cycling transition from qubit level $|1\rangle$ produces no photons. In the right diagram, if the qubit level $|1\rangle$ is occupied, a transition occurs, with subsequent spontaneous emission of a photon, which may be observed. Since the induced transition is cyclic, many laser photon absorptions and spontaneous emissions in a time T result in the detection of several photons, identifying the occupied state unambiguously.

5.5 Coherence

Coherence is the property of maintaining the quantum mechanical phase and amplitude relationships between different states of a superposition. A 2-qubit state $|\psi\rangle$ may be written (without entanglement)

$$|\psi\rangle = 2^{-1}\exp(i\gamma)(|0\rangle + \exp(i\phi)|1\rangle)(|0\rangle_m + \exp(i\phi)|1\rangle_m) \qquad (5.11)$$

in terms of equal superpositions of the internal and motional (subscript m) qubit states. The overall phase factor γ is related to the ion motion, but plays no role in determining the relative populations of the qubit states, since $|\psi\rangle\langle\psi|$ is independent of γ. The relative phase $\phi = (\phi_1 - \phi_0)$ of the internal (logic) qubit states is critically important to the logic operations, however, even though $\langle\psi|\psi\rangle = 1$ is also independent of ϕ. For example, the off-diagonal terms of the density matrix for the internal states

$$|\psi\rangle\langle\psi| = \begin{pmatrix} \rho_{00} & \rho_{01} \\ \rho_{10} & \rho_{11} \end{pmatrix} = 2^{-1}\begin{pmatrix} |0\rangle\langle 0| & \exp(-i\phi)|0\rangle\langle 1| \\ \exp(i\phi)|1\rangle\langle 0| & |1\rangle\langle 1| \end{pmatrix} \qquad (5.12)$$

are called "coherences", and depend on the value of ϕ. For a group of N ions, if the phases ϕ_N of the individual ion amplitudes are distributed randomly, then the coherences tend to average to zero, leaving only the diagonal elements, which do not provide complete information about the quantum state.

  If each ion has the same relative value of ϕ, the off-diagonal elements of the ensemble are non-zero and oscillate periodically in time (since the states $|0\rangle$ and $|1\rangle$ have different energies) without change in amplitude or frequency. Loss of this coherence, or decoherence, is thus a critical consideration for quantum computation, since it reflects a loss of quantum information or entanglement. Fig. 5.19 schematically shows a characteristic coherence loss in a wave. Fidelity of gate operations, cf. Section 4.1 and [35], is a good measure of coherence, since it accounts for changes both in phase and in state amplitudes. Unitary quantum gates, if perfectly executed, will preserve the quantum probability and may change the phase of the coherences in a well-defined way. However, if the relative phases of the $|0\rangle$ and $1\rangle$ levels are changed in an *arbitrary* way, or if quantum probability is lost to states outside of the basis, then the fidelity is decreased and the computation will be corrupted. Fidelity loss occurs whenever the quantum system comes in contact with the "environment", i.e., anything that is not part of the computational basis or is not controlled. The environment is expected to have a large number of degrees of freedom, which cannot be precisely specified. This lack of information is accounted for by averaging over the states of these degrees of freedom, resulting in an evolution of the qubit system, which reduces the quantum superposition of amplitudes to a classical distribution of probabilities. Such contacts with the environment may occur during the execution of gates, or during intervals between gates. Environmental heating of the motional qubit, which can transfer it out of the lowest two motional basis states, is an important consideration for certain confined-ion gate types. Fortunately, many types of qubit errors associated with decoherence can be corrected, but at the cost of increased complexity of the quantum qubit system [35].

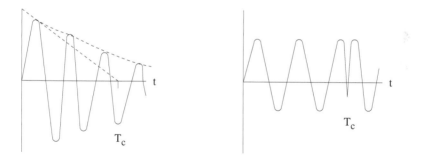

Figure 5.19: (left) A decaying sinusoidal wave form is shown, illustrating a finite coherence time T_C associated with loss of probability. For perfect coherence, the amplitude and frequency would remain constant. (right) A wave with constant amplitude has an interruption in phase at time T_C.

Because of the operation in vacuum, and the isolation of the qubits from everything except electric and magnetic fields, coherence in cold confined ion qubits is expected to be very high. However, experimentally, coherence is found not to be perfect, so its measurement, and the identification and amelioration of the sources of its imperfection, are critical. The NIST group categorizes decoherence into decoherence of the *ion motion*, decoherence of the *internal qubits*, and decoherence by *non-ideal fields* used in logic operations [54].

5.5.1   Coherence of the motional qubit

One of the first critical areas of decoherence to be identified was associated with the CM motion of the cold ions confined in a linear radio frequency trap, primarily between the $n = 0$ and $n = 1$ states. The ion traps used in the study of quantum gates are small, so the electrodes are close to the confined charges. Since in the CM motion, all ions move together creating an oscillating dipole, a relatively large current can be induced in the trap electrodes by capacitive coupling. This current dissipates heat in the resistances of the electrodes. The heat energy must come from the CM motion, damping its amplitude and broadening its frequency width. A maximum coherence time $\tau = 1\,ms$, the time to exchange a motional quantum between $n = 0$ to $n = 1$ of the CM mode, was measured for Be^+ ions in a particular NIST hyperbolic trap [33]. Models used to estimate the damping time based on electrode design, resistivity, etc. predicted coherence times of seconds, so more detailed work was required. However, for two ions, modes other than the CM mode, such as the symmetric stretch mode used for the gate described in Subsection 5.6.5, see Fig. 5.10, were found to be far more coherent, in accord with expectations since to first approximation the induced current in the electrodes is zero. A resonant external field, which might drive this mode, would require a field *gradient* to excite the stretch mode ion motion, since a different field must act on each ion. A uniform field would suffice to excite the CM oscillation.

Heating is sufficient to change the relative populations of the CM normal mode, and motional excitation is well-established in radio frequency traps. In general, the trapping potentials will have some anharmonic character, due to approximations in design, or small displacements from ideal positions, which can lead to unexpected couplings of the ion motion with the radio frequency field which creates the confinement. An exact treatment of the harmonic motion shows that motional sidebands occur at frequencies $j\Omega_{rf} \pm \omega_\zeta$ for each dimension ζ, where j is an integer. Parametric heating at $2\omega_z$ is another resonant heating possibility [28], caused by a potential difference at that frequency between the trap electrodes and the end electrodes. External excitation of the ions at such frequencies was experimentally limited by electrical filters, and since such excitation is resonant, small adjustments in trapping parameters might substantially reduce it, even if accidental resonance with multiples of the trapping frequency occurred in normal operation due to anharmonic potentials. Nevertheless, if the excitation source is Johnson noise similar to that from a hot resistor, with a wide frequency bandwidth, it could still be important in ion heating. Initial estimates indicated that these sources of heating were negligible in the NIST measurements with very cold ions [54], but subsequent measurements under controlled conditions (see Subsection 5.5.4) provided more information.

Heating of ions can arise from elastic collisions between ions due to the Coulomb interaction, which change the momentum of individual ions in the radio frequency trapping field. This heating can be made negligible for very cold ions, when the external potentials are very harmonic, by making sure that the mode frequencies are not sub-harmonics of the trapping frequency Ω_{rf}, and that the secular motional frequencies $\omega_\zeta \ll \Omega_{rf}$. Ion-molecule collisional heating can be reduced by improving the vacuum in the trap.

Another form of ion heating can arise from stray static or slowly varying electric fields which do not have a confinement symmetry, produced by accumulated charges on small electrically neutral patches on conductors or on nearby insulators, or due to different materials or adsorbed gases on electrode surfaces, etc. Surface roughness on a small scale may have an effect. Surface charging effects can be associated with ion production in the trap, which usually is done by electron impact ionization of metal atoms evaporated from a hot wire, while the atoms are passing through the trap structure. The charges and atoms can both strike electrodes, possibly stick and migrate around, and build up over long periods due to the excellent vacuum. The associated fields would result in heating of the confined ions if multiples of the sums and differences of ion motional frequencies equaled the trapping frequency Ω_{rf}. Estimates and tests using deliberately applied static fields did not indicate that this potential problem was serious. Numerical estimates of the effects of fluctuations of patch fields indicated expected coherence times $\tau = 3000s$, well above the observed decoherence times [54].

Despite the discussion in the previous paragraph, the heating rate of the ion motional state was found experimentally to increase over time as Be was deposited onto the trap electrodes [39]. In subsequent work a shadow mask was employed to keep evaporated Be atoms from striking the trap electrodes. This resulted in a coherence time of $10\,ms$ in a very small linear trap, for the time to absorb a motional quantum from the ground state. A *figure of merit*, the ratio of the coherence time to the time to execute an entangling gate, was about 1000. (Cf. Table 2.4 for the coherence quality factor for a single-qubit operation.) For similar traps, an approximate scaling of r_0^{-4} for the heating rate was found, where r_0 is the distance from the ion array to the nearest electrode. This scaling favors large trapping structures. Electroplated gold deposited on the evaporated gold on trap electrodes also led to some coherence improvements, so electrode surface purity and smoothness are positively correlated with motional coherence and low ion heating. Still, excess heating is observed, which does not seem correlated with thermal electronic noise.

Another important possibility for logic operations is mode cross-coupling from static electric field imperfections, leading to heating and decoherence. For N ions in a string, the number of spectator normal modes is $3N - 1$. If these modes are not all cooled to the zero point, energy might be transferred between the CM mode and the spectator modes, leading to heating or at least to decoherence. This has been described theoretically and studied experimentally, and found not to account for the short coherence times observed [54].

Finally, the occasional collision of an ion with a background molecule or atom even in the best of experimental vacuum systems must be considered. Inelastic collisions which destroy the ion by chemical reaction or electron exchange can occur when the ion is in a laser-excited energy state, like a $P_{1/2}$ or $P_{3/2}{}^3$ state. These collision intervals are thousands of seconds at easily-attained pressures in room temperature ultra-high vacuum systems. Elastic collision rates with neutrals are somewhat faster, but thousand second intervals between such collisions are still feasible.

To summarize, the ion heating and coherence problems observed in rf traps are still not fully understood. Coherence times for a single confined ion much heavier than Be^+, such as Hg^+ in the lowest motional CM mode in a similar trap, were found to be as long as 0.15 s [17], so going to heavier qubit ions may help. Since the figure of merit discussed above for Be ions is only 1000, maintaining coherence for long computations will not be feasible for gates which rely on ions remaining in the lowest two motional qubit levels. This condition is relaxed for gates operating at higher temperatures, such as those discussed in Subsections 5.6.4 and 5.6.5. Additional qubit ion cooling, which does not disrupt logic qubit coherence during a computation, is also a feasible consideration.

5.5.2 Coherence of the internal qubit

The internal ion qubit of Be^+ has a measured coherence time longer than 10 minutes [32], when measurements of the hyperfine structure transition frequency were made between specific levels at a magnetic field chosen to enhance coherence (a "field-independent" transition). The linewidth of the transition was less than 1 mHz, at a frequency near 300 MHz. Narrower resonances have been achieved with other ions. However, the qubit levels used in the experimental gates based on hfs levels may not reach this level of coherence.

Radiative decay, in which a photon is emitted in an uncontrolled quantum transition, is a fundamental limit on internal state coherence. For the magnetic dipole radiative decays connecting hyperfine levels, the lifetime for this decay is $> 10^{10}s$ for singly-charged ions, so this is totally negligible. For ions with a qubit composed of a ground state level and a metastable state level, connected by an electric quadrupole transition, lifetimes can range from $\approx 1s$ to 100 s or more. Associated with the qubit level lifetimes T_q are natural line widths for the transition given by $\Gamma = 1/2\pi T_q$.

---

[3]The notation (term symbol) P_J (here $J = 1/2$ and 3/2) here we mean the following: P denotes the second quantum number ℓ in $|n\ell m\rangle$ in Example 2.2.1, i.e., $\ell = P = 2$, and $J = L \pm S$, where $L = \ell$ is the orbit angular momentum quantum number and S is the spin, i.e., J is the total angular momentum quantum number. Here we have $L = 1$ and, thus, $J = 1 \pm 1/2 = 1/2$ and 3/2.

The energy separation of most qubit levels depends on the external magnetic field B, which has an average value B_{ave}. Fluctuations of the field from this value are expected to be a *primary* source of internal state decoherence [54] for a level separation which is directly proportional to B. *Slow* fluctuations in B compared to a single gate operation time T lead to a phase angle

$$\phi(t) = (1/2)(\omega_0 t + (\omega_0/B) \int_0^t (B - B_0(t')dt'). \tag{5.13}$$

In a sequence of operations, the additional phase accumulated under the integral must be kept small, or taken into account. Higher frequency fluctuations of B (with periods shorter than the gate time T) will tend to average away, or to contribute in higher order corrections to an offset to the Rabi frequency which can be taken into account in gate implementation [54]. Fortunately, qubit levels with energy separations which are independent of magnetic field in first order are potentially available, provided that transitions to an auxiliary level are not additionally required. "Spin-echo" laser pulses, which tend to re-phase quantum systems in magnetic fields, have been employed [41].

Electric field energy shifts of qubit levels are generally second order in the field, and are very small at the equilibrium site of the ion. There are *ac* Stark shifts due to time-varying fields, like the ≈ 8 V/cm blackbody radiation field at room temperature, in the infra-red spectral region. In general, an electric field shifts all *hyperfine* levels by the same amount, making Stark effects very small. For optical ground to metastable level transitions, a uniform electric field near 1 V/cm produces relative level shifts of a few mHz. In these transitions, an electric quadrupole shift, which interacts with a field gradient, may also be present. Near-static field gradients due to neighboring ions or stray electric fields with gradients near 10 V/cm^2 are expected to produce relative level shifts as large as 1 Hz, and significantly larger gradients are expected for neighboring ions. However, the effects are expected to be fairly constant, resulting more in correlated energy level shifts rather than decoherence [54].

5.5.3 Coherence in logic operations

Maintaining fidelity in logic operations is a critical task. In Section 5.6, a gate described by an ideal interaction Hamiltonians H_I acts for time T on the system wave function resulting in

$$|\psi(T)\rangle_I = \exp(-iH_I T/\hbar)|\psi(0)\rangle \equiv U(H_I T)|\psi(0)\rangle, \tag{5.14}$$

where $|\psi(0)\rangle$ describes the Hilbert space of the N internal qubits of the system plus the motional qubit and in certain cases any additional levels such as auxiliary levels. The NIST group classifies two categories of decoherence: errors or noise in the gate rotation angles θ and ϕ, and coupling to the environment induced by the gates, caused by imperfections in gate implemen-

tation [54]. An actual attempted implementation of the ideal interaction
Hamiltonian is described by the "practical" propagator $U(H_P T) \approx U(H_I T)$.
The fidelity of the gate can be characterized as

$$F(N) = \langle |\langle U(H_I T)|\psi(0)|U(H_P T)|\psi(0)\rangle|^2 \rangle, \qquad (5.15)$$

where the outer brackets describe the quantum mechanical expectation
value. In this point of view, the fidelity is the probability that the correct
path has been followed in Hilbert space, with $F(N) = 1$ the result for a
practical gate acting perfectly on the N qubits. One can measure the coherence of each state during the operation by using time delays between
the preparation of the state and the application of the gate pulses [54], a
procedure followed experimentally.

A discussion of the effects of possible amplitude and phase errors on
the rotation angles in Hilbert space [54] finds that for *random* errors, the
error terms add only in *quadrature*, i.e., $\varepsilon = \mathcal{O}\left(\sqrt{\sum_{i=1}^{n} \varepsilon_i^2}\right)$. The maximum
number of operations M before the fidelity drops appreciably below one is
given by

$$M_{\max} \cong E^{-2}, \qquad (5.16)$$

where E is the fractional error in amplitude or phase per operation, and the
errors in the angles are assumed small compared to π radians. If the errors
are *systematic* rather than random, they add *coherently*, and accumulate
linearly, so the requirements on systematic errors are more stringent than
on random errors.

The pulse area of a Rabi pulse on a single ion induced by a resonant field
with well-controlled amplitude and duration will vary due to laser power
fluctuations, and by fluctuations in the beam position relative to the ion.
The latter is particularly important to gates in which the ions are individually addressed. Position fluctuations are minimized by rigid mounts for the
optics and trap, and a quadrant detector of the beam position can be used
for feedback control of the beam position. This assumes that the ions of the
array, rather than the beam position, are moved in addressing. Assuming
that fluctuations in laser power and mode dominate, photon shot noise provides the fundamental limit to power stability, but nearly all laser sources
are actually limited by acoustic vibrations. These vibration amplitudes can
be reduced by active power stabilization using feedback, but the quantum
efficiency of the photodetector and the limited laser power directed to the
stabilizer limit the effectiveness of the technique. A calculated example [54]
for 1 Watt laser power at $313\,nm$, a gate time $T = 1\,\mu s$, a detector efficiency
of 0.5, and half the laser beam power used in stabilization give a fractional
power fluctuation $\geq 2.3 \times 10^{-6}$ at the beamsplitter, which is likely larger at
the site of the ion. Fractional laser pulse *duration* errors of 10^{-6} require
timing precision of 10^{-12} s (ps) for a $1\,\mu s$ pulse.

To characterize fluctuations at the site of the ion, the probability $P(0,n)$
of the ion being in the $|0\rangle|n\rangle$ level of the qubit can be measured vs. the

time t of the applied gate pulse. If the laser power is assumed constant for the measurement time t but varies over the time required to make a quantum mechanical average of the measurements at that time, then $\Omega t/2\pi$ Rabi cycles must be measured [54], if the desired relative magnitude of the fluctuations is $\Delta\Omega_{rms}/\Omega$. For a relative measurement of 10^{-4}, about 1500 Rabi cycles are required. Alternatively, if the laser power varies rapidly during a single measurement time t, but slowly over the averaging time, mainly the visibility of the signal (the difference between the maximum and the minimum values of the signal versus time) is affected.

Schneider and Milburn [41] calculate the decoherence for quantum logic operations due to a specific model for phase fluctuations. For optical transitions, the lasers must have the required frequency and phase stability, which may present a problem for long-duration computations, using the stability of present stable lasers [54].

It must be noted that the above discussion refers to constant amplitude Rabi pulses, rather than adiabatic passage between states, such as are produced by 2-photon Raman transitions. On the negative side, the fidelity of adiabatic transfer between ground state hyperfine structure levels is limited by coupling to several excited states during the gate transition, and the relatively high laser power required for Raman transitions can increase decoherence by spontaneous Raman scattering. A numerical estimate for the fractional probability of spontaneous emission during a stimulated Raman carrier π pulse with Rabi frequency Ω between hfs levels of Be$^+$ [52] gives $\approx 10^{-3}$. To successfully incorporate error correction in a long computation, probabilities for errors during a gate operation should be 10^{-4} or smaller [45]. This eliminates Be$^+$ from consideration if Raman pulses are used, but heavier ions have larger fine structure separations and lower rates of spontaneous emission. Comparing various ions otherwise acceptable for comparable gates, only ^{113}Cd$^+$ and ^{199}Hg$^+$ have spontaneous emision rates below $\approx 5 \times 10^{-6}$ (cf. [52, Table 1]). This lower limit is required, since the Rabi frequencies for sideband transitions scale as $\eta\Omega$, increasing the relative transition time by $1/\eta$ compared to carrier transitions. It is possible in principle to make the detuning much larger than the fine structure level separation, $|\Delta| \gg \omega_{FS}$, so that spontaneous emission is dominated by Raleigh scattering rather than off-resonant excitation. The cost of this is that the Raman beam intensities must increase as Δ^2 to keep the Rabi frequency, and hence the gate time, constant.

On the positive side, for π pulses only, the stimulated Raman technique does not depend critically on pulse area, and in general the frequency and phase sensitivity depend on the frequency and phase *differences* between the two laser beams. Since the two beams can be derived from a single laser beam by use of frequency modulators, only the stability of the low frequency modulating source, which can be made very high, is important. The electric fields of the stimulated Raman transitions produce ac Stark shifts of the coupled levels, but if the Stark shifts are the same for each level, their effects cancel. If they are not the same, they can still be made small by suitable choices of the transition parameters such as detuning, polarization, and laser intensity. Nevertheless, there can be *fluctuations* of the Stark shifts, which may be important. Analysis [54] shows that the worst case is for sideband excitation, in which the motional quantum number changes by one unit. Rms (root mean square) phase fluctuations caused by Stark shifts resulting in a frequency offset are worse than Rabi frequency fluctuations by a factor of $1/\eta$ in the Lamb–Dicke limit. However, it is expected that fluctuations in intensity will be dominated by fluctuations in the primary laser beam, which are correlated in the two beams. Under these conditions, with the frequencies of the two beams nearly the same, the phase fluctuations caused by Stark shifts will be *smaller* than those caused by Rabi frequency fluctuations. To the extent that this is the case, stimulated Raman transitions between hyperfine structure levels are viewed as superior in stability to single photon transitions separated by optical frequencies, at least with the present state of laser stability [54].

Laser phase shift stability for Raman transitions depends on *phase differences*, as noted above, but could still depend on *path length* differences of the two beams. Path length differences are more common in transitions coupling the internal and motional qubits, since the two beams arrive at an angle to produce a non-zero wave vector difference Δk. An active path length stabilizer has been described [4].

As more ions are added to a linear string, the high CM vibration frequencies needed for efficient laser cooling and manipulation result in smaller separations between the ions, to the extent that addressing a single ion by tightly focussing the laser beam to approach the diffraction limit becomes experimentally difficult. One possibility is to permit unused ions between the actual qubits, but better alternatives appear to be limiting the length of the ion string, and moving individual ions into the laser beam. This will be discussed further in Section 5.7. Another quantity affected by the number N of ions in the string is the Debye–Waller factors, which reduce the Rabi frequency exponentially due to the large number $3N - 1$ of (uncooled) spectator modes [54]. If the CM mode is used for logic, then the Lamb–Dicke parameter $\eta_N = \eta/N^{1/2}$ holds. So, as for preservation of coherence, it is desirable to cool all spectator modes to the zero point state, or better yet, limit ion string length.

Spectator states are both internal and motional states, and the more of these present, the greater the probability of off-resonant excitation when gates are executed. The lower limit on gate time T is $\approx \omega_{CM}^{-1}$, but approaching this limit may cause broadening which increases spectator excitation.

5.5.4 Studies of decoherence through coupling to engineered reservoirs

An important technique has been tested [34] to study decoherence of quantum states of confined ion qubits by controlled contact with environments designed to produce specific types of decoherence in these superpositions. Coupling between ion states of a mesoscopic scale quantum superposition and an engineered "amplitude" reservoir is made by applying noisy voltages between trap electrodes, which simulate the voltages produced by a hot resistor (or reservoir) which has a controlled temperature and frequency spectrum. For 2-component superposition states, an exponential dependence for the decoherence rate is expected, dependent on the separation of the components in Hilbert space.

Coherent states of the motional qubit are states of a harmonic oscillator which can be represented by a vector $|\alpha\rangle$, with $\alpha = |\alpha| \exp(i\theta)$, where θ is the phase of the oscillator at an initial time $t = 0$, and $|\alpha|$ is the dimensionless amplitude of the motion of that state [21]. Such coherent states lead to motions similar to classical harmonic oscillator trajectories, particularly when many states are superposed. Fig. 5.20 attempts to illustrate this for three harmonic oscillator state functions. The states of the fluctuating reservoir are also taken to be harmonic oscillator states, and the interaction is taken to be proportional to the amplitudes of the motion of the states [34]. By analogy, the "zero point" of the ion oscillation is subject to random displacements produced by the reservoir, resulting in a fluctuating force on the ion system. The rate of decoherence of the ion states is expected to scale as the square of the separation of the two superposed ion states $|\alpha_1 - \alpha_2|^2$, where $|\psi\rangle \propto (|\alpha_1\rangle + |\alpha_2\rangle)$. The coherence remaining after coupling to the reservoir for a time t is [49]

$$C(t) = \exp(-|\alpha_1 - \alpha_2|^2 \xi t), \tag{5.17}$$

where ξ is a constant that describes the degree of coupling between the qubit state and the reservoir. This amplitude reservoir coupling does not depend on the internal state of the ion.

Similarly, for Fock states (cf. Remark 3.2.1) [34], the eigenstates $|n\rangle$ of the number of vibrational units n of the oscillator, the coherent wave function can be written $|\psi\rangle \propto (|n_1\rangle + |n_2\rangle)/2^{1/2}$. This wave function loses

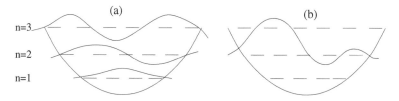

Figure 5.20: (left) The three lowest standing wave states of a quantum harmonic oscillator potential well are sketched. (right) An approximate superposition of the three standing wave states is imagined, in which the "ion wave" is localized off-center. If more states were superposed, and the phases of the states optimally adjusted, then the "ion wave" would oscillate as a well-defined disturbance in the harmonic well, i.e., as a "particle" formed from a coherent superposition of states.

coherence if the modes of the reservoir linearly couple to the energy of the oscillator. Remaining coherence after coupling for time t is

$$C(t) = \exp(-|n_1 - n_2|^2 \kappa t), \qquad (5.18)$$

where the coupling is characterized by the constant κ. Induced random adiabatic variations in the trapping motional frequency ω change the phase of the ion oscillation without changing its energy. This "phase reservoir" couples to superposed Fock states to reduce coherence as described in Eq. (5.18).

A "zero temperature reservoir" [34], based on a bath of laser cooling light and optical spontaneous emission, drives Fock states to $n = 0$, and results in a non-exponential decay of coherence. This is a form of quantum, rather than classical, decoherence.

Experimentally, the coherence of the quantum superpositions were measured using single-atom interferometry. The harmonic motional state was conditionally split into two components, each associated with a different internal ion qubit state, by driving laser Raman transitions. The superposition was perturbed by coupling to a reservoir, and then recombined by reversing the steps that formed it. The internal ion qubit state was measured using the usual method as a function of the relative phase of these creation and reversal steps, resulting in interference fringes. The measurements were averaged over many repetitions, and the contrast of the fringes characterized the amount of remaining coherence $C(t)$.

The measurements of fringe contrast confirmed the theoretical expectations discussed above for a range of superposition sizes $\Delta\alpha$, resulting in a single exponential. The decay constant of fringe visability for the ambient amplitude reservoir agreed within experimental error with the separately measured ion heating rate for the apparatus. Since classical fields produce the decoherence, their effects can in principle be reversed if the information is properly recorded.

5.5.5 Summary of ion coherence

Coherence, which initially was viewed as the great strength of the cold, confined ion technique, can now also be seen as a limiting factor. The time T to execute an entangling quantum gate usually varies as $(\eta\Omega)^{-1}$. The magnitude of Ω is limited by broadening of levels due to short transition times, resulting in the excitation of unwanted states which are not well resolved. If η is kept small, then T becomes large. If η is permitted to approach 1, and the Lamb–Dicke limit ceases to be a good approximation, then precise control of the ion motion is necessary since the Rabi transition rates of the gates depend on the motional states through the Debye–Waller factors. Precise control means excellent ground state cooling of all motional modes. As more ions are added to the string, to accommodate a large scale computation, three more vibrational modes per ion are added, increasing the difficulty of isolating the desired vibrational mode unless T is increased to reduce the uncertainty principle broadening of the levels connected by the transitions during the gate. Larger values of T reduce the number of gates that can be executed during the coherence time of the CM motion, thus reducing the scale of the computation.

Motional coherence times of the ions in the trap, resulting from heating of the motion, are presently limited to about $10\,ms$ for Be ions, but large improvements have recently been achieved, and more seem feasible with improved understanding of the mechanics of decoherence through engineered reservoirs. Ion internal qubit coherence is limited by uncontrolled magnetic field fluctuations, and ultimately by ion collisions with neutral molecules in the vacuum. Coherence in logic operations is limited by Stark shifts produced by the inducing electric fields, by laser intensity fluctuations, by spontaneous emission of a photon during a qubit state transition, or by unwanted coupling between internal and external qubit levels.

Despite the negative impression of these limits, it is important to recall that coherence remains a great strength of the cold, confined ion technique, and that new techniques to study and control it are now available. The limitations on coherence direct the general cold ion concept toward more specific techniques. Instead of a single string of N cold ions, many shorter strings with selective coupling appear more effective. Instead of ions using ground plus metastable levels coupled by single laser transitions, ions with two ground qubit states coupled by stimulated Raman transitions appear favored. Instead of light ions for qubits, heavier Cd^+ ions are preferred. Better control of surface conditions of the trap electrodes is required. Application of periodic excitations to improve coherence, such as laser cooling intervals during gates, or spin-echo re-phasing pulses, appear essential. A concept for large scale quantum computation using this technique has been advanced [27].

5.6 Quantum gates

In the previous sections, the scientific characteristics of a quantum computing technology, as outlined by DiVincenzo [18]; cf. Box 5.1 as well as Box 2.9, have been discussed in relation to the cold, confined ion technique. The important exception is the implementation of the quantum gates, which is now addressed.

5.6.1 General considerations

The quantum gates of cold, confined ions consist of controlled operations, induced by interactions of the ion qubits with laser beams, which affect the time development of the qubit wave functions. The simplest required operations are single qubit operations, which result in wave function rotations for the internal states of a single ion, as illustrated in Fig. 5.21, but leave it in the same motional quantum state. The ion internal qubit is treated in analogy to a spin-$\frac{1}{2}$ particle, well-described by theory, which relates the spin direction of the particle in space to the superposition of quantum basis states associated with the two energy states of the system.

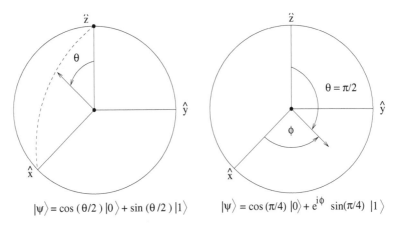

$$|\psi\rangle = \cos(\theta/2)\,|0\rangle + \sin(\theta/2)\,|1\rangle \qquad\qquad |\psi\rangle = \cos(\pi/4)\,|0\rangle + e^{i\phi}\sin(\pi/4)\,|1\rangle$$

Figure 5.21: (left) The Bloch sphere diagram of an electron spin angular momentum vector pointing at angle θ relative to the z axis in the $x-z$ plane. (right) A similar diagram for the spin vector in the $x-y$ plane, at an angle ϕ relative to the x axis. The wave functions associated with the spin directions are written out. See also Appendices C and D.

The 2-qubit operation is the next simplest. Entanglement couples the wave functions of two qubits conditionally. The coupling of the internal ion qubit and the motional qubit, in a way that is *determined by the state of the motional qubit*, is a characteristic example. Fig. 5.22 shows the effect of the entanglement. This entangling operation can be described as a simultaneous rotation of the internal and motional wave functions. These two kinds of operations, single qubit gates and 2-qubit entangling gates, are sufficient to construct a universal quantum gate, such as a phase gate or the CNOT gate, from which a quantum computer of any level of complexity can be synthesized [35]; see Theorems 2.6.1 and 2.6.2. Since scalability and a high degree of coherence have been assumed as given, the research in ion confinement computation has emphasized the demonstration of proposed gates, rather than the execution of algorithms by a finite number of gates. In the following, the scientific basis for single qubit and two qubit operations, and their experimental demonstration, are discussed. Perfect coherence is assumed, unless otherwise noted.

The spin-$\frac{1}{2}$ analogy for two internal ion levels is based on the interaction of the vector magnetic dipole moment μ of an electron with static and time-varying spatially uniform magnetic fields **B**. For the magnetic case, the qubit energy level separation is $\hbar\omega_0 = \mu \cdot \mathbf{B_0}$, where $\mathbf{B_0}$ is the uniform static field conventionally taken to lie along the z-axis. If a uniform magnetic field with amplitude $\mathbf{B_1}$, rotating in time at frequency ω, is applied

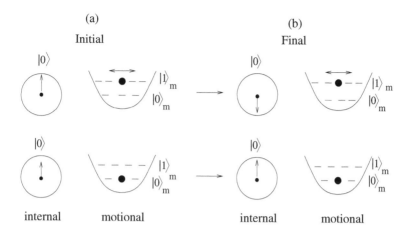

Figure 5.22: (a) The initial internal and motional states of an ion are shown, for two possible states $|1\rangle_m$ (upper) and $|0\rangle_m$ (lower) of the motional (control) qubit, and an internal (target) state $|0\rangle$ in both cases. The subscript m signifies motions as shown in Fig. 5.10.

(b) (upper) The final internal and motional states of the ion are shown. The control (motional) qubit remains in state $|1\rangle_m$, while the internal (target) qubit is flipped from $|0\rangle$ to $|1\rangle$. (lower) The control qubit remains in state $|0\rangle_m$, and the target qubit remains in state $|0\rangle$. The flipping of the target qubit is conditioned on the state of the control qubit.

perpendicular to $\mathbf{B_0}$, the motion of the electron moment μ is most easily understood in a coordinate frame of reference which is rotating at frequency ω about an axis given by the direction of $\mathbf{B_0}$. In this frame of reference, $\mathbf{B_1}$ is at rest and has a direction usually specified by an angle ϕ relative to the x-axis. In the interesting case of resonance, when $\omega = \omega_0$, the uniform field $\mathbf{B_0}$ is effectively zero in the rotating frame, the energy of the electron is $\hbar\Omega = \mu \cdot \mathbf{B_1}$, and μ precesses in time only about $\mathbf{B_1}$. This precession carries μ between the directions associated with the qubit states $|0\rangle$ and $|1\rangle$, at a Rabi frequency Ω. Fig. 5.23 shows the two reference frames.

If the field $\mathbf{B_1}$ is applied for a chosen time $T_{\pi/2}$ (a "$\pi/2$ pulse")[4] and the electron is initially in the state $|0\rangle$, then the precession angle θ is $\Omega T_{\pi/2} = \pi/2$ and the electron is left in the coherent superposition state $|+\rangle = (|0\rangle + |1\rangle)/2^{1/2}$. The rotation of the electron moment in space is correlated with a rotation of the electron wave function by half of the angle, an effect associated with the half-integer spin of the electron. If a "π pulse" is applied,

---

[4]A $\pi/2$ pulse is a high frequency control pulse with a specific duration applied to a qubit, typically rotating the Bloch vector by an angle $\pi/2$ so that that vector is tipped from a pole to the equator, or reversely, on the Bloch sphere. For a π pulse, the rotation is of angle π, called a "qubit flip", changing $|0\rangle$ to $|1\rangle$ or reversely, producing inversion of the qubit level population. See Fig. 5.24. For the Bloch sphere and Bloch vectors, see additional discussions in Fig. 10.5 in Chap. 10 and Appendices D and E.

(a) lab   (b) rotating

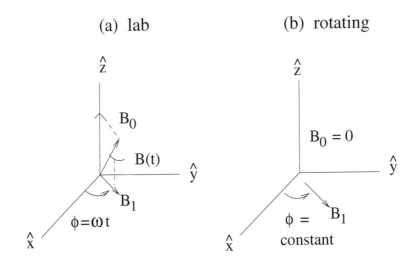

Figure 5.23: Reference frames showing the magnetic fields associated with a magnetic dipole transition of an electron.
(a) In the laboratory frame, a uniform magnetic field B_0 is applied in the z direction, and a much smaller rotating field B_1 in the x-y plane has a direction specified by the angle $\phi = \omega t$ relative to the x axis.
(b) In the rotating frame of reference, at resonance when $\omega = \omega_0$ the z field is transformed to zero, and the angle of the field B_1 is constant in the x-y plane. The motion of an electron spin angular momentum is simply described as a precession about B_1 in the rotating reference frame.

the moment is rotated from the state $|0\rangle$ to the state $|1\rangle$, while a "2π pulse" rotates the spin back to $|0\rangle$, but with a minus sign in the wave function. These pulses are shown in Fig. 5.24. Because of the factor of two difference in wave function and spatial rotation angles, a "4π pulse" is needed to return to the initial state $|0\rangle$ with the same sign.

In the electric analogy, the qubit energy level separation $\hbar\omega_0$ is determined by the ion structure [7]. The analogy for the time dependent field E_1 to the case discussed above is exact for two electronic levels of an ion coupled by an induced electric dipole moment operator $\mathbf{d}$ interacting with a spatially uniform electric field $\mathbf{E_1}$ varying in time with angular frequency ω. The spatial uniformity follows from the rotating wave (or dipole) approximation, see Subsection 3.1.2, in which the wavelength λ of the radiation is much larger than the spatial dimension of the atom, a criterion that applies to tunable laser radiation, and to all lower-frequency electromagnetic waves. The electric dipole operator $\mathbf{d}$ can be expressed as the sum of raising ($\sigma^+ = \tilde{\sigma}_x + i\tilde{\sigma}_y$) and lowering ($\sigma^- = \tilde{\sigma}_x - i\tilde{\sigma}_y$) operators on the internal motion [54], which induce transitions between the two qubit levels $|0\rangle$ and $|1\rangle$:

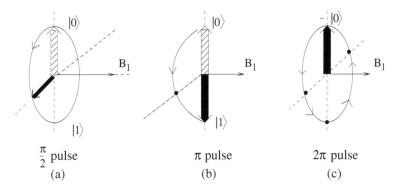

$\frac{\pi}{2}$ pulse π pulse 2π pulse

(a)   (b)   (c)

Figure 5.24: Pulsed rotations of electron spin about a magnetic field B_1 along the y axis (a) θ -90 degrees, (b) $\theta = 180$ degrees, (c) $\theta = 360$ degrees.

$$\mathbf{d} \propto \tilde{\sigma}_x = \frac{1}{2}(\sigma^+ + \sigma^-) \equiv 1/2 \begin{pmatrix} 0 & 1 \\ 0 & 0 \end{pmatrix} + 1/2 \begin{pmatrix} 0 & 0 \\ 1 & 0 \end{pmatrix} = \frac{1}{2}(|1\rangle\langle 0| + |0\rangle\langle 1|). \quad (5.19)$$

(In the above, the Pauli spin matrices σ_x and σ_y as defined in (2.42) and (2.43) are now adapted to $\tilde{\sigma}_x$ and $\tilde{\sigma}_y$ to signify the fact that electrons are spin-1/2 particles.) When these operators act on the ion wave function, they result in a formal mathematical description of a transition, in which the electron transfers from one qubit energy state of the ion to another, while absorbing or emitting a photon to conserve overall energy. This transition is analagous in all respects to the magnetic dipole transition discussed above.

The electric field is the field of a laser beam having polarization ε_L perpendicular to the direction of motion. The field is assumed to be propagating in the z-axis direction:

$$\mathbf{E} = E_0 \boldsymbol{\varepsilon_L} \cos(kz - \omega_L t + \phi) \cong E_0 \boldsymbol{\varepsilon_L} \cos(\omega_L t + \phi), \quad (5.20)$$

where the approximate form follows from the electric dipole approximation $kz \ll 1$, since $k = 2\pi/\lambda$ is small compared to atomic dimensions. The electric field polarization ε_L is expressed in terms of left circular (σ_-) and right circular (σ_+), which can be interpreted as unit vectors in orthogonal directions. The electric vector of left circularly polarized light rotates counterclockwise when viewed back along the beam direction (in imagination), while right circularly polarized light has the opposite helicity. (Such light polarization will be further utilized in Chap. 8 on linear optics.) Linear polarizations are made from superpositions of these orthogonal components. These polarizations are selected experimentally to permit the induction of particular transitions to higher energy levels, essentially specifying the value of a quantum number of the ion.

The Hamiltonian function for the ion can be written approximately as the sum of the Hamiltonian describing the internal ion energies, and an interaction Hamiltonian H_I which describes how the active electron of the ion interacts with external electric and magnetic fields. It is the interaction Hamiltonian which describes the execution of a quantum gate. In the laboratory, the action of exposing the ion to a time-varying field E_1 (or B_1) for time T is described mathematically as a transformation to the rotating coordinate frame discussed above. The development of the electron wave function (or rotation of the electron dipole moment) can be thought of as a rotation of the electron coordinates in a Hilbert space specified by the interaction Hamiltonian. When the external fields are turned off, mathematically a transformation from the rotating coordinate frame to the laboratory frame of reference occurs.

The interaction Hamiltonian appropriate for the qubit levels $|0\rangle$ and $|1\rangle$ is given by (when $\omega_L = \omega_0$):

$$H_I = \hbar\Omega \exp(-i\phi)\sigma^+ + \hbar\Omega \exp(i\phi)\sigma^- \equiv \hbar\Omega \exp(-i\phi)\sigma^+ + h.c. \qquad (5.21)$$

The raising and lowering operators $\sigma^\pm$ induce transitions only between the internal ion levels [54]. Also, $h.c.$ denotes the usual hermitian conjugate operator formed by replacing σ^+ by σ^- and i by $(-i)$. The rotating wave approximation has been used, which means that exponential terms varying at frequencies high compared to ω_0, such as $\exp(\pm i2\omega_L t)$, have been dropped, since they have negligible effect on the resonant interaction; cf. Subsection 3.1.2. The symbol Ω here is the Rabi frequency, which is a measure of the strength E_0 of the electric field in the interaction. It depends on the square root of the laser beam intensity, which is proportional to E_0^2. The relation is given by

$$\Omega = -E_0\langle 1|\mathbf{d} \cdot \varepsilon_{\mathbf{L}}|0\rangle/2\hbar. \qquad (5.22)$$

This interaction equation applies explicitly to electric dipole transitions which are used in ion cooling, or to qubit transitions to an auxiliary level, but also applies by analogy to a higher order magnetic dipole transition to a metastable level which may be one of the qubit levels. The Rabi frequency is the transition rate of the active electron between the internal qubit levels.

It is shown in (3.31) and Subsection 3.1.3 that the unitary quantum gate operator specified in terms of $\theta = \Omega t/2$ and ϕ is

$$\exp(-iH_I t/\hbar) = \begin{pmatrix} \cos(\Omega t/2) & i\exp(-i\phi)\sin(\Omega t/2) \\ i\exp(i\phi)\sin(\Omega t/2) & \cos(\Omega t/2) \end{pmatrix} \qquad (5.23)$$

which has the same form as the *one qubit unitary rotation gate* $U_{\theta,\phi}$ in (3.33), where $\theta = \Omega t/2$ is time-dependent. A particular rotation angle θ corresponds to the application of a time-dependent laser field for a well-defined time; see Fig. 5.25.

When both qubit levels in a single ion are in the electronic ground state, such as two hyperfine structure levels of the ground state of Be^+, then a single qubit rotation described by the Hamiltonian above can be executed by using a radio- or microwave-frequency generator, since the frequency

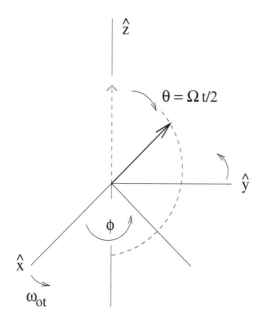

Figure 5.25: The rotation of a single qubit is diagrammed. The qubit states $|1\rangle$ and $|0\rangle$ can be thought of as arrows (or *spins*) lying along the positive and negative z axis, respectively. The action of the interaction Hamiltonian is to rotate the arrow through an angle θ relative to the z axis, at a rate half the Rabi frequency. A rotation through 2π brings the arrow back to its original position, but the wave function acquires a minus sign due to the factor of 2, so a 4π rotation brings the qubit back to its original state. The angle ϕ is measured relative to the x axis of the (rotating) coordinate system.

separation is small. However, it would then not be feasible to execute an entangling operation to couple an internal transition efficiently to the vibrational motion of the ions, since this coupling requires an electric field *gradient* to be effective. The spatial derivative of the electric field of Eq. (5.20) is proportional to the wave number $k = 2\pi/\lambda$, and consequently is very small when the wavelength λ is very large, as it is at microwave and lower frequencies. To make λ relatively small, and k relatively large, optical frequencies are essential. In an electric quadrupole transition to a metastable qubit level in an excited state, the coupling of the moment to the electric field already requires an electric field gradient [12].

Using $\eta = kz_0$ for a single ion, and using the rotating wave approximation while retaining the argument of the cosine function containing kz in Eq. (5.20) for the electric field, the interaction Hamiltonian coupling the internal and vibrational transitions becomes [54]

$$H_I = \hbar\Omega \exp(i\phi)\sigma^+ \exp(-i\phi_L - \omega_0)t[1 + i\eta\{a\exp(-i\omega_z t)$$
$$+ a^+ \exp(+i\omega_z t)\}] + h.c. \tag{5.24}$$

The coordinate z has been interpreted as the displacement of the ion from its equilibrium position in the array. It can be expressed in a quantized operator notation as $z = z_0(a + a^+)$, where a and a^+ are the lowering and raising operators for the normal mode oscillation having angular frequency ω_z. The characteristic spatial part of the wave function for the lowest vibrational level has $z_0 = (\hbar/2m\omega_z)^{1/2}$. The other modes of oscillation are assumed to be cooled to the lowest state, and are neglected. This Hamiltonian (Eq. (5.24)) has operators changing both internal (σ^+, σ^-) and motional (a^+, a^-) states of the ion.

Since the laser frequency ω_L appears in Eq. (5.24), the choice of this frequency determines the coupling between the internal and external coordinates. If, for example, $\omega_L \equiv \omega_0 - \omega_z$, so that the first red side band is excited at resonance, then after substitution for ω_L and a further application of the rotating wave approximation, the interaction Hamiltonian becomes

$$H_I = i\hbar\eta\Omega \exp(i\phi)\sigma^+ a - i\hbar\eta\Omega \exp(-i\phi)\sigma^- a^+. \tag{5.25}$$

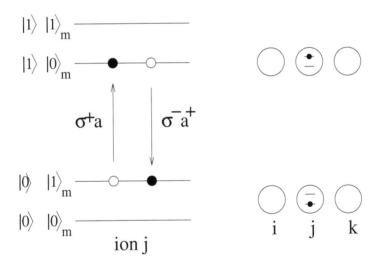

Figure 5.26: The entangling action of a swap gate between the internal and motional qubits is sketched, based on the interaction Hamiltonian controlling the coupling between these qubits. If an ion in its lowest internal state, but in the $n = 1$ vibration state, is excited by a laser to the higher internal state, but the $n = 0$ motional state, then this operation is executed mathematically by the operations σ^+ and a. The adjoint of these operators returns the system to the original state. If the operators act only on ion j of a string, then in the lowest internal state, the string is vibrating, but in the upper internal state, it is in the lowest vibrational level.

The first term of this expression excites the internal qubit (operator σ^+), but simultaneously *de*-excites the CM motional qubit (operator a), while the second term does just the opposite, so this is essentially a qubit swap operation; see Fig. 5.26. The result is entanglement: the conditional connection between the state of the internal qubit of a single ion and the state of the CM motion of the ions acting as a group. Compared to the interaction Hamiltonian of Eq. (5.21), one sees that the effective Rabi frequency of Eq. (5.25) is now $\eta\Omega$, and the phase shift has an additional factor of $\exp(i\pi/2)$ (due to the factor i) in the first term, and $\exp(-i\pi/2)$ in the second (hermitian conjugate) term. For a string of N ions, η is replaced by $\eta/(N^{1/2})$. The 2-qubit rotation matrix analogous to (5.23) applies with these substitutions. Note that the transition frequency $\eta\Omega \ll \Omega$, since η is assumed to be small, and that for an N-ion string, this transition frequency is proportional to $N^{-1/2}$, so "scaling independence" is somewhat compromised.

For two ground state qubit levels, another type of excitation called a *2-photon stimulated Raman transition* [54] is required in order to excite and entangle at optical frequencies, so that k is not negligible and a sufficient field gradient is present. The two laser beams incident on the site of a single ion are assumed to have frequencies ω_{L1} and ω_{L2}, wave vectors $\mathbf{k_1}$ and $\mathbf{k_2}$, and phases ϕ_1 and ϕ_2. The interaction Hamiltonian is still described by Eq. (5.24), but with the substitutions $\Delta k = |\mathbf{k_1} - \mathbf{k_2}|$ replacing k in the electric dipole approximation, ϕ replaced by $\phi_1 - \phi_2$, and ω_L replaced by $\omega_{L1} - \omega_{L2}$. The resonant 2-photon Rabi rate is now

$$\Omega_2 = E_{01}E_{02} \sum_i \langle 1|\mathbf{d} \cdot \boldsymbol{\varepsilon_2}|i\rangle\langle i|\mathbf{d} \cdot \boldsymbol{\varepsilon_2}|0\rangle/(4\hbar^2\Delta_i). \tag{5.26}$$

The summation index i denotes intermediate (virtual) electronic excited levels. The two laser beams are detuned by the frequencies Δ_i from actual ion excited levels, as shown in Fig. 5.27. These excited levels are ideally not populated in the transition, which carries the electron adiabatically between the two ground state hfs qubit levels. This is achieved by making the detunings Δ_i much greater than the natural widths Γ_i of the excited levels, the range of uncertainty of the energy of these levels. The Lamb–Dicke parameter is now $\eta_2 = (\Delta k)z_0$. The resonance conditions for the interaction Hamiltonian Eq. (5.25) are now $(\omega_{L1} - \omega_{L2}) = \omega_0 + \omega_z$ for the blue sideband, and $\omega_0 - \omega_z$ for the red sideband frequencies, respectively, as illustrated in Fig. 5.28. The frequency difference $(\omega_{L1} - \omega_{L2}) = \omega_0$ for the internal qubit transition which does not excite the normal mode oscillation. This is the "carrier" transition, Fig. 5.27. For the carrier transition, the two Raman beams are co-propagating, and the difference frequency is set to the internal ion transition frequency ω_0. It may appear that Δk is always small for stimulated Raman transitions, but since Δk is a vector, a clever choice for the directions of $\mathbf{k_1}$ and $\mathbf{k_2}$ can keep the magnitude of Δk comparable to $\mathbf{k}$, when coupling the internal qubit to the motion.

Because of the detunings, the Rabi frequency $\Omega_2 < \Omega$, but the entangling operation must still be kept short in time. This can be managed by relaxing the condition that $\eta_2 \ll 1$. If $\eta_2 \approx 1$, then the effective Rabi frequency $\eta_2\Omega_2$ which appears in the rotation matrix resulting from the application of

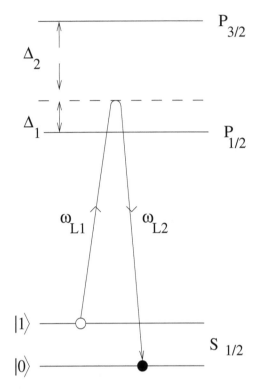

Figure 5.27: A diagram of a 2-photon stimulated Raman transition. An ion in internal qubit state $|1\rangle$ in the ground S state is excited by two laser beams having a frequency difference equal to the qubit frequency separation. Coupling to the excited P states, with detunings Δ_i, permits an adiabatic transfer of the electron from qubit state $|1\rangle$ to qubit state $|0\rangle$ through a detuned *virtual* level without actually exciting the P states, which could result in stimulated emission and loss of coherence. A transition of this type (with different laser frequencies) might only excite the motional qubit, instead of the internal qubit.

Eq. (5.25) is larger, but must be corrected by factors called Debye–Waller factors, in analogy to nuclear effects observed in solids. These Debye–Waller factors depend on the extent in space of the ion wave function as the ion oscillates harmonically in particular quantum levels, with an amplitude *no longer* negligible compared to the wavelength λ of the driving laser beam [54]. In general, these Debye–Waller corrections depend on the normal mode quantum numbers and on η_2, and can be written analytically. For example, if the normal mode remains in the lowest motional energy state $|0\rangle_{CM}$ (where subscript "CM" denotes the CM mode in Fig. 5.10) during the stimulated Raman transition, i.e., the carrier transition, then $\Omega_{2(0,0)} = \Omega \exp(-\eta_2^2)$. Each type of sideband transition now has its characteristic rate, but this property turns out to be useful in executing certain alternative types of quantum gates discussed below.

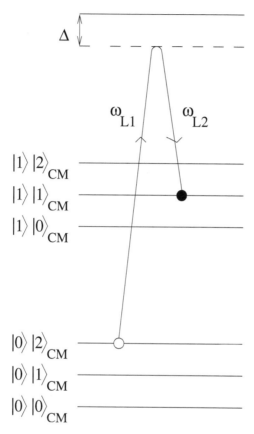

Figure 5.28: A stimulated two photon Raman transition which excites the internal qubit and de-excites the motional qubit, as illustrated. This type of transition is favored for two ground state qubits, since the shorter wavelength of the optical transition increases the electric field gradient, which couples the two qubit types.

   The implementation of quantum gates using Be^+ ions has been done in the following way [33], where the parameters are given for a single ion in a cylindrical trap. The qubit levels were chosen as $^2S_{1/2}$ (cf. footnote 2 of this chapter, here the left superscript 2 denotes "doublet", and S means $L = 0$.) ground state hyperfine structure levels $|F = 2, m_F = -2\rangle \equiv |0\rangle$ and $|F = 1, m_F = -1\rangle \equiv |1\rangle$. The quantum numbers F and m_F refer to angular momentum properties of the hyperfine coupling; they are described in [7] with the heuristic motivation given in Section 2.2 earlier, but can be thought of as level labels here. The frequency separation of the qubit levels was $\omega_0 = 2\pi \times 1.250$ GHz. The auxiliary level $|aux\rangle \equiv |F = 2, m_F = 0\rangle$ was also in the ground state, lying about 2.5 MHz below $|0\rangle$. This internal level separation was produced by the Zeeman effect caused by a weak 0.18 mT uniform magnetic field applied to the trap region. The motional qubits $|0\rangle_m$ and $|1\rangle_m$ were separated by the frequency $\omega_x = 2\pi \times 11.2$ MHz, where the other frequencies in the cylindrical ion trap were $\omega_y = 2\pi \times 18.2$ MHz and $\omega_z = 2\pi \times 29.8$ MHz.

Two Raman beams with a frequency difference of about 1200 MHz, generated from a single laser beam using an acousto-optic modulator (AOM), had Δk directed along the x-axis [33]. The frequency difference was set to a motional sideband frequency $\omega_0 + \omega_x \Delta_n$. An acousto-optic modulator is a crystal, in which density (sound) waves are excited using an external source. These acoustic waves can couple to a laser beam interacting with the modulator, resulting in laser waves having frequencies either augmented or decremented by the acoustic frequency. In the present measurement, the wavelength of each beam was near $313\,nm$, with about 1 mW power. The dominant detuning Δ from the $^2P_{1/2}$ excited level was about 50 GHz. The carrier Rabi frequency was $\Omega = 2\pi \times 140$ kHz. Since $\eta_{2x} \approx 0.2$, $\eta_{2x}\Omega \approx 2\pi \times 30$ kHz for the qubit transition, and about $2\pi \times 12$ kHz for the auxiliary transition. Population detection of the qubits was accomplished by inducing the cycling transition $|0\rangle \Rightarrow |F = 3, m_F = 3\rangle$ of the $2P_{3/2}$ excited level, with detection of a fluorescent photon.

The motional qubits $|0\rangle_m = |0\rangle_x$ and $|1\rangle_m = |1\rangle_x$ were initialized by sideband ion cooling, where the subscript x denotes the coordinate of oscillation of the motional qubit.

5.6.2   Cirac–Zoller CNOT gate

Now that experimental techniques and parameters for implementation have been discussed, attention turns to the unitary universal 2-qubit quantum gates. Cirac and Zoller [13] specified the operations necessary to implement a universal phase gate. A characteristic of gates of this type is that the levels of the internal and motional qubits are populated by resonant laser transitions. A mathematical proof of these operations appears in [9]. A transition to an additional "auxiliary level" of the ion was needed to implement the Cirac–Zoller gate. An additional rotation can be used to transform this phase gate into a CNOT gate. The single qubit rotations, and two qubit entangled rotations were proposed to implement the phase gate, where the two qubits were the internal ion qubit and the CM motional qubit [54]. Detailed mathematical derivations will be given below, following [9].

The gate experimentally implemented [33] by a method similar to that proposed by Cirac and Zoller [13], using the parameters described above, was a 2-qubit controlled NOT gate, which also used the auxiliary level, e.g., $|1\rangle_m|1\rangle \Rightarrow |0\rangle_m|\text{aux}\rangle \Rightarrow -|1\rangle_m|1\rangle$, where the result is a phase shift by π in the qubit wave function [33, 53]. This was carried out in three steps (see Fig. 5.14):

(1) Starting in the qubit levels $|1\rangle_m|1\rangle$, a $\pi/2$ rotation of the internal qubit applied at the carrier frequency created an equal superposition of the

internal qubit levels , but left the motional qubit in the level $|1\rangle_m$.

(2) a 2π rotation applied at the blue sideband frequency between $|1\rangle$ and $|\text{aux}\rangle$ rotated the phase of this part of the superposition by π radians, while returning to the original levels, as shown in the example above. The transition $|1\rangle_m|0\rangle \Rightarrow |1\rangle_m|\text{aux}\rangle$ was not resonant.

(3) A $\pi/2$ phase shift was applied at the carrier frequency, but with a phase shift of π relative to the initial operation. This brought the superposition state $2^{-1/2}(|0\rangle - |1\rangle) \Rightarrow |0\rangle$, with the result that

$$|a\rangle|b\rangle \Rightarrow |a\rangle|a \oplus b\rangle \text{ where } \oplus \text{ denotes addition mod 2.} \quad (5.27)$$

This is the CNOT operation, which for an arbitrary superposition state having general angles θ, ϕ, leaves

$$\cos\theta|0\rangle|0\rangle_m + \exp(i\phi)\sin\theta|1\rangle|0\rangle_m \Rightarrow \cos\theta|0\rangle|0\rangle_m + \exp(i\phi)\sin\theta|1\rangle|0\rangle_m \quad (5.28)$$

uneffected, but changes the part of the wave function in the motional state $|1\rangle_m$:

$$\cos\theta|0\rangle|1\rangle_m + \exp(i\phi)\sin\theta|1\rangle|1\rangle_m \Rightarrow \cos\theta|1\rangle|1\rangle_m + \exp(i\phi)\sin\theta|0\rangle|1\rangle_m. \quad (5.29)$$

The CM motional (control) qubit is left unchanged by the operation, while the internal qubit (target) states are exchanged. The time to execute this gate [33] is $T_{CZ} = 2\pi/\eta\Omega$, set by the slowest (entangling) transition. Had the initial state been $|1\rangle_m|0\rangle$, the same three operations would have resulted in a final state $|1\rangle_m|1\rangle$. For any initial internal state on $|0\rangle_m$, step (2) is non-resonant, and no change occurs (see Fig. 5.29).

Let us now give rigorous derivations of the above. Return to (5.25), but consider a string of N coupled ions. As noted there, for the CM motion, the interaction Hamiltonian coupling the internal qubit states of each ion $j, j = 1, \ldots, N$, to the CM motion can be written as

$$H_j = \frac{i\hbar\eta\Omega}{\sqrt{N}}e^{i\varphi}\sigma_j^+ a - \frac{i\hbar\eta\Omega}{\sqrt{N}}e^{-i\varphi}\sigma_j^- a^+, \quad (5.30)$$

by replacing η in (5.25) with $\eta/[N^{1/2}]$. Noting from (5.19) that $\sigma_j^+ = \begin{bmatrix} 0 & 1 \\ 0 & 0 \end{bmatrix} = |1\rangle_j\langle 0|_j$, $\sigma_j^- = \begin{bmatrix} 0 & 0 \\ 1 & 0 \end{bmatrix} = |0\rangle_j\langle 1|_j$, (5.30) can be written as

$$H_j = \frac{\hbar\eta\Omega}{\sqrt{N}}[|1\rangle_j\langle 0|_j e^{-i\varphi'} a + |0\rangle_j\langle 1|_j e^{i\varphi'} a^+], \quad (5.31)$$

where $\varphi' \equiv -(\frac{\pi}{2} + \varphi)$. From now on, we just rename φ' as φ and write (5.31) as

$$H_j = \frac{\hbar\eta\Omega}{\sqrt{N}}[|1\rangle_j\langle 0|_j e^{-i\varphi} a + |0\rangle_j\langle 1|_j e^{i\varphi} a^+], \quad j = 1, 2, \ldots, N. \quad (5.32)$$

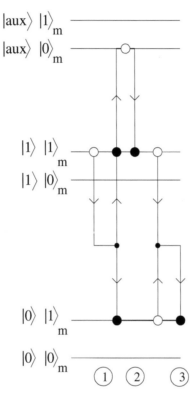

Figure 5.29: The action of a Cirac–Zoller gate is delineated. In step 1, a qubit in logic state $|1\rangle$ (open circle) is excited into a superposition of logic states $|1\rangle$ and $|0\rangle$, while remaining in motional state $|1\rangle_m$. The superposition is represented by black dots. In step 2, the $|1\rangle|1\rangle_m$ state is excited through the internal auxiliary state and back by a 2π pulse, changing the sign of the wave function. In step 3, the superposition (open circles) is recombined by a rotation to the qubit state $|0\rangle|1\rangle_m$ (black dot). If the system alternatively started in motional state $|1\rangle|0\rangle_m$, the transition to the auxiliary level would be non-resonant, and the internal state would not be changed.

Remark 5.6.1. It can be shown that the Rabi frequency Ω_R for the internal qubit transition coupled by a and $a^\dagger$ to the vibration, is equal to $\eta\Omega/\sqrt{N}$, much smaller than that, Ω, in (5.21) or (5.23) for a single qubit (rotation) operation. □

Lemma 5.6.1. *Let H_j be given by (5.32) for $j = 1, 2, \ldots, N$. Then H_j has a family of invariant 2-dimensional subspaces*

$$V_{j,k} \equiv span\{|0\rangle_j|k+1\rangle_{CM}, |1\rangle_j|k\rangle_{CM}\}, \quad k = 0, 1, 2, \ldots, \tag{5.33}$$

in the Hilbert space

$$V_j \equiv span\{|0\rangle_j|k\rangle_{CM}, |1\rangle_j|k\rangle_{CM} \mid k = 0, 1, 2, \ldots\}. \tag{5.34}$$

On the orthogonal complement of all the $V_{j,k}$'s, i.e., $\widetilde{V}_j = \left(span \bigcup_{k=0}^{\infty} V_{j,k}\right)^\perp \subset V_j$, the action of H_j is 0: $H_j|\psi\rangle = 0$ for all $\psi \in \widetilde{V}_j$.

(In the above, the original subscript "m" for motion is henceforth replaced by "CM", the center-of-mass mode.)

Proof. We have

$$\begin{aligned}
H_j|0\rangle_j|k+1\rangle_{CM} &= \frac{\hbar\eta\Omega}{\sqrt{N}}[|1\rangle_j\langle 0|0\rangle_j e^{-i\phi}(a|k+1\rangle_{CM}) \\
&\quad + |0\rangle_j\langle 1|0\rangle_j e^{i\phi}(a^\dagger|k+1\rangle_{CM})] \\
&= \frac{\hbar\eta\Omega}{\sqrt{N}}e^{-i\phi}\sqrt{k+1}\,|1\rangle_j|k\rangle_{CM}, \tag{5.35}
\end{aligned}$$

and, similarly,

$$H_j|1\rangle_j|k\rangle_{CM} = \frac{\hbar\eta\Omega}{\sqrt{N}}e^{i\phi}\sqrt{k+1}\,|0\rangle_j|k+1\rangle_{CM}. \tag{5.36}$$

Thus, $V_{j,k}$ is an invariant 2-dimensional subspace of H_j in V_j. □

Let the time-evolution operator of H_j (depending on ϕ) on V_j be $U_j(t, \phi) = e^{-\frac{i}{\hbar}H_j t}$. Then on the subspace $V_{j,k}$, $k = 0, 1, 2, \ldots, \infty$, with respect to the ordered basis $\{|0\rangle_j|k+1\rangle_{CM}, |1\rangle_j|k\rangle_{CM}\}$, H_j admits the following matrix representation

$$H_j = \hbar\eta\Omega\sqrt{\frac{k+1}{N}}\begin{bmatrix} 0 & e^{-i\phi} \\ e^{i\phi} & 0 \end{bmatrix}. \tag{5.37}$$

Thus, with H_j restricted to $V_{j,k}$, its time-evolution operator is given by

$$\begin{aligned}
U_{j,k}(t, \phi) &\equiv U_j(t, \phi)\big|_{V_{j,k}} = e^{-\frac{i}{\hbar}H_j t}\Big|_{V_{j,k}} \\
&= \begin{bmatrix} \cos(\mathcal{E}_k t) & -ie^{-i\phi}\sin(\mathcal{E}_k t) \\ -ie^{i\phi}\sin(\mathcal{E}_k t) & \cos(\mathcal{E}_k t) \end{bmatrix}, \quad \mathcal{E}_k \equiv \hbar\eta\Omega\sqrt{\frac{k+1}{N}}. \tag{5.38}
\end{aligned}$$

Also, note that

$$U_j(t,\phi)|\psi\rangle = |\psi\rangle \quad \text{for all} \quad |\psi\rangle \in \widetilde{V}_j, \tag{5.39}$$

because the action of H_j on $\widetilde{V}_j$ is annihilation. In physics, these states in $\widetilde{V}_j$ are said to be *off-resonance*.

Next, define

$$H_j^{\text{aux}} = \hbar\eta\Omega[|\alpha\rangle_j\langle 1|_j e^{-i\phi}a + |1\rangle_j\langle\alpha|_j e^{i\phi}a^\dagger], \qquad j = 1, 2, \tag{5.40}$$

where $|\alpha\rangle$ is the auxiliary quantum state $|\text{aux}\rangle$ as indicated in Fig. 5.29. Then, similarly to Lemma 5.6.1, we have the following.

Lemma 5.6.2. *Let H_j^{aux} be given by* (5.40) *for $j = 1, 2, \ldots, N$. Then H_j^{aux} has a family of invariant 2-dimensional subspaces*

$$V_{j,k}^{\text{aux}} \equiv span\{|1\rangle_j|k+1\rangle_{CM}, |\alpha\rangle_j|k\rangle_{CM}\}, \qquad k = 0, 1, 2, \ldots, \tag{5.41}$$

in the Hilbert space

$$V_j^{\text{aux}} \equiv span\{|1\rangle_j|k\rangle_{CM}, |\alpha\rangle_j|k\rangle_{CM} \mid k = 0, 1, 2, \ldots\}. \tag{5.42}$$

On the orthogonal complement $\widetilde{V}_j^{\text{aux}} \equiv \left(span \bigcup_{k=0}^{\infty} V_{j,k}^{\text{aux}}\right)^{\perp} \subset V_j^{\text{aux}}$, the action of H_j^{aux} is 0: $H_j^{\text{aux}}|\psi\rangle = 0$ for all $|\psi\rangle \in \widetilde{V}_j^{\text{aux}}$. □

The time-evolution operator corresponding to H_j^{aux} on V_j^{aux} is denoted as $U_j^{\text{aux}}(t,\phi) = e^{-\frac{i}{\hbar}H_j^{\text{aux}}t}$. Now, restrict H_j^{aux} to the invariant 2-dimensional subspace $V_{j,k}^{\text{aux}}$ with ordered basis $\{|1\rangle_j|k+1\rangle_{CM}, |\alpha\rangle_j|k\rangle_{CM}\}$; its evolution operator has the matrix representation

$$U_{j,k}^{\text{aux}}(t,\phi) \equiv U_j^{\text{aux}}(t,\phi)\big|_{V_{j,k}^{\text{aux}}} = e^{-\frac{i}{\hbar}H_j^{\text{aux}}t}\big|_{V_{j,k}^{\text{aux}}}$$
$$= \begin{bmatrix} \cos(\mathcal{E}_k t) & -ie^{-i\phi}\sin(\mathcal{E}_k t) \\ -ie^{i\phi}\sin(\mathcal{E}_k t) & \cos(\mathcal{E}_k t) \end{bmatrix}, \quad \mathcal{E}_k \equiv \hbar\eta\Omega\sqrt{\frac{k+1}{N}}. \tag{5.43}$$

Here, we again also have

$$U_j^{\text{aux}}(t,\phi)|\psi\rangle = |\psi\rangle \quad \text{for all} \quad |\psi\rangle \in \widetilde{V}_j^{\text{aux}}, \tag{5.44}$$

because H_j^{aux} annihilates the subspace $\widetilde{V}_j^{\text{aux}}$. Using the CM mode as the *bus*, we can now derive the following 2-bit quantum phase gate.

Theorem 5.6.3. *Let $U_j(t,\phi), U_{j,k}(t,\phi), U_j^{\text{aux}}(t,\phi)$ and $U_{j,k}^{\text{aux}}(t,\phi)$ be defined as above satisfying* (5.38)–(5.44) *for $j = 1, 2, \phi = 0$ and $k = 0$. Then for*

$$U \equiv U_1(3T, 0)U_2^{\text{aux}}(2T, 0)U_1(T, 0), \quad T \equiv \frac{\pi}{2\eta\Omega}\sqrt{N}. \tag{5.45}$$

we have

$$U|0\rangle_1|0\rangle_2|0\rangle_{CM} = |0\rangle_1|0\rangle_2|0\rangle_{CM}, \qquad (5.46)$$

$$U|0\rangle_1|1\rangle_2|0\rangle_{CM} = |0\rangle_1|1\rangle_2|0\rangle_{CM}, \qquad (5.47)$$

$$U|1\rangle_1|0\rangle_2|0\rangle_{CM} = |1\rangle_1|0\rangle_2|0\rangle_{CM}, \qquad (5.48)$$

$$U|1\rangle_1|1\rangle_1|0\rangle_{CM} = -|1\rangle_1|1\rangle_1|0\rangle_{CM}. \qquad (5.49)$$

Consequently, by ignoring the last CM-bit $|0\rangle_M$, *we have the phase gate*

$$U = Q_\pi, \quad cf. \ (2.47). \qquad (5.50)$$

Proof. We first verify (5.46):

$$
\begin{aligned}
U|0\rangle_1|0\rangle_2|0\rangle_M &= U_1(3T,0)U_2^{\mathrm{aux}}(2T,0)[U_1(T,0)|0\rangle_1|0\rangle_{CM}]|0\rangle_2 \\
&= U_1(3T,0)U_2^{\mathrm{aux}}(2T,0)[|0\rangle_1|0\rangle_{CM}]|0\rangle_2 \\
&\qquad \text{(by (5.39) because } |0\rangle_1|0\rangle_{CM} \in \widetilde{V}_j \text{ for } j = 1) \\
&= U_1(3T,0)[U_2^{\mathrm{aux}}(2T,0)|0\rangle_2|0\rangle_{CM}]|0\rangle_1 \\
&= U_1(3T,0)[|0\rangle_2|0\rangle_{CM}]|0\rangle_1 \\
&\qquad \text{(by (5.44) because } |0\rangle_2|0\rangle_{CM} \in \widetilde{V}_j^{\mathrm{aux}} \text{ for } j = 2) \\
&= [U_1(3T,0)|0\rangle_1|0\rangle_{CM}]|0\rangle_2 \\
&= |0\rangle_1|0\rangle_{CM}|0\rangle_2 \\
&\qquad \text{(by (5.39) because } |0\rangle_2|0\rangle_{CM} \in \widetilde{V}_j \text{ for } j = 1) \\
&= |0\rangle_1|0\rangle_2|0\rangle_{CM}. \qquad (5.51)
\end{aligned}
$$

Similarly,

$$U|0\rangle_1|1\rangle_2|0\rangle_{CM} = |0\rangle_1|1\rangle_2|0\rangle_{CM}.$$

Next, we verify (5.48):

$$
\begin{aligned}
U|1\rangle_1|0\rangle_2|0\rangle_{CM} &= U_1(3T,0)U_2^{\mathrm{aux}}(2T,0)[U_1(T,0)|1\rangle_1|0\rangle_{CM}]|0\rangle_2 \\
&= U_1(3T,0)U_2^{\mathrm{aux}}(2T,0)[-i|0\rangle_1|1\rangle_{CM}]|0\rangle_2 \\
&\qquad \text{(using (5.38) with } k = 0, \phi = 0 \text{ and } \mathcal{E}_k T = \pi/2) \\
&= U_1(3T,0)\{(-i)[U_2^{\mathrm{aux}}(2T,0)|0\rangle_2|1\rangle_{CM}]|0\rangle_1\} \\
&= U_1(3T,0)\{[-i|0\rangle_2|1\rangle_{CM}]|0\rangle_1\} \\
&\qquad \text{(using (5.43) with } k = 0, \phi = 0 \text{ and } 2\mathcal{E}_k T = \pi) \\
&= -i[U_1(3T,0)|0\rangle_1|1\rangle_{CM}]|0\rangle_2 \\
&= -i[(i)|1\rangle_1|0\rangle_{CM}]|0\rangle_2 \\
&\qquad \text{(using (5.38) with } k = 0, \phi = 0 \text{ and } \mathcal{E}_k T = 3\pi/2) \\
&= |1\rangle_1|0\rangle_2|0\rangle_{CM}. \qquad (5.52)
\end{aligned}
$$

Finally, we verify (5.49):

$$
\begin{aligned}
U|1\rangle_1|1\rangle_2|0\rangle_{\mathrm{CM}} &= U_1(3T,0)U_2^{\mathrm{aux}}(2T,0)[U_1(T,0)|1\rangle_1|0\rangle_{\mathrm{CM}}]|1\rangle_2 \\
&= U_1(3T,0)U_2^{\mathrm{aux}}(2T,0)[-i|0\rangle_1|1\rangle_{\mathrm{CM}}]|1\rangle_2 \\
&\qquad \text{(using (5.38) with } j=1, k=0, \phi=0 \text{ and } \mathcal{E}_k T = \pi/2) \\
&= U_1(3T,0)[(-i)U_2^{\mathrm{aux}}(2T,0)|1\rangle_2|1\rangle_{\mathrm{CM}}]|0\rangle_1 \\
&= U_1(3T,0)[(i)|1\rangle_2|1\rangle_{\mathrm{CM}}]|0\rangle_1 \\
&\qquad \text{(by (5.44) because } |1\rangle_2|1\rangle_{\mathrm{CM}} \in \widetilde{V}_j^{\mathrm{aux}} \text{ for } j=2) \\
&= (i)[U_1(3T,0)|0\rangle_1|1\rangle_{\mathrm{CM}}]|1\rangle_2 \\
&= (i)[(i)|1\rangle_1|0\rangle_{\mathrm{CM}}]|1\rangle_2 \\
&\qquad \text{(using (5.38) with } j=1, k=0, \phi=0 \text{ and } \mathcal{E}_k T = 3\pi/2) \\
&= -|1\rangle_1|1\rangle_2|0\rangle_{\mathrm{CM}}. \tag{5.53}
\end{aligned}
$$

The verifications are complete.　　　　　　　　　　　　　　　　　　　□

The CNOT gate can now be obtained according to (2.48). As a consequence, we have the following.

Theorem 5.6.4. *The quantum computer made of confined ions in a trap is universal.*

Proof. Use (2.48) and (5.50) to deduce Corollary 2.6.3.　　　　　　　□

5.6.3　Wave packet or Debye–Waller CNOT gate

It is also possible to execute an alternative CNOT gate of the Cirac–Zoller type with a single pulse, and not use an auxiliary level [15, 54]. Not requiring an auxiliary level may be important since some candidate ions may not have a level available suitable to this purpose. Also, a level outside the basic qubit invariant Hilbert subspace may potentially allow the loss of some information, decreasing fidelity and causing problems in future error correction schemes. For ions with hyperfine structure, it is possible to make the two qubit levels "magnetic field independent" to a first approximation, but difficult to accomplish simultaneously with a third auxiliary level. This may be important to coherence (see Subsection 5.5.2). The wave packet gate is feasible because of the quantum mechanical wave property of the quantized oscillatory motion, resulting in interactions which are extended in space, dependent on the quantum state.

This alternative CNOT gate can be carried out with a single pulse by using the property that the Debye–Waller corrections to the Rabi frequency depend on the motional quantum level, when the Lamb–Dicke parameter η_2 is not small. It is possible to set η_2 so that the ratio

$$
\Omega_{2(1,1)}/\Omega_{2(0,0)} = (2k+1)/2m \tag{5.54}
$$

holds, where k and m are positive integers, with $m > k \geq 0$. The subscripts of Ω_2 in parentheses refer to values of the quantum numbers of the motional qubit levels, respectively, before and after an operation which causes a transition between the internal ion levels.

If the carrier transition ω_0 is driven for a gate time T, so that $\Omega_{2(1,1)}T = (k+\frac{1}{2})\pi$, then $\Omega_{2(0,0)}T = m\pi$, consistent with the condition on the ratio of the Rabi frequencies, which is set by the mathematics which determines these frequencies. With this condition, the $\pi/2$ carrier pulse swaps the states $|0\rangle|1\rangle_m$ and $|1\rangle|1\rangle_m$ but leaves all motional states involving $|0\rangle_m$ unaffected. Within some phase factors, this is shown to be equivalent to the CNOT gate [54], executed in a single step.

A similar gate has been demonstrated on a single Be$^+$ ion in a linear Paul trap [39]. The ion internal qubit levels were the same as previously described above, but no auxiliary level was used. The detuning Δ of the 2-photon Raman beams from the $^2P_{1/2}$ was $+$ 80 GHz . The motional levels were chosen as $|0\rangle_m = |0\rangle_z$ and $|1\rangle_m = |2\rangle_z$, with the frequency $\omega_z = 2\pi \times 3.4$ MHz. The trapping fields and Raman beams were adjusted to make the Lamb–Dicke parameter $\eta_2 = (\Delta k_z)(\hbar/2\omega_z m)^{1/2} = 0.359$. This choice was made to make the Rabi frequency ratio (dependent on the Debye–Waller factors)

$$\Omega_{2(0,0)}/\Omega_{2(2,2)} = 2/(2 - 4\eta_2^2 + \eta_2^4) \equiv 4/3. \tag{5.55}$$

The Rabi frequency $\Omega_{2(0,0)} = 2\pi \times 92$ kHz. For a wave function superposition

$$|\psi\rangle = \sum_{n=0,2} (b_{0n}|0\rangle + b_{1n}|1\rangle)|n\rangle_{CM}, \tag{5.56}$$

and a gate time T_{DW}, the rotation angles associated with the Rabi frequencies now depend on the initial and final values of the CM oscillator quantum numbers. Suppose the carrier transition is induced with a gate time appropriate to two complete Rabi cycles starting from either internal qubit level, with the motional qubit remaining in the $|0\rangle_z$ level, i.e., $\Omega_{2(0,0)}T = 2\pi$ (called a 4π pulse). Then the same gate duration T produces a 3π pulse $\Omega_{2(2,2)}T = 3\pi/2$, with the motional qubit remaining in the $|2\rangle_z$ level. The internal qubit (target) transition is dependent on the motional (control) qubit state, so a CNOT gate is executed. The transition associated with the control state $|2\rangle_z$ also acquires an additional $-\pi/2$ phase shift, but this can be removed by an additional single qubit rotation before or after the CNOT gate. Fig. 5.30 shows the effect of the gate. The gate time T_{DW} can be expressed as $T_{WP} = \pi/\eta^2\Omega$.

The initialization to the $|0\rangle|0\rangle_z$ state was accomplished as noted earlier, by sideband cooling and optical pumping. A 99.9% probability of occupation of this state was achieved. Detection was carried out with a 200 μs pulse of radiation with σ_- polarization, using the cycling transition described earlier. It was feasible to prepare all four of the basis states specified in Eq. (5.16), by applying appropriate pulses at the carrier, red, or blue sidebands,

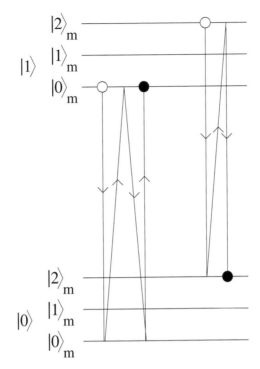

Figure 5.30: The action of the wave packet gate is diagrammed. An ion initially in qubit levels $|1\rangle|0\rangle_m$ (open circle) experiences a 4π pulse at the Rabi frequency associated with state $|0\rangle_m$ which returns it to its original state. During the same time interval, if the ion starts in state $|1\rangle|2\rangle_m$ it makes a transition to state $|0\rangle|2\rangle_m$ by a $3\pi/2$ pulse, because the Rabi frequency associated with state $|2\rangle_m$ has exactly the right value, due to the choice of the Lamb–Dicke parameter η, which depends on trapping parameters. The internal state changes depending on the motional state quantum number, since the Rabi frequencies depend on η.

starting from the initialized state. A 95% accuracy in CNOT logic for each basis state was achieved.

5.6.4 Sørensen–Mølmer gate

An important proposal for a two ion (or two qubit) and multiple qubit gates was made by Sørensen and Mølmer [42]. The cold ion system is basically the same as the one proposed by Cirac and Zoller, but the gate implementation using laser pulses is conceptually different, since the intermediate motional qubit levels are not populated. Now finite (but still small) temperatures of the ions in the array are allowed. The internal and motional qubits of an ion pair in the array are entangled using two laser beams *de-*

tuned by a frequency δ from the first blue and first red motional sidebands of a single ion, $\omega_{L1} = \omega_0 + \omega_{CM} - \delta$ and $\omega_{L2} = \omega_0 - \omega_{CM} + \delta$. One laser beam excites one ion, and the other laser beam excites the other. The ions are coupled through their joint CM oscillatory motion, but the interaction only *virtually* excites the motional quantum states, without populating them. The sum of the two laser frequencies $\omega_{L1} + \omega_{L2} = 2\omega_0$, corresponding to the excitation of the internal qubits of each of the two ions by two interfering transition paths through the vibrational motion from $|0\rangle|0\rangle|n\rangle \Rightarrow |1\rangle|1\rangle|n\rangle$, without changing the vibrational quantum state n. The interaction Hamiltonian is written

$$H_I = \sum_j (\hbar\Omega_j/2)(\sigma_j^+ \exp\{i[\eta_j(a + a^+) - \omega_j t]\} + h.c. \qquad (5.57)$$

For two ions of the same type, the subscripts j which identify the ions may be dropped from the Rabi frequency Ω and the Lamb–Dicke parameter η, but the cooling of the ion motion now requires only that

$$\eta(n + 1)^{1/2} \ll 1, \qquad (5.58)$$

where n is the quantum number of the motional qubit. Since η is small (but not negligible) in this approximation, the terms proportional to η can be removed from the exponential term in Eq. (5.57) by an approximate expansion in η, to make the Hamiltonian appear equivalent to that of Eq. (5.24). By using second order perturbation theory, and restricting the intermediate states to $|0, 1, n + 1\rangle$ and $|1, 0, n - 1\rangle$ (where the first two symbols describe the internal ion qubit states, assumed to be in motional state n, and the terms $n \pm 1$ describe the virtual motional states), it can be shown that the effective Rabi frequency for the interaction Hamiltonian becomes [32, 42]

$$\Omega_{SM} = -(\eta\Omega)^2/2(\omega_{CM} - \delta), \qquad (5.59)$$

where δ is the detuning of the laser addressing ion #1. This expression is independent of the CM oscillation quantum number n, due to interference between the two paths of excitation, so the evolution of the internal qubit states is independent of the quantum number of the motional qubit (subject, of course, to the condition given by Eq. (5.58)). This temperature independence is a significant advantage.

Alternatively, the two ions can be *simultaneously* illuminated by two laser beams having both detunings $\pm\delta$, resulting in two additional interference paths involving the internal states $|0, 1\rangle$ and $|1, 0\rangle$ leading to the same final state; see Fig. 5.31. Under these conditions, Sørensen and Mølmer show that the evolution of all elements of the two qubit internal density matrix operator oscillate with frequency $\Omega_{SM}T/2$, where T is the time of an illuminating laser pulse which produces the maximally entangled state of two ions

$$|\psi\rangle = (2)^{-1/2}(|0, 0\rangle - i|1, 1\rangle). \qquad (5.60)$$

The density operator is $\rho = |\psi\rangle\langle\psi| \oplus \rho_{CM}$. A sequence of operations consisting of single qubit rotations plus the entangling interaction of the bichromatic pulse for time T resulting in a CNOT operation was suggested [32, 42].

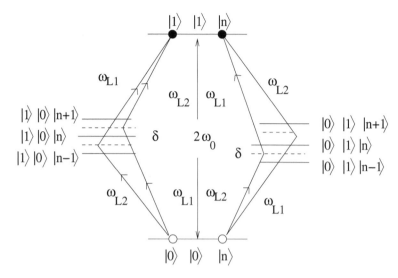

Figure 5.31: The operation of a Sørensen–Mølmer gate is diagrammed. Two logic qubits each in state $|0\rangle$ and motional state $|n\rangle$ are coupled through *virtual* CM motional states by interfering laser excitation paths. This results in both qubits in state $|1\rangle$, with the motional qubit unchanged, and never populated. The lasers have different frequencies, which sum to $2\omega_0$, twice the qubit separation frequency. Both ions are illuminated by both lasers. This operation can be shown to be part of a CNOT gate, which works on warm $(n \neq 0)$ ions in a single step.

This proposal was experimentally investigated [40] for two and four ions in a small linear rf trap, using Be^+ ions in the single ion qubit levels $|0\rangle = |F = 2, m_F = -2\rangle$ and $|1\rangle = |F = 1, m_F = -1\rangle$. The single ion Rabi frequency $\Omega = 2\pi \times 500$ kHz and $\eta = 0.23/N^{1/2}$. Rather than use the CM oscillation modes, which entail some ion heating (see discussion in Section 5.7), the higher frequency symmetric stretch (SS) motional modes were employed, in which alternating ions oscillate out of phase for $N = 2$ and $N = 4$. The symmetric stretch mode frequency was $\omega_{SS} = 3^{1/2}\omega_{CM} = 2\pi \times 8.8$ MHz, and the ions were initialized in the $n = 1$ quantum state. The frequency satisfies $\Omega_{SM} \approx (\eta\Omega)^2/\delta$ for $\omega_{SS} \approx 2\delta$.

The entanglement operation can be completed in a single step for a specific value of $\delta = 2\eta\Omega m^{1/2}$, where m is an integer. Equivalently $\Omega_{SM} = \delta/4m$. Choosing a laser pulse interval $T_{SM} = \pi/2\Omega_{SM} = 2\pi m/\delta$ results in the 2-qubit wave function given by Eq. (5.60). This entanglement is scalable, since the same operation can be used to generate an N-qubit entangled state (with N even). Using $\delta T = 2\pi m$, the time for preparation of the entangled state is $T_{SM} = \pi m^{1/2}/\eta\Omega \approx 6 \times 10^{-6}s$ for two ions, using $m = 1$ and the parameters given above. The entanglement operation was completed for two and for four confined ion qubits. A two ion gate of this type has certain advantages compared to the Cirac–Zoller gate [40]. It does not require precise control of the motional qubit state, nor an auxiliary state; it can be carried out in one operation, and it does not require laser focusing on a single ion.

For a figure of merit for the performance of the gate, the fidelity $F(N)$ for creation of the maximally entangled N-particle state was suggested [40]. Measured fidelities were $F(2) = 0.83 \pm 0.01$ and $F(4) = 0.57 \pm 0.02$ [40]. A fidelity $F(N) > 0.5$ guarantees N particle entanglement [40].

5.6.5 Geometrical phase gate

It is also possible to adjust the phases of the exciting electric fields to couple the internal qubits to different quadrature components such as the position and momentum of the symmetric stretch mode of oscillation of two ion qubits, or so that a commutator term like $[A, B] \equiv AB - BA$ provides the product of such operators. Milburn et al. [31], Sørensen and Mølmer [43], and Wang et al. [50] have proposed alternatives. A phase gate with angle π has been demonstrated between two ion qubits [30].

A classical force $F\sin(\omega t - \delta)$, resonant with the harmonic oscillator frequency, can coherently displace the quantum state $|\psi\rangle$ of the oscillator in phase (position z-momentum p) space [8]. A force that acts for a time τ results in a displacement Δz and Δp which is described by the action of a corresponding displacement operator $D(\alpha) = \exp(\alpha a^+ - \alpha^* a)$ on $|\psi\rangle$, where $\alpha = (2z_0)^{-1}(\Delta z + i\Delta p/m\omega)$ for an oscillator with mass m and spread of the ground state wave function z_0. Since sequential displacements are additive within a phase factor, if state $|\psi\rangle$ is transported around a closed loop in phase space by displacement operators, the phase factors accumulate to give a geometric phase $\Phi = A(z, p, \tau)/\hbar$ to $|\psi\rangle$, where $A(z, p, \tau)$ is the loop area in phase space executed in time τ. If the force *differs* for the two qubit states of each ion in a 2-qubit combination, then *conditional* states are formed, and a logic gate based on the geometric phase can be executed. This has been demonstrated using a state-dependent dipole force generated by laser light [30].

For a pair of ion qubits confined in a harmonic trap, and separated by their Coulomb repulsion, along the linear direction of oscillation there will be a CM mode and a symmetric stretch mode of oscillation. These modes have similar magnitudes of displacements, but in the CM mode the displacements are parallel, while in the stretch mode they are anti-parallel. The frequencies of each mode differ, since in one mode the ions move together, but in the other mode, the ions separate and their Coulomb interaction comes into play. Using two lasers having detunings $\pm\delta$ to illuminate the two ions (the Sørensen–Mølmer technique), the weak field approximation used in Subsection 5.6.3 is relaxed (i.e., $\eta\Omega \ll \omega_{SS} - \delta$, so that linewidths remain small and the vibrational mode levels are not populated in the transition). The remaining conditions are that $\Omega \ll \delta$ and $\Delta\omega = (\omega_{SS} - \delta) \ll \delta$, so that excitation of the vibration modes is small, but not zero. The laser difference frequency $\Delta\omega$ was close to the stretch mode frequency ω_{SS}, which is $3^{1/2}\omega_{CM}$.

The electric field of the laser beams produces a Stark shift of the energies of each internal ion state, associated with an electric dipole force on each ion. The polarizations and frequencies of the laser beams used to produce the stimulated Raman transitions are adjusted to average away the differential energy shifts between the $|0\rangle$ and $1\rangle$ qubit states, but on short time scales, a different force exists on each state, modulated at the frequency $\omega_{SS} + \delta$. The beam directions were chosen to be at right angles to each other, and their wave vector difference $\Delta\mathbf{k}$ was made to coincide with the trap axis. Conditions were chosen to give the interference pattern of the beams the same phase at the position of both ions. When the ions are in the same internal state, the dipole force on each ion is the same, and no differential force exists, so the stretch mode cannot be excited. If the ion internal states are different, a differential force exists, and the symmetric stretch mode is excited. Due to the detuning δ, the driving force does not remain synchronous with the stretch mode frequency, but it becomes synchronous after a time $\tau = 2\pi/\delta$, during which the motion is displaced along a circular path in phase space (see Fig. 5.32). The geometrical phase Φ is the area enclosed by this path, in units of $\hbar$.

To summarize, for this to be an entangling operation, the force on each ion must be dependent on the internal ion qubit level. This has been accomplished experimentally using a level-dependent optical dipole force, generated by tuning the Raman laser beams to have a difference frequency proportional to an ion mode frequency ω_{SS}, with a wave vector difference directed along the trap mode to be excited (such as the trap axis). The dipole force varies sinusoidally at ω_{SS}, but has a different phase for each qubit level, resulting in conditional excitation. The interaction Hamiltonian is given by

$$H_I = \hbar\Omega(\text{level})\{\exp -i(\omega_b - \omega_r)t\}[1 + \eta(a \exp -i\omega_{SS}t + a^+ \exp i\omega_{SS}t)] + h.c.$$
$$(5.61)$$

where the frequency difference $(\omega_b - \omega_r)$ is the frequency difference between the blue and red Raman beams. Note that *no transition of the internal qubit* is induced by this Hamiltonian, only motional qubit transitions. The effective Rabi frequency $\Omega(\text{level})$, which now depends on the internal qubit level, is given by (5.26), multiplied by $\exp i(\phi_b - \phi_r)$. Optical dipole forces are proportional to Stark shifts between levels, which degrade coherence, but the *time-averaged* Stark shift $\chi_0 - \chi_1$ can be made zero for gate times much greater than $2\pi/\omega_{SS}$ and still employ a level-dependent displacement operator for certain choices of the Raman laser beam parameters and detunings. The time scales for having a *different* force on each ion level, modulated at the frequency $\omega_{SS} + \delta$ to excite the symmetric stretch mode, were

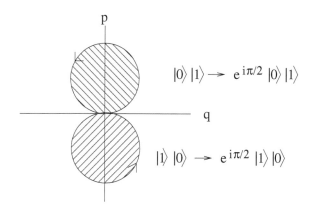

Figure 5.32: The operation of a geometric phase gate on two ions is represented. Optical dipole forces acting on the ion motion (the axial stretch mode) have the same direction for the same internal qubit states, but different directions otherwise. In the latter case, the stretch mode is excited. In phase space, if a closed orbit is execute by the displacements, then a phase shift in the wave function proportional to the enclosed area of the path occurs. System parameters are chosen, so that the states $|0\rangle|1\rangle$ and $|1\rangle|0\rangle$ accumulate $\pi/2$ phase shifts from the areas enclosed with the circular orbits, while the state $|0\rangle|0\rangle$ and $|1\rangle|1\rangle$ experience no net shift because there is no coupling to the stretch mode. Since the phase shifts depend on the phase space paths, and hence on forces which depend on the qubit state, these operations are entangling.

$\Delta^{-1} \ll \tau \ll \omega_{SS}^{-1}$. For Be^+, it was possible to choose parameters such that the Rabi frequencies $\Omega(\text{level})$ in Eq. (5.61) associated with the qubit levels differed by a factor of two, as well as in sign [30]. The dipole force acting to excite the ions in the stretch mode of oscillation with an ion in one qubit level was twice as big, and in the opposite direction, as the force when the ion was in the other qubit level.

The experimental test [30] used the symmetric stretch mode of oscillation along the z axis of two confined ions, with frequency $\omega_{SS} = 3^{1/2}\omega_z + \delta$, where δ is a small detuning. The ion separation was an integer multiple of $2\pi/\Delta k$ for the Raman beams, so the optical dipole force was in the *same* direction on each ion if they occupied the *same* qubit level, but in *opposite* directions when they were in *different* qubit levels. Consequently, the stretch mode was excited when the ions were in internal states $|0\rangle|1\rangle$ or $|1\rangle|0\rangle$, but not otherwise. The detuning δ and displacement pulse duration $2\pi/\delta$ were chosen to complete a circular path in phase space with an area resulting in a phase shift of $\pi/2$ when the qubit states were different. This operator acted equivalently to the product an operator that applies a $\pi/2$ phase shift to qubit level $|1\rangle$ of each ion separately, with a second operator that acts as a π phase gate between the two ions together, i.e., an entangling operation.

The test was reported with measurements of the fidelity [30]. The gate plus single qubit rotations had $97 \pm 2\%$ fidelity for the 2-ion Bell state, Eq. (5.60), while leaving the motional state unchanged. The time required for this operation was $39\,\mu s$. The limit on the fidelity was attributed to 1% fluctuations in both the phase δ, and in the Raman beam intensities. The complete π-phase gate operation occurred in $156\,\mu s$ with the stated fidelity.

5.6.6   Summary of quantum gates

Deterministic quantum gates are executed between qubits, or between single-qubit levels, using laser pulses with well-defined durations. These operations are described by interaction Hamiltonian operators, and occur at a rate described by a Rabi frequency. The Hamiltonians act on internal qubit levels using raising and lowering operators σ^+ and σ^-, and on the CM motional qubit by analogous operators a^+ and a. Two-beam stimulated Raman transitions are used to excite transitions of the internal qubit only (carrier transitions at ω_0), or the motional qubit only (transitions at ω_{CM}), or both (transitions at $\omega_0 + \omega_{CM}\Delta n$), with appropriate directions of laser beams and of the wave vector difference required. An entangling operation together with single qubit rotations acts as a universal quantum gate which changes internal qubit states contingent on the external (motional) qubit state. All operations can be interpreted as rotations of the wave function in Hilbert space.

The CNOT gate proposed by Cirac and Zoller required three laser pulses, including a transition to an auxiliary level outside the minimal basis which coupled to the motional qubit, so the Rabi rate was $\Omega_{CZ} = \eta\Omega$. The simpler wave packet gate required only a single internal qubit transition between well-defined motional states having different Rabi transition rates due to different Debye–Waller factors. The Sørensen–Mølmer CNOT gate uses quantum interference to execute a 2-ion gate without populating motional levels, making it capable of operation at finite temperatures within the Lamb–Dicke regime. The gate rate $\Omega_{SM} \propto (\eta\Omega)^2$, and so is relatively slow. The geometric phase gate uses transitions only between motional qubit levels, which are excited at Rabi frequencies which depend on the internal qubit state.

The operation of entanglement to produce universal quantum gates has progressed well beyond the original proposal of Cirac and Zoller, in terms of simplicity, if not in ease of comprehension. In particular, the two ion and four ion Sørensen–Mølmer gates are a significant advance, which are related theoretically to the geometric phase gate. Other potential mechanisms for gates have been proposed [26], but not yet experimentally verified. The choice of an optimum gate type will depend on a detailed evaluation of the parameters of the final system. In this regard, both the fidelity and the time of execution of the gate, compared to the coherence time available to complete the calculation, are important. In particular, the execution time of the complete geometric phase gate (phase $= \pi$) was about $160\,\mu s$, with a fidelity of about 0.97.

It has been noted that gate times for N ions tend to scale as $N^{1/2}$. The execution times for certain gate types while considering the fidelity of the operation, has been considered theoretically and compared with measurements [47]. A parameter called the recoil frequency of an ion at rest is important in the considerations: it is $\hbar^{-1}$ times the recoil energy per ion, which is undergoing a π pulse. The ion recoils, since a photon is emitted or absorbed, which carries momentum. The recoil energy is given by the equation

$$E_R/\hbar = \hbar(rk_z)^2/2M \qquad (5.62)$$

for an ion with mass M, where r is 1 for a single photon transition, and 2 for a stimulated Raman transition. Single ion rotations require a time $T = 2\pi/\Omega$, which can be made in $\approx 1\mu s$, and are unlikely to reduce fidelity. Gates coupling the internal qubits to the motion, such as swap gates, or some of the CNOT gates, require much longer intervals because of Stark shifts of the ion level separations. When Stark shifts are compensated by laser re-tuning, longer intervals decrease off-resonant excitation of other transitions, which reduce fidelity. The Cirac–Zoller gate requires a time $T_{CZ} = 2\pi/\eta\Omega = 2T_S$ in terms of the time for a swap gate, basically interchanging the excitation of the internal and motional qubit. It is shown that for the swap gate [47]

$$T_S^{-1} \leq 2^{3/2}\varepsilon(E_R\omega_z/2\pi N\hbar)^{1/2} \quad \text{with} \quad \varepsilon \approx \Omega/2^{1/2}\omega_z. \qquad (5.63)$$

where the "imprecision" $\varepsilon = (1 - F)^{1/2}$ is defined in terms of the fidelity F. This gate rate increases proportionally to ω_z, the mode oscillation frequency. Similarly the wave packet gate requires a time $T_{WP} = \pi/\eta^2\Omega$, which turns out to be basically limited by the N-ion recoil frequency, $T_{WP}^{-1} \approx E_R/N\hbar$, with $\varepsilon = 2^{1/2}\eta\Omega/\omega_z$. High fidelity (smaller ε) implies longer gate durations.

It seems clear that the 2-qubit methods related to the Sørensen–Mølmer technique, which does not require the coldest ions or precise laser focusing, are to be preferred, as long as gate execution time and fidelity are acceptable. The phase gate, with no internal qubit transitions, appears to be the best choice for short execution times and high fidelity.

5.7 A vision of a large scale confined-ion quantum computer

A concept of the NIST-Boulder group [27] for overcoming some coherence difficulties summarized in Subsection 5.5.5 is to physically isolate small subsystems or registers of ions that can be operated on independently. These subsystems are then entangled sequentially, or in parallel, so that the large entangled state needed for a large scale quantum computation can be constructed. The subsystems may be transported between nodes of the larger system, for quantum information processing. The proposed architecture consists of an array of interconnected linear ion traps [27]; see Figs. 5.5 and 5.33. Time-dependent potentials applied to control segments of electrodes are used to move ions between nodes of the array. Logic operations between selected ion qubits are applied in the interaction region of an accumulator trap; the ions are then moved to memory locations or to other accumulators. Such arrays allow highly parallel processing and ancilla qubit readout in a separate location, so that logic ions can be shielded from the resonant light emitted in all directions associated with qubit measurement, which is potentially dephasing. The present scheme, called a "quantum charge-coupled device" or QCCD, uses only quantum manipulation techniques that have already been individually demonstrated experimentally, although alternative options may be considered in the future.

A segment of a conceptual quantum QCCD device is shown in Fig. 5.33. It consists of ultra-small, interconnected linear radio-frequency ion traps [27]. By using "dc" control voltages, a few ions can be held in each trap, or moved from trap to trap. In certain designated traps, gates can be implemented between a few ions, while the connections between traps permit information transfer between sets of ions. The speed of the quantum gates, and of the ion transfers, are comparable, both limited by the trap voltages. The qubit ions may be heated by the manipulation, so they are to be cooled sympathetically by a second ion species retained in the interaction region. Using a second ion species for cooling isolates the qubit ions from photons emitted during the cooling process, which might be resonant with the qubit transition if the same ion type were used. The qubits levels may develop uncontrolled relative phase shifts as the ions are moved, and accurate positioning of the ions in the interaction region during a computation may be difficult. These potential problems are to be addressed by decoherence-free encoding [27]; cf. Example 2.9.3.

A decoherence-free subspace (DFS) of two ions is spanned by pairs of ion qubits

$$|0\rangle \equiv |0\rangle_1 |1\rangle_2 \text{ and } |1\rangle \equiv |1\rangle_1 |0\rangle_2, \qquad (5.64)$$

which form the logic qubits $|0\rangle$ and $|1\rangle$. Suppose that the qubit wave function state $|1\rangle_1$ of ion 1 acquires an overall phase α_1 relative to qubit wave function state $|0\rangle_1$ due to its transport to an accumulator, and similarly for ion 2, the wave functions acquire a relative phase α_2 due to this transport. The superposition state $|0\rangle + |1\rangle$ of the logic qubit is then

$$|0\rangle + |1\rangle = \exp(i\alpha_2)|0\rangle_1|1\rangle_2 + \exp(i\alpha_1)|1\rangle_1|0\rangle_2. \qquad (5.65)$$

If $\alpha_1 = \alpha_2$, there is no relative phase shift, the dephasing is collective, and the common phase factor can be disregarded. Only when the phase shifts differ is the dephasing not collective, and decoherence results. The ions are moved in the QCCD structure by electric fields, which may produce unknown Stark phase shifts in each ion wave function, but to the extent that these shifts cancel for similar paths, the logic qubit is unaffected. If a

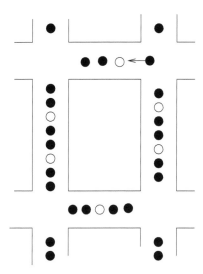

Figure 5.33: A schematic of a portion (or block) of a hypothetical QCCD device, showing the inner electrodes of a set of short, interconnected linear trap segments. The inner electrodes are segmented (not shown), so that slower time-varying forces can confine the ion strings, or move ions at the ends of strings into contact in the horizontal (gate) regions, from the vertical storage (memory) regions. Other electrodes carrying radio-frequency voltages lie above and below the plane shown. Two black dots together form a DFS qubit, while the small open circles represent cooling ions of a different type for each string. Entangling gates are executed in the gate regions, while information is moved by swap gates to and from qubits between the ends of the memory strings.

linear magnetic field B varies over a device with dimensions of 10 cm, but the ions are separated by 10 microns, the dephasing of the logic qubit is reduced by 10^{-4} compared to that for a single ion qubit, assuming that the level separations depend linearly on B. So the logic qubit is robust against decoherence during ion transport in the QCCD [27].

5.8   Trap architecture and performance

A successful experimental test was completed [39] on a trap structure composed of a stack of metalized $200\,\mu m$ thick alumina wafers, created using lithographic techniques. Slots were machined with laser beams, and coated with $0.5\,\mu m$ evaporated gold traces through a shadow mask. Gold was electroplated onto the electrode surfaces to a thickness of $3\,\mu m$, resulting in surfaces smooth at the $1\,\mu m$ scale. A $400\,\mu m$ wide central slot, and a wafer spacing of $360\,\mu m$, established the rf trapping region. Side slots $10\,\mu m$ wide electrically isolated five different control electrodes held at rf ground, used for axial confinement in two possible linear trap sections. Evaporated Be atoms were shielded from striking the trap electrodes by a shadow mask. The trapping rf voltage was 500 V at $\Omega_T = 2\pi \times 230$ MHz. The trap was held in a quartz envelope evacuated to 3×10^{-11} Torr. Be ions were confined, cooled, and studied using previously described techniques. It is expected that this structure could be scaled up to a much more elaborate device.

A key test of the trap was the ion heating rate, which is discussed in Subsection 5.5.1. The heating rate was found to be about one hundred times *lower* than in a similar trap lacking electroplating of the gold surfaces. The heating rate doubled over an initial two weeks of trap operation. This heating is still much faster than is expected for thermal electronic noise, so other unknown causes continue to dominate.

Ions were transferred between the two linear trapping sections by continuously varying the potentials of the five pairs of control electrodes. Transfers were smoothly carried out keeping the secular motional frequencies for each dimension constant. The ions were transferred during a measured characteristic time t_t, held for a transfer time, transferred back, and the motional heating measured. A transfer time near $50\,\mu s$ was found to be adiabatic for all three motional modes, and no ions were lost in over one million successive transfers [39].

Coherence preservation was measured on coherent superpositions of internal ion states after transfer, and without transfer. The results averaged over many trials were the same within $0.2 \pm 0.6\%$, indicating no reduction in internal state coherence associated with ion transfer.

Two confined ions were also separated into different traps, and then brought together again. In $95 \pm 1\%$ of the measurements, one ion was separated out in the transfer, and in the rest of the measurements, either two or no ions were separated. Any number of ions (0, 1, 2 or 3 starting with 3 confined ions) could be picked off and moved to the other trap. In the separation the ions were strongly heated, about 140 ± 70 motional quanta being absorbed, with a 10 ms operational time for executing the separation found to minimize this heating. Most of the heating occurred when the CM frequency of the ions was lowest during the separation operation. From these tests, it was expected that additional ion cooling, likely done sympathetically, would be needed following some ion transport operations [39]. More recently [1], a smaller ion separation electrode was employed, resulting in ion separation with no detectable failure rate, and with substantially reduced ion heating. After separation, the symmetric stretch mode was measured to be in the ground state, and the CM mode had a mean quantum number near 1, permitting a phase gate to be implemented with a fidelity $> 90\%$ even with no ion re-cooling!

5.9 Teleportation of coherent information

Using quantum teleportation, an unknown superposition state of a qubit can be moved from one location in space to another, using entanglement as a resource, by communicating only two *classical* bits between the locations. The method uses halves of a 2-qubit entangled state previously deposited at the two locations. The unknown qubit $|\alpha\rangle$ and one qubit of the 2-qubit entangled state $|\psi\rangle$ are measured in the Bell state basis at the sender location, with the result xy, where xy are two uniformly distributed random bits 00, 01, 10, and 11. The well-known Bell states form a complete basis for all possible 2-qubit states. Based on the non-locality property of quantum mechanics, the second entangled qubit of $|\psi\rangle$ at the receiver location is left in the state $R_{xy}|\alpha\rangle$, where the operation R_{xy} can be expressed using single-qubit Pauli operations I, X, Z, and XZ, where X, Y and Z are, respectively, σ_x, σ_y and σ_z given in (2.42) and (2.43). Once xy is classically communicated, the proper choice R_{xy}^+ from I, X, Z, or XZ can be selected and executed to produce the unknown state $|\alpha\rangle$ at the receiver location [22]. The entangled state, and the information in the original state $|\alpha\rangle$ at the sending location, are destroyed (by measurement) in the process.

Deterministic quantum teleportation of ion qubits has been demonstrated by two groups. In one experiment [38], three Cd^+ ions were stored in a linear Paul trap, with the qubit of ion 1 encoded in a superposition of the ground $S_{1/2}$ state and the metastable $D_{5/2}$ state, which has a lifetime near 1 second. The two qubit states are connected by an electric quadrupole transition with a wavelength near $729\,nm$. The qubit states of the other two Cd ions formed the entangled pair. The teleportation process required only about $2\,ms$, and was verified with a fidelity near 75%. The teleportation was also accomplished with similar fidelity after a delay of $10\,ms$ following the Bell state production.

In the other experiment [1], trapping zones in a linear Paul trap permitted movement and separation of the Be ion qubits. Stimulated Raman transitions were used to implement single-qubit rotations, and also to generate entanglement by implementing a 2-ion phase gate (see Subsection 5.6.4). Auxiliary "spin-echo" pulses were periodically applied to the ions in one trapping region to reduce qubit state dephasing caused by slow variations of the magnetic field and by static field gradients. Ions 1 and 3 were prepared in a Bell singlet state, invariant under global rotations. Consequently, a rotation of all three ions enables any state for the "unknown" ion 2 qubit to be produced, without affecting the other ions. Ions 1 and 2 were placed in one trapping zone, while ion 3 was placed in a different trapping zone. The phase gate was implemented between ions 1 and 2 with fidelity > 90%, and the states of ions 1 and 2 were then individually detected using more ion manipulations in the traps. After obtaining the measurement results, selected unitary operations were applied to ion 3, based on the outcomes to produce the final state. The complete experiment required about $4\,ms$, including $1\,ms$ for cooling the three axial modes to the ground state, two ms to implement ion separations and movements, and three qubit measurements of $0.6\,ms$ each. An average fidelity of 78% for the teleportation process was achieved, well above the threshold of 2/3 for establishing the presence of entanglement in comparisons of the final state with the initial state.

The measurements show that teleportation of ion qubit states can be accomplished on demand, and without post-selection of data, as has been used in other teleportation measurements. This technique appears potentially useful in rapidly moving information over relatively large distances between qubits in a quantum computer.

5.10 Experimental DFS logic gates

Four-ion gates, in which the 4-ion entangled state

$$|\psi\rangle_4 = 2^{-1}\{|0\rangle|0\rangle|0\rangle|0\rangle - i|1\rangle|1\rangle|1\rangle|1\rangle\} \tag{5.66}$$

was demonstrated, have been discussed in Subsection 5.6.4. By considering the states as 2-ion DFS pairs in a decoherence free subspace, a related 4-ion state can be interpreted as an entangled state of two logic qubits, i.e.,

$$|\psi(4)\rangle_2 = 2^{-1}\{|0\rangle|1\rangle|0\rangle|1\rangle - i|1\rangle|1\rangle|1\rangle|1\rangle\}$$
$$= 2^{-1/2}\{|0\rangle|0\rangle - i|1\rangle|1\rangle\}|1\rangle|1\rangle. \tag{5.67}$$

For the logic qubits, this is the 2-ion equivalent of Eq. (5.66), which has also already been produced in a laboratory test [27]. These entangling gates can be used to perform arbitrary rotations of a single logic qubit as well as two logic qubits, so universal quantum logic can be performed. The phase shifts are now 2-ion phase shift *differences*, which depend only on the relative phase of a laser path difference between two ions, rather than the phase developed over macroscopic distance from the laser to the ions. A similar benefit applies to temporal phase fluctuations. The individual ion pairs making up the logic qubit must be initialized in states $|0\rangle|1\rangle$, with readout simply discriminating this state from $|1\rangle|0\rangle$ [27]. The dependence on phase differences means that DFS encoding also does not require clock synchronization between logic gates, required for individual ion qubits in parallel operations occurring in different trap regions.

5.11 Quantum error correction by ion traps

Recall our introduction to QECC in Section 2.9. Quantum error correction is based on encoding a general qubit superposition state $|\psi\rangle = a|0\rangle + b|1\rangle$ (with a, b unspecified) in the form of multiple correlated qubits [45], e.g., for three qubits the encoded qubit is $|\psi\rangle_E = a|000\rangle + b|111\rangle$, by majority voting. Suppose an error process inadvertently reverses the state of individual qubits with some (assumed low) probability p, resulting in erroneous states such as $|\psi\rangle_{EE} = a|010\rangle + b|101\rangle$, where the state of the second qubit has been reversed. By measuring *pairs* of qubits, in a way that information about a *particular* qubit is not revealed, one can determine if the 2-qubit states are the same or not. This produces an unavoidable disturbance (associated with any measurement) but it results in information of a better-defined state that can be more easily corrected if errors have occurred [10]. From sufficient 2-qubit measurements (depending on the number of encoded qubits) any error of the type described can be identified and corrected, without determining the coefficients a or b, which would destroy the coherent information stored in the encoded state.

For additional types of quantum errors, more correlated encoding qubits are required, with five or more qubits essential for correcting general errors. Multiple-qubit Calderbank–Shor–Steane (CSS) stabilizer codes [35], see (2.83) in Section 2.9, are useful for error correction in quantum computations of various sizes.

A three qubit quantum error correcting code (QECC) to protect against spin flip (qubit state) errors has been experimentally demonstrated [10] using 2-photon stimulated Raman laser pulses to manipulate the internal qubit hyperfine states of three previously cooled Be^+ ions confined in a linear, multiple-zone ion trap. This is the first deterministic measurement demonstrating either individual measurement or reinitializing the state of the ancilla qubits. The primary ion qubit was prepared in a state that was encoded into the primary and two ancilla qubits, using a 3-qubit variant of the phase gate discussed in Subsection 5.6.5. Errors were deliberately introduced in all qubits at various rates, and the encoded state was decoded back to the primary single qubit state plus the two ancilla states. The ancilla states were spatially separated in the ion trap from the primary qubit, and measured using state-dependent resonant laser pulses, as previously described. Based on the measurement outcome, the primary qubit state was corrected by applying the appropriate laser pulse. The error correction was verified by comparing the corrected final state to the uncorrected state, and to the initial state.

The quantum error correction is perfect only if at most one qubit experiences an error. The probability of more than one qubit error is $P_{>1} = p^3 + 3p^2(1-p)$ where $p = p(\theta) = \sin^2\theta/2$ for a single qubit error rotation θ (in these measurements). The fidelity of the corrected state is $F(\theta) = 1 - (3/4)|a|^2|b|^2\theta^4 + O(\theta^6)$, increasing quadratically in θ^2 for small errors θ, when $|a|^2|b|^2 \neq 0$, rather than linearly in θ^2 for the uncorrected state.

This measurement is the first step toward demonstrable *fault-tolerant* quantum computation using confined ion qubits.

5.12   Summary of ion quantum computation

A concept for a large scale quantum computer based on cold, confined ions has been presented, called a quantum charge coupled device, that is based on the Cirac–Zoller concept, but overcomes certain experimental limitations in that concept uncovered since its introduction. The QCCD uses an array of small linear rf traps constructed using lithographic techniques. The array is separated into regions devoted to specific operations associated with parallel quantum information processing. Limited numbers of ions are moved between regions, to transfer entangled information by quantum vibrations. Decoherence Free Encoding of ion pairs to form a logic qubit is suggested as a means to overcome or reduce decoherence. Experimental performance of ion transfer, separation, and coherence in a lithographic trap segment, and the use of previously demonstrated 2- and 4-ion gates to form a decoherence free subspace, are discussed. Transferring information over relatively large distances between separated qubits using quantum state teleportation has been demonstrated. The rudiments of a quantum error correction scheme, essential for large scale fault-tolerant quantum computation, have been experimentally demonstrated. Ion cooling using other ion types, or isotopes of the same type, is still essential during operation, but does not degrade qubit manipulations.

5.12.1   Assessment

The development of *experimental* quantum information processing techniques, using cold, condensed ions in arrays as qubits, has been summarized from demonstration of the initial proposal of Cirac and Zoller [13] to a concept for parallel processing in a large number of small, interconnected linear ion traps [27]. The progress of this development has been aided by new theoretical concepts for quantum gates, but driven by the conceptualization and implementation of experimental techniques, while addressing their experimental limitations, largely by the group at NIST-Boulder. The analysis of the present state of the development presented below is based on the work discussed earlier, but its elaboration is the opinion (and estimates) of the book's authors.

5.12.2   Qubits

To carry out a large scale quantum computation, a large number of operations must be executed within the coherence time of the system. The figure of merit of Subsection 5.5.1 is the expected system coherence time compared to the average duration of an entangling quantum operation, or gate. The duration of a 1-bit rotation is given by $T_1 = 2\pi/\Omega$, which can be executed in a time $\leq 1\mu s$ with acceptable fidelity, and hence is negligible compared to 2-qubit gate durations. The time required for a single-operation entangling 2-qubit quantum gate varies with the gate implementation protocol, the properties of the qubit ion, and also depends on the desired fidelity of the operation, which in turn depends on the requirements of error correction for fault-tolerant computation [46]. The threshold for fault-tolerant quantum computing requires relative gate errors $\leq 10^{-4}$.

It is assumed here that the qubit levels will be hyperfine levels in the ground term of an ion, since these levels have the longest potential coherence times. It is also assumed that a decoherence free subspace (DFS) will be used, to enhance the coherence of qubits that are moved about in a larger structure. Each qubit will consist of levels of two ions. Ground state hyperfine levels employed as the qubits require that 2-laser-beam stimulated Raman transitions will be used, because of the short wavelength requirement for gradient coupling of the computational qubit to the motional qubit. This in turn requires the use of relatively heavy ions, since otherwise spontaneous emission during the stimulated Raman gate operation limits coherence times to the point that error correction is unlikely to be successful. Presently Cd^+ appears to be a feasible choice for a qubit, using two ground state hyperfine levels of a suitable isotope [52]. Employment of the geometrical phase gate, which does not require carrier transitions, also reduces spontaneous emission decoherence.

Arguments for the use of 2-photon Raman transitions include the fact that only phase differences and frequency differences of the two Raman beams are important. These differences can be made very stable by using radio-frequency modulation of a single laser beam to produce the two separate beams. The two laser beams may be parallel, or at right angles, when exciting the qubit, depending on the type of gate. Laser beams with essentially the same path length are also less prone to developing phase shifts during transmission. For single photon transitions, the phase stability of the laser itself becomes the limiting factor, and of course any limits set by metastable level lifetimes will set a limit to computational coherence. Use of a DFS also favors Raman beam coherence, since then the phase and frequency differences of the Raman beams *at the site of the two ions* of the logic qubit are the important consideration, canceling path difference effects, etc. The demands placed on laser focusing are reduced when 2-ion qubits are employed. It may be feasible to use four hyperfine levels of a single ion [54] to create the DFS logic qubit, in which case coherence in ion transfer compared to two separate ions, and the task of keeping track of ion identity, would be simplified. On the other hand, to the extent that the frequencies of the two internal qubits differ (rather than, e.g., just the polarizations or wave vector differences of the exciting beams), complicating effects in qubit state detection and coherence might arise. Focusing the laser beams on a single ion would then be essential.

Use of Cd^+ hyperfine levels as the qubit states, compared to Be^+ ions, requires relatively more complex laser procedures. With the hfs of the $^{111}Cd^+$ isotope used for the qubit levels, the $^{116}Cd^+$ isotope can be used as the cooling ion [5]. The 5.2 GHz isotope shift between the resonance frequencies of these two ion masses is viewed as sufficient to make decoherence from spontaneous emission and ac Stark shifts negligible during cooling. This frequency difference is still within the range that electro-optical modulators (EOM) can generate both cooling and computational beams from the same input laser beam, without using additional laser sources [5, 29]. The parameters of the $^{111}Cd^+$ qubit are as follows [5]: the qubit levels are the $^2S_{1/2} |F = 1, m_F = 0\rangle$ and $|F = 0, m_F = 0\rangle$ levels, with a frequency separation $\omega_0 = 2\pi \times 14.53$ GHz. These levels have energies that depend only weakly on magnetic fields. The optical pumping to initialize the qubit, and qubit detection, are accomplished using $214.5\,nm$ radiation from a quadrupled Ti:Sapphire laser operating at $858\,nm$. The cycling transition has σ_- polarization, between $^2S_{1/2}, F = 1 \Leftrightarrow {}^2P_{3/2}, F = 2$.

The stimulated Raman beams for the gates are derived from a $458\,nm$ Nd:YVO$_4$ laser, phase modulated with a resonant EOM at 7.265 GHz, half the frequency separation of the qubit hyperfine levels [29]. The modulated laser frequency is then doubled. The beam is coupled into a build-up cavity with a BBO crystal for sum frequency generation. A comb of frequencies centered at $229\,nm$, separated by the modulation frequency, is produced, with the carrier detuned by 14 THz from the $^2P_{1/2}$ excited state. A Mach–Zehnder interferometer is used to modify the relative phases and amplitudes of the modulated beam in order to drive the desired stimulated Raman transition. Thus two distinct laser systems are required for Cd ion qubits.

5.12.3 Coherence

The present demonstrated coherence time for a ground state hfs transition is at best about 600 seconds, which was likely limited by uncontrolled magnetic field variations or by ion-atom collisions [6]. Recently, during teleportation measurements, short intervals of spin-echo excitation have been used to rephase the qubit states, which dephase due to local magnetic field fluctuations or gradients [1]. At the pressure near 2×10^{-11} Torr achieved in the small linear trap array, intervals between close *elastic* ion-atom collisions should exceed 10^3s, so using a DFS to limit magnetic decoherence, 10^3s seems a not-unreasonable estimate for a feasible internal qubit coherence time in a room temperature vacuum system. A complete calculation by a quantum computer would need to be executed in this time. If ion-molecule collisions ultimately limit coherence, a cryogenic vacuum system can produce lower pressures.

In the above estimate, the coherence of the motional qubit was not considered. At present, the coherence time for a CM normal mode of oscillation is about 10 ms for Be ions, but a CM coherence time for a heavier single Hg ion of $150\,ms$ has been reported [17]. The coherence time is taken here to be the heating time for the absorption of a single quantum between the ground and first excited motional qubit state. There seems no need to preserve phase coherence in the motional qubit for times much longer than the times needed to transfer information from one internal qubit to another through gate operations. For a linear string of ions, this time is just a few gate times, but if a QCCD (Section 5.7) is used for a large scale quantum computer, this time may be as long as an ion separation time, which presently has been reduced to a couple of milliseconds [1]. Ion separation was earlier found to heat the motion substantially, by about 150 quanta on average, with a mean spread of about half that, but improved separation technique has already reduced heating of the CM mode for two ions to an average of one quantum, and no heating of the symmetric stretch mode was observed. This means that using sideband cooling, only a few cooling quanta will be needed to reasonably expect the ion motion to be cooled back down. Assuming this cooling can be done at a rate $\eta\Omega \approx 5 \times 10^6$ photons/s, a cooling time of only about 50 μs is needed. In sum, the present measured motional coherence times are sufficient for expected operations, but since the ion heating is not presently understood, there is much room for improvement, and improvements can be anticipated [39]. In any event, using cooling following some operations is presently essential, but fortunately this cooling can be accomplished in times well less than ion separation operations. The cooling would be done sympathetically using ^{116}Cd ions having a shifted resonance frequency, so that spontaneous emission of cycling photons does not excite the qubit ions, destroying coherence. To avoid requirements of populations remaining in the lowest and first vibrational levels, it seems likely that quantum gates able to function with the vibrational modes of "warm" ions will be used, so perfect cooling to the zero point vibration is not essential.

5.12.4 Gates

It is now appropriate to estimate feasible 2-qubit gate times. Theoretical calculations for swap gates and for wave packet gates, plus measurements using Ca$^+$ ion qubits, have been completed [47]. An imprecision ε is defined in terms of the minimum fidelity F (Eq. (5.63)) as $\varepsilon = (1 - F)^{1/2}$. This definition assumes no undefined imperfection in the execution dynamics, such as noise on beams, etc., nor any difference of the laser-induced gate from an ideally defined gate, but rather an imprecision due to unavoidable effects occurring in the planned execution of the gate, not included in the Hamiltonian. Examples are Stark shifts, or off-resonance radiation coupling out of the computational basis. For a swap gate, the maximum execution rate is found to be limited by [47]

$$(T_S)^{-1} \leq 2^{3/2}\varepsilon\{(E_R/Nh)(\omega_z/2\pi)\}^{1/2} \quad \text{with} \quad \epsilon \approx \Omega/2^{1/2}\omega_z, \qquad (5.68)$$

where E_R/h is the r-photon recoil frequency, with $E_R = (r\hbar k_z)^2/2M$ and an N ion string. The number r is a positive integer and for a stimulated Raman transition, $r = 2$. The rate for a wave packet gate, induced on the carrier transition, is a factor of two smaller than (5.68), assuming that $\Omega \ll \omega_z$. For the wave packet gate, $\varepsilon = 2^{1/2}\eta\Omega/\omega_z$. The CNOT wave packet gate is executed by driving the carrier transition with a single pulse, and so should be relatively fast and free from Stark shifts, as discussed in Subsection 5.6.3.

For the wave packet gate in [15] using one Be ion, $\eta = 0.359, \omega_z/2\pi = 3.4$ MHz, $\Omega/2\pi = \Omega_{0,0}/2\pi \exp(-\eta_s^2) = 104$ kHz and $E_R/h = 452$ kHz [15]. From this one calculates $\varepsilon = 1.6 \times 10^{-2}$ and $T_{WP} = 36\mu s$ using Eq. (5.68) (divided by 2). The parameters require $T_{WP(m)} = 11\mu s$ and using an estimate that $< .05\%$ of the population moved outside the basis by off-resonant excitation [15], $\varepsilon_m = 2.3 \times 10^{-2}$, in approximate agreement with the theory. For two Cd^+ DFS qubits, with the same values for all parameters except $E_R/h = 1.5 \times 10^5$, T_{WP} is $60\,\mu s$ with $\varepsilon = 1.6 \times 10^{-2}$. The swap gate limit is twice as long, for the same ε. Duration limits on other types of gates are expected to scale similarly from the qubit durations for Be ions, and to be proportional to ε as well.

Execution of a geometric 2-ion phase gate is formally the same [30] as execution of a Sørensen–Mølmer gate discussed in Subsection 5.6.4. Individual ion addressing is not required during the gate, reducing laser focus requirements. The ions may be separated to execute single ion gates. The accumulated phase during the gate does not depend on the exact starting state distribution, path shape, or time of execution, but rather on the accumulated path area $A/\hbar$. This means that the ions need not be cooled to the motional ground state for accurate gate operation, as noted above.

The geometrical phase gate uses the same interaction strength as the Cirac–Zoller and Sørensen–Mølmer gates, so for a given laser intensity, its speed is about the same [30]. To achieve a given fidelity, as discussed above for swap and wave packet gates, the phase gate is an improvement, since spin flips are not required for gate execution. Limits to fidelity due to off-resonant spin-changing carrier transitions are absent. The main source of gate error has been found to be fluctuations in δ, (the relative detuning from the motional levels), and fluctuations in the Raman beam intensity, both near 1%. If frequency drift and intensity relative errors can be reduced to about 10^{-3} in future work, the expected relative gate error is 10^{-4} with comparable gate speeds. An error probability of 10^{-4} is the asymptotic threshold for fault-tolerant quantum computation, see Subsection 5.12.5 below.

5.12.5 Computation

With estimates of feasible gate times and coherence times, the efficacy of a quantum calculation using confined ions can be considered. Estimates of (error-free) memory used in a quantum computer, and the computation time, have been made in order to execute Shor's algorithm (Subsection 2.7.5) to find the prime factors of a K-bit number N [2]. The calculation is based on NOT and CNOT gates, using $O(K^3)$ gate operations. Including scratch qubits, $5K + 1$ logic qubits were found to be required.

About $Q = 72K^3$ elementary gate operations must be executed to evaluate the modular exponential function $x^a \bmod N$ (where a is a number with

$\approx 2K$ bits), while only a nearly negligible $K(2K - 1)$ 2-qubit gates plus 2K single qubit rotations were needed for the 2K-qubit Fourier transform. These are the two main parts of Shor's factoring algorithm.

To factor a 9 decimal digit number, $K = 30$ logical ion qubits in a single string are required to store the number. About $5K + 1 = 151$ qubits in a single string would be needed for the overall computation, undergoing at least $Q = 2 \times 10^6$ gate operations overall. Estimating that all of these operations are 2-qubit entangling gates, executing them in series would require about 300 s, assuming a $150\,\mu s$ gate duration using a geometrical phase gate (which depends on the choice of ion (Cd^+) and ε through calculation fidelity). Measurement times for Cd^+ qubits at the end of the calculation require about $200\,\mu s$ each, contributing negligibly to the total time if performed largely in parallel. Compared to the $\approx 10^3 s$ of the expected coherence time, a computation of this size appears feasible, neglecting problems associated with the long qubit string. However, if the QCCD architecture (Section 5.7) is used, the time to transport and separate qubits must be also included [39], and of course no provision for error sensing or correction has been made.

Assuming a DFS with two ions per logic qubit, at least 300 ions would need to be stored for the $5K + 1 = 151$ qubits. These qubits would be both memory and gate qubits, and since the manipulating operations heat the qubits, additional cooling ions are essential. Suppose a string of about 6 logic qubits are used for memory, with a single cooling qubit (of a different isotope or ion type) per string. Two logic qubits interacting in a gate region will also require one cooling ion. One might estimate there will be about 23 memory regions and 23 gate regions, with an additional ≈ 50 cooling ions for a total of about 350 ions.

The present segment length for a trap region is about 0.4 cm, so a square array of micro-traps less than 3 cm on a side is sufficient. Suppose each 2-qubit gate requires two transport and two ion separation operations. Presently, an adiabatic transport requires about $50\,\mu s$, and ion separation times are now about $2\,ms$. Let us optimistically assume that ion separation times can be reduced to about $200\,\mu s$ by further experimental work. Then an additional $600\,\mu s$ would be required per gate operation (including $100\,\mu s$ for two qubit re-coolings). The total calculation time (for all operations in series) becomes about 1500 s, exceeding a system coherence time. An assumed dramatic experimental reduction in ion separation time was clearly crucial. Benefits may be achieved using information transfer based on teleportation, but the separated entangled states will need to be physically distributed.

For this factoring calculation, $Q \approx 2 \times 10^6$ elementary gate operations were required on the $K \approx 30$ logic qubits of the number N without error, an optimistic assumption indeed. A relative precision $(KQ)^{-1} = 2 \times 10^{-7}$ is needed to successfully complete the calculation [45]. Error correction procedures are clearly necessary to produce meaningful results.

Steane has assumed a hierarchical computer design to analyze fault-tolerant quantum error corrections [50]. He assumes *memory groups* ≤ 10 qubits, with gates between any qubit pair. End qubits communicate with nearest neighbors. About ten such groups form a *block*. This structure so far is similar to that of the proposed QCCD [27]. Each block is coupled to other blocks by another physical mechanism, which allows communication over relatively long distances, such as teleportation. Switching networks may be needed to allow each block to communicate with all the others [46].

It is feasible to encode a general unknown qubit state in a set of five other qubits, in a way that the originally stored state can be precisely recovered following any unknown change in the state of one of the qubits. Fault tolerant methods have a good probability of success, even when all operations are imperfect [45].

Now assume a quantum computer for the previously discussed computation, but with error correction. The computer is assumed to have $K = 500$ logic qubits encoded in K/k blocks [45], where k is the number of logic qubits per block. Each block has n physical qubits. There are an additional $2n_{rep}$ ancilla blocks of n physical qubits, and $2n_{rep}$ sets of verification bits containing $(n+k)/2$ physical qubits. Steane has run Monte Carlo evaluations of various error correcting codes, including the Hamming $\{7, 1, 3\}$ and Golay $\{23, 1, 7\}$ CSS codes, which have a general utility for large scale-ups. He used $n_{rep} = 1$, the number of pairs of ancilla blocks per data block that can be prepared in parallel. The total number of logic qubits is now $N_q = (n + n_{rep}(3n + k))K/2k \approx 4500$ for a scale-up of only 9. Gates act in parallel during most gate executions of the generation and verification network. It is assumed that the same relative error rate $\gamma = 10^{-4}$ for errors in rotations, 2-qubit gates, initialization of the qubits, and qubit measurement. If the memory error rate is assumed to be $\gamma/10$, which is likely optimistic, since two swap gates must be used to transfer data between memory qubits, then from [45, Fig. 5b], the BCH $\{127, 43, 13\}$ code seems essential for KQ near 10^8.

Since the gates for the error correction often act in parallel, the increase in computation time may not be all that large. The computer architecture would now require about 380 trap segments, and be a square about 8 cm on a side, still convenient for vacuum work. Clearly, fault-tolerant error correction will greatly increase the size of any computation.

5.12.6 Summary

In summary, it appears that the presently *demonstrated* cold ion quantum computing technology (with the exception of trap operations separating ions) has approached the requirements for a large scale quantum computation. The fidelity of some types of quantum gates is marginal at present, but the literature often blames laser power stability for the shortcomings, which could presumably be significantly improved over that available with the present scientific lasers by a determined engineering effort. One may anticipate further improvements in ion motional coherence, and in the time and coherence of manipulation of ions within the QCCD storage devices, but the laboratory time scale to achieve these required improvements cannot be estimated, since basic insights are still lacking and improved techniques are unpredictable. Communication between widely separated parts of the computer is essential for computation with teleportation the likely means, but the details are not yet specified. On the other hand, experimental quantum computing research is only about nine years old, and impressive progress has been achieved. But it should be kept in mind that the overall confined ion-laser cooling research, which forms the backbone of this effort, has been developed over a longer period of twenty-five years. The laboratory development of the work discussed here, coupled with innovations in theory, shows that the choice of an eventual quantum computing technology should be based on experiment, rather than only on theory.

5.12.7 Outlook

It seems clear that with reasonable approximations of the existing technology, cold ions in traps could potentially be used for limited large-scale fault-tolerant computations ($KQ \approx 10^6$). The present bottlenecks are the movement of ions required by the proposed architecture of the QCCD computer, and the stability of laser pulses. On the other hand, this type of architecture is compatible with a hierarchical computer structure [50], which seems essential as the scale of computation increases. Such a hierarchical structure requires some type of communication between separated blocks of information storage and manipulation, which may be supplied by the recent demonstrations of teleportation.

Other gates have been proposed [26], but not experimentally evaluated, using the Raman sideband technique. A recently proposed 2-qubit gate [20] is claimed to use a sequence of fast laser pulses to execute a phase gate on stored ion qubits, with a duration $T = 1.08 \times 2\pi/\omega_z$. If these gates do not exhibit appreciable loss of fidelity, then gate times can be reduced by a factor of 100 or more from present values. However, the limits to computation time would still be set by the much longer times to transport and separate ions in the QCCD, unless dramatic improvements are forthcoming. The figure of merit can be increased to about 10^9 only if all parts of the general gate operation can be completed in about a μs. This would permit calculations with KQ of about 10^{13} (neglecting error correction) in a single device of reasonable physical size in room temperature vacuum. It still falls well short of KQ needed for the important task of factoring a 130 digit number into primes, however.

At present, cold confined ions provide a valuable and interesting technology for quantum computers, which appears worthy of much further study. This technology permits theoretical estimates of success, since it is well-defined and understood. Major new advances will be necessary to reach the ultimate goals of quantum computation on very large numbers.

References

[1] M.D. Barrett, J. Chiaverini, T. Schaetz, J. Britton, W.M. Itano, J.D. Jost, E. Knill, C. Langer, D. Liebfried, R. Ozeri, and D.J. Wineland, *Nature* **429** (2004), 737.

[2] D. Beckman, A.N. Chari, S. Devabhaktuni, and J. Preskill, Efficient networks for quantum factoring, *Phys. Rev.* **A54** (1996), 1034.

[3] J.C. Bergquist, J.J. Bollinger, W.M. Itano, C.R. Monroe, and D.J. Wineland, (eds)., *Trapped ions and laser cooling, Vol. IV*, NIST Technical Notes **1380** (U. S. Department of Commerce, 1996); plus earlier and later volumes of this series.

[4] J.C. Bergqusit, W.M. Itano, and D.J. Wineland, *Frontiers in Laser Spectroscopy*, T.W. Hänsch and M. Inguscio, (eds.), (North-Holland, Amsterdam, 1994) p. 359.

[5] B.B. Blinov, L. Deslauriers, P. Lee, T. Madsen, R. Miller, and C. Monroe, Sympathetic cooling of trapped Cd^+ isotopes, *Phys. Rev.* **A65** (2002), 040304.

[6] J.J. Bollinger, D.J. Heinzen, W.M. Itano, S.L. Gilbert, and D.J. Wineland, *IEEE Trans. Instrum. Measurement* **40** (1991), 126.

[7] B.H. Bransden and C.H. Joachain, *Physics of Atoms and Molecules*, (Longman, London, 1983).

[8] P. Carruthers and M.M. Nieto, Coherent states and the forced quantum oscillator, *Am. J. Phys* **7** (1965), 537.

[9] G. Chen, D.A. Church, B.-G. Englert, and M.S. Zubairy, Mathematical models of contemporary elementary quantum computing devices, *CRM Proceedings and Lecture Notes* **33** (2003), Amer. Math. Soc., Providence, Rhode Island, 77–116.

[10] J. Chiaverini, D. Liebfried, T. Schaetz, M.D. Barrett, R.B. Blakestad, J. Britton, W.M. Itano, J.D. Jost, E. Knill, C. Langer, R. Ozeri, and D.J. Wineland, Realization of quantum error correction, *Nature* **432** (2004), 602.

[11] D.A. Church, Storage ring ion trap derived from the linear quadrupole radio-frequency mass filter, *J. Appl. Phys.* **40** (1969), 3421.

[12] D.A. Church, *Physics Reports* **228** (1993), 253.

[13] J.I. Cirac and P. Zoller, Quantum computation with cold, trapped ions, *Phys. Rev. Lett.* **74** (1995), 409.

[14] H.G. Dehmelt, *Adv. At. Mol. Phys.* **3** (1967), 53.

[15] B. DeMarco, A. Ben-Kish, D. Liebfried, V. Meyer, M. Rowe, B.M. Jelenkovic, W.M. Itano, J. Britton, C. Langer, T. Rosenband, and D.J. Wineland, Experimental demonstration of a controlled-NOT wavepacket gate, *Phys. Rev. Lett.* **89** (2002), 267901.

[16] R.G. DeVoe, Elliptical ion traps and trap arrays for quantum computation, *Phys. Rev.* **A58** (1998), 910.

[17] F. Diedrich, J.C. Bergquist, W.M. Itano, and D.J. Wineland, *Phys. Rev. Lett.* **62** (1989), 403.

[18] D.P. Divincenzo, The physical implementation of quantum computation, in *Scalable Quantum Computers*, S.L. Braunstein and H.K. Lo, (eds). (Wiley-VCH, Berlin, 2001) pp. 1–11.

[19] E. Fischer, *Z. Phys.* **156** (1959), 1.

[20] J.J. Garcia-Ripoll, P. Zoller, and J.I. Cirac, Fast and robust 2-qubit gates for scalable ion trap quantum computing, quant-ph/0306006 v1.

[21] R.J. Glauber, Coherent and incoherent states of the radiation field, *Phys. Rev.* **131** (1963), 2766.

[22] D. Gottesman and I.L. Chuang, *Nature* **402** (1999), 390.

[23] L. Gruber, J.P. Holder, J. Steiger, B.R. Beck, H. DeWitt, J. Glassman, J.W. McDonald, D.A. Church, and D. Schneider, Evidence for highly charged ion Coulomb crystallization in mixed, strongly-coupled plasmas, *Phys. Rev. Lett.* **86** (2001), 636.

[24] S. Ichimaru, Strongly coupled plasmas, high density classical plasmas, and degenerate electron liquids, *Rev. Mod. Phys.* **54** (1982), 1017.

[25] W.M. Itano and D.J. Wineland, Laser cooling of ions stored in harmonic and Penning traps, *Phys. Rev.* **A25** (1982), 35.

[26] D. Jonathan and M.B. Plenio, Light-shift-induced quantum gates for ions in thermal motion, *Phys. Rev. Lett.* **87** (2001), 127901.

[27] D. Kielpinski, C. Monroe, and D.J. Wineland, Architecture for a large-scale ion-trap quantum computer, *Nature* **417** (2002), 709–711.

[28] L. Landau and E.M. Lifshitz, *Mechanics*, Trans. by J. B. Sykes and J. S. Bell (Addison-Wesley, Reading, MA, 1960) pp. 93–95.

[29] P.J. Lee, B.B. Blinov, K. Brickman, L. Deslauriers, M.J. Madsen, R. Miller, D.L. Moehring, D. Stick, and C. Monroe, Atomic qubit manipulation with an electro-optic modulator, quant-ph/0304188 v1.

[30] D. Liebfried, B. DeMarco, V. Meyer, D. Lucas, M. Barrett, J. Britton, W.M. Itano, B. Jelenkovic, C. Langer, T. Rosenband, and D.J. Wineland, Experimental demonstration of a robust, high-fidelity geometric two ion-qubit phase gate, *Nature* **422** (2003), 412.

[31] G.J. Milburn, S. Schnieder, and D.F. James, Ion trap quantum computing with warm ions, *Fortschr. Physik* **48** (2000), 801.

[32] K. Mølmer and A. Sørensen, *Phys. Rev. Lett.* **82** (1999), 1835.

[33] C. Monroe, D.M. Meekhof, B.E. King, W.M. Itano, and D.J. Wineland, *Phys. Rev. Lett.* **75** (1995), 4714.

[34] C.J. Myatt, B.E. King, Q.A. Turchette, C.A. Sackett, D. Kielpinski, W.M. Itano, C. Monroe, and D.J. Wineland, *Nature* **403** (2000), 269.

[35] M.A. Nielsen and I.L. Chuang, *Quantum Computation and Quantum Information*, (Cambridge University Press, Cambridge, 2000).

[36] W. Paul, Electromagnetic traps for charged and neutral particles, *Rev. Mod. Phys.* **62** (1990), 531–540.

[37] M.G. Raizen, J.M. Gilligan, J.C. Bergquist, W.M. Itano, and D.J. Wineland, *Phys. Rev.* **A45** (1992), 6493.

[38] M. Riebe, H. Haeffner, C.F. Roos, W. Haensch, J. Benheim, G.P.T. Lancaster, J.W. Koerber, C. Becher, F. Schmidt-Koler, D.F.V. James, and R. Blatt, *Nature* **429** (2004), 334.

[39] M.A. Rowe, A. Ben-Kish, B. DeMarco, D. Liebfried, V. Meyer, J. Beall, J. Britton, J. Hughes, W.M. Itano, B. Jelenkovic, C. Langer, T. Rosenband, and D.J. Wineland, Transport of quantum states and separation of ions in a dual rf ion trap, *Quant. Inform. Computation* **4** (2002), 257.

[40] C.A. Sackett, D. Kielpinski, B.E. King, C. Langer, V. Meyer, C.J. Myatt, M. Rowe, Q.A. Turchette, W.M. Itano, D.J. Wineland, and C. Monroe, Experimental entanglement of four particles, *Nature* **404** (2000), 256.

[41] S. Schneider and G.J. Milburn, *Phys. Rev.* **A57** (1998), 3748.

[42] A. Sørensen and K. Mølmer, Quantum computation with ions in thermal motion, *Phys. Rev. Lett.* **82** (1999), 1971.

[43] A. Sørensen and K. Mølmer, Entanglement and quantum computation with ions in thermal motion, *Phys. Rev.* **A62** (2000), 022311.

[44] A. Steane, *Appl. Phys.* **B64** (1997), 623.

[45] A.M. Steane, Overhead and noise threshold of fault tolerant quantum error correction, quant-ph/0207119 v4.

[46] A. Steane, Quantum efforts corrected, *Nature* **432** (2004), 560.

[47] A. Steane, C.E. Roos, D. Stevens, A. Mundt, D. Liebfried, F. Schmidt-Kaler, and R. Blatt, Speed of ion trap quantum-information processors, *Phys. Rev.* **A62** (2000), 042305.

[48] Q.A. Turchette, D. Kielpinski, B.E. King, D. Liebfried, D.M. Meekhof, C.J. Myatt, M.A. Rowe, C.A. Sackett, C.S. Wood, W.M. Itano, C. Monroe, and D.J. Wineland, Heating of trapped ions from the quantum ground state, *Phys. Rev.* **A61** (2000), 063418.

[49] D.F. Walls and G.J. Milburn, *Quantum Optics*, Springer, Berlin, 1994.

[50] X. Wang, A. Sørensen, and K. Mølmer, Multibit gates for quantum computing, *Phys. Rev. Lett.* **86** (2001), 3907.

[51] D.T. Wilkinson and H.R. Crane, *Physical Review* **130** (1963), 852.

[52] D.J. Wineland, M. Barrett, J. Britton, J. Chiaverini, B. DeMarco, W.M. Itano, B. Jelenkovic, C. Langer, D. Liebfried, V. Meyer, T. Rosenband, and T. Schaetz, Quantum information processing with trapped ions, quant-ph/0212079 v2 (2003).

[53] D.J. Wineland and W.M. Itano, Laser cooling of atoms, *Phys. Rev.* **A20** (1979), 1521.

[54] D.J. Wineland, C. Monroe, W.M. Itano, D. Liebfried, B.E. King, and D.M. Meekhof, Experimental issues in coherent quantum state manipulation of trapped atomic ions, *J. Res. Natl. Inst. Stand. Technol.* **103** (1998), 259.

Chapter 6

Quantum Logic Using Cold, Confined Atoms

- Atoms trapping and interactions

- Optical lattices and atom chips

- Various designs for qubits and universal quantum gates

- Data tables

- Assessment

6.1 Introduction

The principal arguments for using cold, confined atoms to implement quantum computation are essentially the same as those for using cold, confined ions: *long coherence times, interactions that are theoretically well-defined, high isolation of the qubits* and *scalability*. It is reasonable to suppose that perturbations of an atomic system should be even less than in an ionic system, since ion charges interact strongly with the surrounding neutral environment, while neutral systems interact very weakly with that environment. Of course, this weak interaction between neutrals also presents challenges to the entanglement of two qubits, which is required to implement a quantum gate. Nevertheless, there has been progress both theoretically and experimentally in logic implementation.

The perturbation-free confinement of qubits in well-defined locations is a necessity for quantum logic using atoms. Atoms can be cooled to low temperatures using lasers in the same way that ions are cooled, and there are additional laser cooling techniques that are more effective with atoms. This has resulted in atoms cooled to micro-Kelvin temperatures, where they exist in the ground state of confining wells. Recent successes with Bose-Einstein condensation (BEC) of atoms into a coherent, cold multi-atom quantum state have aided the loading of arrays of atom traps. There is now also interest in the employment of condensates in quantum logic, but the progress in this topic is poorly defined at present.

Atom confinement and manipulation are accomplished in several rather different ways. Arrays of traps for cold atoms have been created using laser standing waves [20]. The relative positions and separation of these traps can be adjusted by changes in the angles or polarizations of the laser beams, while the depth of the traps is controlled by the laser intensity. Another technique confines atoms near the surface of a microchip [16, 41]. There, cold atoms can be confined, transported, and reflected by electric and magnetic fields created by microstructures that are carrying currents and voltages. A conceptually separate way of performing similar functions using laser light is to use micro-fabricated optical elements [4]. Micro-focused laser beams can be combined with microscopic fields near surfaces to form traps and to manipulate atoms to execute quantum gates.

The execution of single atom quantum gates can be accomplished using microwave or laser fields, and several rather different techniques for coupling two atoms have been proposed for entangled 2-qubit gates.

In the following sections, the interactions of atoms with radiation or fields that are used to confine or transport cold atoms are briefly described, followed by descriptions of the existing optical and microchip technologies that are used to implement the atom confinement and desired operations. With this background, several theoretical proposals to implement entangling gates are briefly discussed. These gates are based on particular interactions of cold atoms at absolute temperature $T = 0$, and also for finite T. Experimental progress in implementing gates is currently lagging behind theory; this is why gate operations are modeled mainly theoretically. Finally, potential problems are addressed, some data tables and a brief assessment of progress are presented.

6.2   Preparation and detection

Due to the electric neutrality of atoms, their coherent manipulation usually involves interactions much more feeble than in the case of ions. It is thus not surprising that typical trapping potentials are substantially shallower than what could be easily achieved with charged species. Loading an atom trap but with particles from the Maxwellian low-energy tail of a thermal sample would lead to phase-space densities too low for most experiments and it has thus become mandatory to develop cooling techniques suitable for neutral atoms[1]. It is not our intention here to describe the often quite subtle physical details of the many methods in existence, inasmuch

---

[1]If the particle source is a thermal beam, then the atoms do not only need to be cooled, but also first to be stopped. Cf. [2] for a discussion of such methods.

as from the point of view of a quantum informatics practitioner, sufficiently
cool atoms are a factual prerequisite and not a problem to be solved anew.
Nonetheless, it is our opinion that whoever works in the field should at least
be familiar with the basic notions involved. Their importance can hardly be
overestimated. The so-called Doppler cooling of atoms was first proposed by
two Nobel prize winners, Hänsch (2005) and Schawlow (1981), in their seminal paper [23]. The essence of the method is similar to the ionic version
already described in Chap. 5 (Subsection 5.2.2). The reader may leap to
Section 6.8 and see Table 6.3 there, whose fifth and sixth columns provide
a feeling for the minimal temperatures and average velocities that can be
achieved by applying the Doppler cooling concept to neutrals. Precise experimental tests, however, soon demonstrated that the actually measured
temperatures of laser cooled samples could be much lower than the values
obtainable by the Doppler cooling process alone. It was immediately clear
that a much more efficient—yet unpredicted—cooling process was taking
place. Comparison with more elaborate theoretical models demonstrated
that it was the interaction of polarization gradients with the atoms' magnetic (Zeeman) sublevel structure which caused this unexpected cooling.
Indeed, these new phenomena allow one to cool atoms down to temperatures corresponding to the recoil velocity suffered by the atom under emission of a single resonant photon. Under these circumstances, the atomic de
Broglie wavelength becomes of the order of the employed laser wavelength.
The last two columns again of Table 6.3 list the values of temperatures and
velocities at this (so-called) recoil limit. This pioneering work was honored
with the 1997 physics Nobel prize (Chu, Cohen–Tannoudji and Phillips; cf.
also Subsection 3.3.1). The three review papers [8, 10, 40] by the laure-
ates provide excellent introductions into the subject and into its fascinating history. They also contain discussions of various approaches that lead to
sample temperatures well below the recoil-limit ("sub-recoil cooling"). New
methods have been added since and the field is in constant progress. Even
the sideband-cooling technique that had long been in use with ions (based
on laser Raman scattering, as described in Chap. 5), became available for
pre-cooled atoms trapped in strongly detuned optical lattices..

Achieving temperatures low enough to load but the ground state of traps
intended for quantum information processing is one aspect of the problem.
The controlled coherent transfer of atoms from a cooling/precooling region
to the proper trapping area can be achieved by carefully monitoring and
adiabatically varying external "atom optical" potentials, which are mostly
of the sort discussed in the next section. For example, microscopic traps on
atom chips can be loaded starting from a magnetic trap at a few hundred
microns above the chip and then ramping up currents on the chip to shift
the trap towards the surface [16]. For a number of gates to be discussed
below, atom traps must be loaded with exactly one atom per site. This has
been achieved experimentally in tightly focused laser beams [43] and in optical lattices [1]. One essentially monitors the amount of light scattered by
the atoms, which is proportional to the number of atoms, while atoms are

slowly leaking out of the trap, leaving it ultimately one by one. By cali-
brating the amount of scattered light, the trap parameters can be changed
to increase the trap lifetime for the last atom, just after the before-the-last
atom leaves the trap. Detecting single atoms is possible by collecting the
light they emit or scatter [13]. Trapping sites can be individually resolved
even if their spacings are below the wavelength of the emitted light (i.e.,
beyond the usual resolution of microscopes) by applying, e.g., a magnetic
field gradient that shifts the emission frequencies in a position-dependent
way so that spatial information is encoded in the spectrum [45]. On an
atom chip, optical fibers can be used to form micron-sized cavities where
the atom-light interaction is enhanced so that detection at the single-atom
level becomes feasible [28].

6.3   Atom interactions with external fields

The interactions of an atomic electric dipole moment d with electric (E)
fields, or a magnetic dipole moment μ with magnetic (B) fields enables cold
atom confinement in laser beams, in space, or near surfaces, and also the
guiding and reflection of moving atoms. These dipole interactions are far
weaker than those of a separated charge with the corresponding fields, and
consequently provide better isolation and coherence for the atomic state.
On the other hand, the confining potential wells are shallow, so the atom
temperature must certainly be small. Fig. 6.1 compares electric and magnetic interactions in atoms.

6.3.1   Electric interaction

An electric dipole moment d consists of opposite-signed charges separated
by a small distance. There are no observed permanent electric dipole moments in atoms; however, an electric field will induce a dipole moment in an
atom. The electric dipole interaction energy can be written $H_{el} = -d\cdot E(r,t)$,
where the field is evaluated at the position r of the center of charge. This
interaction Hamiltonian is actually an approximation, the "electric dipole"
or "long wavelength" approximation, and it works well into the ultraviolet
spectral region. This is because the wavelength λ of the electric field of
a laser beam is much greater than the induced dipole charge separation,
which is smaller than the atomic radius. The observed induced dipole mo-
ment $\langle d(t)\rangle = \alpha(\omega)E(r,t)$ in a weak field, where $E(r,t) = E(r)e^{-i\omega t} + c.c.$ for
a time-dependent laser electric field. Based on lowest order perturbation
theory, the polarizability tensor is

$$\alpha(\omega) = \sum_{n\neq g} \frac{2\omega_{ng}}{\hbar(\omega_{ng}^2 - \omega^2)} \langle g|d|n\rangle\langle n|d|g\rangle. \tag{6.1}$$

In this equation, g denotes the ground state and n denotes excited states,
with $\hbar\omega_{ng}$ the energy difference between these states. The quantities $\langle g|d|n\rangle$

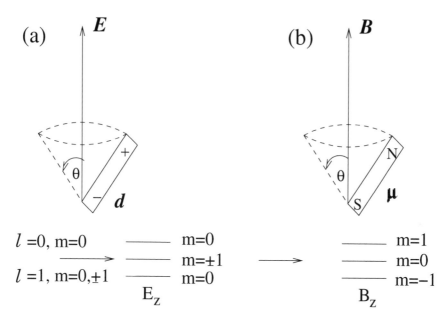

Figure 6.1: An *induced* atomic electric dipole moment d interacting with the electric field E (a) is analogous to a *permanent* atomic magnetic dipole moment μ interacting with a magnetic field B (b). The linear Stark effect describes the energy splitting of the $n = 2$ state of atomic hydrogen in an electric field, while the Zeeman effect describes the splitting of the level in a magnetic field. This splitting is illustrated by the horizontal lines at the bottom of the figure.

are matrix elements which contain the details of the specific atomic wave functions $|g\rangle$ or $|n\rangle$ and the electric dipole moment. Of interest here, $\alpha(\omega)$ is a scalar for the spherical ground state characteristic of alkali atoms (i.e., a tensor proportional to the unit matrix). Note that if the laser frequency $\omega < \omega_{ng}$ (red detuning), then α is positive, while if $\omega > \omega_{ng}$ (blue detuning), then α is negative. In general, those particular excited states n dominate the interaction that lead to the smallest denominator for a chosen laser frequency ω.

The induced, time-averaged interaction called the ac Stark shift of the atomic ground state is found from second order perturbation theory to be

$$V_{el}(\boldsymbol{r}) = -\langle \boldsymbol{d}(t) \cdot \boldsymbol{E}(\boldsymbol{r}, t)\rangle/2 = -\text{Re}[\alpha(\omega)]|E^2(\boldsymbol{r})|. \tag{6.2}$$

Since the force $F_e = -\nabla V_{el}(\boldsymbol{r})$, an inhomogeneous electric field can be used to guide, reflect, or trap atoms. Depending on the sign of $\alpha(\omega)$, atoms will be attracted to an intensity maximum (red detuning) or minimum (blue detuning) of the electric field. Of course, the electric field can also be static ($\omega = 0$), an extreme case of red detuning, which attracts atoms toward the intensity maximum of the electric field. Optical traps have recently been reviewed in [19].

Example 6.3.1. *DC Stark effect.* Strong electric dipole moments can be induced by a static electric field if atomic wavefunctions with opposite parity are nearly degenerate in energy (the so-called dc Stark effect). A simple example is illustrated in Fig. 6.1 for the hydrogen atom: the states with principal quantum number $n = 2$ and angular momentum quantum number $\ell = 0$ (s-state) and $\ell = 1$ (p-state) are very close in energy. The interaction of the electron with the applied electric field E_z (along the z-axis) is described by

$$H_{\text{Stark}} = -eE_z z = -eE_z r \cos \theta, \tag{6.3}$$

where $-E_z z$ is the electrostatic potential corresponding to this field (assumed homogeneous across the atomic extension) and $z = r \cos \theta$ the z-coordinate of the electron relative to the nucleus. Among the four states in the $n = 2$ manifold, $|2s, m = 0\rangle$ (singlet) and $|2p, m\rangle$ ($m = -1, 0, +1$) (triplet), (see Box 7.2 in Chap. 7 for explanations of singlets and triplets) H_{Stark} has nonzero matrix elements only between

$$\langle 2s0|H_{\text{Stark}}|2p0\rangle = \langle 2p0|H_{\text{Stark}}|2s0\rangle = V_{sp} = DE_z, \tag{6.4}$$

as can be easily checked from the hydrogen wave functions given in Example 2.2.1. The coefficient D is of the order of the electron charge times the atomic radius, ea_0.

Let us denote E_s and E_p the energy eigenvalues of the hydrogen Hamiltonian H without the Stark interaction.[2] The total Hamiltonian is represented in the ordered basis $\{|2s, 0\rangle, |2p, -1\rangle, |2p, 0\rangle, |2p, 1\rangle\}$ by the matrix

$$H + H_{\text{Stark}} = \begin{bmatrix} E_s & 0 & V_{sp} & 0 \\ 0 & E_p & 0 & 0 \\ V_{sp} & 0 & E_p & 0 \\ 0 & 0 & 0 & E_p \end{bmatrix} \tag{6.5}$$

whose eigenvalues are

$$\left\{ E_p, \frac{E_s + E_p}{2} \pm \frac{1}{2}\sqrt{(E_s - E_p)^2 + 4V_{sp}^2} \right\}, \tag{6.6}$$

the first one being doubly degenerate (with eigenstates $|2p, \pm 1\rangle$).

If $E_s = E_p$ (or, alternatively, in the asymptotic limit $|V_{sp}| \gg |E_s - E_p|$), the energy shifts $\pm|V_{sp}|$ of the states that split off the $n = 2$ manifold are linear in the applied electric field. This case is illustrated in Fig. 6.1, column (a). Such a linear shift is typical for a resonantly coupled pair of states, the resonance condition in a static field being that the pair of states is degenerate in energy. The eigenstates of the Stark Hamiltonian (6.5) in this case are given by the symmetric and antisymmetric superpositions:

$$\frac{|2s0\rangle \pm |2p0\rangle}{\sqrt{2}}. \tag{6.7}$$

---

[2]E_s and E_p actually coincide in the simplest model for the hydrogen atom, see Example 2.2.1. However, relativistic corrections (called spin-orbit interactions, see Section 2.2) lift this degeneracy.

These states have a "permanent dipole moment" in the sense that the expectation value $\langle d_z \rangle = \langle -er \cos \theta \rangle \neq 0$: one finds a number of the order of $D \sim ea_0$. Note that a superposition of states with different angular momentum quantum number ℓ (more precisely, with different parity) is required for a permanent dipole.

In the opposite case $|E_s - E_p| \rightarrow \infty$, the energy shift is proportional to $\pm |V_{sp}|^2$, hence quadratic in the field. This corresponds to the regime where Eq. (6.2) applies.

We finally mention that the DC Stark shift provides a simple example of an energy shift that depends on the "internal" (here: electronic) state of the atom. This provides the possibility, e.g., to generate a phase shift $\varphi = \pm |V_{sp}|\tau/\hbar$ that is conditioned on the internal state and can be controlled by the duration τ the electric field is applied. This is an important ingredient for quantum gate operations. Another technique is based on a spatially inhomogeneous electric field where the two states (6.7) are subject to opposite forces. This can be used to bring a first atom in contact with a second one, conditioned on the internal state of the first one. This principle is used in many 2-atom gates, as we illustrate in more detail in Section 6.6.   □

6.3.2 Magnetic interaction

Atoms can have a permanent magnetic dipole moment $\boldsymbol{\mu}$, which in the ground state of an alkali atom is due to the electron and nuclear magnetic moments. The energy of the magnetic interaction is $H_m = -\boldsymbol{\mu}\cdot\boldsymbol{B}(\boldsymbol{r}, t)$, which is the Zeeman effect. Of course, induced time-varying magnetic moments may also be present, but static interactions dominate. In a weak external static magnetic field, $\boldsymbol{\mu} = \mu_{\text{eff}}\boldsymbol{F}$, where $\boldsymbol{F}$ is the vector total of the electronic and nuclear angular momentum. For atomic hydrogen and alkali metal atoms in the ground state, $F = I \pm 1/2$, where I is the nuclear spin, and $\mu_{\text{eff}} \approx \pm\mu_B/(I + 1/2) + \mu_I$. Here, $\mu_B = e\hbar/2m_e$ is the Bohr magneton and $\mu_I \sim (m_e/m_p)\mu_B$ the magnetic moment of the nucleus (m_p being the proton mass), typically three orders of magnitude smaller than μ_B. The external field is weak when the Zeeman effect is small compared to the hyperfine interaction between the electronic and nuclear magnetic moments ($\sim 2\pi\hbar\times$ GHz); see Example 6.3.2 below. As a typical order of magnitude, we quote $\mu_B \approx 2\pi\hbar \times 1.4\,\text{MHz/G}$.

Magnetic traps are usually designed such that the atoms move sufficiently slowly so that their magnetic moment follows the changing direction of the trapping field. In this "adiabatic regime", the magnetic quantum number m_F with respect to a local quantization axis along $\boldsymbol{B}$ is conserved. Taking the expectation value in this spin state, the magnetic interaction energy leads to the effective potential $\langle H_m \rangle = -\mu_{\text{eff}}m_F|\boldsymbol{B}(\boldsymbol{r})|$ for the motion of the atoms. If $\mu_{\text{eff}}m_F > 0$, atoms are attracted to the maximum of the magnetic field, i.e., they are "strong field seekers", while they are weak field seekers if the opposite is true. The atomic hyperfine ground state $m_F = F$

(μ_{eff} negative) is particularly interesting for coherent trapping, since the potential energy is already a minimum, no energy can be released during atomic collisions between very cold atoms. The latter must therefore be elastic. When $F \geq 3/2$, more than one magnetic sublevel of the same hyperfine manifold can be confined in a magnetic atom trap, so these sublevels could perform as a potential information qubit. Magnetic atom mirrors and waveguides as well as traps have been constructed [16, 26].

Example 6.3.2. *Hyperfine and Zeeman interactions—Breit-Wigner formula.* The Hamiltonian for an atom with electron spin S, zero orbital angular moment (an s-state), and nuclear spin I in a magnetic field $\boldsymbol{B} = B_z \boldsymbol{e}_z$ is of the form

$$
\begin{aligned}
H &= H_{\mathrm{hf}} + H_{\mathrm{Zeeman}} & (6.8) \\
H_{\mathrm{hf}} &= A \boldsymbol{S} \cdot \boldsymbol{I} & (6.9) \\
H_{\mathrm{Zeeman}} &= -\mu_B B_z \left(g_S S_z + g_I I_z \right), & (6.10)
\end{aligned}
$$

where the constant A characterizes the coupling strength between the nuclear and electron spins (the so-called hyperfine interaction), μ_B is the Bohr magneton, and $g_{S,I}$ are the so-called Landé factors for the electron and nuclear spins. For the electron spin, $g_S = 2$, up to corrections from quantum electrodynamics; for the nuclear spin, the ratio between electron mass and proton mass enters, $g_I \approx m_e/m_p \approx 1/1836$. The energy eigenvalues for the Hamiltonian (6.10) are given by the Breit-Wigner formula (see Fig. 6.2)

$$
\begin{aligned}
E_m(B_z) &= -\frac{A}{4} \pm \frac{1}{2} \left(E_{\mathrm{hf}}^2 - \frac{2\mu B_z m E_{\mathrm{hf}}}{I + 1/2} + \mu^2 B_z^2 \right)^{1/2} - \mu B_z g m, \\
& \quad m = -I + 1/2, \ldots I - 1/2, & (6.11) \\
E_{\mathrm{hf}} &= A(I + 1/2), & (6.12) \\
\mu &= \mu_B(g_S - g_I), & (6.13) \\
g &= \frac{g_I}{g_S - g_I}. & (6.14)
\end{aligned}
$$

Here, I is the quantum number for the magnitude of the nuclear spin, the electron spin has magnitude $S = 1/2$, and m is a magnetic quantum number that runs in integer steps through the indicated range. The positive (negative) sign connects to the hyperfine manifold with $F = I + 1/2$ ($F = I - 1/2$), respectively. In the case $|m| = I + 1/2$, the energy shift is given by

$$
\begin{aligned}
E_m(B_z) &= -\frac{A}{4} + \frac{E_{\mathrm{hf}}}{2} - \frac{\mu B_z m}{2(I + 1/2)} - \mu B_z g m, \\
& \quad m = \pm(I + 1/2). & (6.15)
\end{aligned}
$$

The derivation of this formula, that we give below, illustrates some of the principles that are used in atom manipulation with external fields. □

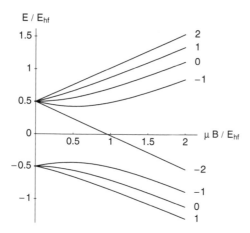

Figure 6.2: Energy eigenvalues of the hyperfine states of an atom in a magnetic field, given by the Breit-Wigner formula (6.11) for $I = 3/2$. The numbers on the right give the total angular momentum quantum number m. The curves starting from $E = E_{\text{hf}}/2$ at zero field ($m = -2, \ldots + 2$) correspond to the hyperfine manifold with $F = 2$. The manifold $F = 1$ corresponds to the lower curves labelled $m = -1, 0, +1$.

We first introduce the total spin of the atom, $\boldsymbol{F} = \boldsymbol{S} + \boldsymbol{I}$. This operator allows us to write the hyperfine interaction in the form

$$2\boldsymbol{S} \cdot \boldsymbol{I} = \boldsymbol{F}^2 - \boldsymbol{S}^2 - \boldsymbol{I}^2. \tag{6.16}$$

This is useful to evaluate the hyperfine energy eigenvalues because the Hilbert space can be decomposed into a direct sum of subspaces where the operators $\boldsymbol{F}^2, \boldsymbol{S}^2, \boldsymbol{I}^2$ act like a simple multiplication by $F(F+1)$, $S(S+1) = 3/4, I(I+1)$, respectively. The subspaces are labelled by F, the total angular momentum, that can take two possible values $I \pm 1/2$. This and other useful properties of the spin operators can be found in any textbook on quantum mechanics and atomic physics [21, 36] and are summarized in Box 5.1. The hyperfine eigenvalues are thus given by

$$
\begin{aligned}
A\boldsymbol{S} \cdot \boldsymbol{I} &= \frac{A}{2} \left((I \pm 1/2)(I \pm 1/2 + 1) - 3/4 - I(I+1) \right) \\
&= -\frac{A}{4} \pm \frac{A}{2}(I + 1/2), \tag{6.17}
\end{aligned}
$$

depending on the value of $F = I \pm 1/2$. Their difference is the hyperfine splitting E_{hf} given by Eq.(6.12).

As a second step, we re-write the Zeeman interaction in the form

$$
\begin{aligned}
\mu_B &\left(g_S S_z + g_I I_z \right) \\
&= \mu_B[(g_S - g_I)S_z + g_I(S_z + I_z)] \\
&= \mu S_z + \mu g F_z, \tag{6.18}
\end{aligned}
$$

where the second term involving F_z commutes with the hyperfine interaction (6.16), but S_z does not. (This can be checked from the commutation relations given in Box 6.1.)

We consider here a spin operator $\mathbf{J} = (J_x, J_y, J_z)$. Its components satisfy the commutation relations

$$[J_k, J_l] = i \sum_n \epsilon_{kln} J_n , \tag{6.19}$$

where ϵ_{kln} is the totally antisymmetric tensor (Levy-Civita symbol) with $\epsilon_{xyz} = 1$. As a consequence of these relations, the squared spin $\mathbf{J}^2$ commutes with any of the spin components: $[J_k, \mathbf{J}^2] = 0$. One can thus find simultaneous eigenvectors and eigenvalues for $\mathbf{J}^2$ and J_z, say. The eigenvalues of $\mathbf{J}^2$ are given by $J(J+1)$ where J is a non-negative half-integer or integer, $J = 0, 1/2, 1, 3/2, \ldots$ For a given value of J, there exists a $(2J+1)$-dimensional sub-space spanned by the vectors $|m\rangle$ with $m = -J$, $-J+1, \ldots + J$ on which the spin operators can be represented by $(2J+1) \times (2J+1)$ matrices. The simplest case is $J = 1/2$, the spin-1/2 system where the Pauli matrices (adjusted by a factor of 1/2) occur:

$$J_x \mapsto \frac{1}{2} \begin{pmatrix} 0 & 1 \\ 1 & 0 \end{pmatrix}, \quad J_y \mapsto \frac{1}{2} \begin{pmatrix} 0 & -i \\ i & 0 \end{pmatrix}, \quad J_z \mapsto \frac{1}{2} \begin{pmatrix} 1 & 0 \\ 0 & -1 \end{pmatrix}. \tag{6.20}$$

In the general case, the action of the spin operators is most easily computed using the so-called ladder operators. We define them here by

$$J_\pm \equiv \frac{1}{\sqrt{2}} (J_x \pm i J_y), \tag{6.21}$$

and get the following basic relations

$$[J_z, J_\pm] = \pm J_\pm, \tag{6.22}$$
$$[J_+, J_-] = J_z, \tag{6.23}$$
$$\mathbf{J}^2 = J_z^2 + J_+ J_- + J_- J_+. \tag{6.24}$$

The commutator (6.22) implies that when $|m\rangle$ is eigenvector to J_z with eigenvalue m, the vector $J_\pm|m\rangle$ is also an eigenvector with the eigenvalue $m \pm 1$: this explains the name "ladder operators" for $J_\pm$. Using the relations (6.23) and (6.24), one can show that

$$J_+|m\rangle = \left(\frac{(J-m)(J+m+1)}{2} \right)^{1/2} |m+1\rangle, \tag{6.25}$$

$$J_-|m\rangle = \left(\frac{(J+m)(J-m+1)}{2} \right)^{1/2} |m+1\rangle, \tag{6.26}$$

where an arbitrary phase factor was conventionally set to unity. In particular, $J_+|J\rangle = 0$ and $J_-|-J\rangle = 0$.

Box 6.1 Spin algebra and eigenstates

We finally use the raising and lowering operators $S_\pm$, $I_\pm$, as defined in Box 6.1 to write the hyperfine interaction as

$$H_{\mathrm{hf}} = A\left(I_+ S_- + I_- S_+ + I_z S_z\right). \tag{6.27}$$

We can now identify 1- and 2-dimensional subspaces in which the Hamiltonian $H_{\mathrm{hf}} + H_{\mathrm{Zeeman}}$ can be diagonalized. They are spanned by the product states $|m_I, m_S\rangle$, $m_I = -I, \ldots + I$, $m_S = -1/2, +1/2$ that are simultaneous eigenstates of I_z and S_z. The first quantum number gives the eigenvalue for I_z, the second one the eigenvalue for S_z. They are also eigenstates of the total angular momentum F_z with an eigenvalue $m = m_I + m_S$. The eigenvalue m will be used below to classify the subspaces.

The states $|I, +1/2\rangle$ and $|-I, -1/2\rangle$ $(m = \pm(I + 1/2))$ are already eigenstates and span the 1-dimensional subspaces. This follows from the fact that $I_+ |I, +1/2\rangle = 0$ and $S_+ |I, +1/2\rangle = 0$. The remaining terms are easily evaluated to give the energy eigenvalues (6.15).

The values $m = -I + 1/2, \ldots I - 1/2$ correspond to a family of 2-dimensional subspaces. Let us consider the ordered basis $\{|m-1/2, +1/2\rangle, |m+1/2, -1/2\rangle\}$ and compute the matrix representation of $H_{\mathrm{hf}} + H_{\mathrm{Zeeman}}$. The Zeeman Hamiltonian gives, for example,

$$\mu B_z \left(S_z + g F_z\right) |m + 1/2, -1/2\rangle = \mu B_z \left(-1/2 + gm\right) |m + 1/2, -1/2\rangle,$$

as both operators act on their eigenvectors. For the hyperfine interaction, we need to compute, for example,

$$I_- S_+ |m + 1/2, -1/2\rangle = \frac{1}{2} (I + m + 1/2)^{1/2} (I - m + 1/2)^{1/2} |m - 1/2, +1/2\rangle$$

$$= \frac{1}{2} \left[(I + 1/2)^2 - m^2\right]^{1/2} |m - 1/2, +1/2\rangle,$$

using Eqs. (6.25), (6.26) from Box 6.1. Here, the coupling between the two basis vectors appears. The terms involving the operator product $I_z S_z$ is easy since we are dealing with the eigenvectors of I_z and S_z. Putting everything together, we get the following 2×2 matrix:

$$H = -\frac{A}{4} - \mu B_z g m +$$

$$+ \frac{1}{2} \begin{bmatrix} Am - \mu B_z & A\left[(I + 1/2)^2 - m^2\right]^{1/2} \\ A\left[(I + 1/2)^2 - m^2\right]^{1/2} & -Am + \mu B_z \end{bmatrix}, \tag{6.28}$$

where it is understood that the first two terms are proportional to the unit matrix. The eigenvalues are easily computed: they are the sum of the first two terms plus the eigenvalues of the third term which are:

$$\pm \frac{1}{2} \left[(Am - \mu B_z)^2 + A^2((I + 1/2)^2 - m^2)\right]^{1/2}$$

$$= \pm \frac{1}{2} \left[A^2(I + 1/2)^2 - 2Am\mu B_z + \mu^2 B_z^2\right]^{1/2}.$$

Given the definition (6.12) for E_{hf} in terms of A, this leads to Eq. (6.11). □

The Breit-Wigner formula (6.11) implies the following design rules for qubit manipulation with neutral atoms: if hyperfine states are used as qubits, a magnetic field can lead to energy and phase shifts with different signs and magnitudes. Conversely, to protect the qubit from uncontrolled magnetic field fluctuations, pairs of states can be chosen that are shifted in the same way. Such pairs correspond to opposite values of m splitting off from different hyperfine states, up to small corrections due to the nuclear magnetic moment. In fact, by linearizing Eq. (6.11), one finds a Zeeman-like shift for the two hyperfine manifolds of the form $E_m(B_z) - E_m(0) = \mu_{\text{eff}} m B_z$ where $\mu_{\text{eff}} = \mu_B/(I + 1/2)$ ($\mu_{\text{eff}} = -\mu_B/(I + 1/2)$) for the manifold $F = I + 1/2$ ($F = I - 1/2$), respectively. It is possible to compensate for the nuclear correction by using a static magnetic field B_* such that $\partial E_m^+(B_*)/\partial B_z = \partial E_{-m}^-(B_*)/\partial B_z$, where the full dependence on B_z is taken into account. A typical value for this "magic field" in the alkali atom Rubidium is a few gauss which is easily achieved in experiments. A magnetic trap with $|B| = B_*$ at the minimum can trap atoms in both internal states, provided the energy shift is such that it increases with the magnetic field.

Finally, with a time-dependent magnetic field in addition to a static field, one can induce resonant couplings between the split levels. This is the basic principle of magnetic resonance spectroscopy. The resonant frequency ω_{res} is determined by the energy difference between the levels, $\hbar\omega_{\text{res}} = E_m(B_z) - E_{m'}(B_z)$. This can be used for qubit manipulation as well (flipping qubits, qubit-dependent phases), as explained in Subsection 6.5.1. In the limit of an off-resonant excitation, the coupled levels are shifted by an amount proportional to $\pm|\Omega|^2/(E_m(B_z) - E_{m'}(B_z) - \hbar\omega_{\text{res}})$ where Ω is the "magnetic Rabi frequency" proportional to the amplitude B_1 of the time-dependent field. The field polarization must be chosen such that the magnetic moment has a nonzero matrix element $\langle m|\boldsymbol{\mu} \cdot \boldsymbol{B}_1|m'\rangle$ between the pair of qubit states. This shift is position-dependent if B_1 varies in space, and can be used to implement state-dependent potentials for the atomic center-of-mass motion.

6.3.3   Cold, controlled atom–atom interactions

For cold bosonic atoms, s-wave scattering dominates in atom-atom collisions. This scattering has been proposed for use in the execution of certain 2-atom quantum gates. s-wave scattering is described by a contact potential of the form [6]:

$$U(\boldsymbol{r}_1 - \boldsymbol{r}_2) = (4\pi a_s \hbar^2/m)\delta(\boldsymbol{r}_1 - \boldsymbol{r}_2), \tag{6.29}$$

where δ is the Dirac delta function and a_s is the s-wave scattering length, a measure of the strength and sign of the contact interaction. In a spin-triplet collision for ^{85}Rb, the spin triplet s-wave scattering length $a_{st} = -369a_0$ while for ^{87}Rb $a_{st} = +106a_0$, both in units of the Bohr radius a_0. Fig. 6.3 shows a definition of the sign of the s-wave scattering length. If the colliding atoms remain in the transverse vibrational ground state of the confining well (frequency $\omega_\perp$) during the collision interaction, the 1-dimensional collision interaction is

$$U(x_1 - x_2) = 2a_{st}\hbar\omega_\perp\delta(x_1 - x_2),$$

as shown by Olshanii [39] where also corrections due to the confinement are derived.

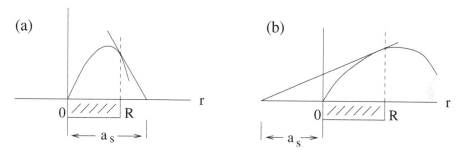

Figure 6.3: *S*-wave scattering lengths a_s in cold atomic collisions result from phase shifts of the long-wavelength interatomic *s*-wave function $\Psi(r)$. The scattering lengths are defined by the slope of the wave function at the edge R of the radial potential well describing the interaction. When the slope is zero, the scattering length is infinite. (a) $a_s > 0$ (b) $a_s < 0$.

6.4   Atom trapping

The previous sections have shown that both time-averaged electric and magnetic interactions in non-uniform time-varying fields can lead to atom confinement at potential minima or maxima. For static fields, there is no absolute potential minimum in charge-free space, so an additional interaction of another type must be employed to achieve atom confinement away from the static field source. Additionally, if a magnetic field drops to zero in the atom well, then so-called non-adiabatic transitions can flip the atom angular momentum, resulting in atom loss from the trap, so zero B field regions must be avoided. More generally, the traps are designed such that the atoms move sufficiently slowly so that their atomic moments (electric or magnetic) can adjust to the changing direction of the trapping field. This "adiabatic regime" is typically ensured when the trap oscillation frequency is much smaller than the energy splitting between adjacent internal states [3, 46].

6.4.1   Optical lattices

The ac Stark shift potential given by (6.2) will confine cold alkali atoms in a far-off-resonance optical lattice. The optical lattice is a set of periodic potentials spaced near $\lambda/2$, created by a set of interfering laser beams. A lattice can exist in one to three dimensions. A 3-dimensional (3D) optical lattice, using lasers detuned greatly to the blue to minimize photon scattering, traps atoms at the nodes. Transverse counter-propagating beams along the x and y directions confine atoms in tubes oriented along the z direction, while longitudinal counter-propagating beams produce standing waves within each tube. For example [5], laser intensities of 100 W/cm^2 for all beams with $\lambda \leq 852\,nm$ can be used to confine Cs atoms. The transverse and longitudinal beams are respectively detuned to blue wavelengths by $\Delta_\perp/2\pi = 120$ GHz and $\Delta_{||}/2\pi = 2 \times 10^{12}$ Hz. This results in potential wells with oscillation frequencies $\omega_\perp = 2\pi \times 200$ kHz and $\omega_{||} = 2\pi \times 50$ kHz (see (6.32) below, for an approximate relation to laser power). The Lamb–Dicke parameter η provides a measure of the resolution of the oscillation frequency of the atom in the potential well from an internal transition frequency of the atom. It is also a measure of atom localization in the lowest state of the separate elliptical (approximately harmonic) wells. $\eta_x = kx_0 \approx 0.1$ and $\eta_z = kz_0 \approx 0.2$ correspond to lowest state atom localizations of $\lambda/60$ and $\lambda/30$, respectively, in the perpendicular and parallel directions of the elliptical well.

In the optical lattice, polarization gradients can be achieved and be used to displace two sets of atoms trapped in different Zeeman states relative to each other. This effect is based on an atomic polarizability that is no longer a scalar. The effective potential can be written in form of an effective magnetic field whose magnitude is proportional to the helicity (i.e., the degree of circular polarization) of the field. If the longitudinal beams are linearly polarized, their interference creates a field polarization that changes periodically from right circular via a sequence of linear, left circular, and linear back to right circular, depending on the relative phase shift between the two beams. The optical lattice potential contains wells at the locations of right and left circular polarization that trap Zeeman states $|m\rangle$ with opposite sign. By changing the angle θ between the polarizations of the longitudinal beams, the distance δz between minima of the right circularly polarized and left circularly polarized wells can be varied according to $\delta z = (\lambda/2\pi)\arctan(\tan\theta/2)$. As $\theta \to 0$ adiabatically, $\delta z \to 0$ and two neighboring atoms are moved into the same linearly polarized well. As $\theta \gg \pi/2$, the neighboring atom wells can be translated over long distances, to permit interaction with other atoms. This enables controlled atom-atom interactions to be implemented. Significant progress towards this direction has been achieved by working with a Bose-Einstein condensate of atoms loaded in an optical lattice of this type [18]. The atom-atom interaction then leads to a suppression of fluctuations in the number of atoms per trapping site, called a phase transition to a Mott insulator [49]. This state implements the "initialized" state of a quantum register. Multi-atom entangled states have been created in a reversible way by shifting neighboring atoms in a state-dependent way [69].

6.4.2 Focused laser traps

Assume a Gaussian-focused laser beam propagating along the z-direction with power P_L and a beam waist diameter w at the focus. The optical dipole potential (6.2), with a scalar polarizability for simplicity, is then near the focus proportional to

$$V_{el}(\mathbf{r}) \quad \propto \quad \frac{\exp\left(-2(x^2 + y^2)/[w(z)]^2\right)}{[w(z)]^2} \tag{6.30}$$

$$w(z) \quad = \quad w(1 + z^2/z_R^2)^{1/2}, \qquad z_R \equiv \pi w^2/\lambda. \tag{6.31}$$

Expanding this to second order around $\mathbf{r} = \mathbf{0}$, one finds a harmonic trap with a trapping frequency that scales like [19]:

$$\omega_\perp \approx \left(\frac{\text{Re } \alpha(\omega)P_L}{\varepsilon_0 M c w^4}\right)^{1/2}, \qquad \omega_\| \approx \omega_\perp(\lambda/w). \tag{6.32}$$

Here the atom mass is M. Typical trapping frequencies can be in the range of $10 - 10^3$ Hz. The well depths are smaller than in the optical lattice, since the beam waist w is $\gg \lambda$. The trap is harmonic as long as the atoms are confined radially (longitudinally) within a region much narrower than w (much smaller than z_R).

Crossing "hollow" laser beams, having transverse modes with intensity minima near the center, results in an isotropic potential well when the lasers are detuned to the blue, or a beam splitter, when detuned to the red [29].

One- or 2-dimensional lattices of laser beam foci can be created with many micrometers separation using an array of microlenses. This has been called an optical microtrap array [4]. This technique has the advantage over an optical lattice that the potential wells are sufficiently separated so that single confined atoms can be unambiguously addressed by a single laser beam, for qubit manipulation. However, the traps have also been used to confine hundreds of atoms [14] using red-detuned lasers. The traps can be moved relatively in space by changing the angle of incidence of the laser beam. Cylindrical lenses produce linear potentials that can be used as atom waveguides, and four crossed linear guides result in an atom interferometer geometry. This microlens concept is scalable in the number of traps, and provides some dynamic control of the arrangement of the array; see Fig. 6.4.

A 2-dimensional 50×50 array of spherical microlenses with focal lengths of $625\,\mu m$ and separation $125\,\mu m$ produced an array of foci with $125\,\mu m$ separation and a spot size of $7 \pm 2\,\mu m$ (at e^{-2} radius of intensity) [14]. A red-detuned beam of $100-200\,\text{mW}$ of linearly polarized laser light detuned 0.2 to $2\,nm$ below the $780\,nm(F = 3)\ 5S \leftrightarrow (F = 4)5P$ transition of Rb was imaged on a *magneto optical trap* (MOT) to produce an array of dipole traps for the cold atoms in the MOT. The traps had a potential depth corresponding to an atom temperature of about 1 mK, and contained up to 10^3 atoms each. A

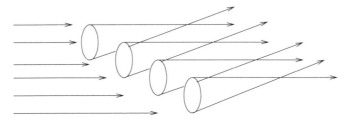

Figure 6.4: A 1-dimensional array of microlenses establishing an array of focus points for laser light. With red-detuning of the laser light, optical dipole traps for atoms are created at these foci, which can act as a qubit register.

1-dimensional array of microtraps could also be made by illuminating only one row of the microlens array.

When an optical lattice of traps is created from the interference of laser standing waves, as discussed in Subsection 6.4.1, the potential minima are spaced by about $\lambda/2$, and the trap depths are about a factor of ten larger than for laser focus traps (w is replaced by λ in (6.32)). Optical traps on a scale below an optical wavelength can also be achieved using a planar waveguide formed above a blue-detuned evanescent wave outside a transparent substrate [17]. A laterally structured field, produced by laser light diffracted by nanostructures deposited in the transparent substrate, can produce these traps [32], as can sub-micrometer electrodes deposited on the substrate [44]. Time-dependent potentials are created by changes in electrode voltages.

6.4.3 Static magnetic traps

The separation of magnetic traps is determined by the size of the structures that produce the magnetic field, either by magnetization of some medium like a tape, or by electric currents. With structure sizes in the micrometer range or below, tight confinement of atoms can be achieved.

A waveguide for atoms in a weak-field-seeking atomic state is created from the azimuthal magnetic field of a straight filament current I superimposed on an external bias field B_b perpendicular to the current. These fields cancel at a distance $y_t = \mu_0 I / 2\pi B_b$ from the filament, where μ_0 denotes the permeability of the vacuum. Around the zero of field at vector position (x_t, y_t, z), the magnetic field vector is approximately $B(r) \approx (B_b/y_t)(y - y_t, x - x_t, 0)$, so $|B|$ increases linearly with distance relative to the field minimum, with a gradient B_b/y_t; see Fig. 6.5. A smaller trapping distance and a steeper gradient is achieved by reducing the current I, but since $B \to 0$, the adiabatic approximation no longer holds. By rotating the bias field so that a component $B_b \cos\theta$ lies in the z direction, the minimum $|B| \neq 0$ is feasible, to overcome this adiabaticity problem. The transverse oscillation frequency in the trap is given by the equation

$$\omega_\perp = \left(\frac{-\mu_{\text{eff}} m_F B_b \sin^2 \theta}{M y_t^2 |\cos \theta|} \right)^{1/2}. \tag{6.33}$$

The result is in the range 100 kHz to 1 MHz for $y_t = 1$ to 10 micrometers. The atoms are not confined along the filament axis, but if the current wire is bent, a longitudinal confinement is achieved to hold them [16].

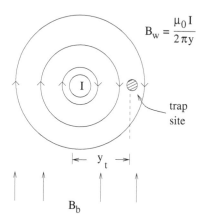

Figure 6.5: The fields of a current-carrying wire B_w and a uniform bias field B_b cancel at a position y_t, to produce a magnetic trap in the x-y plane above the surface of a microchip. The surface could be located at $y = 0$, with the wire deposited on top.

The magnetic trap built from a current-carrying wire is easily integrated into a solid substrate, leading to the concept of the "atom chip". It opens the road to "integrated atom optics" to design devices for quantum manipulation of stored single atom and multi-atom samples. Field generating structures like wires, electrodes, and optical fibers are built on a semi-conductor surface, for example, adapting micro-fabrication technology such as micro-electronics and micro-optics. Atom chips offer large electric and magnetic field gradients near microscopic conductors, and tight confinement with deep potential wells and large energy level separations for trapped atoms. Micrometer distance scales are feasible, permitting strong qubit coupling and fast gate operations. In addition, qubits can be encoded in different degrees of freedom, like hyperfine states and atom trap vibrational levels.

A trap based on a magnetic mirror (see Subsection 5.2.1, Chap. 5) is another alternative. A linear array of magnetic quadruples is created when a homogeneous bias field is superimposed on a mirror field with the proper symmetry. Atoms can be translated by adiabatically rotating the bias field [4].

6.5 Qubits and gates

A quantum register based on atoms can be implemented using atomic hyperfine states or the eigenstates of the effective confining potential, which to a good approximation are harmonic oscillator states for small quantum numbers. If these states differ in their atomic hyperfine or magnetic quantum number, they are considered internal qubits, but if they differ in their center-of-mass motion, then they are considered external qubits. Both qubit types have long confinement and coherence times, permitting unitary operations to be performed with laser pulses, either single qubit or 2-qubit operations. The 2-qubit operations require controlled interactions between pairs of atoms. Eventually, the results of the gate operations must be measured.

The simplest 2-qubit gates are the CNOT gate: one qubit is flipped depending on the state of the other qubit, the phase gate Q_η: only the state $|11\rangle$ acquires a phase shift η, while the others remain unchanged, and the "square root of SWAP" ($U_{sw}^{1/2}$) gate. It can be shown that any of these 2-qubit gates can actually be expressed in terms of any other one, combined with single-qubit operations. (See Eq. (2.48) for an example.) Applying Theorem 2.6.2, we have:

Theorem 6.5.1. *A register of N qubits implemented with neutral atoms performs as a universal quantum computer if the following conditions hold:*
(1) to any qubit, arbitrary 1-qubit gates can be applied, and
(2) to any pair of qubits, one of the three 2-qubit gates CNOT, Q_η, or $U_{sw}^{1/2}$ can be applied. □

6.5.1 Single qubit gates based on internal states

The magnetic sublevels of hyperfine structure (hfs) manifolds of atomic ground states are stable. Atoms in all such sublevels are simultaneously confined in electrostatic and optical traps. For $F \geq 3/2$, at least two sublevels are confined in magnetic traps. The internal qubit states $|0\rangle \equiv |F, m_F = -F\rangle$ and $|1\rangle \equiv |F + 1, m_F = +F\rangle$ are particularly important, since they are both subject to the same potential in all trap types (see Fig. 6.2), and do not dephase in a fluctuating magnetic field [47]. Also, the $|0\rangle$ state is the absolute ground state in a weak magnetic field, and consequently is less perturbed by any atomic collisions.

An arbitrary qubit state has a time evolution given by:

$$|\Psi(t)\rangle = c_0|0\rangle + c_1 \exp(-i\omega_{\text{hfs}}t)|1\rangle, \qquad (6.34)$$

where ω_{hf} is the hyperfine structure splitting frequency between states F and $F + 1$. This frequency is the order of GHz in alkali atom ground states. The relative phase can be controlled by locking the qubit levels to an atomic clock, used as a local oscillator. Both qubit states have the same magnetic moment, so their relative phase is not changed by a spatially-varying static magnetic field.

Qubit state amplitudes are changed by a microwave pulse at the hfs frequency, or by a Raman laser transition when two lasers with a difference frequency of ω_{hfs} illuminate the atom in the trap. Both cases are described by the interaction Hamiltonian

$$H_1(t) = \hbar\{\Omega_1(t)e^{-i\omega_{hfs}t}|1\rangle\langle 0| + \Omega_1^*(t)e^{+i\omega_{hfs}t}|0\rangle\langle 1|\}. \tag{6.35}$$

Here $\Omega_1(t)$ is the complex Rabi frequency, proportional to the amplitude of the external time-dependent fields, which describes the transition rate between qubit states. A field pulse of suitable time duration implements any single qubit unitary Rabi rotation operation (see Chap. 3, Section 3.1).

6.5.2 Single qubit gates based on external states

The stationary states of the trap potential have the same internal atomic quantum numbers, so external field fluctuations affect them relatively little. Usually the trap ground state and first excited state, separated by the trap oscillation angular frequency ω, are chosen (see (6.32) and (6.33) for trap frequencies). Appropriate cooling techniques initially prepare the atoms in the ground state. An interaction that depends on the atomic displacement $\pm u$ from the trap center can be used to implement single qubit gates. A spatially constant force F (in a magnetic or electric field gradient) with an interaction Hamiltonian

$$H_I = -\boldsymbol{F} \cdot \boldsymbol{u} = -\boldsymbol{F} \cdot \hat{\boldsymbol{u}} r_0 (a + a^+) \tag{6.36}$$

is the simplest example. The unit vector $\hat{u}$ is directed along one of the principal axes of the confining potential. The qubit states are specified by the trap ground state $|0\rangle$ and the state $|1\rangle = a^+|0\rangle$, where a^+ and a are the usual raising and lowering operators for the harmonic well levels. The radius of the atomic wave function in the ground state $r_0 = (\hbar/2M\omega)^{1/2}$ can be about $100\,nm$. It can be shown that (6.36) results in an interaction Hamiltonian similar to that given by (6.35), which performs an analog of an internal qubit rotation on the external qubit. This rotation is based on a time-dependent force $F(t)\cos(\Omega t + \varphi)$, where $F(t)$ is an envelope function and $\Omega \approx \omega$. Excitation of the undesired $|1\rangle \rightarrow |2\rangle$ transition in the well can be avoided by making the trap slightly anharmonic.

A qubit rotation can also be based on an adiabatic deformation of the trapping potential, which splits this potential into a double well, but leaves the atom in the ground state. At the maximum splitting, a short force pulse gives a phase difference 2θ to the two parts of the split atom wave function, which are separated by a distance $\Delta \gg r_0$. This pulse couples the nearly degenerate lowest even and odd parity states of the double well. A superposition state

$$|0\rangle \rightarrow \cos\theta|0\rangle - ie^{i\varphi}\sin\theta|1\rangle \tag{6.37}$$

results as the atom wave function recombines when the double wells are merged, with the even and odd states mapping to $|0\rangle$ and $|1\rangle$ respectively. An accumulated frequency difference between the odd and even states of the double well determines the phase φ [22, 27].

6.6 Controlled 2-qubit gates

Quantum logic gates involving two atomic qubits must overcome the problem of the short range coherent interaction of neutral atoms, while maintaining atom confinement and suppressing decoherence. The main challenge is to perform the gate sufficiently fast compared to typical decoherence and relaxation times. This may require elaborated techniques to transfer atoms into other electronic states where atom-atom interactions are stronger (leading to faster gate operations). Additional complications arise when successive interactions with different atoms are required. Some interesting and resourceful schemes to achieve 2-atom gates have been proposed.

6.6.1 Cold, controlled collisions in an optical lattice

This 2-atom gate is based on atom confinement in ground states of 3-dimensional wells of an optical lattice, as discussed in Subsection 6.4.1. Key to the success of this gate is the suppression of spontaneous emission of photons by the atoms, by operating the lattice far off the atomic resonance, and performing the atomic manipulations quickly compared to the residual photon scattering rate [5]. The photon scattering rate $\Gamma_{sc} = s\Gamma/2$, where the saturation parameter s is proportional to the excited state population. The spontaneous emission rate $\Gamma = k^3 |d_{eg}|^2 / 3\pi\varepsilon_0\hbar$, where e and g denote excited and ground states coupled by the dipole matrix element d_{eg}. If atoms are spaced at distances $r_{12} \ll \lambda$, then the near field dipole-dipole interaction between atoms 1 and 2 varies as $V_{dd} \sim \langle d_1\rangle\langle d_2\rangle/r_{12}^3$. For weak excitation, the dipole expectation value $d \sim s^{1/2}d_{eg}$. The ratio of the atom interaction energy to the scattering rate scales as $\kappa \sim V_{dd}/\hbar\Gamma_{sc} \sim (kr_{12})^{-3}$, so if $r_{12} \ll \lambda$ as for two atoms in the samel well of an optical lattice, the photon scattering is negligible compared to the coherent dipole-dipole interaction [5].

By changing the angle between longitudinal laser polarizations, atoms trapped in neighboring wells can be brought into the same linearly polarized well, as discussed in Subsection 6.4.1. Interatomic collisions, as described by the contact potential (Subsection 6.3.3), are becoming relevant at these short distances and have been employed to entangle and disentangle multi-atom states [69]. Alternatively, amplified dipole-dipole interactions can be initiated by a "catalysis laser" that excites the atom dipoles from s to p states for a short time at frequency ω_c, as shown in Fig. 6.6. A conditional operation such as CNOT is performed by a weak π-polarized laser field, which has a frequency that can excite an atom only if it is in the logical

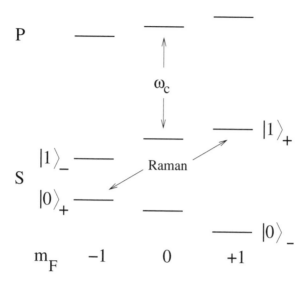

Figure 6.6: Schematic diagram of S and P states (not to scale) of an alkali atom with half-integer nuclear spin. The catalysis laser excites the interacting dipoles with S-P resonant frequency ω_c, slightly shifting the $|1\rangle_- \otimes |1\rangle_+$ state of two atoms ($-$ and $+$) in the same well. Raman lasers excite a π pulse on the target ($+$) atoms, leaving the control ($-$) atoms unaffected.

state $|1\rangle_\pm$, defined for atoms labeled $+$ and $-$ by two ground state hyperfine levels with $F = I + 1/2$, $m_F = \pm 1$. The dipole-dipole interaction V_{dd} causes a level shift only for the $|1\rangle_- \otimes |1\rangle_+$ 2-qubit state. A Raman π-pulse on this shifted state $|1\rangle_- \otimes |1\rangle_+ \leftrightarrow |1\rangle_- \otimes |0\rangle_+$ transition achieves the CNOT. The $|0\rangle_\pm$ states are defined by the hyperfine levels $F = I - 1/2$, $m_F = \mp 1$. No other transition is driven since an external magnetic field splits the m_F levels, and particular polarizations of the Raman laser beams are used.

The motional state of the atoms, assumed to remain in $|0\rangle_m$, determine the atomic wave function overlap with the dipole-dipole potential V_{dd}. The wave function overlap in the $|0\rangle_m$ state is maximized by a spherical well, but is adequate for gate operation in the well specified in Subsection 6.4.1. The dipole-dipole interaction depends both on the internal electronic state (tensor form of the interaction) and the motional state (atomic wave function overlap with the dipole-dipole potential). Photon scattering from the lattice produces a homogeneous linewidth $\Gamma_{sc}/2\pi \approx 4$ Hz, while $V_{dd}/\hbar \approx 5$ kHz.

6.6.2 Molecular interactions in an optical lattice

In this gate, qubits are represented in internal atomic states in atoms confined in a large regular array of an optical lattice, or alternatively a microchip (see Subsection 6.4.3). An auxiliary set of "marker atoms" are to be used between different lattice sites containing the qubits [6]. These marker atoms

(1) address atomic qubits by "marking" a single lattice site by atomic interactions, and

(2) they are also messenger qubits, which transport quantum information between different sites of the optical lattice and entangle distant qubits.

To move marker atoms, only the global laser parameters generating the optical superlattice need be changed. The time scale of the movements in the lattice is the order of the oscillation period of the confining lattice potential. Since the optical lattice is created very far off resonance,

(1) decoherence due to spontaneous emission is strongly suppressed, and

(2) the atomic qubit states can be chosen corresponding to a clock transition, i.e., with quantum numbers such that the frequency connecting the levels is insensitive to stray magnetic fields.

Resonant molecular interactions based on *magnetic or optical Feshbach resonances* are used for the interaction between the qubit and marker atoms. The resonant properties result in interactions over distances comparable to the trap spacing in the optical lattice, so the time scale of the operations is approximately the same as for lattice transport. The resonant molecular interactions can be dependent on the internal (atom) qubit state, resulting in a mechanism to entangle marker and atomic qubits and perform swap operations.

The model for the optical superlattice is a 1-dimensional periodic potential with one register atom R_i every second lattice site. Marker atoms can sit in the ground state of some of the remaining sites. The lattice control parameters are globally changed to create a periodic array of double-well structures with different well depths, so the marker atom can be transferred from its initial (ground state) site into the first excited state of a neighboring well, while the register atom remains in its ground state. From there, the marker atom can be moved to the ground state of an empty neighboring site, or back to its original position. All of the marker atoms undergo the same, parallel movement.

While occupying a common lattice site, the marker and qubit atoms are coupled by the strong, resonant molecular Feshbach interaction controlled by an external magnetic or optical field [6]. The Feshbach resonance occurs[3] when a molecular bound state $|n_\beta\rangle$ of a particular collision channel

---

[3]We explain a *Feshbach resonance* briefly by way of an example. Suppose an electron scatters off of helium ion He+. The incoming electron can excite the He+ (helium ion) to an $n = 2$ state (where n is the principal quantum number) and if it does not have enough energy it can be temporarily captured into a resonance state to form doubly excited He, where both electrons of the helium atom are excited. This is a resonance state because, of course, the two electrons

β having a well-defined vibrational quantum number ν crosses the dissociation threshold for another state having the same quantum numbers, whose energy ε_β is being changed by an external magnetic field. The collision channel β corresponds to the relative motion of the two atoms of the molecule, and the hyperfine angular momentum quantum numbers (specifying qubit states) of these two colliding atoms; see Fig. 6.7. The resonance energy ε_β varies almost linearly with the magnetic field B, with a slope s_β which depends on the collision channel. As B is linearly ramped, the ground state of the relative trap vibrational motion is adiabatically connected to the resonant state. Close to resonance, the scattering length $A_\beta(B) = A_{bg}(1 - \Delta B/(B - B_\beta))$ where B_β is the resonant value of the magnetic field B, ΔB is the width of the resonance, and A_{bg} is the background (nonresonant) scattering length. The effective Hamiltonian can be written as

$$H = \varepsilon_\beta(B)|n_\beta\rangle\langle n_\beta| + \sum_v [v\hbar\omega|v\rangle\langle v| + V_v^\beta|v\rangle\langle n_\beta| + h.c.] . \tag{6.38}$$

In this equation, the trapped eigenstates of the relative motion are designated v (for vibration), and a harmonic trap with frequency ω is assumed. The resonance couplings are $V_v^\beta = 2\hbar\omega[(4v + 3)^{1/2}A_{bg}(m\omega/\hbar)^{1/2}\Delta B(s_\beta)/\pi\hbar\omega]$. Accurate values of the resonance parameters ΔB, B_β, and A_{bg} are available from both theory and measurement [34]. Atoms prepared in their relative motion ground state in the well are transferred into the bound molecular state whose energy depends on the B field. This process is reversible. The resonance induced energy shift causes a 22-particle phase to appear only for a particular 2-qubit computational basis state associated with β, while all other qubit states are not affected. A phase gate or SWAP gate is executed. The phase is acquired during the Feshbach ramp affecting only the state $|0,0\rangle$, while a SWAP gate requires an additional laser field that couples the state shifted by the molecular interaction to the states $|10\rangle$ or $|01\rangle$.

In a simulation, ^{87}Rb clock states $|0\rangle = |F = 1, m_F = 0\rangle$ and $|1\rangle = |F = 2, m_F = 0\rangle$, and an auxiliary state $|a\rangle = |F = 1, m_F = 1\rangle$ were chosen, so that the qubit states were not strongly dependent on magnetic field, i.e., $m_F = 0$, while the $|a\rangle$ state energy depended linearly on B. Two resonances at fields $B_{0x} = 386$ G and $B_{00} = 407$ G were calculated, with respective widths $\Delta_{0x} = 5.7$ mG and $\Delta_{00} = 16$ mG. The resonant scattering length of ^{87}Rb depended on β for collisions in 2-atom channels $|0x\rangle$ and $|00\rangle$.

The potential wells can be manipulated with a fidelity of 99.99%, for non-adiabatic atom transfer between lattice sites in a time of $200\,\mu s$ for Rb

---

can later exchange energy again and one electron will be ejected.

Typical Feshbach resonances have certain simple properties. The first is that the phase shift of the spatial wave function increases by π over an energy range covering the resonance. The resonance has an energy width that depends on the coupling between the channels. The "probability" for being in the resonance peaks at the resonance energy; the maximum of this probability as a function of energy is inversely proportional to the energy width of the resonance.

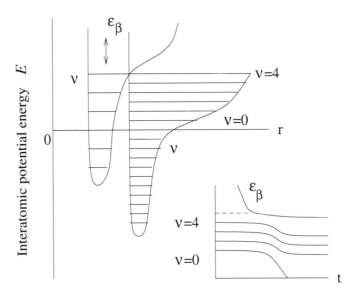

Figure 6.7: The two curves represent the interaction potential energy for two scattering channels of two atoms confined in a trap in the lowest vibration level ν. The resonant energy ε_β around a Feshbach resonance varies linearly with magnetic field. This resonant energy couples adiabatically with the atomic vibrations in the well, as shown in the insert plot, with the $\nu = 0$ level coupling to the resonant state. The atoms are transformed into the molecular bound state $(E < 0)$ which has an energy dependent on the magnetic field B. The transformation is reversible.

and $32\,\mu s$ for Na atoms. For readout of the qubit state, a SWAP operation to a marker atom followed by transport of this atom to a different lattice region for individual measurement suffices.

6.6.3   Cold atom collisions in state-dependent potentials

The goal of this proposal [7] is to implement a 2-atom phase gate, i.e., if two *internal* states of the atoms represent $|0\rangle$ and $|1\rangle$ for one qubit, then the 2-qubit state $|1\rangle|1\rangle \rightarrow -|1\rangle|1\rangle$ while all other qubit state combinations remain unchanged. This is to be accomplished by *state-selective* switching of the trapping potential, such that only atoms in qubit state $|1\rangle$ (a hyperfine structure state) would undergo a collisional interaction and experience a phase shift. This was to be achieved by suddenly lowering and then later raising a potential barrier between the wells confining the two atoms. The atom collision is a coherent interaction described by a pseudopotential proportional to the s-wave scattering length [6]; see Subsection 6.3.3. The collision results in a phase shift of the wave function for each atom in state $|1\rangle$. The amount of phase developed is controlled by the number of collisions, as well as the interaction strength.

For two separated atoms in the same internal quantum state, the internal state factorizes from the motional part of the total wave function. The motional part separates into center-of-mass and relative motion parts. If each atom is in state $|1\rangle$ the potential barrier drops, and the two atoms oscillate in a single well, in which collisions occur during an interaction time τ, as shown in Fig. 6.8. The atoms eventually return to their original position in separate traps, with an interaction phase that is adjusted to $\pm\pi$ by a proper choice of interaction time and trap parameters. If both atoms are in state $|0\rangle$, the potential remains unchanged and there can be no collisions of the atoms in the separated wells. For atoms in *different* states, the states do not factorize into internal and motional parts, and the center-of-mass and relative motions do not decouple. There is a phase shift ϕ_{01} which must be numerically evaluated, but which is eventually made negligible by proper choice of parameters.

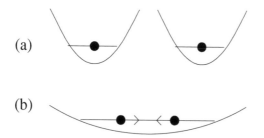

Figure 6.8: (a) At $t < 0$, the potentials confining the two atoms are the same for both internal atom states $|0\rangle$ and $|1\rangle$. (b) For $t > 0$, the potentials remain the same for the internal states $|0\rangle$, but are changed to a single well for internal states $|1\rangle$, in which the atoms collide. The well is eventually returned to its original double form, with a single atom in each well. Only atoms in state $|1\rangle$ experience the collisional phase shift.

In a suggested physical implementation, the Stark shift induced by an external electric field results in a confining energy independent of the atom hyperfine structure level. The magnetic dipole moment in a hyperfine state $|F, m_F\rangle$ and an external static magnetic field give an energy which depends on the quantum number m_F of the atomic spin state. Together, the two effects produce a trapping potential that has a magnitude and shape dependent on the internal atomic state.

Atomic mirrors, for example, created by the sinusoidal magnetization of a conventional programmed videotape, plus an external bias magnetic field, result in a magnetic trapping potential with non-zero periodic potential minima lying fractions of a micrometer above the tape surface [26]. Trapping frequencies can range from tens of kHz to several MHz. Microscopic electrodes nanofabricated on the atom mirror surface can produce the electric potentials needed to implement the 2-atom gate. For ^{87}Rb atoms in the

5 $S_{1/2}$ level, the hyperfine states $|0\rangle = |F, m_F\rangle = |1, -1\rangle$ and $|1\rangle = |2, 2\rangle$. Simulations [7] show that the collision interaction phase shift for both atoms in qubit state $|1\rangle$ increases in steps when the atoms collide at the center of the common well. Both atoms in qubit state $|0\rangle$ remain in separate wells. For atoms in *different* states, the phase shifts due to the collisions occur in larger steps when the atoms are near turning points. If the potential minimum of the well for state $|0\rangle$ is displaced along a transverse direction from the well for state $|1\rangle$, collision interactions will not occur when the states are different, and the collision-induced phase shift occurs only for both atoms in state $|1\rangle$. At zero temperature, the maximum simulated gate fidelity was found to be $F(0) = 0.99$, and even at $T \approx 2\hbar\omega/k_B \approx 3\,\mu K$, the fidelity $F(T) \approx 0.96$. Gate operation times were ≈ 0.4 ms.

6.6.4 Quantum gates based on interactions between Rydberg states

A fast phase gate for neutral atoms has been proposed by Jaksch et al. [30]. It entangles atoms by strong interactions of permanent dipole moments of laser-excited Rydberg (high principal quantum number n) states in a constant electric field. The gate operation times are set by the time scale of the laser excitation of the 2-atom interaction energy: this time is much shorter than a trap oscillation period. The gate is insensitive to atom temperature, or to small variations in atom-atom separation.

For hydrogen atoms, with degenerate Rydberg states with the same principal quantum number n, an electric field $\boldsymbol{E}$ along the z-axis splits the energy levels by $\Delta U_{nqm} = 3nqea_0E_z/2$, where quantum number $q = n - 1 - |m|$, $n - 3 - |m| \ldots - (n - 1 - |m|)$ is called the parabolic quantum number, m is the usual magnetic projection quantum number, e is the charge unit, and a_0 is the Bohr radius. These Stark states have permanent electric dipole moments $\boldsymbol{d} = d_z\hat{z} = 3nqea_0z/2$. In alkali atoms, these effects occur somewhat differently in detail, but still apply. The simpler case of hydrogen atoms is considered further. For two atoms 1 and 2 at fixed relative positions $\boldsymbol{r}$, prepared in Stark eigenstates by laser excitation to a common value of m, the dipole-dipole interaction potential

$$V_{\text{dip}}(\boldsymbol{r}) = \frac{1}{4\pi\varepsilon_0}\left[\frac{\boldsymbol{d}_1 \cdot \boldsymbol{d}_2}{|\boldsymbol{r}|^3} - \frac{3(\boldsymbol{d}_1 \cdot \boldsymbol{r})(\boldsymbol{d}_2 \cdot \boldsymbol{r})}{|\boldsymbol{r}|^5}\right] \tag{6.39}$$

describes the interaction. In the limit that the energy splitting between two Stark states produced by the electric field $E_z\hat{z}$ is $\gg V_{\text{dip}}$, then the diagonal terms of V_{dip} produce an energy shift, while the non-diagonal terms couple adjacent m manifolds $(m, m) \to (m \pm 1, m \mp 1)$ with each other. If the initial Stark eigenstate $|n, q = n - 1, m = 0\rangle$ is chosen, then this coupling is suppressed, and the energy shift becomes $U(R) = -9[n(n - 1)^2](a_0/R)^3(e^2/4\pi\varepsilon_0a_0) \propto n^4$. This large energy shift, for atoms separated by R, is used for entanglement.

The qubits are internal ground state hyperfine structure levels $|g\rangle_j$ and $|e\rangle_j$ for each atom $j = 1, 2$, of which only the states $|g\rangle_j$ are coupled by laser excitation to a given Rydberg Stark eigenstate $|r\rangle_j$ ($j = 1, 2$). This can be accomplished exploiting angular momentum selection rules for electric dipole transitions. The Hamiltonian of the system can be separated into two parts: one part describes the motion and trapping of atoms, while the second part describes the internal atom states. To execute a phase gate on the internal states, the gate operation time Δt must be much shorter that any external state evolution time, to avoid entanglement of the internal and external degrees of freedom. The spatial width of the atomic wave function is assumed to be much smaller than the atom separation R.

Expanding (6.39), $V_{\text{dip}}(\hat{r}) = U(R) - F(\hat{r}_z - R) + \dots$ with $F = 3U(r)/R$. $U(R)$ gives the internal energy shift if both atoms are excited to state $|r\rangle$, while the second term gives the mechanical force on the atoms due to V_{dip}, which contributes to the motional part of the Hamiltonian. Assume that the atoms can be individually addressed by the exciting laser beam, with a Rabi frequency Ω_j to excite atom j from the ground state $|g\rangle_j$ to the Rydberg state $|r\rangle_j$. Consider the regime $\Omega_j \ll U(R)$ and zero laser detunings $\delta_j = 0$. The three steps to execute the gate are now

(1) apply a π pulse to atom 1, swapping the ground state $|g\rangle$ to the Rydberg state $|r\rangle$;

(2) apply a 2π pulse to atom 2 (tune the "area" of the pulse similarly to step 1); and

(3) apply a π pulse to atom 1.

This pulse sequence results in a sign change of the wave function for atoms initially in $|ge\rangle$ or $|eg\rangle$. The state $|ee\rangle$ is not affected. The state $|gg\rangle$ is excited, and accumulates a phase shift that executes the phase gate, as visualized in Fig. 6.9. The gate operation time is $\Delta t = 2\pi/\Omega_1 + 2\pi/\Omega_2$, and excited state loss of population from $|r\rangle$ due to spontaneous decay is only a few percent.

This same gate can be performed adiabatically, with the advantage that individual addressing of the two atoms is not required: $\Omega_{1,2}(t) \equiv \Omega(t)$ and identical laser detunings $\delta \neq 0$ are used. The gate operation time Δt needs to be about ten times longer to be adiabatic. If alternatively $\Omega_j \gg U(R)$ is chosen, an undesirably large mechanical force F occurs in the non-adiabatic gate.

Gate fidelity is estimated to be high, with the probability of exciting a trap state *without* changing the internal state of the atoms $< 2.4 \times 10^{-3}$ in an adiabatic approximation with $\Delta t = 100/\Omega$. The optical potential experienced by the atom in the Rydberg state $|r\rangle$ may be different from that for the qubit states $|g\rangle$ and $|e\rangle$, permitting motional excitation during the gate. This probability is found to be $< 3.9 \times 10^{-3}$ in an analytical approximation [30].

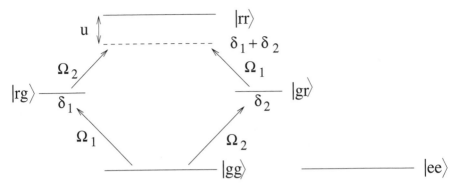

Figure 6.9: The two atom ground state $|gg\rangle$ is coupled to the Rydberg states $|rg\rangle$ and $|gr\rangle$ by laser excitation with Rabi frequencies $\Omega_1(t)$ and $\Omega_2(t)$. A π pulse is applied to atom 1 to bring the system to the state $i|rg\rangle$. A 2π pulse applied to atom 2 (in terms of the unperturbed states) is detuned from $|rr\rangle$ by the interaction energy u, and accumulates a trivial phase shift. A second π pulse applied to atom 1 results in $e^{i\pi}|gg\rangle$, with a phase $\phi = \pi$, to complete the phase gate.

6.6.5 Quantum gates for qubits implemented in motional states

A 2-atom quantum gate based on adiabatically interacting two bosonic atoms in different microtraps has been described [15], with the qubit implemented in the motional states of the confined atoms, i.e., $|0\rangle$ and $|1\rangle$ are the ground and first vibrational states of the (harmonic) trap, respectively. Tunneling and cold, controlled collisions are important interactions when the traps are close together, as diagrammed in Fig. 6.10. The dynamics depend strongly on the particular state of motion of the atoms. After the traps are separated, the gate has been implemented successfully, provided only one atom is left in each trap. Both ^{85}Rb and ^{87}Rb are considered, which have, respectively, negative and positive scattering lengths in cold collisions (cf. Subsection 6.3.3). A quantum gate can be performed in about 20 ms, using optical microtraps [14].

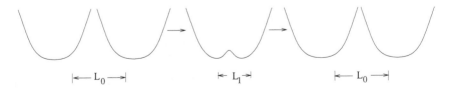

Figure 6.10: Originally, with traps separated by L_0, atoms in each trap do not interact. When the traps are brought together, tunneling and cold collisions occur to produce a SWAP gate. The traps are separated at the appropriate time to leave one atom in each separated well.

When the atom traps are well-separated, single qubit quantum gates can be executed using two laser pulses in a Raman configuration on either trapped atom (see Subsection 6.5.2). As the traps are moved together, quantum tunneling gives a probability that both atoms exist in the same trap, so the qubits are not well defined, and bosonic statistics become important. If the trap motion is well controlled, when the traps are separated there will again be a single atom in each trap (although not necessarily the same atom as initially). In the adiabatic limit (slowly transformed trap), the only allowed transitions are those initially degenerate in energy and which, at short distances, are coupled by tunneling or cold atomic collisions. In the presence of an interaction, the resonant couplings are $|01\rangle \leftrightarrow |10\rangle$, suggesting a SWAP gate, and a second coupling which must be suppressed or controlled by adjusting the interaction so that the final state is not populated. By letting the resonant coupling $|01\rangle \leftrightarrow |10\rangle$ act for one half of the time required for a complete SWAP, one can naturally implement the so-called $(\text{SWAP})^{1/2}$ gate which is imprimitive (contrary to the ordinary SWAP). In fact, two $(\text{SWAP})^{1/2}$ gates and single qubit operations implement a CNOT gate. A simulation of gate operation with a gate duration of about 17 ms and with $\omega_\| = 1.25 \times 10^4 s^{-1}$ gave a fidelity $F > 0.9997$, or an error rate $< 0.1\%$ per gate operation.

In the simulation, the minimum distance between the traps was $1\,\mu m$ (presently achieved using optical microtraps). Laser powers were $1-10$ mW per trap, with Rb atoms confined along the laser beam direction with $\omega_\|$ in the range 10^4 to $10^5 s^{-1}$ and $\omega_\perp$ 10 to 100 times higher. The lifetime of atoms in the traps is 100 to 1000 ms, and coherence is limited by spontaneous scattering of lattice photons, with one atom scattering about 50 photons per coherence time. This raises the total qubit error rate to about 3%. For single site addressing before and after gates, trap separations near $20\,\mu m$ are desired. Overall, including spontaneous scattering during gate times as long as 40 ms, qubit error rates between 5% and 12% were estimated for a practical test.

6.6.6 Quantum phase gate on an atom chip

Atom chips offer a highly integrated and scalable environment with multiple trapping sites for atoms and the possibility of swapping qubits between different degrees of freedom, each with two energy levels. In one particular example [9], a phase gate for a pair of Rb atoms trapped on a chip is analyzed. One degree of freedom stores information while another processes it. As "storage qubit", a pair of hyperfine states is used, while the phase gate operations employ two vibrational levels of the atom trap.

On an atom chip, ^{87}Rb $5S_{1/2}$ ground state hyperfine levels $|F = 2, m_F = 1\rangle$ and $|F = 1, m_F = -1\rangle$ have the same magnetic moment at a "magic" magnetic field $B_M = 3.23$ G [33], used as an offset field. These states then act like clock states $|F, m_F = 0\rangle$ for atoms in free space, exhibiting quadratic Zeeman and Stark shifts in stray fields. Alternatively, for calculations, the qubit can be encoded on external motional states of the confined atom, such as the ground and first excited states in a harmonic trap, or the left and right states of a double well, with the atoms in the same internal state.

Denoting clock state information storage qubits as $|0\rangle$ and $|1\rangle$, and vibrational state operation qubits as $|g\rangle$ and $|e\rangle$, a phase gate can be implemented by collisions [9]. Information is SWAPed from the vibrational logic states to the internal states for storage between gates. This is accomplished by 2-photon adiabatic Raman sideband excitation in the Lamb–Dicke regime (trap oscillation amplitude < laser wavelength), based on wavelengths near $800\,nm$. The desired sideband excitation $|1g\rangle \leftrightarrow |0e\rangle$ requires $\Delta k \approx 1.43$ cm^{-1}, a small linewidth, to prevent the undesired $|0g\rangle \leftrightarrow |1e\rangle$ transition. In this way the information is transferred between the internal and external qubit states.

A double well configuration is preferred for the 2-atom phase gate. The two qubits are trapped in two adjacent magnetic wells, and interact via tunneling through the barrier. The logic states are the ground and first excited vibrational states in each potential well—these states are linear superpositions of the eigenstates of the trapping potential. The ground state must be deep in the well to avoid tunneling, while the excited state must interact to develop the correct value of the phase ϕ of the gate. The gate requires complete revival of all initial 2-atom states, and fulfillment of the phase condition $\phi = \pi = \phi_{ee} + \phi_{gg} - 2\phi_{ge}$. Simulations show a near complete revival of the wave functions after 16.25 ms, with fidelity > 0.99, with the state $|gg\rangle$ stationary on this time scale.

6.7   Coherence properties of atom gates

Coherence of the wave functions of isolated atom internal hyperfine structure states is expected to be quite high, since uncontrolled atom collisions are rare, and collisions of cold atoms do not greatly perturb the hyperfine structure. This is one of the main advantages of confined atomic qubits. On the other hand, stray fields or field gradients can decohere the state functions, particularly through the magnetic interaction, if the qubits are moved from one location to another. In some gate schemes, it is possible to choose "clock" states as internal atom qubits, which interact only quadratically with external fields. The energy separation of these states, and hence any phase difference developed between them, is relatively immune to external field variations while the qubits are moved from site to site. On atom chips, hyperfine levels may have the same magnetic moment at so-called "magic" fields, which can be used as bias fields for an array of trap sites. In this environment, such levels are similarly phase coherent, as discussed in Subsection 6.6.6. However, the chip surfaces radiate thermal magnetic fields that can flip the magnetic moment in a resonant process. This may transform a weak-field seeking atom into a high-field seeker that is subsequently expelled from the trap. The corresponding trap lifetimes are proportional to the magnetic energy density at the Zeeman splitting frequency, which shows some power law dependence on the distance to the chip surface [16, 24, 25]. Typical lifetimes are in the range of at least few seconds at distances of a few microns.

Other coherence problems are associated with the implementation of particular gate schemes. Gates based on excitation to high quantum number n Rydberg states [30], described in Subsection 6.6.4, have potential decoherence associated with use of the dipole-dipole interaction, which scales as n^4 and so is strongest for high n. Rydberg states can be ionized by blackbody radiation, or by trapping laser fields. Spontaneous emission of radiation from the Rydberg states may also occur. These loss processes result in lower gate fidelity, and loss of unitarity in the gate operations. However, blackbody radiation ionization becomes negligible for quantum numbers $n < 20$, and spontaneous emission and field ionization both decrease as n^{-3}. Consequently, an optimal choice of n minimizes decoherence by these mechanisms. It is also possible to excite a trap state, without changing the internal state of the atoms. The probability of this has been evaluated in a simulation to be $< 2.4 \times 10^{-3}$, and a much smaller value is obtained in an analytic approximation [30].

Transport of qubits in the execution of gates or in the transfer of information is usually associated with heating of the atomic motion, which must be re-cooled to perform further gates. Re-cooling using a laser results in resonance radiation, which can excite other qubits if absorbed. It is preferable to re-cool a qubit in a remote location, but the qubit is likely to re-heat when it is returned. A better alternative is to use a different isotope as a cooling agent, which can be distributed in the lattice of trapping sites, or even a different type of atom. The spontaneous emission of radiation associated with cooling is then shifted in wavelength from that absorbed by a qubit atom. Perhaps the best solution is to use another cooling means. Cooling by immersion of atoms in superfluid helium has been proposed [11].

When molecular interactions are used in gates (Subsection 6.6.2), the superlattice is implemented by a far-off-resonance optical lattice, which suppresses decoherence due to the scattering of "lattice" photons by off-resonant qubit atoms. In this scheme, the marker atoms are always confined in a trap during transport. The register (internal) qubits never become entangled with the motional (gate) qubits, eliminating a major source of decoherence during gate operations. Clock states can be used for the internal qubits. Other lattice schemes may not permit a large degree of detuning of the lattice photons, which normally preserves coherence by reducing photon scattering.

When motional states are used as qubits, with adiabatic interactions as in Subsection 6.6.5, non-adiabatic effects produce an expected error rate $\leq 1\%$. Shorter gate durations rapidly increase the error rate, populating unwanted states. Doubly occupied trap states become resonant with singly-occupied states, and parameters of the system must be chosen to avoid this degeneracy. Coherence is mostly limited by spontaneous scattering of lattice photons at intervals near 10 ms, contributing a 2% error rate. Fidelity is very sensitive to the minimum distance of the atoms, since tunneling between wells has an exponential dependence on distance.

Traps must be loaded with exactly one atom per site for gates based on controlled cold collisions [7] (Subsection 6.6.3), and readout should be performed without removing atoms from the trap. Coherent atom transfer between sites is essential for this gate scheme. In an optical lattice trap (Subsection 6.6.1) with blue-detuning, there is some residual photon scattering from the atoms, and there is also the possibility that the intense blue-detuned laser beam will be resonant with a molecular potential [31], resulting in uncontrolled atom collisions. Long range molecular potentials may affect the atom-atom interaction, particularly when the atoms are in the same well. The qubits are generally spaced closer than the wavelength λ, complicating qubit readout using a laser due to the diffraction limit of the focused laser spot.

It seems clear that the details of gate operation are important to the limits to computation set by coherence. Storage time for an individual atom in present traps seems to be limited to about a second. However, this area of research is relatively young, so it is too early to form definite opinions about the ultimate limits which may be set by decoherence in an actual computational device.

6.8   Recapitulation: atomic data tables for atoms and ions

We can now go over the techniques of trapping of ions (Chap. 5) and that of atoms (addressed in this chapter). Most of such experiments are actually performed on a very limited number of chemical elements. The main reason for this restriction is related to the need to have spectral properties simple enough to permit efficient cooling according to the principles described in the main text. Although the list of candidate elements is continually expanding and even tough spectra such as that of Erbium are being tamed for the purpose of laser cooling [35], here we will list but the standard examples used in most approaches and which, due to the much less involved experimental requirements, should remain the candidates of choice also in the near future.

6.8.1   Ions

This subsection should really belong to Chap. 5 but we place it here for the expediency of recapitulation, together with atom trapping. Elements in the second column of the periodic table (e.g., Be, Mg, Ca, Sr and Ba) have two valence electrons. When ionized they lose one of these and become isoelectronic to the contiguous alkali metals. Second column ions are popular

Ion	$\lambda_1\,(nm)$	$\lambda_2\,(nm)$	$\lambda_3\,(nm)$	$\lambda_4\,(nm)$	$\lambda_5\,(nm)$	$\lambda_6\,(nm)$	$\lambda_7\,(nm)$
Be^+	313	313	-	-	-	-	-
Mg^+	280	279	-	-	-	-	-
Ca^+	397	393	866	850	854	732	729
Sr^+	422	408	1092	1004	1033	687	674
Ba^+	493.41	455.40	649.69	585.37	614.17	2052	1760
Zn^+	206	202.6	-	-	-	-	-
Cd^+	226.5	214.5	-	-	-	-	-
Hg^+	194	164	-	-	-	-	-

Table 6.1: Transition wavelengths (in nm) of a representative list of ionic species employed in ion-trap quantum-information experiments. Consult Fig. 6.11 to identify the individual electronic transitions. The wavelengths are given in terms of the energies $E_{e,g}$ of the upper and lower states by $2\pi\hbar c/\lambda_i = E_e - E_g$.

because of this feature—-the single remaining valence electron is a simple system that allows a straightforward application of standard laser cooling methods. Excellent candidates are also provided by the second column side-group elements Zn, Cd and Hg. We do not include in our table the more or less fortuitously convenient Yb^+ ion nor the elements from the third main group (i.e., B, Al, Ga, In and Tl). The level scheme depicted in Fig. 6.11 defines the transition wavelengths λ_j listed in Table 6.1. For more details, consult [42]. We remark that the choice for a specific transition is often made with regards to the more mundane problem of commercially available and stable, tunable laser systems.

Standard selection rules apply: dominant (i.e., electric dipole) transitions are "allowed" for $\Delta\ell = \pm 1$, for example, between $\ell = 0$ (i.e., "S") and $\ell = 1$ (i.e., "P") states or between $\ell = 1$ and $\ell = 2$ (i.e., "D") states. See Box 3.1. The transitions between S and D states are "forbidden" as electric dipole transitions, but they can occur as much less probable electric quadrupole transitions (involving two photons). Spontaneous decay times from optically allowed states are typically in the order of 10^{-8} s. The spontaneous decay time from D states, on the other hand, can be as large as 83 s (resp. 37 s) in the case of the Ba infrared $D_{5/2} \to S_{1/2}$ (resp. $D_{3/2} \to S_{1/2}$) transitions. (For this reason, such states are called *metastable*.) In view of the typical time scales involved in the coherent manipulation of quantum information, the $D_{5/2} - S_{1/2}$ transition—when available—provides a convenient means to represent a qubit. The two states can be detected efficiently by illuminating the ion with laser light resonant on the $S_{1/2} - P_{1/2}$ transition: the laser photons can be absorbed by the $S_{1/2}$ state only, and if one is absorbed, the ion is getting de-excited from the $P_{1/2}$ state within a few 10^{-8} s and emits a so-called "fluorescence photon" that can be detected. The process continues cyclically, producing in fact a strong fluorescence. This does not happen for the $D_{5/2}$ state because the conditions for absorption are not satisfied: the ion remains "dark". The two qubit states can also be coupled directly (even if weakly) by a laser to perform arbitrary qubit rotations, similar to the procedure discussed in Subsection 6.5.1.

Whenever metastable D-states are either nonexistent or impractical, ground-state hyperfine splittings can alternatively be employed to represent qubits. Their energy splittings are in the microwave domain and the qubit can be manipulated via microwave pulses or with the Raman technique. Since their origin is linked to the properties (in particular the spin) of the atomic nucleus, they are isotope specific. Table 6.2 lists some representative examples of these hyperfine splittings.

6.8.2 Atoms

The preferred candidates for pure optical trapping of *neutral* atoms are typically hydrogenic (the alkali metals Li, Na, K, Rb and Cs) or metastable noble (see Table 6.3 for the corresponding transitions). In order to keep decoherence produced by spontaneous decays to a minimum, for the pur-

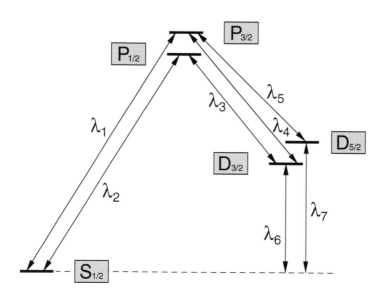

Figure 6.11: Optical transition labeling used in Table 6.1. The hyperfine splitting of the states (see Table 6.2) is not shown.

Ion isotope	$S_{1/2}$ (GHz)	$P_{1/2}$ (GHz)	$P_{3/2}$ (GHz)	$\Gamma[P_{3/2}]$ (MHz)	$\Gamma[P_{1/2}]$ (MHz)
$^9\text{Be}^+$	1.25	-	-	20	-
$^{25}\text{Mg}^+$	1.8	-	-	43	-
$^{43}\text{Ca}^+$	3.2	-	-	23	-
$^{87}\text{Sr}^+$	5	-	-	23	-
$^{135}\text{Ba}^+$	7.2	-	-	19	15
$^{137}\text{Ba}^+$	8	1.5	-	-	-
$^{111}\text{Cd}^+$	14.5	2	0.8	60	-
$^{113}\text{Cd}^+$	15.3	2	0.8	60	-
$^{199}\text{Hg}^+$	40.5	6.9	-	-	70

Table 6.2: Representative list of isotopes used in ion-based quantum informatics and corresponding hyperfine energy splittings (in GHz) of S and P levels. The last two columns quantify the decay rates from the $l = 1$ states $P_{3/2}$ and $P_{1/2}$. The splitting frequencies ν_{hfs} are given in terms of the energies $E_{e,g}$ of the upper and lower hyperfine states by $2\pi\hbar\nu_{\text{hfs}} = E_e - E_g$.

pose of trapping, these transitions are driven far off-resonance, so that the atoms essentially remain in their ground (or effective-ground, in the case of metastables) state. For this reason, it is unnecessary here to graphically present a more detailed picture of all level scheme details. Quantum information is usually stored in the hyperfine structure of the respective ground states, if existent. (A splitting is absent in the case of spinless nuclei, as is the case for the main isotope ^{52}Cr of chromium.) Table 6.4 lists the hyperfine splittings for the stable isotopes of the alkali atoms quoted in Table 6.3. This Table 6.4 also gives data to determine the interaction with static magnetic and electric fields, as discussed in Section 6.3: the nuclear spin determines an effective magnetic potential and the static polarizability an electric potential. More details can be found in Ref. [42].

6.9 Assessment

At present the experimental implementation of cold, confined atom quantum logic substantially lags behind the theoretical progress in this area. On the other hand, the recent development and tests of many basic experimental techniques using cold atoms, which are occurring at a steady rate, has spurred the theoretical development. This happy convergence of experiment and theory promises to lead to continued rapid progress in this area of research. As discussed in Section 6.6, a number of clever 2-atom quantum gate designs have been proposed and theoretically modeled. These gates are implemented using a variety of interactions, such as cold, controlled atom collisions, tuned Feshbach resonances of 2-atom molecular levels, interaction of atomic Rydberg states excited by laser beams, and gates executed using the motional states of atoms confined in single or double wells. These gate alternatives, when modeled, provide plausible estimates of gate times and fidelities.

The experimental basis on which this theory rests consists of the cooling of atoms to very low temperatures using laser beams and evaporation, followed by the control of atomic motion and the confinement of these ultracold atoms in shallow electromagnetic wells. Internal ground state hyperfine structure states of the atom can act as highly coherent qubit states. The motional states of the atoms in the confining wells may also be used as qubit states, particularly when 2-atom gates are to be executed. These kinds of qubits have the advantage of being theoretically well-described and understood, so that predictions of gate behavior and optimization can be made.

An example of a means of confinement is *optical lattices*, arising from the interference of standing coherent electromagnetic waves produced by lasers. Confinement by the lattice fields implies an interaction with the atom quantum states, but the perturbations of the atom states can be made weak by tuning the lattice laser wavelengths far from any atomic resonance. A consequence is a low scattering rate of lattice photons, which

Atom	Transition	λ (nm)	τ (ns)	T_d (μK)	v_d (cm/s)	T_r (μK)	v_r (cm/s)
He*	$2^3S_1 \rightarrow 2^3P_2$	1083.03	98.04	38.96	28.44	4.08	9.20
	$2^3S_1 \rightarrow 3^3P_2$	388.86	106.83	35.75	27.25	31.6	25.6
Ne*	$3s[3/2]_2 \rightarrow 3p[5/2]_3$	640.22	19.5	196	28.4	2.32	3.09
Ar*	$4s[3/2]_2 \rightarrow 4p[5/2]_3$	811.53	30.2	126	16.2	0.73	1.23
Kr*	$5s[3/2]_2 \rightarrow 5p[5/2]_3$	811.29	28	136	11.6	0.35	0.59
Xe*	$6s[3/2]_2 \rightarrow 6p[5/2]_3$	881.94	34	112	8.4	0.188	0.34
Li	$2^2S_{1/2} \rightarrow 2^2P_{3/2}$	670.78	27.1	141	41.1	6.1	8.57
Na	$3^2S_{1/2} \rightarrow 3^2P_{3/2}$	588.99	16.2	236	29.2	2.40	2.95
K	$4^2S_{1/2} \rightarrow 4^2P_{3/2}$	766.49	26.4	145	17.5	0.83	1.33
Rb	$5^2S_{1/2} \rightarrow 5^2P_{3/2}$	780.03	27	141	11.7	0.37	0.60
Cs	$6^2S_{1/2} \rightarrow 6^2P_{3/2}$	852.11	30.52	125	8.8	0.20	0.35
Cr	$4a^7S_3 \rightarrow 4z^7P_4$	285.21	2.0	1910	81	9.7	5.8

Table 6.3: Representative list of neutral atom species employed for pure-optical (i.e., induced dipole-force) trapping. In the leftmost column, the asterisks mean that these atoms are in a metastable state, not in their true ground state. Here λ is the corresponding transition wavelength and τ the upper state transition lifetime (corresponding transition linewidth $\Gamma = 2\pi/\tau$). The transition wavelengths differ only slightly between the different isotopes of the same element. The Doppler and recoil limit temperatures achievable by laser cooling with the respective transitions are denoted T_d and T_r. At the recoil limit, the atom de Broglie wavelength equals to the light wavelength. To illustrate the orders of magnitude of velocities involved, we also include in this table the Doppler velocity $v_d = [h\Gamma/(2m)]^{1/2}$ and the recoil velocity $v_r = h/(m\lambda)$, where h is Planck's constant and m the atomic mass. (This table is adapted from (2).)

Isotope	ν_{hfs} (MHz)	I	α (Å^3)
^{6}Li	228.2	1	24
^{7}Li	803.5	3/2	24
^{23}Na	1772	3/2	24
^{39}K	462	3/2	44
^{40}K	1286	4	44
^{41}K	254	3/2	44
^{85}Rb	3036	5/2	47
^{87}Rb	6835	3/2	47
^{133}Cs	9192.6	7/2	60

Table 6.4: The hyperfine splittings of the ground state of selected alkali atoms (second column). The third column gives the nuclear spin I and determines the effective magnetic moment in a weak magnetic field: the allowed values for the total spin are $F = I \pm 1/2$. For a given magnetic quantum number $m_F = -F, \ldots, F$, the magnetic interaction energy in a weak field is $V_{\text{mag}}(m_F) \approx \mp \mu_B |B| m_F / (I + 1/2)$ where $|B|$ is the magnitude of the magnetic field and the Bohr magneton $\mu_B = 2\pi\hbar \times 1.4\,\text{MHz/G} = k_B \times 68\,\mu\text{K/G}$. (For a more careful discussion, see Subsection 6.3.2.) The fourth column gives the electric polarizability α in units of Å^3. In a given static electric field with magnitude $|E|$, the electric potential is $V_{\text{el}} = 2\pi\varepsilon_0\alpha|E|^2$ which corresponds to $2\pi\hbar \times 84\,\text{kHz} = k_B \times 4.1\,\mu\text{K}$. for $\alpha = 1\,\text{Å}^3$ and $|E| = 1\,\text{V/}\mu\text{m}$. (These data are adapted from Refs. (19, 37).)

tend to disrupt the atom qubit states in an uncontrolled way. The relative position of lattice sites can be manipulated by changing laser beam parameters, to facilitate interactions between individual atoms in neighboring lattice sites, and then transporting information to other sites. An experimental problem associated with this technique is the addressing of individual atoms with laser beams to execute single qubit gates, since the lattice sites are spaced near $\lambda/2$, below the diffraction limit. There may also be problems with reading out information from individual atoms for the same reason, and with filling all sites with exactly one atom per site.

Some of these problems are alleviated by using microscopic atom traps near semiconductor surfaces, generated by microstructures producing strong electric and magnetic field gradients. Optical effects can also be used to transport, confine, and control atoms near the surface of these devices. The combination of microchip techniques with miniaturized optical components to manipulate atoms has resulted in the "atom chip" concept. Atom chips seem to hold much promise for future developments in the area of control and manipulation of atoms to execute quantum information procedures, since regular arrays of traps can be produced, with sufficient separation for laser manipulation of individual atoms, and with trap depths which can be adjusted easily to interact with the confined atoms. Different types of atoms can be introduced to facilitate atom cooling, and bias magnetic fields can be used to adjust internal atomic state energies to minimize decoherence during atom transport on the chip.

New experimental techniques continue to appear, such as Bose–Einstein condensates of atoms loaded into optical lattices, with proposals for quantum logic gates quickly following. An advantage of this technique lies in the ability to produce a coherent atomic state, which potentially can be used to place a single atom in each lattice site [18], difficult to accomplish by other methods. An alternative is also possible with fermionic atoms [48].

In sum, one has the impression of a vibrant field with advances rapidly appearing in theory as well as experimentation. At present, the implementation of quantum gates using confined atoms is experimentally behind this topic using confined ions, but the theory of gate operation appears to be on a comparable level in both areas. It is far too early to form any judgments about the eventual viability of quantum computation using confined atoms, since the parameters that eventually will constrain a practical implementation are still poorly defined.

References

[1] W. Alt, D. Schrader, S. Kuhr, M. Müller, V. Gomer, and D. Meschede, *Phys. Rev. A* **67** (2003), 033403.

[2] K.G.H. Baldwin, *Contemp. Phys.* **46** (2005), 105.

[3] T.H. Bergeman, P. McNicholl, J. Kycia, H. Metcalf, and N. Balazs, *J. Opt. Soc. B* **6** (1989), 2249.

[4] G. Birkl, F.B.J. Buchkremer, R. Dumke, and W. Ertmer, *Opt. Commun.* **191** (2001), 67.

[5] G.K. Brennen, C.M. Caves, P.S. Jessen, and I.H. Deutsch, *Phys. Rev. Lett.* **82** (1999), 1060.

[6] T. Calarco, U. Dorner, P.S. Julienne, C.J. Williams, and P. Zoller, *Phys. Rev. A* **70** (2004), 102306.

[7] T. Calarco, E.A. Hinds, D. Jaksch, J. Schmiedmayer, J.I. Cirac, and P. Zoller, *Phys. Rev. A* **61** (2000), 022304.

[8] S. Chu *Rev. Mod. Phys.* **70** (1998), 685.

[9] M.A. Cirone, A. Negretti, T. Calarco, P. Krüger, and J. Schmiedmayer, *Eur. Phys. J. D* **35** (2005), 165.

[10] C.N. Cohen-Tannoudji, *Rev. Mod. Phys.* **70** (1998), 707.

[11] A.J. Daley, P.O. Fedichev, and P. Zoller, *Phys. Rev. A* **69** (2004), 022306.

[12] J. Dalibard and C. Cohen-Tannoudji, *J. Opt. Soc. Am. B* **6** (1989), 2023.

[13] I. Dotsenko, W. Alt, M. Khudaverdyan, S. Kuhr, D. Meschede, Y. Miroshnychenko, D. Schrader, and A. Rauschenbeutel, *Phys. Rev. Lett.* **95** (2005), 033002.

[14] R. Dumke, M. Volk, T. Müther, F.B.J. Buchkremer, G. Birkl, and W. Ertmer, *Phys. Rev. Lett.* **89** (2002), 097903.

[15] K. Eckert, J. Monpart, X.X. Yi, J. Schliemann, D. Bruss, G. Birkl, and M. Lewenstein, *Phys. Rev. A* **66** (2002), 042317.

[16] R. Folman, P. Krüger, J. Schmiedmayer, J.H. Denschlag, and C. Henkel, *Adv. At. Mol. Opt. Phys.* **48** (2002), 263.

[17] H. Gauck, M. Hartl, D. Schneble, H. Schnitzler, T. Pfau, and J. Mlynek, *Phys. Rev. Lett.* **81** (1998), 5298.

[18] M. Greiner, O. Mandel, T. Esslinger, T.W. Hänsch, and I. Bloch, *Nature* **415** (2002), 39.

[19] R. Grimm, M. Weidenmüller, and Y.B. Ovchinnikov, *Adv. At. Mol. Opt. Phys.* **42** (2000), 95.

[20] G. Grynberg and C. Robilliard, *Phys. Rep.* **355** (2001), 335.

[21] H. Haken and H.C. Wolf, *The Physics of Atoms and Quanta, Advanced Texts in Physics*, 6th ed. (Springer, Heidelberg-Berlin, 2000).

[22] W. Hänsel, J. Reichel, P. Hommelhoff, and T.W. Hänsch, *Phys. Rev. A* **64** (2001), 063607.

[23] T.W. Hänsch and A.L. Schawlow *Opt. Commun.* **13**, (1975), 68.

[24] D.M. Harber, J.M. McGuirk, J.M. Obrecht, and E.A. Cornell, *J. Low Temp. Phys.* **133** (2003), 229.

[25] C. Henkel, P. Krüger, R. Folman, and J. Schmiedmayer, *Appl. Phys. B* **76** (2003), 173.

[26] E.A. Hinds and I.G. Hughes, *J. Phys. D: Appl. Phys.* **32** (1999), R119.

[27] E.A. Hinds, C.J. Vale, and M.G. Boshier, *Phys. Rev. Lett.* **86** (2001), 1462.

[28] P. Horak, B.G. Klappauf, A. Haase, R. Folman, J. Schmiedmayer, P. Domokos, and E.A. Hinds, *Phys. Rev. A* **67** (2003), 043806.

[29] O. Houde, D. Kadio, and L. Pruvost, *Phys. Rev. Lett.* **85** (2000), 5543.

[30] D. Jaksch, J.I. Cirac, P. Zoller, S.L. Rolston, R. Cote, and M.D. Lukin, *Phys. Rev. Lett.* **85** (2000), 2208.

[31] P.D. Lett, *Ann. Rev. Phys. Chem.* **46** (1995), 423.

[32] G. Lévêque, C. Meier, R. Mathevet, B. Viaris, J. Weiner, and C. Girard, *Eur. J. Phys. AP* **20** (2002), 219.

[33] H.J. Lewandowski, D.M. Harber, D.L. Whitaker, and E.A. Cornell, *Phys. Rev. Lett.* **88** (2002), 070403.

[34] A. Marte, T. Volz, J. Schuster, S. Dürr, G. Rempe, E.G.N. van Kampen, and B.J. Verhaar, *Phys. Rev. Lett.* **89** (2002), 283202.

[35] J.J. McClelland and J.L. Hansen, *Phys. Rev. Lett.* **96** (2006), 143005.

[36] A. Messiah, *Mécanique quantique*, Vol. II, Dunod, Paris, 1995, new edition.

[37] R.W. Molof, H.L. Schwartz, T.M. Miller, and B. Bederson, *Phys. Rev. A* **10** (1974), 1131.

[38] N.R. Newbury, C.J. Myatt, E.A. Cornell, and C.E. Wieman, *Phys. Rev. Lett.* **74** (1995), 2196.

[39] M. Olshanii, *Phys. Rev. Lett.* **81** (1998), 938.

[40] W.D. Phillips, *Rev. Mod. Phys.* **70** (1998), 721.

[41] J. Reichel, W. Hänsel, P. Hommelhoff, and T.W. Hänsch, *Appl. Phys. B* **72** (2001), 81.

[42] J.E. Sansonetti and W.C. Martin, *NIST Handbook of Basic Spectroscopic Atomic Data*. Available online at
http://physics.nist.gov/PhysRefData/Handbook/index.html

[43] N. Schlosser, G. Reymond, and P. Grangier, *Phys. Rev. Lett.* **89** (2002), 023005.

[44] J. Schmiedmayer, *Eur. Phys. J. D* **4** (1998), 57.

[45] D. Schrader, I. Dotsenko, M. Khudaverdyan, Y. Miroshnychenko, A. Rauschenbeutel, and D. Meschede, *Phys. Rev. Lett.* **93** (2004), 150501.

[46] C.V. Sukumar and D.M. Brink, *Phys. Rev. A* **56** (1997), 2451.

[47] P. Treutlein, P. Hommelhoff, T. Steinmetz, T.W. Hänsch, and J. Reichel, (2003).

[48] L. Viverit, C. Menotti, T. Calarco, and A. Smerzi, *Phys. Rev. Lett.* **93** (2004), 110401.

[49] W. Zwerger, *J. Opt. B: Quantum Semiclass. Opt.* **5** (2003), 59.

Chapter 7

Quantum Dots Quantum Computing Gates

- Properties and fabrication of quantum dots

- Three basic designs of quantum dots computing gates

- Universality of quantum gates

- Measurements

7.1 Introduction

Quantum dot(s) (QD) fabrication is a major segment of contemporary nanotechnology. Presently, there are three major proposals for designing quantum computing gates based on the QD technology:

(1) electrons trapped in microcavity;

(2) spintronics;

(3) biexcitons.

In this chapter, we will survey the designs and show mathematically how they, in principle, will generate 1-bit rotation gates as well as 2-bit entanglement and, thus, provide a class of universal quantum gates. Some physical attributes and issues related to their limitations, decoherence and measurement are also discussed. This chapter is a slightly redacted form of a tutorial paper by Chen, et al. [11]. We gratefully acknowledge the **copyright permission from World Scientific Publishing Co. Pte. Ltd., Singapore**, for the reproduction of text and figures throughout this chapter.

Building devices to store and process computational bits quantum-mechanically (qubits) is a challenging problem. In a typical field-effect transistor

(FET) in an electronic computer chip, $10,000$ to $100,000$ electrons participate in a single switching event. It is impossible to isolate, out of such a complex system, two quantum mechanical states that would evolve coherently to play the role of a qubit.

QD devices, including diode lasers, semiconductor optical amplifiers, IR (infrared) detectors, mid-IR lasers, quantum-optical single-photon emitters, etc., are being developed and considered for a wide variety of applications. In this chapter, we hope to elucidate the connection between the physics of QD and the basic mathematics of quantum gate operations.

7.1.1 QD properties and fabrication: from quantum wells, wires to quantum dots

We begin by introducing what QD are. QD consist of nano-scale crystals from a special class of semiconductor materials, which are crystals composed of chemical elements in the periodic groups II-VI, III-V, or IV-IV. The size of QD ranges from several to tens of nanometers (10^{-9} m) in diameter, which is about 10-100 atoms. A QD can contain from a single electron to several thousand electrons since the size of the quantum dot is designable. QD are fabricated in semiconductor material in such a way that the free motion of the electrons is trapped in a quasi-zero dimensional "dot". Because of the strong confinement imposed in all three spatial dimensions, QD behave similarly to atoms and are often referred to as artificial atoms or giant atoms.

When a free electron is confined by a potential barrier, its continuous spectrum becomes discretized. In particular, the gap between two neighboring energy levels increases as the length where the free electron moves decreases. A similar thing happens to solid state. If the motion of electrons in the conduction band or that of the hole in the valence band is limited in a small region with a scale such as the De Broglie wavelength or a phase-coherence length,[1] then the conduction band or the valence band is split into subbands or discrete levels depending on the dimensionality of the confined structure. Such is the case when a material with a lower bandgap is confined within a material with a higher bandgap. More efficient recombination of electron-hole pairs can be achieved by incorporation of a thin layer of a semiconductor material, with a smaller energy gap than the cladding layers, to form a double heterostructure. As the active layer thickness in a double heterostructure becomes close to the De Broglie wavelength (about $10\,nm$ for semiconductor laser devices) or the Bohr exciton radius for lower dimensional structures, and as the motion of the electron is restricted within such a very small regime, energy quantization or momentum quantization is observed and quantum effects become apparent.

---

[1]If an electron travels far enough to be scattered by impurities or other electrons, it will lose its phase coherence (cf. the footnote on quantum coherence in Subsection 7.1.2). This is the *dephasing* we have been talking about all along. Here the length in which an electron travels and yet can keep its phase coherent is called the phase coherence length.

Therefore the electron states are not continuous but discrete. This phenomenon is known as the *size quantization effect*. A proposal for QD-based QC utilizing various energy levels of one or several electrons in QD by confining those QD in a *microcavity* will be studied in Section 7.2.

In natural bulk semiconductor material, the overwhelming majority of electrons occupy the valence band. However, an extremely small percentage of electrons may occupy the conduction band, which has higher energy levels. The only way for an electron in the valence band to be excited and be able to jump to the conduction band is to acquire enough energy to cross the *bandgap*. If such a jump or transition occurs, a new electric carrier in the valence band, called a *hole*, is generated. Since the hole moves in the opposite direction to the electron, the charge of a hole is regarded as positive. The pair of raised electron and hole is called an *exciton*. The average physical separation between the electron and the hole is called the *exciton Bohr radius*. Exciton moves freely in bulk semiconductor. However, an exciton is trapped by high energy barriers as an electron is. The size quantization effect is optically observable. A proposal utilizing excitons' (and also biexcitons') energy levels for QC will be the topic of discussion in Section 7.4.

If the device length is smaller than the phase coherence length of the electron or exciton Bohr radius, the energy levels are discrete and the size quantization effect is observed. Since the energy levels are discrete, the 3-dimensional energy band becomes lower-dimensional depending on the number of confinement directions. If there is only one directional length of device shorter than the phase coherence length, the device is regarded as a 2-dimensional device, called a *quantum well*. The phase coherence length of a quantum well is about $1.62\,\mu m$ for GaAs and about $0.54\,\mu m$ for Si at low temperature. However, since the phase coherence length depends on impurity concentration, temperature, and so on, it can be modified for electronic applications. The exciton Bohr radius of GaAs is about 13 nanometers.

There are two approaches to fabricate nano-scale QD: *top-down* and *bottom-up*. The semiconductor processing technologies, such as metal organic chemical vapor deposition, molecular beam epitaxy and e-beam lithography, etc., are used in the top-down approach. Surface and colloid chemistry, such as self-assembly, vapor-liquid-solid techniques, are used in the bottom-up approach.

There are many methods to synthesize QD in the bottom-up approach, e.g., chemical reactions in colloidal solutions, long time annealing in solid state, chemical vapor deposition on solid surface, wet or dry etching of thin film on solid surface, etc. Even though, through colloidal solutions crystalline QD are synthesized relatively economically and conveniently in the bottom-up approach, as compared with the top-down approach, the alignment of QD is a very serious problem for applications to QC even though such crystalline QD have been made into a few optical devices, such as optical sensors or field effect diodes, and biomedical apparatus.

As mentioned above, several semiconductor processing technologies can be applied to QD fabrication in the top-down approach. Usually, a quantum well is the starting point of the quantum dot fabrication. Thus, let us first describe the technology of the quantum well fabrication.

By molecular beam epitaxy and metal organic chemical vapor deposition techniques, an ultra-thin single crystalline layer can be deposited on a bulk substrate. The development of these advanced epitaxy techniques makes it possible to fabricate quantum wells with a very fine boundary. There are two types of quantum wells. One is formed by depositing several single crystalline layers through molecular beam epitaxy, or through the metal organic chemical vapor deposition technique. The other is by depositing single crystalline layers with modulated impurity concentration. The former is usually chosen for optoelectronic devices such as laser where electrons and holes need to be confined at the same time, and the latter is for electronic devices where only either the electron or the hole needs to be confined.

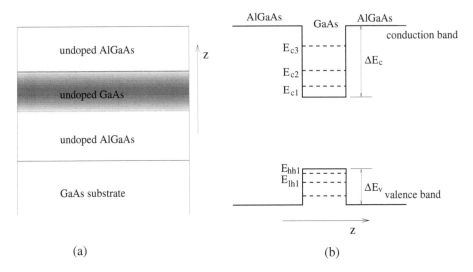

(a)   (b)

Figure 7.1: An AlGaAs/GaAs quantum well structure: (a) cross-section and (b) conduction and valence band profile. The energy barrier heights of the conduction band and of the valence band are ΔE_c and ΔE_v, respectively. E_{ci} is the lowest energy of the i-th conduction subband for $i = 1, 2$ and 3, E_{hh1} is the highest energy of the first heavy hole subband, and E_{lh1} is the highest energy of the light hole subband.

A typical application of quantum wells with single crystalline layers is AlGaAs/GaAs quantum well laser as shown in Fig. 7.1. An Aluminum Gallium Arsenide (AlGaAs) layer is deposited on a GaAs substrate and then a GaAs layer of thickness less than $100 \, nm$ is deposited on the AlGaAs layer, and AlGaAs layer is again deposited on the GaAs layer. The energy band profile shown in Fig. 7.1 is rotated by $90°$. The thickness of the GaAs quantum well is about or less than the exciton Bohr radius and size quantization can be observed clearly. The electron and hole in the GaAs quantum well are confined by the AlGaAs energy barrier, respectively. The laser wave length is dependent on the thickness of GaAs quantum well in the AlGaAs/GaAs quantum well laser.

Remark 7.1.1. We briefly mention here some properties of semiconductor materials that are used in the fabrication of QD.

Silicon oxide (SiO_2) has a low dielectric constant and is easily created by the oxidation of silicon; most of electronic processors are made of silicon and silicon oxide. Such technology has been developed by the silicon industry since the early 1960s. Nowadays silicon electronic devices of about $50\,nm$ in size can be fabricated in mass production.

GaAs is one of direct semiconductors for which a transition from the valence band to the conduction band does not require change of momentum for electrons. It can be used as a photodetector.

GaAs/AlGaAs heterostructures can make electrons have very high mobility at low temperature. The high mobility lengthens electrons' phase-breaking mean free path. Therefore, coherent transport can be observed in GaAs/AlGaAs devices of μm scale at low temperature.

Bulk CdS absorbs visual light of the yellow-green wavelength, and its resistivity decreases with increasing illumination. CdS can be used as a photodetector and window material in solar cells. □

Another quantum well application is the *high electron mobility transistor*. Fig. 7.2 shows the schematic of the cross-section of a high electron mobility transistor and its conduction band profile. The impurities in the doped AlGaAs layer provide electrons to the undoped GaAs layer, and the space charges and excess electrons bend the conduction band as shown in Fig. 7.2(b). Therefore, a quantum well is formed at the boundary of the undoped AlGaAs and the undoped GaAs. Since Si-based metal-oxide-semiconductor FET has a structure similar to high electron mobility transistor, quantum wells can be formed under certain conditions.

Fig. 7.3 shows how quantum wires and QD are fabricated from quantum wells in a top-down approach. One method is to remove part of a quantum well by etching, and the other is to apply an electric field above the quantum well. The former is used to confine electron and hole at the same time, while the latter is to confine only an electron or a hole since the electric field prevents electron (or hole) from forming beneath the metal electrode. To etch the semiconductor material or to deposit the electrode, a nano-scale pattern has to be transferred by lithography. Various lithography technologies have been developed, such as nano-imprinting, AFM-STM lithography, dip-pen nanolithography, etc. Nevertheless, electron-beam lithography is the most widely used currently.

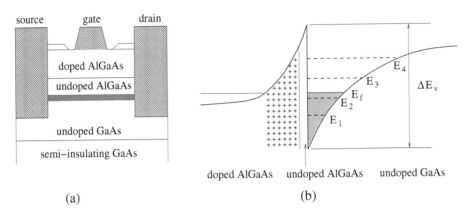

(a)   (b)

Figure 7.2: A high electron mobility transistor based on the modulating doping:
(a) cross-section and (b) conduction band profile of doped and undoped AlGaAs layers and undoped GaAs layer. The energy barrier
of the conduction band is ΔE_c. The quantum well lies just below the
undoped AlGaAs layer in (a), and lies to the right of the undoped
AlGaAs in (b). The "bend" in the conduction band is caused by the
excess electrons in the undoped GaAs layer and the spatial charges
in the doped AlGaAs layer (as marked by "+"). The energy levels
in the quantum well are discrete with regard to the dependence of
variable in the longitudinal direction.

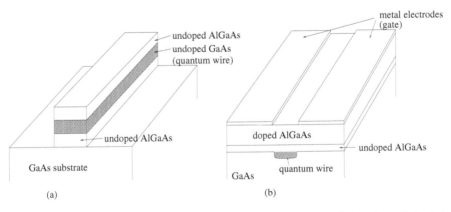

(a)   (b)

Figure 7.3: A quantum wire fabricated from a quantum well structure: (a) etching of a portion of a quantum well structure and (b) deposition of
metal gate on the top layer. The voltage applied to the metal gate
depletes electrons underneath the gate, and electrons can gather
underneath the place where the gate is not deposited.

7.1.2 QD-based single-electron devices and single-photon sources

In Subsection 7.1.1, QD fabrications have been discussed. In most of the proposals to be discussed in the next few sections, QD-based QC are known to be single-electron devices. Recent advances in epitaxial growth technology have lead to confinement of single electrons in semiconductor QD. In QD-based "single-electron transistors" (SET), the position of a single electron governs the electrical conductance. However, the same factors that make single-electron detection simple also complicate construction of a quantum computer based on sensing an electron's position. Charged electrons are easily de-localized by stray electric fields due to Coulomb interaction, and electrons placed in delicate entangled quantum states rapidly lose quantum coherence. The localization of a single dot can be achieved either by advanced epitaxial growth techniques or by using novel optical manipulation techniques such as near-field optical probe.

It was first predicted in 1938 that any two materials with different lattice constants would result in the formation of islands instead of flat layers beyond a critical thickness [36]. The growth of first strain induced islands were reported by Goldstein, *et al.* [15] in 1985 where InAs islands were formed on GaAs.[2] These islands can have sizes in the range of a few nanometers and can confine charge carriers both in the conduction band and in the valence band. Whatever we use the QD system for and whatever the fabrication technology we use, there will always be a statistical distribution of QD size and composition. This statistical distribution in turn produces inhomogeneous broadening of the QD optical response such as transition frequencies: This favors the distinction of one qubit from the others since the energy-domain discrimination is facile. Access to a specific qubit is achieved by positioning the excitation probe beam spot onto the desired location where a number of qubits with different frequencies can be accessed. Access to specific qubits can therefore be achieved by position selective addressing combined with frequency discrimination.

In addition, single-electron devices have a unique mechanism known as the *Coulomb blockade* which is different from size quantization. Single electron tunneling occurs at the ultra-small junction. Electron *cannot* pass through the ultra-small junction due to electrostatic charging energy, which is the Coulomb blockade. Only when the electrostatic charging energy can be lowered by electron tunneling, can a single electron then tunnel through the ultra-small junction, called a single electron junction. Quantitatively, when the capacitance of the junction is much smaller than $e^2/k_B T$, where e is the absolute charge of electron, k_B is the Boltzmann constant and T is the temperature, single-electron tunneling is observed. Fig. 7.4 shows a schematic of a single-electron tunneling device. It is built on a quantum well structure. There are two single electron junctions and one QD. If there

---

[2]This strain-induced-QD-growth method belongs to the bottom-up approach

is no electron in the QD, only a single electron can be at the QD by the single electron tunneling. Various single-electron devices have been introduced such as the SET aforementioned, single-electron box, single-electron turnstile, etc.

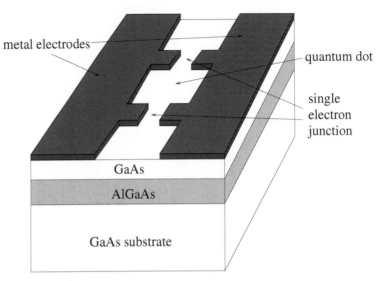

Figure 7.4: A split-dual gate single-electron tunneling device. The voltage applied to the gate leads to single-electron junctions at the narrow channels and a quantum dot between the narrow channels. The capacitance of the junctions and the quantum dot size are dependent on the gate voltage.

Next, we consider QD as single photon generation source. Single photons were first generated in a completely different kind of quantum dots, colloidal quantum dots, which are synthesized in solutions [24]. These dots tend to suffer from blinking and bleaching, thus, improvements in their stability is required if practical devices are to be built with these dots. Their properties are currently closer to those of molecules than to those of Stranski-Krastanow QD.[3] But because of the advantages of Stranski-Krastanow-grown QD, most research has concentrated on epitaxially grown QD. However, one advantage of colloidal dots over epitaxially grown dots is that they still emit efficiently at room temperature.

The recombination of an electron-hole pair leading to the emission of a photon with a specific energy is uniquely determined by the total charge configuration of the dot [20]. If a QD is optically pumped with a pulsed laser leading to the creation of several electron-hole pairs in the dot, then it

---

[3]Briefly speaking, Stranski-Krastanow QD are grown by utilizing the strain caused by the lattice mismatch between the QD layer and the substrate during molecular beam epitaxy to form QD islands which are in the shape of a cylinder, pyramid, or truncated pyramid, with dimensions about exciton Bohr radius (see, e.g., [4, 26, 27, 41]).

Figure 7.5: Scanning-electron microscope image of a micropost microcavity with a top diameter of $0.6\,\mu m$ and a height of $4.2\,\mu m$.

is possible to spectrally isolate the single photon emitted by recombination of the last electron-hole pair [14]. QD offer several advantages as *sources for single photons*. They have large oscillator strengths, narrow spectral linewidths, high photon yield, and excellent long-term stability. The materials used to make QD are compatible with mature semiconductor technologies, allowing them to be further developed and integrated with other components. The usefulness of most QD single-photon sources, though, is limited by their low efficiencies. The dots radiate primarily into the high-index substrates in which they are embedded, and very few of the emitted photons can be collected. The source efficiency can be increased by placing a dot inside a microscopic optical cavity. Perhaps the most practical microcavities for this purpose are microscopic posts etched out of distributed-Bragg reflector (DBR) microcavities [34]. See Figs. 7.5 and 7.6. Light escaping from the fundamental mode of a micropost microcavity is well approximated by a Gaussian beam, and can thus be efficiently coupled into optical fibers, detectors, or other downstream optical components. Q as high as 10^4 together with a mode volume as small as 1.6 cubic optical wavelengths has been achieved. This translates to a nearly 100% efficiency for a single-photon source.

7.1.3 A simple quantum dot for quantum computing

We now illustrate a simple QD experiment as follows. Single quantum dots for quantum computation can be localized at the tip of strain induced self-assembled structures such as Stranski-Krastanow growth mode of GaN quantum dots grown on AlN layers. A uniquely large hexagonal GaN pyramid (cf. Fig. 7.7) is self-assembled on the AlN cap on the surface of the GaN

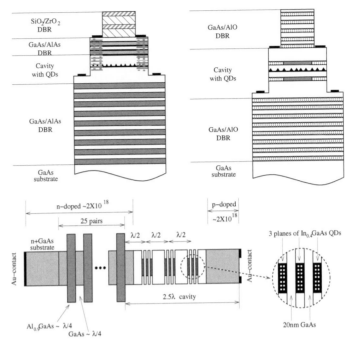

Figure 7.6: QD in a DBR (distributed-Bragg reflector), excerpted from (17) and (18).

QD layers with a radius of curvature no more than $300\,nm$. The faces of the pyramids are the $\{10\bar{1}1\}^4$ planes as evidenced by the angle between the inclined edge and the base of the pyramid. The measured angle of around 58-60° is in good agreement with the calculated angle of 58.4° using the GaN lattice parameters of $c = 5.185$ Å and $a = 3.189$ Å. The formation of the pyramids indicates that the $\{10\bar{1}1\}$ surfaces are self-assembled preferentially compared to the $\{000\bar{1}\}$ surface. Thus it can be inferred that $\{10\bar{1}1\}$ surfaces have the lowest surface potential with respect to the self-assembly process. The tip of the pyramid is very sharp with a diameter measured to be less than $2\,nm$. It is observed that GaN quantum dots are localized at the tip of the pyramid as demonstrated by the strong *excitonic* properties.

Near-field optical spectroscopy can be used for quantum computation as this probing technique is highly selective and has been utilized for exciting a single quantum dot system. Fig. 7.8 shows the exciton emission from quantum dots localized at the tip of the pyramid. A comparison of the far-field and near-field spectrum shows that the emission from quantum confined states in a quantum dot is significantly blue-shifted compared to the bulk GaN states. In GaAs based quantum dots the linewidth of emis-

---

[4]This (and similar ones later) notation is called *Miller indices*, representing the plane on which the GaN lattices $\{10\bar{1}1\}$ is grown (GaN is a wurtzite structure). A reference for Miller indices may be found in [42].

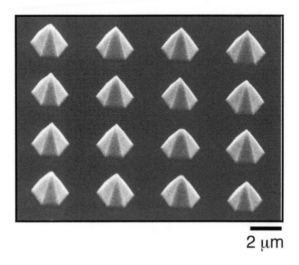

2 μm

Figure 7.7: GaN pyramids were selectively grown in $4\,\mu m$ period and $2\,\mu m$ square openings in a grid-like pattern. Three period of InGaN MQW structures ($30\,nm$) were grown selectively on top of hexagonal GaN pyramids.

sion from single quantum dots has been observed to be less than a few μeV. In GaN based quantum dots the linewidth is broader due to larger longitudinal optical (LO) phonon scattering rate and electron effective mass which leads to homogeneous broadening.

A periodic array of such GaN based pyramids shown in Fig. 7.9, has been fabricated by Arakawa's group at University of Tokyo [37], which can be used for quantum computation based on lateral coupling of the dots by using a near-field optical probe.

7.1.4 Spintronics

Spintronics is applicable to a QD-base QC, a proposal to be addressed in Section 7.3 of this chapter. Spintronics is spin-based electronics with a spin degree of freedom added to the conventional charge-based electronic devices. Electron has a half-spin angular momentum. Therefore, there are two states of electron spin: spin-up and spin-down. Spintronics distinguishes spin-up electron current from spin-down electron current while the charge-based electronics does not. Therefore, the electron spin can be made to carry information in spintronics. In 1988, it is reported in [3] that the resistance of a material is dependent on the magnetic moment alignment of the ferromagnetic layer, which is known as the *giant magnetoresistive* (GMR) effect. A simple GMR-based application is a spin valve as shown in Fig. 7.10. The top anti-ferromagnetic layer fixes the magnetic moment of the upper ferromagnetic layer. The lower ferromagnetic layer can change

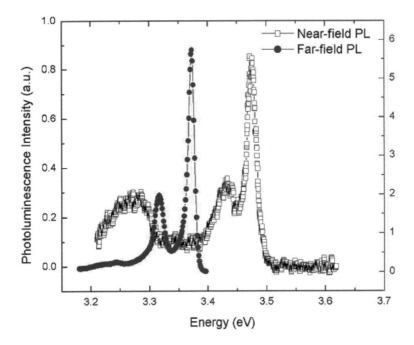

Figure 7.8: Far-field and near-field photoluminescence (PL) spectrum from GaN pyramid. The high energy emission at 3.48eV observed in the near-field limit is from the quantum dot localized at the top of the pyramid, where as the far-field emission is dominated by the photoluminescence from *bulk* GaN states. The near-field PL spectrum signifies various energy levels (i.e., eigenstates) of the quantum dot, which can be used to represent a single qubit for quantum computation.

the direction of the magnetic moment in the presence of an external magnetic field. The resistance of the conductive layer varies with the external magnetic field due to the GMR effect. Spintronics is the type of device which merges electronics, photonics and magnetism. It is currently developed at a rapid pace.

7.1.5   Three major designs of QD-based quantum gates

QD designs allow for tunable bandgap through the choices of QD sizes, shapes and semiconductor materials. For quantum gate logic operations, as mentioned in the preceding paragraphs, one can utilize *energy levels*, *spins*, or *excitonic levels* of confined electrons in quantum dots. At present, there exist three major designs of the quantum-dots based QC, due to

(i) Sherwin, Imamoglu and Montroy [33]: The idea is similar to a cavity-QED design [10, 31] by trapping single electrons in quantum dot mi-

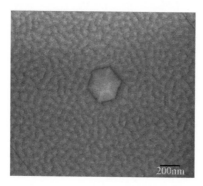

Figure 7.9: SEM image showing the surface morphology and self-assembly of a hexagonal pyramid shape GaN structure with $\sim 300\,nm$ diameter.

crocavities;

(ii) Loss and DiVincenzo [23]: It utilizes electron spins and their interactions via the electromagnetic effect of tunneling;

(iii) Piermarocchi, Chen, Dale, and Sham [29]: It is based on coherent optical control of two electron-hole pairs (called a *biexciton*) confined in a single QD. Efforts are being made to couple two or more QD in order to make this design scalable.

7.1.6 Universality of 1-bit and 2-bit gates in quantum computing

According to Theorem 2.6.2 in Chap. 2, in order to ensure universality, it is sufficient to have all the 1-bit gates $U_{\theta,\phi}$ in (2.46), plus any single 2-bit gate that is imprimitive. These gates are derived from applying laser coherent control pulses and/or tunneling voltage gates in QD. In the next few sections, we will provide the physical background and the mathematical equation governing these QD systems.

This chapter is organized as follows:

(1) In Section 7.2, we study the microcavity approach of Sherwin, *et al.* [33].

(2) In Section 7.3, we show the "spintronics" model following Loss and DiVincenzo [23].

(3) In Section 7.4, we describe the biexciton model of Sham, *et al.* [29].

In each section, issues related to decoherence and measurement will also be discussed.

We need to note that in addition to the three major proposals (i), (ii), and (iii), described above, many researchers have proposed other QD-based entanglement schemes which can also work as universal quantum gates. We

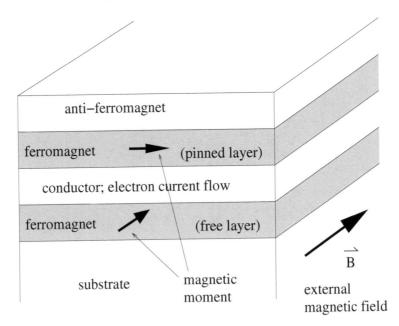

Figure 7.10: A spin valve structure, which is used to select desired electron spin current by an external magnetic field. The magnetization of the pinned layer is insensitive to the magnetic field, but that of the free layer can be changed. When the magnetic field aligns with the magnetic moments in the ferromagnetic layers, the resistance of the conductor reaches minimum. When the magnetic field anti-aligns with the magnetic moments, the resistance reaches maximum.

cannot cite all such schemes exhaustively here due to our limited knowledge and resources.

Another different application of QD to computing, called the *quantum dot cellular automata* (QCA) approach, utilizes special properties of quantum decoherence in QD to construct electronic logic gates. Information concerning QCA may be found on the website [48]. A good expository reference may be found in [40].

7.2   Electrons in quantum dots microcavity

The quantum computer proposed by Sherwin, *et al.* [33], is a collection of QD contained in a 3-D microcavity. Each QD contains exactly one electron. The two lowest electronic states are used to encode $|0\rangle$ and $|1\rangle$, respectively. The third energy level $|2\rangle$, as an auxiliary state, is utilized to perform the conditional phase shift operation; nevertheless, it doesn't directly encode any information. In addition, each QD is addressed by a pair of gate electrodes. Voltage pulses can be applied to control the energy levels of the QD via the Stark effect,[5] in particular, the energies E_{01} and E_{02} of the 0-1 and 0-2 transitions. The microcavity has a fundamental resonance with frequency ω_c. There is also a continuous-wave laser with a fixed frequency ω_l (different from ω_c) through one side of the cavity. The key technique to manipulate the state of the QD is to tune E_{01} and E_{02} with appropriate voltage pulses (through the gate electrodes) such that resonances with $\hbar\omega_c$, $\hbar\omega_l$, and $\hbar\omega_l + \hbar\omega_c$ are achieved. Coupling of different QD is done via microcavity mode photons acting as the data bus. See the schematic in Fig. 7.11. Here, the QD configuration is similar to the QD in DBR microcavities in Subsection 7.1.2. The difference is that we use the cavity-mode photon confined in the cavity to communicate back and forth between QD, while in the application of Subsection 7.1.2, the emphasis was to enhance the efficiency of generating cavity-mode photons emitted from the DBR microcavities for the utilization by downstream optical devices.

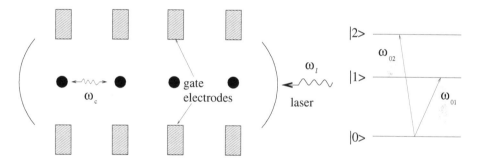

Figure 7.11: The black dots represent the locations of quantum dots where the electrons are confined. The cavity photon with frequency ω_c serves as the data bus to couple two quantum dots. The figure on the right describes the three lowest energy levels utilized to perform qubit operations. The transition energies from $|0\rangle$ to $|1\rangle$ and $|0\rangle$ to $|2\rangle$ are given by $\hbar\omega_{01}$ and $\hbar\omega_{02}$, respectively, which can be tuned by the voltage pulse e.

We may write down the following Hamiltonian to describe a QD interacting with cavity photons and the laser fields:

---

[5] For semiconductor material, the research of electroabsorption effects near the semiconductor band edges dates back to 1950s. These include the interband photon-assisted tunneling or Franz-Keldysh effects [13] and the exciton absorption effects [12]. With the appearance of quantum-confined structures, the optical absorption in quantum wells or quantum dots has been found to exhibit a dramatic change by exciton effect with an applied external electric field. This is caused by the so-called Quantum Confined Stark Effect (QCSE) [25] which led to the development of integrated electroabsorption devices at room temperature.

$$H = \hbar\omega_c \hat{a}_c^+ \hat{a}_c + E_{10}(e)\hat{\sigma}_{11} + E_{20}(e)\hat{\sigma}_{22} + \underbrace{\hbar g_{01}(e)\{\hat{a}_c^+ \hat{\sigma}_{01} + \sigma_{10}\hat{a}_c\}}_{\mathcal{J}_1}$$

$$+ \underbrace{\hbar\Omega_{l,01}(e)\{\hat{\sigma}_{01}e^{i\omega_l t} + \hat{\sigma}_{10}e^{-i\omega_l t}\}}_{\mathcal{J}_2}$$

$$+ \underbrace{\hbar g_{12}(e)\{\hat{a}_c^+ \hat{\sigma}_{12} + \hat{\sigma}_{12}\hat{a}_c\} + \hbar\Omega_{l,12}(e)\{\hat{a}_c^+ \hat{\sigma}_{12}e^{i\omega_l t} + \hat{\sigma}_{21}\hat{a}_c e^{-i\omega_l t}\}}_{\mathcal{J}_3}, \quad (7.1)$$

where $\hat{a}_c$ is the cavity mode annihilation operator, g_{ij}'s are the vacuum Rabi frequencies, $\hat{\sigma}_{ij} = |i\rangle\langle j|$ for $i, j \in \{0, 1, 2\}$ are the transition operators, and e is the voltage pulse. Let e_c, e_l, and e_{l+c} be the proper heights of e such that $E_{10}(e_c) = \hbar\omega_c$, $E_{10}(e_l) = \hbar\omega_l$, and $E_{20}(e_{l+c}) = \hbar\omega_l + \hbar\omega_c$, respectively. The energies of the cavity mode photon, $\hbar\omega_c$, and the laser photon $\hbar\omega_l$ should be sufficiently separated in order to have a satisfactory resonance performance. Typically $\hbar\omega_c$ is 11.5 meV, and $\hbar\omega_l$ is 15 meV. The different effects of applying voltage pulses $e \approx e_c$, e_l, and e_{l+c} can be explained from this Hamiltonian. The meanings of individual terms in (7.1) are explained in Box 7.1.

In order to simplify the analysis of these three primary resonant cases, we have adopted the following assumptions:

1. The time durations for the rise and fall of the voltage pulses e are relatively short, so that ideal heights are quickly reached to ensure that the targeted resonances dominate the overall state vector evolution.
2. The changes to the Hamiltonian H by the voltage pulses e is adiabatic, so that unwanted transitions between $|0\rangle$ and $|1\rangle$ induced by ramping electric fields are minimized.
3. The ac Stark shifts in the energy levels of the QD caused by the laser field are neglected.
4. The effect from the terms which do not satisfy resonance conditions is also neglected.

There are a total of 11 terms in the Hamiltonian H in (7.1). Their origins can be explained as follows:

(1) The first three terms are called the *unperturbed Hamiltonian*, i.e., the Hamiltonian without interaction, where $\hbar\omega_c \hat{a}_c^+ a_c$ represents the energy of the cavity, and $E_{10}(e)\hat{\sigma}_{11} + E_{20}(e)\hat{\sigma}_{22}$ represents the energy of the 3-level atom. Note that $E_{10}(e) = \hbar\omega_{01}$ and $E_{20}(e) = \hbar\omega_{02}$. (For the sake of simplicity, the vacuum energy of the cavity is omitted as it is a constant, and the energy of the atom at level $|0\rangle$ is taken to be 0.)

Interaction terms: $\mathcal{J}_1, \mathcal{J}_2$ and $\mathcal{J}_3$:

(2) The interaction between the atom and the cavity between the two levels $|0\rangle$ and $|1\rangle$ is modeled by $\mathcal{J}_1$. The single-mode cavity field stimulates a Rabi oscillation between $|0\rangle$ and $|1\rangle$, where $g_{01}(e)$ is the vacuum Rabi frequency. Term $\hat{a}_c^+\hat{\sigma}_{01}$ describes the transition process when the atom jumps from $|1\rangle$ to $|0\rangle$ and a cavity photon is created, while $\hat{\sigma}_{01}\hat{a}_c$ describes the reverse process.

(3) $\mathcal{J}_2$ contains the laser-atom interaction terms in a semi-classical form, where the electric field is not quantized (as the electric field induced by the laser is strong), with $\Omega_{\ell,01}$ as the Rabi frequency.

(4) $\mathcal{J}_3$ is a combination of the effects of $\mathcal{J}_1$ and $\mathcal{J}_2$, except that it now describes the transition between levels $|0\rangle$ and $|2\rangle$.

Box 7.1 Meanings of the terms in the Hamiltonian (7.1)

7.2.1 Resonance, 1-bit and CNOT gates

The details of the three primary resonant cases for (7.1) are:

(i) $e \approx e_c$:

The first interaction term, $\hat{H}_1 = \hbar g_{01}(e)\{\hat{a}_c^+\hat{\sigma}_{01} + \sigma_{10}\hat{a}_c\}$ in H, cf. the term $\mathcal{J}_1$ in (7.1), dominates because of the resonance at $\omega_{10}(e) \approx \omega_c$. If the QD is in state $|1\rangle$ or if there is one photon in the cavity, i.e., the cavity state is $|1\rangle_c$, the qubit will undergo vacuum Rabi oscillations with frequency g_{01}:

$$|0\rangle|1\rangle_c \rightarrow \cos(g_{01}t)|0\rangle|1\rangle_c - ie^{i\phi}\sin(g_{01}t)|1\rangle|0\rangle_c; \qquad (7.2)$$

$$|1\rangle|0\rangle_c \rightarrow \cos(g_{01}t)|1\rangle|0\rangle_c - ie^{-i\phi}\sin(g_{01}t)|0\rangle|1\rangle_c, \qquad (7.3)$$

where $\phi = 0$. The above defines a rotation between state vectors $|0\rangle|1\rangle$ and $|1\rangle|0\rangle$, which is very similar to the 1-bit gate (2.46). We denote it as $U_{\theta,\phi}^{(1)} \equiv U_{\theta,\phi}(|0\rangle|1\rangle, |1\rangle|0\rangle)$ with $\theta = g_{01}t$, and $\phi = 0$.

(ii) $e \approx e_l$:

The second interaction term, $\hat{H}_2 = \hbar\Omega_{l,01}(e)\{\hat{\sigma}_{01}e^{i\omega_l t} + \hat{\sigma}_{10}e^{-i\omega_l t}\}$ in H, cf. the term $\mathcal{J}_2$ in (7.1), dominates. The state vector will rotate between $|0\rangle$ and $|1\rangle$ with Rabi frequency $\Omega_{l,01}$. We denote it as $U_{\theta,\phi}^{(2)} \equiv U_{\theta,\phi}(|0\rangle, |1\rangle)$ with $\theta = \Omega_{l,01}t$:

$$|0\rangle \rightarrow \cos(\Omega_{l,01}t)|0\rangle - ie^{i\phi}\sin(\Omega_{l,01}t)|1\rangle; \qquad (7.4)$$

$$|1\rangle \rightarrow \cos(\Omega_{l,01}t)|1\rangle - ie^{-i\phi}\sin(\Omega_{l,01}t)|0\rangle. \qquad (7.5)$$

(iii) $e \approx e_{l+c}$:

The Rabi oscillation between $|0\rangle$ and $|2\rangle$ involving both cavity and laser photons will dominate the resonance behavior. From the $\mathcal{J}_3$ term in (7.1), after reduction we obtain the following effective Hamiltonian to describe these 2-photon processes:

$$H_2 = \hbar\tilde{\Omega}(e)\{\hat{a}_c^+ \hat{\sigma}_{02}e^{i\omega_l t} + \hat{\sigma}_{20}\hat{a}_c e^{-i\omega_l t}\}. \tag{7.6}$$

If the cavity contains one photon, i.e., in state $|1\rangle_c$ and the QD is in state $|0\rangle$, then it rotates between $|0\rangle$ and $|2\rangle$ with frequency $\tilde{\Omega}(e_{l+c})$. We denote it as $U_{\theta,\phi}^{(3)} \equiv U_{\theta,\phi}(|0\rangle, |2\rangle)$ with $\theta = \tilde{\Omega}(e_{l+c})t$:

$$|0\rangle \rightarrow \cos(\tilde{\Omega}(e_{l+c})t)|0\rangle - ie^{i\phi}\sin(\tilde{\Omega}(e_{l+c})t)|2\rangle \tag{7.7}$$

$$|2\rangle \rightarrow \cos(\tilde{\Omega}(e_{l+c})t)|2\rangle - ie^{-i\phi}\sin(\tilde{\Omega}(e_{l+c})t)|0\rangle \tag{7.8}$$

Now let's address how to implement the CNOT (Controlled-NOT) gate in such a system. With respect to the ordered basis $\{|00\rangle, |01\rangle, |10\rangle, |11\rangle\}$, the matrix representation of the CNOT gate is

$$\begin{bmatrix} 1 & 0 & 0 & 0 \\ 0 & 1 & 0 & 0 \\ 0 & 0 & 0 & 1 \\ 0 & 0 & 1 & 0 \end{bmatrix}.$$

This task is achieved by first constructing the conditional phase shift $|00\rangle \rightarrow |00\rangle$, $|01\rangle \rightarrow |01\rangle$, $|10\rangle \rightarrow |10\rangle$, and $|11\rangle \rightarrow -|11\rangle$. The procedure is described in the following (by starting with a cavity with no photons):

1. Apply a "π" pulse with height e_c and duration $\pi/(2g_{01})$ on the control bit. This pulse implements the rotation $U_{\pi/2,\pi}(|0\rangle|1\rangle, |1\rangle|0\rangle) = \begin{bmatrix} 0 & i \\ i & 0 \end{bmatrix}$, where a negative sign representing a global phase is omitted.

2. Apply a "2π" pulse with height e_{l+c}, phase 0, and duration $\pi/\tilde{\Omega}(e_{l+c})$ on the target bit. This pulse implements the rotation $U_{\pi,0}(|0\rangle, |2\rangle) = \begin{bmatrix} -1 & 0 \\ 0 & -1 \end{bmatrix}$.

3. Apply a "π" pulse with height e_c and duration $\pi/(2g_{01})$ on the control bit. Again, this pulse implements the rotation $U_{\pi/2,\pi}(|0\rangle|1\rangle, |1\rangle|0\rangle) = \begin{bmatrix} 0 & i \\ i & 0 \end{bmatrix}$.

These three steps will yield the quantum phase gate

$$Q_\pi = Q_\eta|_{\eta=\pi} \equiv \begin{bmatrix} 1 & 0 & 0 & 0 \\ 0 & 1 & 0 & 0 \\ 0 & 0 & 1 & 0 \\ 0 & 0 & 0 & e^{i\eta} \end{bmatrix}|_{\eta=\pi}.$$

Theorem 7.2.1. *The above procedure implements the desired conditional phase shift to achieve the quantum phase gate Q_π.*

Proof. Following the previous discussion, we can derive the system states after each operation, starting from different initial states. The composite operation is $U_{\pi/2,\pi}^{(1)} U_{\pi,0}^{(3)} U_{\pi/2,\pi}^{(1)}$:

$$|00\rangle|0\rangle_c \rightarrow \quad |00\rangle|0\rangle_c \rightarrow \quad |00\rangle|0\rangle_c \rightarrow \quad |00\rangle|0\rangle_c \quad (7.9)$$

$$|01\rangle|0\rangle_c \rightarrow \quad |01\rangle|0\rangle_c \rightarrow \quad |01\rangle|0\rangle_c \rightarrow \quad |01\rangle|0\rangle_c \quad (7.10)$$

$$|10\rangle|0\rangle_c \rightarrow \quad i|00\rangle|1\rangle_c \rightarrow \quad -i|00\rangle|1\rangle_c \rightarrow \quad |10\rangle|0\rangle_c \quad (7.11)$$

$$|11\rangle|0\rangle_c \rightarrow \quad i|01\rangle|1\rangle_c \rightarrow \quad i|01\rangle|1\rangle_c \rightarrow \quad -|11\rangle|0\rangle_c \quad (7.12)$$

Ignoring the cavity bit $|0\rangle_c$, we have realized the conditional phase shift Q_π on two QD. $\qquad \square$

This quantum phase gate Q_π satisfies (2.47) and is, therefore, a universal 2-bit gate. From Q_π, we can easily derive the CNOT gate by applying voltage pulses with height e_l, utilizing two more "$\frac{\pi}{2}$" and "$\frac{3\pi}{2}$" pulses with duration $\frac{\pi}{4\Omega_{l,01}}$ and $\frac{3\pi}{4\Omega_{l,01}}$, and phase $-\frac{\pi}{2}$, i.e.,

$$\text{CNOT} = U_{3\pi/4,-\pi/2}^{(2)} Q_\pi U_{3\pi/4,-\pi/2}^{(2)}. \quad (7.13)$$

7.2.2 Decoherence and measurement

Decoherence as well as dissipation are major problems in the physical implementation of quantum computers. Any realistic quantum computer will have some interaction with its environment, which causes decoherence (decay of the off-diagonal elements of the reduced density matrix) and dissipation (change of populations of the reduced density matrix) [19]. Because the data is encoded in the electronic states and the coupling of different QD is carried out via microcavity mode photons in this QD scheme, decoherence of both electronic states and cavity photons must be considered. There are many interactions which may cause decoherence. Some of them come from device imperfection, which can be minimized by precision engineering. Examples of imperfections include the emission of freely propagating photons, interactions with the fluctuations in the potentials of the gate electrodes, and inhomogeneity of quantum dots. These problems can be tackled by using very high quality cavities, making the gate electrodes out of superconductor, and individual calibration of each quantum dots. Also, though the cavity photon loss—a dissipation effect—is inevitable in the long run, we can extend the lifetime of a cavity photon by using cavities made from materials such as ultrapure Si. However, there are other sources of decoherence which are more "essential" and hard to get rid of, such as the relaxation of an electron from state $|1\rangle$ to $|0\rangle$ by emission of an acoustic phonon, and the "pure dephasing" of electronic states by the interaction with acoustic phonons. If we assume that these are the dominant source of decoherence, it is estimated that within the decoherence time several thousands of CNOT operations can be safely performed. Nevertheless, limited by the technological complexity and lack of experimental results, accurate conclusions about the decoherence in this scheme still remain to be drawn.

The final stage of the quantum computing process is the readout of the states of the qubits. A tunable antenna-coupled intersubband terahertz (TACIT) photon detector is proposed for this task. Terahertz photons with frequency close to the absorption frequency of the TACIT detector are efficiently detected. The absorption frequency can be tuned via the Stark effect similar to the way the transition energies of the QD are tuned. At the readout stage, we can tune this frequency to the fundamental resonance frequency ω_c of the cavity. Under the cavity resonance, if the qubit is in state $|1\rangle$, it will undergo Rabi oscillation with frequency g_{01} and emit a photon at the time $\frac{\pi}{2g_{01}}$, which is immediately detected; otherwise no photon will be seen. This way we can deduce whether the original state of the QD is $|0\rangle$ or $|1\rangle$.

7.3   Coupled electron spins in an array of quantum dots

7.3.1   Electron spin

The electron spin is a "natural" representation of a qubit since it comprises exactly two levels. Unlike charge (energy-level) states in an atom or quantum dot, there are no additional degrees of freedom into which the system could "leak". Another great advantage of spins as compared to charge qubits is that in typical semiconductor materials like GaAs or Si, the time over which the spin of a conduction-band electron remains phase coherent can be several orders of magnitude longer than the corresponding charge decoherence times. Of course these numbers have to be compared with the time it takes to perform an elementary gate operation. Even with this being considered, single spins seem to be very well suited as qubits. The transverse decoherence time T_2, which is most relevant in the context of quantum computing, is defined as the characteristic time over which a single spin which is initially prepared as a coherent superposition of "spin-up" and "spin-down" coherently precesses about an external magnetic field. The transverse dephasing time $T_2^* \leq T_2$ of an ensemble of spins in n-doped GaAs can exceed 100 ns, as demonstrated by optical measurements [16], while switching times are estimated to be on the order of 10-100 ps. The longitudinal (energy) relaxation time T_1 determines how long it takes for a non-equilibrium spin configuration to relax to equilibrium. T_1 can be much longer than T_2 (and particularly long in confined structures), but while suppression of spin relaxation is necessary for quantum computation, it is not sufficient.

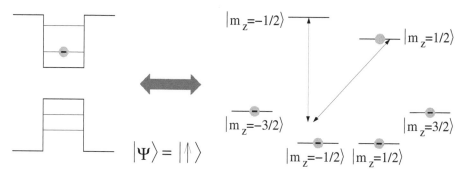

Figure 7.12: Schematic for achieving qubits in multi-level QD using the electron-spin orbits.

There are two main schemes for achieving qubits in quantum dots using electron spin:

1. Single-qubit rotations: In principle, spin-flip Raman transition could rotate the electron spin in $\tau_{gate} \sim 10\,ps \ll \tau_{decoh} \sim 1\,\mu s$.

2. Two-qubit gates: the real challenge in most schemes. In this case, the spin decoherence during gate operation is a problem.

Spintronics requires the fabrication of ferromagnetic nanostructures that at room temperature can transport spin-polarized carriers, and which can be assembled into addressable hierarchies on a macroscopic chip. Most efforts have been directed towards the mixing of transition-metal atoms (such as Ni, Fe and Mn, which have permanent magnetic moments) into semiconductor devices based on periodic table group II-VI (such as CdS) or III-V (GaAs) compound semiconductor. Superstructures consisting of alternating ferromagnetic/diamagnetic, metallic/oxide thin films have also received attention; like spin valves, spin-polarized currents can be injected into them and transported. An all-electrically controlled quantum dot array can be used for switching qubits.

Recently, a new class of diluted magnetic semiconductor based on III-V system is being studied due to its large intrinsic magnetic dipole moment. Gd-doped GaN materials are reported to have a strong intrinsic spin dipole moment. The tunneling in quantum dot based diluted magnetic semiconductor can also be enhanced by using a nanoscale electrode on a diluted magnetic semiconductor system.

7.3.2 The design due to D. Loss and D. DiVincenzo

In this section, we study questions related to the spintronics design [23]. The basics of the Loss and DiVincenzo scheme is quite mathematically elegant. For a linear array of quantum dots, see Fig. 7.13, a single electron is injected into each dot. The electron's spin up and down constitute a single qubit. Each quantum dot is coupled with its (two) next neighbors through gated tunneling barriers. The overall Hamiltonian of the array of coupled quantum dots as given in [9] is

$$H = \sum_{j=1}^{n} \mu_B g_j(t) \boldsymbol{B}_j(t) \cdot \boldsymbol{S}_j + \sum_{1 \leq j < k \leq n} J_{jk}(t) \boldsymbol{S}_j \cdot \boldsymbol{S}_k, \qquad (7.14)$$

where the first summation denotes the sum of energy due to the application of a magnetic field $\boldsymbol{B}_j$ to the electron spin at dot j, while the second denotes the interaction Hamiltonian through the tunneling effect of a gate voltage applied between the dots, and $\boldsymbol{S}_j, \boldsymbol{S}_k$ are the spin of the electric charge quanta at, respectively, the j-th and k-th quantum dot.

Eq. (7.14) is a 1-dimensional *Ising (chain) model*. Ernst Ising (1900–1998) introduced the famous Ising model in 1925 consisting of a linear chain or (2- and 3-dimensional) lattices of spins with the following two characteristics:

(1) each of the spin variables takes on "up" and "down" positions;

(2) only pairs of nearest neighboring spins are coupled.

The Ising model is important in the modern theory of phase transitions and, more generally, of cooperative phenomena.

Due to the universality of the collection of all 2-bit gates, we see that Ising chains are useful in quantum computing. For example, we will see an Ising model again in (9.82) and several other places with superconducting quantum devices.

Box 7.2 Ising models

Quantum dots themselves may be viewed as artificial atoms as both manifest similar behaviors. Coupled quantum dots, in this connection, may be considered to a certain extent as *artificial molecules* [23]. Thus, Burkard, Loss and DiVincenzo applied naturally the Heitler–London and Hund–Mulliken methods in molecular quantum chemistry to evaluate the

---

[6]The *Zeeman effect*, discovered by P. Zeeman in 1896, refers to the splitting of spectral lines by a magnetic field as noted in Section 2.2. For a quantum particle with spin lying in a magnetic field, interaction between its spin magnetic moment with the magnetic field causes its energy levels to split into several levels depending on the spin and the angular momentum quantum number. For QD, the discrete energy levels, the transitions between those levels, and the associated spectral lines discussed so far we have implicitly assumed that there are no magnetic fields influencing the quantum dots. If there are magnetic fields (due to the local magnetized layer) present, the electronic energy levels are split into a larger number of levels and the spectral lines are also split.

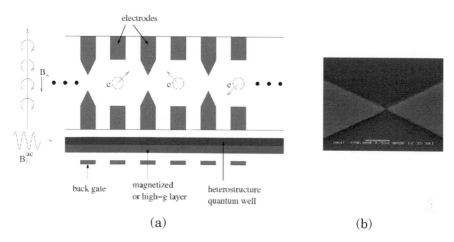

(a)   (b)

Figure 7.13: (a) A linear array of laterally coupled quantum dots, as given in
Burkard, Engel and Loss (8). Each circle represents a quantum dot,
where the arrow represents the spin-$\frac{1}{2}$ ground state of the confined
electron. The electrodes (dark gray) confine single electrons to the
dot regions (circles). The electrons can be moved by electrical gating into the magnetized or high-magnetic dipole moment layer to
produce locally different Zeeman splittings.[6] Alternatively, such local Zeeman fields can be produced by magnetic field gradients as,
e.g., produced by a current wire (indicated on the left of the dot-
array). Since every dot-spin is subject to a different Zeeman splitting,
the spins can be addressed individually, e.g., through ESR (electron-
spin resonance) pulses of an additional in-plane magnetic ac field
with the corresponding Larmor frequency (ω_L). Such mechanisms
can be used for single-spin rotations and the initialization step. The
exchange coupling between the dots is controlled by electrically
lowering the tunnel barrier between the dots. (b) A nanofabricated
bow-tie shaped electrode on Gd-doped GaN/AlN semiconductor
quantum dots for spintronics.

"exchange energy" J, which in terms of our notation in (7.20) later in Subsection 7.3.3, is

$$J = \frac{\hbar}{2}\omega(t).$$

J is a function of B, E and a, among others, where

$$J = J(B, E, a), \tag{7.15}$$

with

$B = $ the magnetic field strength,

$E = $ the electric field strength, and

$a = $ tunneling barrier height or, equivalently, inter-dot distance, by varying which we will be lead to, respectively, the effects of wave function suppression, level detuning, and the suppression of tunneling between the dots [23]. The determination of $w(t)$ or, equivalently, J, is important. Technologically, the tailoring, design and implementation of the control pulse $w(t)$ are also perhaps the most challenging.

The coupling between two quantum dots consists of the usual Coulomb repelling potential between the two electrons located within each dot and, in addition, a quartic potential

$$V(x, y) = \frac{m\omega_0^2}{2} \left[\frac{1}{4a^2}(x^2 - a^2)^2 + y^2 \right]. \tag{7.16}$$

to model the effect of tunneling. See Fig. 7.14(b). Using the Heitler–London approach (likening the coupled quantum dots as the H_2 dimer), Burkard, Loss and DiVincenzo obtained the exchange energy as

$$J = \frac{\hbar\omega_0}{\sinh(2d^2(2b - \frac{1}{b}))} \left[c\sqrt{b} \left(e^{-bd^2} I_0(bd^2) - e^{d^2(b-\frac{1}{b})} I_0 \left(d^2 \left(b - \frac{1}{b} \right) \right) \right) \right.$$
$$\left. + \frac{3}{4b}(1 + bd^2) + \frac{3}{2}\frac{1}{d^2} \left(\frac{eBa}{\hbar\omega_0} \right)^2 \right]. \tag{7.17}$$

The above result given in [9] is very commendable but its calculation is lengthy. Its derivations require special techniques and carefulness but details were not available in [9]. We will fill in such mathematical technical details in Appendices B and C.

7.3.3   Model of two identical laterally coupled quantum dots

For the model given by Loss and DiVincenzo in [9], the underlying assumptions leading to the main result (7.17) are itemized below:

(1) The *geometry* of the two coupled dots is described in Fig. 7.14(a). The electron confinement is based on single GaAs heterostructure quantum dots formed in a 2DEG (2-dimensional electron gas). The directions of the electric and magnetic fields are indicated in Fig. 7.14(a), where

$$\boldsymbol{B} = Be_z, \quad \text{due to the vector potential } \boldsymbol{A}(x, y, 0) = \frac{B}{2}(-ye_x + xe_y) \tag{7.18}$$

$$\boldsymbol{E} = Ee_x \tag{7.19}$$

(2) The quartic potential (7.16) for tunneling, see Fig. 7.14(b), was motivated by the experimental fact from [38] that the spectrum of single dots in GaAs is well described by a parabolic confinement potential, e.g., with $\hbar\omega_0 = 3meV$ [9, 38]. (The quartic potential (7.16) separates into two harmonic wells centered at $x = \pm a$.) The constant a, the half interdot distance, satisfies

$$a \gg a_B,$$

where $a_B = [\hbar/(m\omega_0)]^{1/2} =$
the effective Bohr radius of a single isolated harmonic well,

μ_B :   is the Bohr magneton;
$g_j(t)$:   is the effective g-factor;
$\boldsymbol{B}_j(t)$:   is the applied magnetic field;
$J_{jk}(t)$:   the time-dependent exchange constant (see [10] and the references therein), with $J_{jk}(t) = 4t_{jk}^2(t)/u$, which is produced by the turning on and off of the tunneling matrix element $t_{ij}(t)$ between quantum dots i and j, with u being the charging energy of a single dot. Moreover, $J_{jk}(t) \equiv 0$ if $|j - k| > 1$.

Note that for

$$\boldsymbol{S}_j = \sigma_x^{(j)}\boldsymbol{e}_x + \sigma_y^{(j)}\boldsymbol{e}_y + \sigma_z^{(j)}\boldsymbol{e}_z, \qquad j = 1, 2, \ldots, n,$$

and

$$\boldsymbol{B}_j(t) = b_x^{(j)}(t)\boldsymbol{e}_x + b_y^{(j)}(t)\boldsymbol{e}_y + b_z^{(j)}(t)\boldsymbol{e}_z, \qquad j = 1, 2, \ldots, n,$$

where

$$\boldsymbol{e}_x = \begin{bmatrix} 1 \\ 0 \\ 0 \end{bmatrix}, \quad \boldsymbol{e}_y = \begin{bmatrix} 0 \\ 1 \\ 0 \end{bmatrix}, \quad \boldsymbol{e}_z = \begin{bmatrix} 0 \\ 0 \\ 1 \end{bmatrix}$$

and $\sigma_x^{(j)}, \sigma_y^{(j)}$ and $\sigma_z^{(j)}$ are the standard Pauli spin matrices (at dot j):

$$\sigma_x^{(j)} = \begin{bmatrix} 0 & 1 \\ 1 & 0 \end{bmatrix}, \quad \sigma_y^{(j)} = \begin{bmatrix} 0 & -i \\ i & 0 \end{bmatrix}, \quad \sigma_z^{(j)} = \begin{bmatrix} 1 & 0 \\ 0 & -1 \end{bmatrix},$$

the dot products are defined by

$$\boldsymbol{S}_j \cdot \boldsymbol{S}_k = \sigma_x^{(j)}\sigma_x^{(k)} + \sigma_y^{(j)}\sigma_y^{(k)} + \sigma_z^{(j)}\sigma_z^{(k)},$$
$$\boldsymbol{B}_j(t) \cdot \boldsymbol{S}_j = b_x^{(j)}(t)\sigma_x^{(j)} + b_y^{(j)}(t)\sigma_y^{(j)} + b_z^{(j)}(t)\sigma_z^{(j)}.$$

From the universal quantum computing point of view, as the collection of 1-bit and 2-bit quantum gates are universal, it is sufficient to study a system with *only two coupled quantum dots*, whose Hamiltonian may now be written as [8, 10]

$$H(t) \equiv \frac{\hbar}{2}[\boldsymbol{\Omega}_1(t) \cdot \boldsymbol{\sigma} + \boldsymbol{\Omega}_2(t) \cdot \boldsymbol{\tau} + w(t)\boldsymbol{\sigma} \cdot \boldsymbol{\tau}], \tag{7.20}$$

followed by rewriting the notation

$$\boldsymbol{S}_1 = \boldsymbol{\sigma}, \quad \boldsymbol{S}_2 = \boldsymbol{\tau}; \quad \mu_B g_j(t)\boldsymbol{B}_j(t) = \frac{\hbar}{2}\boldsymbol{\Omega}_j(t); \quad j = 1, 2; \quad J_{12}(t) = \frac{\hbar}{2}w(t).$$

The $\Omega_1(t), \Omega_2(t)$ and $w(t)$ are the *control pulses*. Thus, varying $\Omega_1(t)$ and $\Omega_2(t)$ will generate complete 1-bit Rabi-rotation gates for the first and second qubits, respectively [10]. However, in order to generate the *entangling* controlled-not (CNOT) gate or a quantum phase gate, both being 2-bit gates, the coupling term $w(t)\boldsymbol{\sigma} \cdot \boldsymbol{\tau}$ in (7.20) is indispensable. Therefore,

(3) The Coulomb interaction between the two electrons is described by

$$C = \frac{e^2}{\kappa|\boldsymbol{r}_1 - \boldsymbol{r}_2|}, \qquad \boldsymbol{r}_1 = x_1\boldsymbol{e}_x + y_1\boldsymbol{e}_y, \boldsymbol{r}_2 = x_2\boldsymbol{e}_x + y_2\boldsymbol{e}_y. \qquad (7.21)$$

Here we assume that the screening length λ satisfies

$$\lambda/a \gg 1.$$

(4) The ratio between the Zeeman splitting (due to the magnetic field B) and the relevant orbital energies (see item (5) below) is small for all values of B of interest here. The spin-orbit effect can be neglected as

$$H_{\text{spin-orbit}} = \left(\frac{w_0^2}{2mc^2}\right) \boldsymbol{L} \cdot \boldsymbol{S}$$

$$\left(c = \sqrt{\frac{\pi}{2}} \frac{e^2}{\kappa a_B} \frac{1}{\hbar w_0} \approx 2.4 \text{ for } \hbar w_0 = 3\text{meV}, \quad a_B = \sqrt{\frac{\hbar}{mw_0}}\right) \qquad (7.22)$$

is of the magnetide

$$H_{\text{spin-orbit}}/(\hbar w_0) \approx 10^{-7}.$$

Consequently, the dephasing effects by potential or charge fluctuations can couple only to the charge of the electron, instead of the "holes."

Under conditions (1)–(4) above, the total orbital Hamiltonian of the coupled system may be given as

$$H_{\text{orb}} = h_1 + h_2 + C, \qquad (7.23)$$

where

$$h_j = \frac{1}{2m}\left|\boldsymbol{p}_j - \frac{e}{c}\boldsymbol{A}(\boldsymbol{r}_j)\right|^2 + ex_jE + V(\boldsymbol{r}_j), \text{ for } j = 1, 2. \qquad (7.24)$$

(5) Assume further the cryogenic condition $kT \ll \hbar w_0$, so we need only consider the two lowest orbital eigenstates of the orbital Hamiltonian H_{orb}, which are, respectively, the (symmetric) *spin-singlet* and the (antisymmetric) spin-triplet, see Box 7.3. A perturbation approximation then leads to the effective Heisenberg spin Hamiltonian

$$H_s = J\boldsymbol{S}_1 \cdot \boldsymbol{S}_2 \qquad \text{(cf. } J \text{ in (7.15))}$$

$J \equiv \epsilon_t - \epsilon_s =$ the difference between the triplet and singlet energies.

(7.25)

A self contained account for the derivation of J involves rather technical mathematical analysis of the Fock–Darwin Hamiltonians and states, and

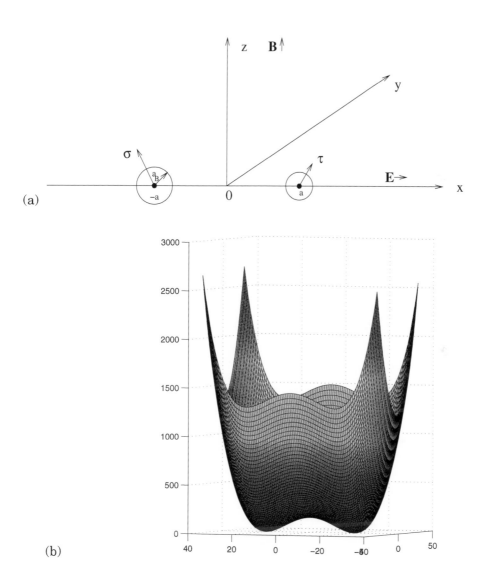

Figure 7.14: (a) Geometry of the two identical coupled quantum dots. The two electrons are confined to the (x, y)-plane. Electron spins are denoted by σ and τ. (b) Profile of the quartic potential given by (7.16).

clever simplifications of the various integrals in the exchange energy. We put together such work in Appendices B and C near the end of the paper.

The universality of the Loss–DiVincenzo QD quantum gates can now be presented. We first show how to choose the control pulse $\Omega_1(t)$ in order to obtain the 1-bit unitary rotation gate $U_{\theta,\phi}$ in (2.46).

For a 2-electron system, because electrons are Fermions, the combined wave function (i.e., the spatial wave function ψ times spin S) must be antisymmetric with respect to the permutation (or exchange) of particle 1 and 2 in order to satisfy the Pauli exclusion principle. Thus, if ψ is symmetric, S must be antisymmetric, and vice versa.

A single electron has spin 1/2. For two electrons, the combined (total) spin angular momentum can be either 0 or 1:

total spin angular momentum 0: $\dfrac{1}{\sqrt{2}}(|\uparrow\downarrow\rangle - |\downarrow\uparrow\rangle)$

(antisymmetric)

total spin angular momentum 1: $\dfrac{1}{\sqrt{2}}(|\uparrow\downarrow\rangle + |\downarrow\uparrow\rangle), |\uparrow\uparrow\rangle, |\downarrow\downarrow\rangle$

(symmetric)

The total spin 0 state is a singlet while the total spin 1 states are triplets.

Box 7.3 Singlets and triplets

Theorem 7.3.1. *([10, pp. 111-112]) Let $\phi, \theta \in [0, 2\pi]$ be given. Denote $e(\phi) = \cos\phi e_x + \sin\phi e_y + 0e_z$ for the given ϕ. Let $U_{1,\Omega_1}(t)$ be the time evolution operator corresponding to the quantum system*

$$i\hbar\frac{\partial}{\partial t}|\psi(t)\rangle = H(t)|\psi(t)\rangle, \quad T > t > 0, \quad cf.\ H(t)\ in\ (7.20) \tag{7.26}$$

where the pulses are chosen such that

$$\Omega_1(t) = \Omega_1(t)e(\phi), \quad \Omega_2(t) = 0, \quad w(t) = 0, \quad t \in [0, T], \tag{7.27}$$

with $\Omega_1(t)$ satisfying

$$\int_0^T \Omega_1(t)dt = 2\theta, \quad for\ the\ given\ \theta. \tag{7.28}$$

Then the action of $U_{1,\Omega_1}(t)$ on the first qubit satisfies

$$U_{1,\Omega_1}(t) = U_{\theta,\phi}, \text{ the 1-bit unitary rotation gate (2.46)}. \tag{7.29}$$

Proof. We have

$$
U_{\theta,\phi} = \begin{bmatrix} \cos\theta & -ie^{-i\phi}\sin\theta \\ -ie^{i\phi}\sin\theta & \cos\theta \end{bmatrix}
$$

$$
= \cos\theta\mathbf{1} - ie^{-i\phi}\sin\theta\left(\frac{\sigma_x - i\sigma_y}{2}\right) - ie^{i\phi}\sin\theta\left(\frac{\sigma_x - i\sigma_y}{2}\right)
$$

$$
= \cos\theta\mathbf{1} - i\sin\theta\cos\phi\sigma_x - i\sin\theta\sin\phi\sigma_y
$$

$$
= \cos\theta\mathbf{1} - i\sin\theta(\cos\phi\sigma_x + \sin\phi\sigma_y)
$$

$$
= \cos\theta\mathbf{1} - i\sin\theta e(\phi)\cdot\sigma
$$

$$
= e^{-i\theta e(\phi)\cdot\sigma}, \tag{7.30}
$$

noting that in the above, we have utilized the fact that the 2×2 matrix

$$
e(\phi)\cdot\sigma = \begin{bmatrix} 0 & \cos\phi - i\sin\phi \\ \cos\phi + i\sin\phi & 0 \end{bmatrix} \tag{7.31}
$$

satisfies $(e(\phi)\cdot\sigma)^{2n} = \mathbf{1}$ for $n = 0, 1, 2, \ldots$, where $\mathbf{1}$ is the 2×2 identity matrix.

With the choices of the pulses as given in (7.27), we see that the second qubit remains steady in the time-evolution of the system. The Hamiltonian, now, is

$$
H_1(t) = \frac{\hbar}{2}\Omega_1(t)e_1(\phi)\cdot\sigma \tag{7.32}
$$

and acts only on the first qubit (where the subscript 1 of $e_1(\phi)$ denotes that this is the vector $e(\phi)$ for the first bit). Because $\Omega_1(t)$ is scalar-valued, we have

$$
H_1(t_1)H_1(t_2) = H_1(t_2)H_1(t_1) \quad \text{for any} \quad t_1, t_2 \in [0, T]. \tag{7.33}
$$

Thus

$$
U_{1,\Omega_1}(T) = e^{-\frac{i}{2}\int_0^T \Omega_1(t)e_1(\phi)\cdot\sigma \, dt}
$$

$$
= e^{\left[-\frac{i}{2}\int_0^T \Omega_1(t)dt\right]e_1(\phi)\cdot\sigma}
$$

$$
= e^{-i\theta e_1(\phi)\cdot\sigma}, \qquad \text{(by (7.28))} \tag{7.34}
$$

using (7.30). The proof is complete. $\qquad\square$

We may define U_{2,Ω_2} in a similar way as in Theorem 7.3.1.

Next, we derive the 2-bit quantum phase gate Q_π and the CNOT gate. This will be done through the square root of the swap gate U_{sw}:

$$
U_{sw}(|ij\rangle) = |ji\rangle, \quad \text{for } i, j \in \{0, 1\}. \tag{7.35}
$$

Theorem 7.3.2. *([10, pp. 110-111]) Denote by $U(t)$ the time evolution operator for the quantum system (7.20) for the time duration $t \in [0, T]$. Choose $\Omega_1(t) = \Omega_2(t) = 0$ in (7.20) and let $w(t)$ therein satisfies*

$$
\int_0^T w(t)dt = \frac{\pi}{2}. \tag{7.36}
$$

Then we have $U(T) = -e^{\pi i/4} U_{\text{sw}}$, i.e., $U(T)$ is the swapping gate (with a nonessential phase factor $-e^{\pi i/4}$.)

Proof. By assumptions, we have now

$$H(t) = \omega(t)\boldsymbol{\sigma} \cdot \boldsymbol{\tau}/2. \tag{7.37}$$

Since $\omega(t)$ is scalar-valued, we have the commutativity

$$H(t_1)H(t_2) = H(t_2)H(t_1), \quad \text{for any} \quad t_1, t_2 \in [0, T]. \tag{7.38}$$

Therefore

$$U(T) = e^{-i\int_0^T H(t)dt/\hbar} = e^{\left[-\frac{i}{2}\int_0^T \omega(t)dt\right]\boldsymbol{\sigma}\cdot\boldsymbol{\tau}}$$

$$= e^{-i\phi\boldsymbol{\sigma}\cdot\boldsymbol{\tau}} \qquad\qquad \left(\phi \equiv \frac{1}{2}\int_0^T \omega(t)dt\right)$$

$$= \cos(\phi\boldsymbol{\sigma}\cdot\boldsymbol{\tau}) - i\sin(\phi\boldsymbol{\sigma}\cdot\boldsymbol{\tau}), \tag{7.39}$$

where $e^{-i\phi\boldsymbol{\sigma}\cdot\boldsymbol{\tau}}$, $\cos(\phi\boldsymbol{\sigma}\cdot\boldsymbol{\tau})$ and $\sin(\phi\boldsymbol{\sigma}\cdot\boldsymbol{\tau})$ are 4×4 matrices. Since

$$\boldsymbol{\sigma}\cdot\boldsymbol{\tau} = \begin{bmatrix} 1 & 0 & 0 & 0 \\ 0 & -1 & 2 & 0 \\ 0 & 2 & -1 & 0 \\ 0 & 0 & 0 & 1 \end{bmatrix}$$

has a 3-fold eigenvalue $+1$ (triplet) and a single eigenvalue (singlet) -3, the associated projection operators can be easily found to be

$$P_1 = \frac{1}{4}(31 + \boldsymbol{\sigma}\cdot\boldsymbol{\tau}) \text{ and } P_2 = \frac{1}{4}(1 - \boldsymbol{\sigma}\cdot\boldsymbol{\tau}); \quad P_j P_k = \begin{cases} 0, & j \neq k, \\ P_j, & j = k. \end{cases} \tag{7.40}$$

Therefore, for any sufficiently well-behaved function, including polynomials and analytic functions, of $\boldsymbol{\sigma}\cdot\boldsymbol{\tau}$,

$$f(\boldsymbol{\sigma}\cdot\boldsymbol{\tau}) = f(1)\frac{1}{4}(31 + \boldsymbol{\sigma}\cdot\boldsymbol{\tau}) + f(-3)\frac{1}{4}(1 - \boldsymbol{\sigma}\cdot\boldsymbol{\tau})$$

$$= \frac{1}{4}[3f(1) + f(-3)]\mathbf{1} + \frac{1}{4}[f(1) - f(-3)]\boldsymbol{\sigma}\cdot\boldsymbol{\tau},$$

according to the spectral theorem. As an application, consider $f(x) = x^2$ and we find

$$(\boldsymbol{\sigma}\cdot\boldsymbol{\tau})^2 = 31 - 2\boldsymbol{\sigma}\cdot\boldsymbol{\tau}. \tag{7.41}$$

Thus, from (7.39) and (7.40), we obtain

$$U(T) = e^{-i\phi\boldsymbol{\sigma}\cdot\boldsymbol{\tau}} = e^{-i\phi}\cdot\frac{1}{4}(31 + \boldsymbol{\sigma}\cdot\boldsymbol{\tau}) + e^{-3i\phi}\cdot\frac{1}{4}(1 - \boldsymbol{\sigma}\cdot\boldsymbol{\tau}). \tag{7.42}$$

With a little manipulation, (7.42) becomes

$$U(T) = e^{i\phi} \left[\cos(2\phi)\mathbf{1} - i\sin(2\phi)\frac{1 + \boldsymbol{\sigma} \cdot \boldsymbol{\tau}}{2} \right]$$
$$= e^{i\phi}[\cos(2\phi)\mathbf{1} - i\sin(2\phi)U_{sw}], \tag{7.43}$$

by the fact that

$$U_{sw} = \begin{bmatrix} 1 & 0 & 0 & 0 \\ 0 & 0 & 1 & 0 \\ 0 & 1 & 0 & 0 \\ 0 & 0 & 0 & 1 \end{bmatrix} = \frac{1}{2}(1 + \boldsymbol{\sigma} \cdot \boldsymbol{\tau}).$$

Choosing $\phi = \pi/4$, we obtain the desired conclusion. $\square$

Corollary 7.3.3. *([10, pp. 110-111]) The square roots of the swapping gate,* $U_{sw}^{1/2}$, *are*

$$U_{sw}^{1/2} = \frac{e^{\pm\pi i/4}}{\sqrt{2}}(1 \mp iU_{sw}). \tag{7.44}$$

Proof. From (7.43), we first obtain

$$U_{sw} = ie^{-\frac{\pi i}{4}}U(T). \tag{7.45}$$

Then use $\phi = \pm\pi/8$ in (7.43) to obtain

$$U_{sw}^{1/2} = (ie^{-\frac{\pi i}{4}})^{1/2}e^{\pm\pi i/8} \left[\frac{1}{\sqrt{2}}(1 \mp iU_{sw}) \right] \tag{7.46}$$

and the desired conclusion. (Note that these two square roots of U_{sw} reflect the choices of $\sqrt{1} = 1$ and

the square root of $-1 = \pm i$

for the square roots of the eigenvalues of U_{sw}.) $\square$

Corollary 7.3.4. *([10, p. 112]) The quantum phase gate* Q_π *is given by*

$$Q_\pi = (-i)U_{1,\Omega_1^{(2)}}U_{2,\Omega_2}U_{sw}^{1/2}U_{1,\Omega_1^{(1)}}U_{sw}^{1/2}, \tag{7.47}$$

where

$$\begin{cases} \int \Omega_1^{(1)}(t)\,dt = -\pi e_{1z}, \\ \int \Omega_1^{(2)}(t)\,dt = \pi e_{1z}/2, \\ \int \Omega_2(t)\,dt = -\pi e_{2z}/2, \end{cases} \tag{7.48}$$

and e_{1z}, e_{2z} *denote the* e_z *vector of, respectively, the first and the second qubit.*

Remark 7.3.1. In order to realize this succession of gates, only one of the $\Omega(t)$ in (7.48) is nonzero at any given instant t, with the duration when $\Omega_1^{(1)}(t) \neq 0$ earlier than that when $\Omega_2(t) \neq 0$, and that when $\Omega_1^{(2)}(t) \neq 0$ even later. Earliest is the period when $w(t) \neq 0$ for the first $U_{\text{sw}}^{1/2}$, and another period when $w(t) \neq 0$ is intermediate between those when $\Omega_1^{(1)}(t) \neq 0$ and $\Omega_2(t) \neq 0$.

Proof. Define

$$U_{\text{XOR}} \equiv e^{\frac{\pi i}{4}\sigma_z} e^{-\frac{\pi i}{4}\tau_z} U_{\text{sw}}^{1/2} e^{i\frac{\pi}{2}\sigma_z} U_{\text{sw}}^{1/2}, \tag{7.49}$$

with $U_{\text{sw}}^{1/2} = \frac{e^{-\frac{\pi}{4}i}}{\sqrt{2}}(1 + iU_{\text{sw}})$ chosen from (7.44). Then it is straightforward to check that

$$\begin{aligned}
U_{\text{XOR}}|00\rangle &= |00\rangle(i), \quad U_{\text{XOR}}|01\rangle = |01\rangle(i), \\
U_{\text{XOR}}|10\rangle &= |10\rangle(i), \quad U_{\text{XOR}}|11\rangle = |11\rangle(-i),
\end{aligned} \tag{7.50}$$

so that

$$\begin{aligned}
U_{\text{XOR}} &= i(|00\rangle\langle00| + |01\rangle\langle01| + |10\rangle\langle10| - |11\rangle\langle11|) \\
&= iQ_\pi.
\end{aligned} \tag{7.51}$$

$\square$

From the quantum phase gate Q_π, we again obtain the CNOT gate as in (7.13).

As a final comment of this section, we note that the two quantum dots in coupling are assumed to be identical. However, the state-of-the-art of fabrication of quantum dots with uniform size and characteristics is far from being perfected in contemporary technology. A more refined mathematical treatment for the modeling of two non-identical quantum dots in coupling is needed.

7.3.4   More details of the QD arrangements: laterally coupled and vertically coupled arrays

The discussion so far in this section is geared toward laterally coupled QD. Let us now give some details of quantum dot array arrangement. In coupled quantum dots, there exists the combined action of the Coulomb interaction and the Pauli exclusion principle. Two coupled electrons in absence of a magnetic field have a spin-singlet ground state, while the first excited state in the presence of strong Coulomb repulsion is a spin triplet. (Recall the discussions of singlet and triplet in the proof of Theorem 7.3.2.) Higher excited states are separated from these two lowest states by an energy gap, given either by the Coulomb repulsion or the single-particle confinement. For lateral coupling, the dots are arranged in a plane, at a sufficiently small distance, say $2a$, cf. (7.15)–(7.17), such that the electrons can tunnel between the dots (for a lowered barrier) and an exchange interaction J between the two spins is produced. Lateral coupling amongst quantum dots lying in a single plane can be achieved by two different techniques. First by controlling the material system, and then by having spatial correlation between adjoining dots that can lead to splitting of eigenstates within a single dot into symmetric and antisymmetric states. Or secondly by using a near-field probe that can induce an electromagnetic coupling between neighboring QD. Fig. 7.15 shows laterally coupled GaN/AlN single period QD system probed by near-field optical spectroscopy. In the absence of tunneling between the dots we still might have direct Coulomb interaction left between the electrons. However, this has no effect on the spins (qubit), provided the spin-orbit coupling is sufficiently small, which is the case for s-wave electrons in III-V semiconductors with unbroken inversion symmetry (this would not be so for hole-doped systems since the hole has a much stronger spin-orbit coupling due to its p-wave character). Finally, the vanishing of J in (7.14) or (7.15) can be exploited for switching by applying a constant homogeneous magnetic field to an array of quantum dots to tune J to zero (or close to some other desirable value). Then, for switching J on and off, only a small gate pulse or a small local magnetic field is needed.

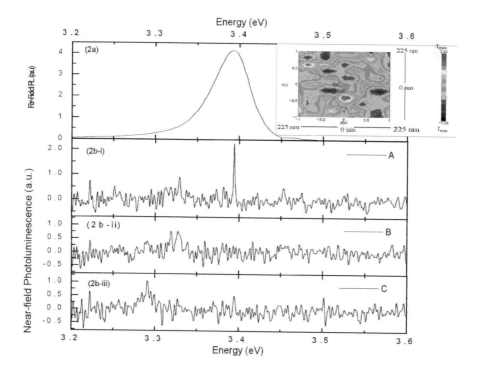

Figure 7.15: Near-field photoluminescence emission from laterally coupled quantum dots.

Fig. 7.15(a) shows the far-field PL (photonluminescence) spectrum, which is broadband and includes an ensemble effect of the emission from various quantum dots. Fig. 7.15(b) shows the emission from various regions A, B and C as mapped in spatial PL map shown above. The emissions from individual dots are observed from the narrow PL spectrum, some of which are inhomogeneously broadened due to lateral coupling. A greater control of laterally coupled quantum dots can be achieved by using patterned substrates so that the electrons in a confined quantum dot can be controlled by tailoring the eigenfunction of the conduction subbands [28].

The arrangement of vertically tunnel-coupled QD has also been studied by Burkard, Loss, and DiVincenzo [9]. The mathematical modeling is quite similar to the laterally coupled case so we omit it here and refer the interested readers to [9]. Neogi, *et al.* [27] considers such a setup of the dots, which has been produced in multilayer self-assembled QD as well as in etched mesa heterostructures.

To gain insight on the spatial variation of GaN QD and quantum wells on the emission intensity and linewidth, cross-sectional transmission electron microscopy (TEM) was performed. Samples were processed in a dual-beam SEM/FIB[7] (FEI Nova 600) using a Ga ion beam accelerating voltage of 5kV, followed by examination in a Tecnai F20 analytical HRTEM.[8] A near vertical correlation of the GaN dots $\sim 30\,nm$ in width is observed from STEM[9]-HAADF[10] image (not shown here), with some dot assemblies being correlated at an angle slightly off vertical. It is also observed that the width of these dots and their period correspond to the surface texture observed in SEM image (Fig. 7.9).

The vertical correlation provides a lower radiative recombination lifetime and higher emission efficiency due to tunneling via the vertically connected dots, especially in case of thin AlN spacer layers. Fig. 7.16 shows strong emission from a single cluster of QD from a $450\,nm \times 450\,nm$ area.

Switching of the spin-spin coupling between dots of different size can be achieved by means of varying external electric and magnetic fields. The exchange interaction[11] is not only sensitive to the magnitude of the applied fields, but also to their direction. It has been predicted that an in-plane magnetic field $B_\parallel$ suppresses J exponentially; a perpendicular field in laterally coupled dots has the same effect. A perpendicular magnetic field $B_\perp$ reduces on one hand the exchange coupling between identically sized dots and on the other hand, for different dot sizes, increasing $B_\perp$ amplifies the exchange coupling J until the electronic orbitals of the various QD are magnetically compressed to approximately the same size. A perpendic-

---

[7]SEM: scanning electron microscope; FIB: focused ion-beam.

[8]HRTEM: high-resolution transmission electron microscope.

[9]STEM: scanning transmission electron microscope.

[10]high-angle annular dark-field.

[11]The exchange interaction is the "off-diagonal" effect of the Coulomb forces acting on two interacting particles (here, the two electrons and their spins in two neighboring QD). This exchange interaction is characterized by the quantity J given in Appendix B.

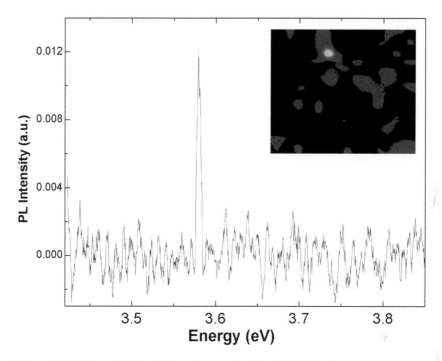

Figure 7.16: Near-field emission from a single GaN/AlN QD excited by a He-Cd laser at $325\,nm$. The scanning area is $450\,nm \times 450\,nm$.

ular electric field $E_\perp$ detunes the single-dot levels, and thus reduces the exchange coupling; the very same finding was made for laterally coupled dots and an in-plane electric field [9].

7.3.5 Decoherence and measurement

The main source of decoherence for this model is the environmental fluctuation. For electron spin, it is the fluctuation of the ambient magnetic field. This effect can be modeled as the linear interaction between the electron spin and the environment modeled as a set of harmonic oscillators (called a Caldeira-Leggett-type model). For a 2-spin system, the interaction Hamiltonian is $H_{int} = \lambda \sum_{i=1,2} \boldsymbol{S}_i \cdot \boldsymbol{b}_i$, where $\boldsymbol{S}_i$ is the angular momentum of the electron spin and the jth component of $\boldsymbol{b}_i$, $b_i^j = \sum_\alpha g_\alpha^{ij}(a_{\alpha,ij}^+ + a_{\alpha,ij})$, is the fluctuating field. Here $a_{\alpha,ij}^+$ and $a_{\alpha,ij}$ are, respectively, the creation and annihilation operators of the magnetic field with mode α, respectively, and g_α^{ij} is the amplitude. The unperturbed harmonic-oscillator Hamiltonian of the magnetic field can be written as $H_B = \sum \omega_\alpha^{ij} a_{\alpha,ij}^+ a_{\alpha,ij} + \frac{1}{2}$, where the subscript "B" represents "bath", i.e., the environment, and ω_α^{ij} is the frequency of the mode α of the field. Interest is focused on the density matrix of the system, which is obtained by tracing the bath from the total density matrix $\bar{\rho}(t)$, $\rho(t) = \text{Tr}_B \bar{\rho}(t)$. Dynamics of $\bar{\rho}(t)$ satisfies the von Neumann equation (cf. (4.13)):

$$\frac{\partial \bar{\rho}(t)}{\partial t} = -i[H, \bar{\rho}(t)] = -i\mathcal{L}\bar{\rho}(t), \qquad (7.52)$$

where H is the overall Hamiltonian including the unperturbed system Hamiltonian H_s:

$$H = H_s + H_{int} + H_B, \qquad (7.53)$$

and $\mathcal{L}$ is the corresponding Liouvillian super operator; see Subsection 4.3.4.

A detailed computation of the effect of the decoherence on the quantum system when the swap gate or 1-bit rotation gate are applied is given in [23]. That result shows that the decoherence time has order of 1.4 ns while the time needed for one logic gate operation has order of 25 ps, which is very satisfactory for quantum computation. The reference model used in the computation has an exchange constant $J = 80\,\mu\text{eV}$ and the ambient magnetic fields are assumed to be at thermal equilibrium.

Measurement of a single spin in a quantum dot is obviously difficult because of the weak signal and strong background noise. Elegant schemes are needed to overcome this difficulty. One could utilize a switchable tunneling with which the electron tunnels into a super-cooled paramagnetic dot (PM) before the measurement. Then it nucleates a ferromagnetic domain whose magnetization can be measured by conventional methods. Fidelity of successful measurement with this method is expected to be about 75% when the magnetization in the upper hemisphere is interpreted as a spin-up state of the electron. Another scheme utilizes a switchable valve or barrier which is only transparent to electrons of the spin-up state. When a spin is to be measured, this switchable valve only let the spin-up electron pass to another quantum dot (called a measure dot). Then a nanoscale single-electron electrometer can be used to detect the presence of electron in the measure dot. If an electron is found there, it is in spin-up state; otherwise, it is in spin-down state.

7.3.6   New advances

Recent advances in nanofabrication has greatly facilitated the development of spin-based devices. The accurate estimation of the QD size distribution in the nanoscale limit is critical to the optimization of the radiative emission rate, device design and fabrication of integrated quantum dot based quantum computational structures. The focused ion-beam (FIB) lithography technique overcomes the diffraction-limited spot size occurring in convention photolithography. FIB offers advantages over the conventional photo-processing technique in precision and fine resolution, and extends the capability of photoprocessing to the nanometer region. FIB aided processing of semiconductor optoelectronic devices and definite progress has been made recently in the areas of photoprocessing and photofabrication with a higher degree of precision and resolution [2, 39]. The patterning and machining of electrodes with nanoscale resolution over quantum dot structures in the FIB allows for the unique opportunity of simultaneously preparing cross-sectional TEM at sub-micron spatial resolution.

We also mention that many solid-state implementations for quantum computing have been proposed subsequently [22], including superconducting qubits, nuclear spins of donor atoms in silicon, and charge qubits in QD.

7.4 Biexciton in a single quantum dot

Piermarocchi, Chen, Dale and Sham [29] propose to utilize the robustness of the elementary excitation of the electrons in semiconductor nanostructure quantum dots, i.e., the excitons. An exciton consists of a conduction-band electron and a valence-band hole. This electron-hole pair may be likened to a hydrogen atom, which has an orbiting electron and a nucleus with one proton carrying a positive charge corresponding to the *hole*. In an undoped quantum dot, the optically excited electron-hole pair feels the presence of a large number of atoms (in the order of 10^5-10^6) in the dot and the effect may be well characterized by the static dielectric constant and the electron-hole's effective mass [32]. Thus, the exciton works in the same way as excitations in "giant atoms". A single quantum dot can hold multiexciton complexes containing many interacting excitons.

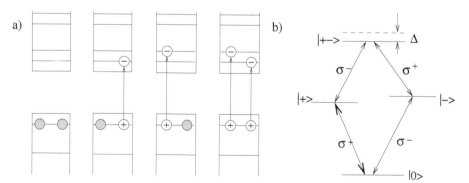

Figure 7.17: This schematic illustrating biexciton transitions is adapted from (21). Here ⊖ represents an electron and ⊕ a hole. (a). The four quantum states considered, from left to right, the ground state $|0\rangle$, the excited exciton state $|+\rangle$, the excited exciton state $|-\rangle$ and the biexciton state $|-+\rangle$, which could be utilized to encode $|00\rangle$, $|01\rangle$, $|10\rangle$ and $|11\rangle$, respectively. (b). Diagram of the energy levels of the four states and their transitions induced by optical pulses, where σ^+ represents left-polarized light and σ^- represents right-polarized light. The binding energy is $\Delta = \epsilon_+ + \epsilon_- - \epsilon_{-+}$.

Piermarocchi, *et al.* [29] consider two electron-hole pairs, each with two confined energy levels, inside a single QD; see Fig. 7.17. The underlying assumptions for the biexciton model considered in this section according to [29] are:

1. The lifetime of the biexciton and exciton is large enough for the quantum operations involved, and no unintended states will be introduced. Thus, the system evolves in a desirable subspace. This also means that only the optically active states are considered.

2. The size of the dot considered is about $40 \times 35 \times 5\,nm^3$. The electronic levels considered are the first two states derived from the localization of the s-like conduction band states, carrying a spin $\pm\frac{1}{2}$, and the hole levels are derived from the the localization of the states in the p-like valence-band heavy hole, carrying a $\pm\frac{3}{2}$ total spin in the direction of the QD growth axis.

3. Only the Coulomb interaction between the carriers which conserve their conduction or valence band indices is taken into account. The electron-hole exchange is neglected, whose energy is of order μeV and only affects the fine structure of the excitonic levels.

According to the above assumptions, the Hamiltonian of the four level biexciton system under optical field with σ^+ polarization can be written as

$$H^+ = \epsilon_+|01\rangle\langle 01| + \epsilon_-|10\rangle\langle 10| + \epsilon_{-+}|11\rangle\langle 11|$$
$$+ \frac{1}{2}\{\Omega_+(t)|01\rangle\langle 00| + f\Omega_+(t)|11\rangle\langle 10| + \text{h.c.}\} \qquad (7.54)$$

where ϵ_+, ϵ_-, and ϵ_{-+} are, respectively, the unperturbed energies of the states $|01\rangle$, $|10\rangle$, and $|11\rangle$ from $|00\rangle$, and

$$\Omega_+ = d_+ E_+(t) \qquad (7.55)$$

represents a time dependent Rabi energy provided by the electric field, with d_+ being the dipole moment of the exciton state $|+\rangle$; f is a correction factor to the dipole moment due to the Coulomb interaction. The electric field $E_+(t)$ normally has a Gaussian shape: $E_+(t) = \varepsilon_+ e^{-(t/\epsilon)^2} e^{i(\omega_+ t + \phi)}$. If a series of pulses are applied instead of one, $E_+(t)$ is the sum of several pulses and $\Omega_+(t)$ can be expressed as

$$\Omega_+(t) = \sum_j d_+ E_{+,j}(t - \tau_j), \qquad (7.56)$$

where τ_j is the center of the jth pulse. The Hamiltonian H^+ may be rewritten in matrix form

$$H^+ = \begin{pmatrix} 0 & \Omega_+^*(t)/2 & 0 & 0 \\ \Omega_+(t)/2 & \epsilon_+ & 0 & 0 \\ 0 & 0 & \epsilon_- & f\Omega_+^*(t)/2 \\ 0 & 0 & f\Omega_+(t)/2 & \epsilon_{-+} \end{pmatrix} \qquad (7.57)$$

with respect to the ordered basis $|00\rangle, |01\rangle, |10\rangle, |11\rangle$. The two isolated block form of the matrix in (7.57) implies that the whole state space can be divided into two invariant subspaces. One is spanned by $|00\rangle$ and $|01\rangle$, or the ground state $|0\rangle$ and excited state $|+\rangle$; another is spanned by $|10\rangle$ and $|11\rangle$, or the excited sate $|-\rangle$ and biexciton state $|-+\rangle$. The Hamiltonian H^+ introduces Rabi rotation within each subspace, which will be shown in next subsection (cf. (7.68)). Similarly, the Hamiltonian of the system under σ^- polarized optical field reads

$$
\begin{aligned}
H^- &= \epsilon_+ |01\rangle\langle 01| + \epsilon_- |10\rangle\langle 10| + \epsilon_{-+} |11\rangle\langle 11| \\
&+ \frac{1}{2}\{\Omega_-(t)|10\rangle\langle 00| + f\Omega_-(t)|11\rangle\langle 01| + h.c.\}
\end{aligned} \tag{7.58}
$$

and the associated matrix form is

$$
H^- = \begin{pmatrix}
0 & 0 & \Omega_-^*(t)/2 & 0 \\
0 & \epsilon_+ & 0 & f\Omega_-^*(t)/2 \\
\Omega_-(t)/2 & 0 & \epsilon_+ & 0 \\
0 & f\Omega_-(t)/2 & 0 & \epsilon_{-+}
\end{pmatrix}. \tag{7.59}
$$

We can also write H^- in a similar block matrix form as H^+ in (7.57) by reordering the four basis, while H^- introduces Rabi rotations within subspace spanned by $|00\rangle$ (the ground state $|0\rangle$) and $|10\rangle$ (the excited state $|-\rangle$); and another subspace spanned by $|01\rangle$ (the excited state $|+\rangle$) and $|11\rangle$ ($|-+\rangle$). This is easy to understand because exciton states $|+\rangle$ and $|-\rangle$ are created by optical fields with different polarizations. The fact that H^+ and H^- share the same form after reordering the basis ensures that we only have to focus on one of them. The evolution operator, or propagator, generated by H^- can be obtained easily and similarly after we know the propagator generated by H^+.

7.4.1 Derivation of the unitary rotation matrix and the conditional rotation gate

The key to realize a logic gate, or unitary transformation, of this quantum system is to choose appropriate optical pulses. Before the derivation of the various logic gates, we first convert H^+ into its interaction picture. We separate H^+ into the unperturbed and interaction parts $H^+ = H_0^+ + H_I^+$, where

$$
H_0^+ = \epsilon_+ |01\rangle\langle 01| + \epsilon_- |10\rangle\langle 10| + \epsilon_{-+} |11\rangle\langle 11|; \tag{7.60}
$$

$$
H_I^+ = \frac{1}{2}\{\Omega_+(t)|01\rangle\langle 00| + f\Omega_+(t)|11\rangle\langle 10| + h.c.\}. \tag{7.61}
$$

Standard deduction leads to the Hamiltonian in the interaction picture

$$
\begin{aligned}
V^+ &= e^{iH_0^+ t} H_I^+ e^{-iH_0^+ t} \\
&= \frac{1}{2}\{\Omega_+(t)e^{i\epsilon_+ t}|01\rangle\langle 00| + f\Omega_+(t)e^{i(\epsilon_+ - \Delta)t}|11\rangle\langle 10| + h.c.\} \tag{7.62}
\end{aligned}
$$

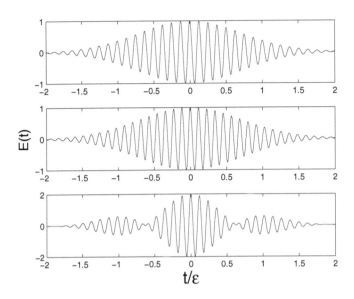

Figure 7.18: A schematic illustration of the control pulses applied for logic gate. The top and middle curves have the same Gaussian shape but different frequencies. The bottom curve is a composite pulse which is the sum of the above two pulses. For simplicity, all the angles are set to 0. Also, $\varepsilon_0 = \varepsilon_1 = 1$ and $\epsilon_0 = \epsilon_1 = \epsilon$. While x-axis represents time, y-axis represents the amplitude of the electric field.

where $\Delta = \epsilon_+ + \epsilon_- - \epsilon_{-+}$ is the biexciton binding energy, which is not zero because of the interaction between the two excitons. Choosing the optical pulse to be $E_+(t) = \varepsilon_0 e^{-(t/\epsilon_0)^2} e^{-i\epsilon_+ t - i\phi_0} + \varepsilon_1 e^{-(t/\epsilon_1)^2} e^{-i(\epsilon_+ - \Delta)t - i\phi_1}$, a composite bi-chromatic phase-locked pulse shown in Fig. 7.18, we obtain the Hamiltonian with this pulse, in the interaction picture as

$$
\begin{aligned}
\mathbf{V}^+ = {} & \frac{d_+}{2}(\varepsilon_0 e^{-(t/\epsilon_0)^2} e^{-i\phi_0} + \varepsilon_1 e^{-(t/\epsilon_1)^2} e^{i\Delta t - i\phi_1})|01\rangle\langle00| \\
& + \frac{d_+}{2}(\varepsilon_0 e^{-(t/\epsilon_0)^2} e^{-i\Delta t - i\phi_0} + \varepsilon_1 e^{-(t/\epsilon_1)^2} e^{-i\phi_1})|11\rangle\langle10| + h.c. \quad (7.63)
\end{aligned}
$$

where we have used the usual rotating wave and dipole approximations, and f is set to 1 for convenience. The pulse has an envelope shape $e^{-(t/\epsilon)^2}$ with width 2ϵ, and its spectra has two peaks centered at ϵ_+ and $\epsilon_+ - \Delta$. See Fig. 7.18. To determine and approximate the unitary evolution operator, we need the Magnus expansion

$$
\begin{aligned}
U^+ &= \mathcal{T}e^{-\frac{i}{\hbar}\int_{-\infty}^{\infty} \mathbf{V}^+(t)dt} \\
&= e^{-\frac{i}{\hbar}(\hat{H}^0 + \hat{H}^1 + \dots)}, \quad (7.64)
\end{aligned}
$$

where $\mathcal{T}$ is the time ordering operator, and

$$\hat{H}^0 = \int_{-\infty}^{\infty} \mathbf{V}^+(t)dt,$$

$$\hat{H}^1 = \int_{-\infty}^{\infty} dt_2 \int_{-\infty}^{\infty} dt_1 [\mathbf{V}^+(t_2), \mathbf{V}^+(t_1)],$$

$$\cdots \qquad \cdots$$

The two integral limits are set to be infinity, because the amplitude of the electric field decreases to zero exponentially fast. Simulation has shown that the first term already has good accuracy, and the evolution operator can be obtained by the approximation

$$U^+ = e^{-\frac{i}{\hbar}\int_{-\infty}^{\infty}\mathbf{V}^+(t)dt}$$

$$= e^{-iA}, \tag{7.65}$$

where

$$A = \frac{1}{\hbar}\int_{-\infty}^{\infty}\mathbf{V}^+(t)dt$$

$$= \frac{1}{2}\begin{pmatrix} 0 & e^{i\phi_0}\theta_0+\delta_1 e^{i\phi_1} & 0 & 0 \\ e^{-i\phi_0}\theta_0+\delta_1 e^{-i\phi_1} & 0 & 0 & 0 \\ 0 & 0 & 0 & \delta_0 e^{i\phi_0}+\theta_1 e^{i\phi_1} \\ 0 & 0 & \delta_0 e^{-i\phi_0}+\theta_1 e^{-i\phi_1} & 0 \end{pmatrix} \tag{7.66}$$

in matrix form. Here, for simplicity, we have introduced new variables $\theta_i = \frac{d+\varepsilon_i}{\hbar}\int_{-\infty}^{\infty}e^{-(t/\epsilon_i)^2}dt$ and $\delta_i = \frac{d+\varepsilon_i}{\hbar}\int_{-\infty}^{\infty}e^{-(t/\epsilon_i)^2}e^{i\Delta t}dt = \frac{\epsilon_i\sqrt{\pi}d+\varepsilon_i}{\hbar}e^{-(\epsilon\Delta/2)^2}$, $i = 0, 1$. When the pulse is sufficiently flat, $\epsilon\Delta \ll 1$, δ_i is almost zero, and A is further reduced to an even simpler form

$$A = \frac{1}{2}\begin{pmatrix} 0 & e^{i\phi_0}\theta_0 & 0 & 0 \\ e^{-i\phi_0}\theta_0 & 0 & 0 & 0 \\ 0 & 0 & 0 & e^{i\phi_1}\theta_1 \\ 0 & 0 & e^{-i\phi_1}\theta_1 & 0 \end{pmatrix}, \tag{7.67}$$

which agrees with the result obtained by the area theorem.[12] It is also worth noting that A is composed of two independent subsystems, which gives great convenience in computation. A general form of the propagator U^+ in matrix form can derived as

$$U^+ = e^{-iA}$$

$$= \begin{pmatrix} \cos(\theta_0/2) & ie^{i\phi_0}\sin(\theta_0/2) & 0 & 0 \\ ie^{-i\phi_0}\sin(\theta_0/2) & \cos(\theta_0/2) & 0 & 0 \\ 0 & 0 & \cos(\theta_1/2) & ie^{i\phi_1}\sin(\theta_1/2) \\ 0 & 0 & ie^{-i\phi_1}\sin(\theta_1/2) & \cos(\theta_1/2) \end{pmatrix}. \tag{7.68}$$

---

[12]Assuming θ is the angle the Bloch vector rotates, and $\Omega(t)$ is the Rabi frequency, then the *area theorem* says that

$$\theta = \int_{-\infty}^{t}\Omega(t_1)dt_1,$$

i.e., θ is equal to the area below the frequency curve.

The matrix in (7.68) is useful in the design of the pulses for the 1-bit rotations and the conditional rotation gates.

Theorem 7.4.1. *In the above four-level biexciton quantum system (within the accuracy of the area theorem), polarized optical pulses can be used to realize unitary 1-bit rotation and the conditional rotation matrices.*

Proof. When $\theta_0 = \theta_1 = \theta$, from (7.68) we obtain the y-rotation of the first qubit by setting $\phi_0 = \phi_1 = \pi/2$, denoted by $R_{1y}(\theta)$, and the x-rotation $R_{1x}(\theta)$ by setting $\phi_0 = \phi_1 = 0$. If $\theta_0 = 0$, a rotation of first qubit when the second qubit is in state $|1\rangle$ is obtained, denoted by $C_{1,2}^{\text{ROT}}(\theta, \phi)$:

$$
C_{1,2}^{\text{ROT}}(\theta, \phi) = \begin{pmatrix} 1 & 0 & 0 & 0 \\ 0 & 1 & 0 & 0 \\ 0 & 0 & \cos(\theta/2) & ie^{i\phi}\sin(\theta/2) \\ 0 & 0 & ie^{-i\phi}\sin(\theta/2) & \cos(\theta/2) \end{pmatrix}. \tag{7.69}
$$

Another conditional rotation triggered when the second qubit is in state $|0\rangle$ can be obtained with $\theta_1 = 0$, denoted by $C_{1,\bar{2}}^{\text{ROT}}(\theta, \phi)$:

$$
C_{1,\bar{2}}^{\text{ROT}}(\theta, \phi) = \begin{pmatrix} \cos(\theta/2) & ie^{i\phi}\sin(\theta/2) & 0 & 0 \\ ie^{-i\phi}\sin(\theta/2) & \cos(\theta/2) & 0 & 0 \\ 0 & 0 & 1 & 0 \\ 0 & 0 & 0 & 1 \end{pmatrix}. \tag{7.70}
$$

The proof of the rotations and conditional rotations of the second qubit is similar. We only need to apply the σ^- polarized field instead of the σ^+ polarized field. The result can be obtain by just reordering the basis. □

The derivation of the above theorem also provides blocks to construct a CNOT gate, and we have following corollary.

Corollary 7.4.2. *A CNOT gate can be simulated with the one qubit rotation and conditional rotation gates obtained in above theorem, thus the proposal is universal.*

Proof. We will only investigate the CNOT gate controlled by the second qubit. Using the same notation in the last theorem, we can construct a z-rotation operation of the second qubit,

$$
R_{2z}(\theta) = R_{2y}(\pi/2)R_{2x}(\theta)R_{2y}(-\pi/2) \tag{7.71}
$$

and a special case,

$$
R_{2z}(\pi/2) = R_{2y}(\pi/2)R_{2x}(\pi/2)R_{2y}(-\pi/2)
$$

$$
= \begin{pmatrix} e^{-\pi/4} & 0 & 0 & 0 \\ 0 & e^{-\pi/4} & 0 & 0 \\ 0 & 0 & e^{\pi/4} & 0 \\ 0 & 0 & 0 & e^{\pi/4} \end{pmatrix}. \tag{7.72}
$$

Above z-rotation can be used in a quantum phase gate Q_π up to a phase shift:

$$Q_\pi = R_{2z}(\pi/2)R_{1y}(\pi/2)R_{1x}(\pi/2)C_{1,\bar{2}}^{ROT}(-\pi/2)C_{1,2}^{ROT}(\pi/2)R_{1y}(-\pi/2)$$

$$= e^{-i\pi/4} \begin{pmatrix} 1 & 0 & 0 & 0 \\ 0 & 1 & 0 & 0 \\ 0 & 0 & 1 & 0 \\ 0 & 0 & 0 & -1 \end{pmatrix}. \tag{7.73}$$

Now we can construct the CNOT gate with the second qubit as the control qubit using above phase shift gate and a conditional rotational gate combined as

$$C_{1,2}^{NOT} = Q_\pi C_{1,2}^{ROT}(\pi, -\pi/2)$$

$$= e^{-i\pi/2} \begin{pmatrix} 1 & 0 & 0 & 0 \\ 0 & 1 & 0 & 0 \\ 0 & 0 & 0 & 1 \\ 0 & 0 & 1 & 0 \end{pmatrix}, \tag{7.74}$$

up to a phase factor. This ends the proof.   □

7.4.2 Decoherence and measurement

The life time of an exciton is critical to quantum information procession. It constrains how many logical operations can be performed before the system loses coherence. Theoretical and experimental results [35, 5] show that the life time spans from 10ps to about 1ns, depending on the confinement and temperature. The source of decoherence comes from spontaneous emission caused by ambient fluctuation and other factors. The shape of the pulse should be carefully designed to reduce the spillover of unwanted excited states, such as unlocalized exciton states with energy levels close to that of the localized excitons. During the operations, heating effect may also accelerate the decoherence.

Research on coherent nonlinear optical spectroscopy of the individual excitons provides a method to probe a single exciton in a semiconductor quantum dot [6, 35]. When a weak probe optical pulse $E(t)$ is applied on the quantum dot, an induced nonlinear optical polarization field is created. The polarization field is homodyne detected with the transmitted field, generating a signal. The integral of this signal provides a differential transmission signal proportional to the inversion of the exciton, i.e., the difference of probability amplitudes of the excited and ground states.

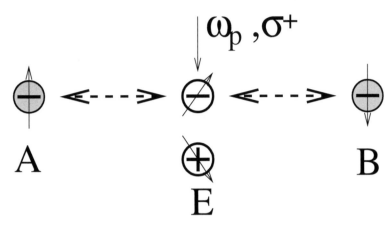

Figure 7.19: A schematic of the optical induced spin-spin interaction. A and B are two electron spins in two neighboring quantum dots; E is the photo-generated unlocalized exciton which interacts with A and B. The optical field has frequency w_p and σ^+ polarization.

7.4.3 Proposals for coupling of two or more biexciton QD

It is not realistic to set up more qubits by exciting more excitons in a single quantum dot. To make the above proposal *scalable*, other strategies to couple two or more QD are necessary. Potential candidates include microcavity [33] as already discussed in Section 7.2, where the cavity mode photon is used as an agent to connect two QD, and a linear support [7], where the quantum phonon state of a big molecule or nano-rod serves as the data bus. Piermarocchi, *et al.* [30] also suggest using pure optical control to couple spins in two neighboring QD. The main control method is to use an external optical field with σ^+ polarization which creates an unlocalized electron-hole pair, i.e., an exciton. This photo-generated electron-hole pair has a fixed angular momentum configuration corresponding to the polarization of the field, with electron spin state $-\sigma(= -1/2)$ and hole spin state $\sigma(= 3/2)$. The density of the optical field is kept below the lowest discrete exciton state. The electron spin in the photo-generated exciton interacts with the two electron spins from the two neighboring quantum dots and serves as an agent to couple them together. After reasonable simplification, the interaction Hamiltonian between the two spins can be shown to be

$$H_S = -2J_{12}\boldsymbol{S}^1 \cdot \boldsymbol{S}^2, \tag{7.75}$$

where $\boldsymbol{S}^1$ and $\boldsymbol{S}^2$ are the two electron spins from the two quantum dots, and J_{12} is an constant. Fig. 7.19 gives a schematic of this proposal. It is an interaction term similar to that of NMR except that the constant J_{12} is controllable.

7.5 Conclusions

Physical implementations of qubits using QD are fundamentally limited by the interaction of the qubits with their environment and the resulting decoherence. These interactions of the qubits set the maximum time of coherent operation and an upper bound for the number of quantum gate operations to be applied on a single qubit; therefore understanding the origin of decoherence is critical to control or reduce it, in order to implement quantum logic gates. Because of their strong localization in all directions, electrons confined in quantum dots are strongly coupled to longitudinal optical (LO) vibrations of the underlying crystal lattice. If the coupling strength exceeds the "continuum width", the energy of keeping the LO phonons delocalized, a continuous Rabi oscillation of the electron arises, that is, an everlasting emission and absorption of one LO phonon. As a result, electron-phonon entangled quasi-particles known as polarons form; these play a substantial role in the rapid decoherence of the spin-based quantum dot qubits. The decoherence time for an exciton typically ranges from $20\,ps$ to $100\,ps$, which is considerably shorter than the decoherence times of nuclear or electron spin. This is a problem since gate operations take approximately $40\,ps$ to perform. However, implementing ultrafast (femtosecond) optoelectronics may eventually enable us to bypass this problem. Read-out on the QD can be achieved by placing the excitation and probe beam spots on a specific location where a number of qubits with different excitonic frequencies can be accessed. The somewhat randomized distribution of the QD size and composition allow qubits with different excitation frequencies to exist, making it easier to identify specific qubits by singling out the different frequencies.

There are many interesting or useful websites on QD or quantum complexity in general, maintained by individual researchers or research centers. We mention just a few of them here:

(i) [43], of the Centre for Quantum Computation, at Oxford University, U.K., contains comprehensive, current information and activities in quantum computation.

(ii) [44], of the Los Alamos National Laboratory, U.S., contains perhaps the most current papers and manuscripts on any quantum-related topics.

(iii) [45], Quantum Dot Group homepage, maintained by a group of European (mainly French) researchers.

(iv) [46], maintained by researchers in Oxford and Cambridge Universities, U.K., on optical quantum dots.

(v) [47], maintained by Dr. L. Kouwenhoven of the Technical University of Delft in the Netherlands.

(vi) [48], maintained by researchers in the Electrical Engineering Department of the University of Notre Dame, U.S., is the website of quantum cellular automata.

References

[1] Abramowitz and I. Stegun, *Handbook of Mathematical Functions*, 9th printing, Dover, New York, 1970.

[2] J. Ahopelto, Emerging Nanopatterning Methods, Int. Conference on Ultimate Lithography and Nanodevice Engineering, 2004, Agelonde, France.

[3] M.N. Baibich, J.M. Broto, A. Fert, F. Nguyen Van Dau, F. Petroff, P. Eitenne, G. Creuzet, A. Friederich, and J. Chazelas, Giant magnetoresistance of (001)Fe/(001)Cr magnetic superlattices, *Phys. Rev. Lett.* **61** (1988), 2472.

[4] P. Bhattacharya, *et al.*, In(Ga)As/GaAs self-organized quantum dot lasers: DC and small-signal modulation properties, *IEEE Trans. Electron Devices* **46** (1999), 871–883.

[5] D. Birkedal, K. Leosson, and J.M. Hvam, Long lived coherence in self-assembled quantum dots, *Phys. Rev. Lett.* **87** (2001), 227401.

[6] N.H. Bonadeo, G. Chen, D. Gammon, D.S. Katzer, D. Park, and D.G. Steel, Nonlinear nano-optics: probing one exciton at a time, *Phys. Rev. Lett.* **81** (1998), 2759–2762.

[7] K.R. Brown, D.A. Lidar, and K.B. Whaley, Quantum computing with quantum dots on quantum linear supports, *Phys. Rev. A* **65** (2001), 012307.

[8] G. Burkard, H-A. Engel, and D. Loss, Spintronics, quantum computing and quantum communications in quantum dots, in *Fundamentals of Quantum Information* W. Dieter Heiss, (ed.), Springer-Verlag, Berlin and Heidelberg, 2002, 241–265.

[9] G. Burkard, D. Loss, and D. DiVincenzo, Coupled quantum dots as quantum gates, *Phys. Rev. B* **59** (1999), 2070–2078.

[10] G. Chen, D.A. Church, B-G. Englert, and M.S. Zubairy, Mathematical models of contemporary elementary quantum computing devices, in *Quantum Control: Mathematical and Numerical Challenges* A.D.

Bandrauk, M.C. Delfour and C. Le Bris, (eds.), CRM Proc. & Lecture Notes, Vol. 33, Amer. Math. Soc., Providence, R.I., 2003, 79–117.

[11] G. Chen, Z. Diao, J.U. Kim, A. Neogi, K. Urekin and Z. Zhang, *Int. J. Quantum Information*, **4** (2006), 233–296.

[12] R.J. Elliot, *Phy. Rev.* **108, 1348** (1957).

[13] W. Franz, *Z. Naturforsch.13a* **484** (1958).

[14] J.M. Gerard and B. Gayral, Strong Purcell effect for InAs quantum boxes in 3-dimensional solid-state microcavities, *J. Lightwave Technol.* **17** (1999), 2089–2095.

[15] L. Goldstein, F. Glas, J.Y. Marzin, M.N. Charasse, and G. Le Roux, *Appl. Phys. Lett.* **47** (1985), 1099.

[16] J.M. Kikkawa and D.D. Awschalom, Resonant spin amplification in n-Type GaAs, *Phys. Rev. Lett.* **80** (1998), 4313.

[17] I.L. Krestnikov, N.A. Maleev, A.V. Sakharov, A.R. Kovsh, A.E. Zhukov, A.F. Tsatsul'nikov, V.M. Ustinov, Zh.I. Alferov, N.N. Ledentsov, D. Bimberg, and J.A. Lott, 1.3 m resonant-cavity InGaAs/GaAs quantum dot light-emitting devices, *Semicond. Sci. Technol.* **16** (2001), 844–848.

[18] A.A. Lagatsky, E.U. Rafailov, W. Sibbett, D.A. Livshits, A.E. Zhukov, and V.M. Ustinov, Quantum-dot-based saturable absorber with p-n junction for mode-locking of solid-state lasers, *IEEE Photonics Technology Letters* **17** (2005), 294–296.

[19] R. Landauer, *Proc. R. Soc. London, Ser. A* **454, 305** (1998).

[20] L. Landin, *et al.*, Optical studies of individual InAs quantum dots in GaAs: few-particle effects, *Science* **280** (1998), 262–264.

[21] X. Li, Y. Wu, D. Steel, D. Gammon, T.H. Stievater, D.S. Katzer, D. Park, C. Piermarocchi, and L.J. Sham, An all-optical quantum gate in a semiconductor quantum dot, *Science* **301** (2003), 809–811.

[22] *Fortschr. Phys.* **48** (2000), Special issue on *Experimental Proposals for Quantum Computation*, H.K. Lo and S. Braunstein (eds.), Wiley-VCH, Berlin.

[23] D. Loss and D. DiVincenzo, Quantum computation with quantum dots, *Phys. Rev. A* **57** (1998), 120–126.

[24] P. Michler, A. Imamoglu, M.D. Mason, P.J. Carson, G.F. Strouse, and S.K. Buratto, Quantum correlation among photons from a single quantum dot at room temperature, *Nature* **406** (2000), 968.

[25] D.A.B. Miller et al., *Phy. Rev. Lett.* **53, 2173** (1984).

[26] A. Neogi, B.P. Gorman, H. Morkoç, T. Kawazoe, and M. Ohtsu, Near-field optical spectroscopy and microscopy of self-assembled GaN/AlN nanostructures, *Appl. Phys. Lett.* **86** (2005), 43103.

[27] A. Neogi, H. Morkoç, T. Kuroda, A. Tackeuchi, T. Kawazoe, and M. Ohtsu, Exciton localization in vertically and laterally coupled GaN/AlN quantum dots, *Nano Letters* **5** (2005), 209–212.

[28] A. Neogi, H. Yoshida, T. Mozume, N. Georgiev, and O. Wada, Intersub-band transitions and ultrafast all-optical modulation using multiple InGaAs-AlAsSb-InP coupled double-quantum-well structures, *IEEE J. Selected Topics in Quantum Electronics* **7** (2001), 710–717.

[29] C. Piermarocchi, P. Chen, Y.S. Dale, and L.J. Sham, Theory of fast quantum control of exciton dynamics in semiconductor quantum dots, *Phys. Rev. B* **65** (2002), 075307.

[30] C. Piermarocchi, P. Chen, L.J. Sham, and D.G. Steel, Optical RKKY interaction between charged semiconductor quantum dots, *Phys. Rev. Lett.* **89** (2002), 167402.

[31] A. Rauschenbeutel, G. Nogues, S. Osnaghi, P. Bertet, M. Brune, J.M. Raimond, and S. Haroche, Coherent operation of a tunable quantum phase gate in cavity QED, *Phys. Rev. Lett.* **83** (1999), 5166–5169.

[32] L.J. Sham and T.M. Rice, Many-particle derivation of the effective-mass equation for the wannier exciton, *Phys. Rev.* **144** (1966), 708–714.

[33] M.S. Sherwin, A. Imamoglu, and T. Montroy, Quantum computation with quantum dots and terahertz cavity quantum electrodynamics, *Phys. Rev. A* **60** (1999), 3508–3514.

[34] G.S. Solomon, M. Pelton, and Y. Yamamoto, Single-mode spontaneous emission from a single quantum dot in a 3-dimensional microcavity, *Phys. Rev. Lett.* **86** (2001), 3903.

[35] T.H. Stievater, X. Li, D.G. Steel, D. Gammon, D.S. Katzer, D. Park, C. Piermarocchi, and L.J. Sham, Rabi oscillations of excitons in single quantum dots, *Phys. Rev. Lett.* **87** (2001), 133603.

[36] I.N. Stranski and L. Krastanow, Stizungsberichte d. mathem.-naturw. Kl., *Abt. IIb* **146** (1938), 797.

[37] K. Tachibana, T. Someya, S. Ishida, and Y. Arakawa, Selective growth of InGaN quantum dot structures and their microphotoluminescence at room temperature, *Appl. Phys. Lett.* **76** (2000), 3212.

[38] S. Tarucha, D.G. Austing, T. Honda, R.J. van der Hage, and L.P. Houwenhoven, *Phys. Rev. Lett.* **77** (1996), 3613; L.P. Kouwenhoven, T.H. Oosterkamp, M.W.S. Danoesastro, M. Eto, D.G. Austing, T. Honda, and S. Tarucha, *Science* **278** (1997), 1788.

[39] J. Vukovic, Photonic crystal structures for efficient localization of ex-traction of light, Ph.D. Thesis, Cal. Inst. of Technology, Pasadena, California, 2002.

[40] D.O.S. Wei, Quantum computing with quantum dots, UROPS Paper (Module PC3288), Dept. of Physics, National University of Singapore, 2003.

[41] F. Widmann, J. Simon, N.T. Pelekanos, B. Daudin, G. Feuillet, J.L. Rouviere, and G. Fishman, Giant piezoelectric effect in GaN self-assembled quantum dots, *Microelectronics J.* **30** (1999), 353–356.

[42] http://britneyspears.ac/physics/crystals/wcrystals.htm

[43] http://www.qubit.org

[44] http://xxx.lanl.gov/archive/quant.ph

[45] http://pages.ief.u-psud.fr/QDgroup/quantuminformation.html

[46] http://www.nanotech.org/?path=Research/Optical/sQuantum/sDots

[47] http://vortex.tn.tudelft.nl/grkouwen/kouwen.html

[48] http://www.nd.edu/~qcahome

Chapter 8

Linear Optics Computers

- Classical optics and optical circuits

- Quantum optics and qubits

- The method of Knill, Laflamme and Milburn

- Quantum teleportation

The use of light as a physical medium to perform complex mathematical transformations has a long history. The possibility to efficiently implement quantum algorithms using this purely electromagnetic "material", however, has only recently been recognized [39]. Our goal in this chapter is to make the non-specialist familiar both with the basic methods and with the particular language used in this context. Although we have tried to compile a more or less comprehensive and updated bibliographical list, this collection is mainly aimed to help readers rapidly acquire most of the pertinent literature in an efficient way. This chapter is not a review, but rather an introduction to the subject. Instead of presenting the most recent and elaborate ideas, here we will provide a basic introduction for the newcomer. It will contain most of the ingredients necessary for a deeper study of the literature and for the development of new ideas. Simple situations are better suited for the comprehension of fundamental concepts, and it is in this spirit that our choice of illustrative examples has been made. Special attention has also been paid to clarify the oftentimes confusing, heterogeneous use of notational conventions and terminology by different authors.

Linear optics computers come in two flavors. There are those that exclusively exploit the classical properties of light. Like any ordinary, i.e., classical computer, they can in principle also be used to simulate quantum algorithms. This approach will be presented in Section 8.1. Quantum correlations are at the heart of quantum informatics. Classical simulations thereof are inherently inefficient in the computer-scientific sense. Section 8.2 is devoted to the efficient use of quantum properties of light for the purpose of

347

data manipulation. A brief summary and outlook is given at the end of the chapter.

8.1 Classical electrodynamics— Classical computers

In classical—as opposed to quantum—physics, the behavior of electric and magnetic fields is described in a unified way by a set of coupled equations known as the Maxwell equations. Light is an electromagnetic phenomenon and, as such, its position- and time-dependent electric and magnetic field vectors satisfy a wave equation subject to boundary conditions defined by the presence of metallic mirrors, dielectric surfaces, absorbing materials, etc. The most general solution of the wave equation may be expressed as a linear superposition of eigensolutions of the corresponding Helmholtz equation. These particular solutions are called the *optical modes* of the system. Throughout this chapter we are concerned with light whose propagation is concentrated along beams having very well defined directions in space. This property allows directions in space to be used as unambiguous labels to identify individual beams, and in the simplest case it is commonly referred to just as moving "to the left" (L) or "to the right" (R), as we will do later. Unless special experimental care is taken, beams are usually made of a superposition of several optical modes.

It is standard nowadays to transmit light not only through the air or vacuum, but also through *optical fibers.* Since long rolls of such fibers may be placed on top of an optical table, light pulses may easily travel over distances of thousands of kilometers. Optical fibers are used to delay or even temporarily store light pulses until they are needed for a certain task. The geometric properties of free optical modes, as produced for instance by a laser light source, and the bound modes of a (multi-modal) optical fiber are usually not perfectly compatible. An *optical (mode) coupler* is used to link the various subsystems of an optical experiment, but *mode mismatches* are difficult to avoid and represent an important source of optical loss and, in the sense of information theory, computational errors.

As a direct consequence of the electromagnetic wave equation, the electric (as well as the magnetic) field vector is always orthogonal to a light beam's propagation direction. The two dimensions of this orthogonal plane represent a degree of freedom called *polarization.* Since optical tables are usually horizontal, it is customary (though of course not necessary) to represent the electric field's vector as a linear superposition of its *vertical* (V) and *horizontal* (H) polarization components. This property, too, can be used to encode information in a light beam, in addition to the already mentioned path encoding [77]. Other possible ways to encode information in a classical light beam (which have been experimentally demonstrated) will be discussed at the end of this section.

In order to read out information from a light beam, one uses a set of *photo-detectors*. These measure (i.e., produce an electric signal which is more or less linearly related to) the *intensity* of the incident light, which is proportional to the square of its electric field *amplitude*. Path-encoded information can be read-out directly using such devices. Information stored in the polarization degree of freedom, however, is as invisible to such a detector as it is to our eyes. In order to obtain a *polarization-sensitive detector*, we may filter out either the horizontal or the vertical polarization beam component by placing a corresponding *polarizer* (a polarizing filter) in front of the detector. A less invasive technique makes use of *polarizing beam splitters* (PBSs). These optical devices, when correctly placed into a beam path, transmit the horizontal polarization component of the beam and reflect the vertically polarized one. Two detectors, one at each exit of the PBS, are then able to separately determine the intensity of each polarization component.

In addition to amplitude and polarization, a vector wave also has a *phase*. The exploitation of this phase degree of freedom for information encoding is the essence of the whole field of interferometry. The only way to retrieve phase information is by suitably transforming it into amplitude information which can then be read-out using conventional photo-detectors. We remark that throughout our discussion, phase angles will be expressed in radians. Although electric (and magnetic) field amplitudes are real-valued quantities, it is customary and computationally convenient to express phases using the (complex-valued) exponential $e^{i\phi}$. Since the wave equation is linear, we may just define the actual electric field vector to be the real (or the imaginary) part of the quantities expressed through our formulas.

8.1.1 Light beam manipulation with four degrees of freedom

Let us proceed to transform these ideas into a mathematical formalism. To be concrete, we will consider $N = 4$ degrees of freedom. This number is both large enough to demonstrate all relevant ideas and small enough to explicitly write down the involved matrices without losing the overview. From our treatment, it will become clear how situations with $N > 4$ should be handled. Four degrees of optical freedom can be physically obtained by considering two alternative beam propagation directions ("right" and "left"), each with two possible polarization directions ("vertical" and "horizontal"). A general state of such a system can then be expressed as a linear superposition of the following four basis states,

$$V_R \equiv \begin{bmatrix} 1 \\ 0 \\ 0 \\ 0 \end{bmatrix} \quad \begin{pmatrix} \text{Vertically polarized beam} \\ \text{moving to the right} \end{pmatrix}, \tag{8.1}$$

$$H_R \equiv \begin{bmatrix} 0 \\ 1 \\ 0 \\ 0 \end{bmatrix} \quad \begin{pmatrix} \text{Horizontally polarized beam} \\ \text{moving to the right} \end{pmatrix}, \qquad (8.2)$$

$$V_L \equiv \begin{bmatrix} 0 \\ 0 \\ 1 \\ 0 \end{bmatrix} \quad \begin{pmatrix} \text{Vertically polarized beam} \\ \text{moving to the left} \end{pmatrix}, \qquad (8.3)$$

$$H_L \equiv \begin{bmatrix} 0 \\ 0 \\ 0 \\ 1 \end{bmatrix} \quad \begin{pmatrix} \text{Horizontally polarized beam} \\ \text{moving to the left} \end{pmatrix}. \qquad (8.4)$$

A convenient graphical representation, adapted to our choice of basic alternatives, makes use of light beams drawn at angles of $\pm 45°$. Fig. 8.1, e.g., shows the performance (and the standard symbol) of the simplest of all optical components, the *mirror*. Its action on an impinging light beam can be described by the matrix

$$M \equiv \begin{bmatrix} 0 & 0 & 1 & 0 \\ 0 & 0 & 0 & 1 \\ 1 & 0 & 0 & 0 \\ 0 & 1 & 0 & 0 \end{bmatrix}, \qquad (8.5)$$

which indeed correctly transforms R-moving beams into L-moving ones and vice-versa, without affecting the polarization:

$$\left. \begin{matrix} MV_R = V_L \qquad MH_R = H_L \\ MV_L = V_R \qquad MH_L = H_R \end{matrix} \right\} . \qquad (8.6)$$

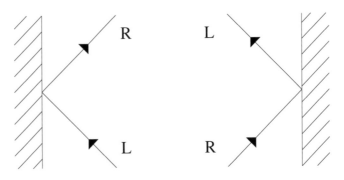

Figure 8.1: Symbolic representations of a mirror acting on left- (L) and right- (R) travelling beams.

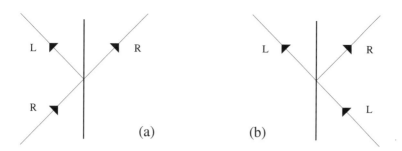

Figure 8.2: Symbolic representations of a beam splitter acting on rays coming from (a) the left, and (b) from the right.

A (non-polarizing, i.e., conventional) *beam splitter* separates an incoming beam into a transmitted component, which continues propagating in the same direction as the one incoming, and a reflected one, which behaves as if it has reflected off a mirror. Evidently, the total amount of transmitted plus reflected light should be equal to the total amount of light reaching the beam splitter at its entrance. A symmetric (or *balanced*) beam splitter divides the incoming beam into a pair of outgoing beams of equal intensity. Fig. 8.2 illustrates its action, which can be conveniently described by the matrix

$$B \equiv \frac{1}{\sqrt{2}} \begin{bmatrix} 1 & 0 & i & 0 \\ 0 & 1 & 0 & i \\ i & 0 & 1 & 0 \\ 0 & i & 0 & 1 \end{bmatrix}. \tag{8.7}$$

Phase conventions have several variances, depending on the specific application and individual preferences. In our own phase convention, a V_R beam that passes a symmetric beam splitter will be represented by the beam vector

$$BV_R = (V_R + iV_L)/\sqrt{2}, \tag{8.8}$$

which describes a pair of still vertically polarized beams moving partly to the right and partly to the left. The most general representation of a beam vector in terms of the four basis states V_R, H_R, V_L and H_L may be written as

$$a_{VR}V_R + a_{HR}H_R + a_{VL}V_L + a_{HL}H_L, \tag{8.9}$$

where the expansion factors a_{VR}, a_{HR}, a_{VL} and a_{HL} are proportional to the respective electric field amplitudes. We assume them to be normalized in such a way that $|a_{VR}|^2 + |a_{HR}|^2 + |a_{VL}|^2 + |a_{HL}|^2 = 1$. Because of this normalization, the squared norm of each component represents its relative intensity. Since we are studying only idealized situations and are not concerned with optical component imperfections, we consider only intensity-preserving interactions, which translate mathematically to norm-preserving (i.e., unitary) matrices, such as M and B.

We have already mentioned the action of a polarizing beam splitter, depicted in Fig. 8.3 and described by the matrix

$$P \equiv \begin{bmatrix} 0 & 0 & i & 0 \\ 0 & 1 & 0 & 0 \\ i & 0 & 0 & 0 \\ 0 & 0 & 0 & 1 \end{bmatrix}. \tag{8.10}$$

One possible way of using this transformation is by interpreting it as a logic gate, where the polarization state controls that of the motional. For whenever the incoming beam is H polarized, the direction of motion (be it L or R) remains unchanged, while a V polarization at the input causes R and L to interchange at the output. This gate is reversible and is our *controlled-NOT gate* ("CNOT"). Its usual matrix representation is (2.44) but now with respect to the ordered basis $\{V_R, H_R, V_L, H_L\}$ is

$$\begin{bmatrix} 0 & 0 & 1 & 0 \\ 0 & 1 & 0 & 0 \\ 1 & 0 & 0 & 0 \\ 0 & 0 & 0 & 1 \end{bmatrix}, \tag{8.11}$$

which can be obtained from (8.10) by correcting the imaginary entries using the propagation (8.13) and polarization (8.15) phase shifters to be introduced below. Polarizing beam splitters are ubiquitous in the final detection stage of any circuit using polarization-encoding. Incidentally, a V-polarizing filter placed in the right- or in the left-travelling beam is represented by the non-unitary matrices

$$\begin{bmatrix} 1 & 0 & 0 & 0 \\ 0 & 0 & 0 & 0 \\ 0 & 0 & 0 & 0 \\ 0 & 0 & 0 & 0 \end{bmatrix} \quad \text{and} \quad \begin{bmatrix} 0 & 0 & 0 & 0 \\ 0 & 0 & 0 & 0 \\ 0 & 0 & 1 & 0 \\ 0 & 0 & 0 & 0 \end{bmatrix}, \tag{8.12}$$

respectively, with H-polarizing ones being obtainable similarly. Such polarizers would be used in the detection stage, if only the information from one of the channels is actually required.

Phase shifts between two optical paths can be simply introduced by changing the optical path length of one of them. The matrix describing an arbitrary phase shift of φ_R (φ_L) in the right (left) beam is given by

$$F(\varphi_R, \varphi_L) \equiv \begin{bmatrix} \exp(i\varphi_R) & 0 & 0 & 0 \\ 0 & \exp(i\varphi_R) & 0 & 0 \\ 0 & 0 & \exp(-i\varphi_L) & 0 \\ 0 & 0 & 0 & \exp(-i\varphi_L) \end{bmatrix}. \tag{8.13}$$

Our graphical representation of such a (propagation) *phase shifter* is shown in Fig. 8.4(a). The reader shall be warned, though, that optical component symbology is notoriously heterogeneous in the literature.

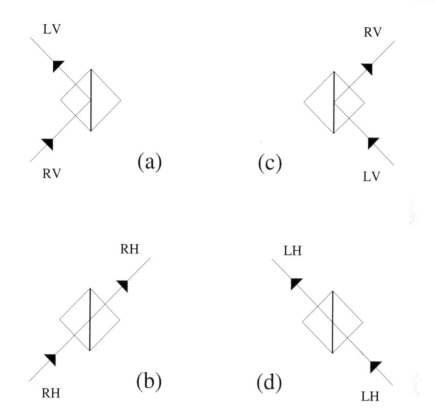

Figure 8.3: The usual symbols for a polarizing beam splitter (PBS) in four different situations, e.g., acting on an incoming
(a) vertically polarized R-beam,
(b) horizontally polarized R-beam,
(c) vertically polarized L-beam and
(d) horizontally polarized L-beam.

Some crystals are *birefringent*, i.e., they transmit H- and V-polarized light with different velocities. The crystal direction along which the velocity is larger is called the *fast axis*. A *wave plate* (or *retardation plate*) is a thin slice of such a material. If a beam of arbitrary polarization traverses a wave plate of thickness d, a phase difference ϕ between the H and the V components will accumulate, and this phase difference is proportional to d. Wave plates are omnipresent in optics, and they are made in two sizes, viz., *half-wave plates* with $\phi = \pi/2$ (Fig. 8.4(b)) and *quarter-wave plates* with $\phi = \pi/4$ (Fig. 8.4(c)). By rotating the axis of such a crystal slice by an angle $\vartheta/2$, an extremely versatile polarization-handling device is obtained. Its matrix representation

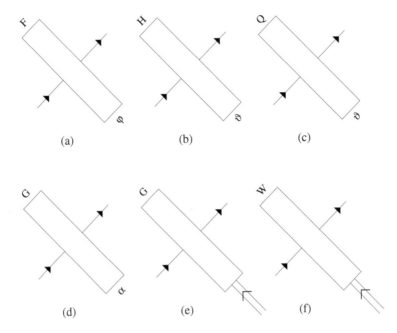

Figure 8.4: Symbols chosen in this chapter for the representations of
(a) a phase shifter,
(b) a half-wave plate,
(c) a quarter-wave plate,
(d) a polarization rotator,
(e) a Faraday rotator, and
(f) a Pockels cell.

$$W_{2\phi_R,2\phi_L}(\vartheta_R/2, \vartheta_L/2) \equiv$$

$$\begin{bmatrix} \cos\phi_R - i\sin\phi_R\cos\vartheta_R & -i\sin\phi_R\sin\vartheta_R & 0 & 0 \\ -i\sin\phi_R\sin\vartheta_R & \cos\phi_R + i\sin\phi_R\cos\vartheta_R & 0 & 0 \\ 0 & 0 & \cos\phi_L - i\sin\phi_L\cos\vartheta_L & -i\sin\phi_L\sin\vartheta_L \\ 0 & 0 & -i\sin\eta_L\sin\vartheta_L & \cos\phi_L + i\sin\phi_L\cos\vartheta_L \end{bmatrix}$$

$$(8.14)$$

contains the *polarization phase shifter*

$$W_{2\phi_R,2\phi_L}(0,0) = \begin{bmatrix} \exp(-i\phi_R) & 0 & 0 & 0 \\ 0 & \exp(i\phi_R) & 0 & 0 \\ 0 & 0 & \exp(-i\phi_L) & 0 \\ 0 & 0 & 0 & \exp(i\phi_L) \end{bmatrix} \quad (8.15)$$

as a special case. Up to an overall phase, a half-wave plate rotated by an angle $\vartheta_{R(L)}/2 = 45°$,

$$W_{\pi,0}(\pi/4, 0) = \begin{bmatrix} 0 & -i & 0 & 0 \\ -i & 0 & 0 & 0 \\ 0 & 0 & 1 & 0 \\ 0 & 0 & 0 & 1 \end{bmatrix} \tag{8.16}$$

or

$$W_{0,\pi}(0, \pi/4) = \begin{bmatrix} 1 & 0 & 0 & 0 \\ 0 & 1 & 0 & 0 \\ 0 & 0 & 0 & -i \\ 0 & 0 & -i & 0 \end{bmatrix}, \tag{8.17}$$

respectively, performs the interchange $V_{R(L)} \leftrightarrow H_{R(L)}$, and is the polarization-analogue of a mirror. Unlike a spatial mirror, which acts on the V and H polarizations simultaneously, however, we can choose to place such a $\pi/4$-rotated half-wave plate either into the R or into the L beam (or also into both simultaneously, if desired, thus becoming the exact polarization counterpart of a conventional mirror). It becomes possible, therefore, to create a CNOT gate in which the motional degree of freedom R/L controls the polarizational one, simply by introducing either $W_{\pi,0}(\pi/4, 0)$ into an R beam or, respectively, $W_{0,\pi}(0, \pi/4)$ into an L beam [75]. Half-wave plates rotated by 45° thus behave essentially as polarization-analogues of a polarizing beam splitter. The reader may wish to create a custom-made optical device by designing a system that realizes the *swap* (or, *swapping*) *gate* (cf. (2.45))

$$\begin{bmatrix} 1 & 0 & 0 & 0 \\ 0 & 0 & 1 & 0 \\ 0 & 1 & 0 & 0 \\ 0 & 0 & 0 & 1 \end{bmatrix}. \tag{8.18}$$

A device which rotates the polarization plane by an arbitrary angle α is called a *polarization rotator* (cf. Fig. 8.4(d)). The matrix representation $G(\alpha_R, \alpha_L)$ of a pair of polarization rotators in the R- and the L-beams reads

$$G(\alpha_R, \alpha_L) \equiv \begin{bmatrix} \cos\alpha_R & -\sin\alpha_R & 0 & 0 \\ \sin\alpha_R & \cos\alpha_R & 0 & 0 \\ 0 & 0 & \cos\alpha_L & -\sin\alpha_L \\ 0 & 0 & \sin\alpha_L & \cos\alpha_L \end{bmatrix}. \tag{8.19}$$

Since optical fibers usually preserve the polarization state of the beam that they are transmitting, a controlled twist of such a fiber permits an elementary realization of a polarization rotator. If the rotation angle has to be controlled in real-time, it can be achieved by using *Faraday-rotators* which rely on a magneto-optic effect that allows to control the rotation angle α with an external magnetic field (cf. Fig. 8.4(e)). By placing a polarization rotator between two quarter-wave plates tilted by 45° in opposite directions, we obtain a practical implementation of the polarization phase shifter in (8.15):

$$W_{\frac{\pi}{2},\frac{\pi}{2}}\left(\frac{\pi}{4},\frac{\pi}{4}\right)G(\alpha_R,\alpha_L)W_{\frac{\pi}{2},\frac{\pi}{2}}\left(-\frac{\pi}{4},-\frac{\pi}{4}\right)=W_{2\alpha_R,2\alpha_L}(0,0), \qquad (8.20)$$

without requiring any specially fabricated wave-plate of appropriate thickness. A wave-plate of effectively controllable thickness is in fact available and is called a *Pockels cell* (cf. Fig. 8.4(f)). It allows to control ϕ via an external electric field and can be used as an extremely fast actuator for real-time applications.

It has long been known that any unitary 2×2 matrix (and therefore, any possible manipulation of the polarization degree of freedom) can be physically realized by surrounding a half-wave plate with two appropriately tilted quarter-wave plates (see, e.g., [77]). During the 1990s it was realized that in fact any N dimensional unitary matrix can be represented in terms of the optical elements that we have introduced so far [69]. The recursive proof of this assertion is constructive, so that it provides an explicit recipe for creating an optical circuit that realizes any given unitary transformation. For a detailed treatment of our present $N = 4$ case, with several concrete examples of possible physical realizations, see [14].

In order to present an example of an interesting optical circuit, let us introduce two more building blocks that are frequently encountered in the literature. On the one hand, there is the polarization analogue of a conventional beam splitter, which can be simply obtained by tilting a half-wave plate by an angle of $22.5°$. If placed into an R-beam, it will produce

$$W_{\pi,0}(\pi/8,0)=\begin{bmatrix} \frac{-i}{\sqrt{2}} & \frac{-i}{\sqrt{2}} & 0 & 0 \\ \frac{-i}{\sqrt{2}} & \frac{i}{\sqrt{2}} & 0 & 0 \\ 0 & 0 & 1 & 0 \\ 0 & 0 & 0 & 1 \end{bmatrix}. \qquad (8.21)$$

The matrix $W_{0,\pi}(0,\pi/8)$ correspondingly describes a polarization-state splitter in the L-beam.

On the other hand, we have the so called *Mach–Zehnder interferometer* (see Fig. 8.5). This fundamental device consists of a beam splitter followed by two mirrors placed in such a way that the reflected beams merge again at a location where a second beam splitter is carefully positioned. Since this configuration is extremely sensitive to relative phase shifts between the two *interferometer arms* (i.e., internal trajectories) of the circuit, we include a phase shifter (8.13) between the beam splitters. It may phenomenologically describe an unavoidable path-length mismatch arising during the build-up of the apparatus, or an adjustable control-knob for its operation. The corresponding matrix representation is given by

$$Z(2\varphi) \equiv BMF(\varphi,\varphi)B = i\begin{bmatrix} \cos\varphi & 0 & -\sin\varphi & 0 \\ 0 & \cos\varphi & 0 & -\sin\varphi \\ \sin\varphi & 0 & \cos\varphi & 0 \\ 0 & \sin\varphi & 0 & \cos\varphi \end{bmatrix}. \qquad (8.22)$$

Due to its wide applicability, and in spite of being actually a composed system, it is often treated as a basic building unit. A Mach–Zehnder interferometer that has been adjusted in such a way that an incoming R (or L) beam exits the device again in its R (or L) state (possibly modulo a global phase constant) is said to be *balanced*, a condition which is equivalent to $Z(0) = i$.

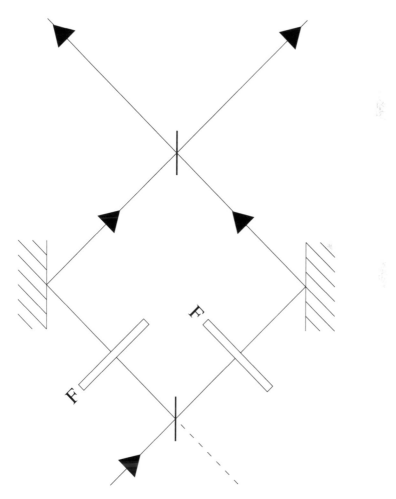

Figure 8.5: Ray trajectories in a Mach-Zehnder interferometer.

8.1.2 Optical circuits and examples

The most common starting point of a typical light circuit may, e.g., be a V_R beam produced by a laser source and an appropriate polarizing filter. As a second step, one usually requires all possible alternatives (four of them in our present context) to stand on equal footing. An unbiased superposition may be the "average" state $(V_R + H_R + V_L + H_L)/2$. It can be produced, quite intuitively, by placing two splitters in series. First, one which splits the V_R beam with respect to its inner (polarization) degree of freedom $[W_{\pi,0}(\pi/8,0)]$ and second, one which (irrespective of the beam's internal degree of freedom) acts as a splitter in space (B). Using two (not really that relevant) phase shifters $F(\pi/2, \pi/2)$, we may correct the phase factors, so that (cf. Fig. 8.6(a))

$$F(\pi/2, \pi/2)BF(\pi/2, \pi/2)W_{\pi,0}(\pi/8,0) \begin{bmatrix} 1 \\ 0 \\ 0 \\ 0 \end{bmatrix} = \frac{1}{2} \begin{bmatrix} 1 \\ 1 \\ 1 \\ 1 \end{bmatrix}. \qquad (8.23)$$

Quite generally, a device which transforms any state of the standard basis (8.1)–(8.4) into equal-weight superpositions thereof is called a *Walsh–Hadamard gate*.

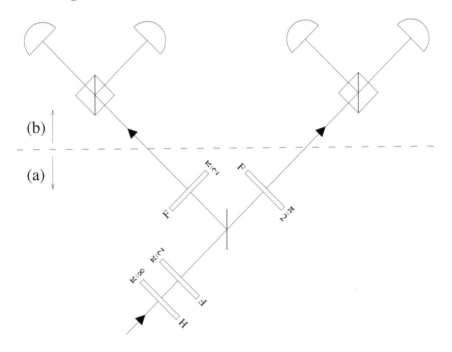

Figure 8.6: (a) Preparation of an "unbiased" optical state, and
(b) A typical, polarization-discerning detection scheme.

A pair of polarizing beam splitters, one in the R and one in the L beam, and four detectors behind their exit ports, as shown in Fig. 8.6(b), represent a typical detection scheme. It is able to determine the squared norm of all four expansion coefficients (8.9). The individual phases of a_{VR}, a_{HR}, a_{VL} and a_{HL}, however, cannot be obtained this way. It would certainly be handy to have an optical system capable of distinguishing the four different states

$$\frac{1}{2}\begin{bmatrix} -1 \\ 1 \\ 1 \\ 1 \end{bmatrix}, \quad \frac{1}{2}\begin{bmatrix} 1 \\ -1 \\ 1 \\ 1 \end{bmatrix}, \quad \frac{1}{2}\begin{bmatrix} 1 \\ 1 \\ -1 \\ 1 \end{bmatrix}, \quad \frac{1}{2}\begin{bmatrix} 1 \\ 1 \\ 1 \\ -1 \end{bmatrix}, \qquad (8.24)$$

which for the described scheme all look alike, since they only differ by a π shift in one of their components. The goal is met simply by introducing a Mach–Zehnder interferometer $Z(\pi/2)$ and a pair of phase shifters $W_{\pi,\pi}(-\pi/8, \pi/8)$ before the detection stage. A pair of mirrors is required to redirect the beams to the Mach–Zehnder inputs. By applying this scheme to the four possible states (8.24),

$$Z(\pi/2)\, M\, W_{\pi,\pi}(-\pi/8, \pi/8)\, \frac{1}{2}\begin{bmatrix} -1 & 1 & 1 & 1 \\ 1 & -1 & 1 & 1 \\ 1 & 1 & -1 & 1 \\ 1 & 1 & 1 & -1 \end{bmatrix} = \begin{bmatrix} 0 & 0 & -1 & 0 \\ 0 & -1 & 0 & 0 \\ 0 & 0 & 0 & 1 \\ -1 & 0 & 0 & 0 \end{bmatrix}, \qquad (8.25)$$

we conclude that we have achieved what we intended: to each phase-marked vector in (8.24) there will be one and only one corresponding detector receiving light. Phase information has been bijectively mapped onto intensity information. In fact, this is even more than what is needed in order for such a mapping to be useful. In principle, it would be enough if to every phase-shifted input line there corresponds one and only one detector which receives a signal that is different from the other $N-1$ signals, but without these signals necessarily vanishing. The algorithm developed by Grover [24], cf. Box 2.5, allows to achieve this precisely. In the $N = 4$ case, the *Grover algorithm* essentially corresponds to the process decribed by (8.25). For $N > 4$, the algorithm iterates the basic sequence in (8.25) a number of times whose complexity can be shown to grow as $O(N^{1/2})$ in the large-N limit [24]; see Subsection 2.7.4 of Chap. 2.

If one considers the expansion coefficients of either one of the vectors (8.24) as the elements of a list of length N, of which the one element with a different quality—a phase factor in this case—is to be found by an appropriate search algorithm, then Grover's algorithm seems to perform well: instead of taking roughly $O(N)$ steps to identify the marked element, it does so in only $O(\sqrt{N})$ steps. Yet, what it actually does so efficiently is to transform one type of marking (one which is not visible to a common detector) into another one (which is easily detectable). The output list again consists of N items, one and only one of which has a different quality—an intensity factor in this case—which has to be identified. And this still takes the usual number of steps, which is invariably proportional to N. We note that an experiment that realizes the scheme described by (8.25) has actually been performed by Kwiat *et al.* [42].

A more recent experiment [31] demonstrates how the four states $\Psi^{(\pm)} \equiv (H_R \pm V_L)/\sqrt{2}$ and $\Phi^{(\pm)} \equiv (V_R \pm H_L)/\sqrt{2}$ can be produced in a simple way using a half-wave plate, a polarizing beam splitter and a series of controllable phase shifters. These four *Bell states* form an orthonormal basis and play an important role in many algorithms. The controlled creation and detection of Bell states is essential, e.g., to implement the $N = 4$ cryptography scheme proposed in [3] and further discussed in [31]. After first preparing a horizontally polarized R beam, that experiment can be realized by the circuit shown in Fig. 8.7,

$$
W_{2\phi_R, 2\phi_L}\left(\frac{\pi}{4}, \frac{\pi}{4}\right) F(\varphi_R, \varphi_L)\, P\, W_{\pi, 0}\left(\frac{\pi}{8}, 0\right)
\begin{bmatrix} 0 \\ 1 \\ 0 \\ 0 \end{bmatrix}
= \frac{1}{\sqrt{2}}
\begin{bmatrix} \sin\phi_R\, e^{i\varphi_R} \\ i\cos\phi_R\, e^{i\varphi_R} \\ \cos\phi_L\, e^{-i\varphi_L} \\ -i\sin\phi_L\, e^{-i\varphi_L} \end{bmatrix}.
\tag{8.26}
$$

A possible set of parameters can be chosen as

$$
\begin{cases}
\phi_R &= \phi_L = 0 \\
\varphi_R &= -\pi/2 \\
\varphi_L &= 0
\end{cases}
$$

for $\Psi^{(+)}$,

$$
\begin{cases}
\phi_R &= \phi_L = 0 \\
\varphi_R &= -\pi/2 \\
\varphi_L &= \pi
\end{cases}
$$

for $\Psi^{(-)}$,

$$
\begin{cases}
\phi_R &= \phi_L = \pi/2 \\
\varphi_R &= 0 \\
\varphi_L &= -\pi/2
\end{cases}
$$

for $\Phi^{(+)}$ and

$$
\begin{cases}
\phi_R &= \phi_L = \pi/2 \\
\varphi_R &= 0 \\
\varphi_L &= \pi/2
\end{cases}
$$

for $\Phi^{(-)}$. A polarizing beam splitter oriented at $45°$ can be used to conveniently detect the four Bell states. In [31] an equivalent combination of a half-wave plate and a polarizing beam splitter is used instead.

8.1.3 Complexity issues of LOCC and alternatives

Our choice so far to consider only a set of two binary alternatives was very convenient for elucidation purposes. The relative simplicity of this case also explains why early experiments were restricted to $N = 4$ linearly independent states. The observation that $4 = 2^2 = 2 \times 2 = 2 + 2$ may, however, misleadingly suggest that the addition of one more binary alternative just

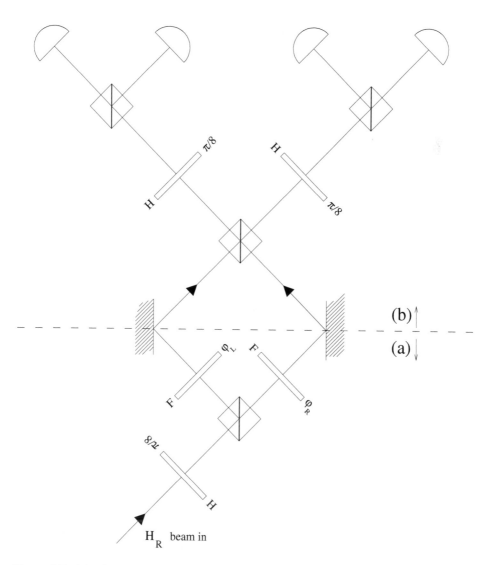

Figure 8.7: (a) Optical Bell-state preparation starting with a horizontally-polarized R-beam, and
(b) its measurement.

corresponds to adding two more beams to the device. This, of course, is wrong. If, following a convention proposed in [75] and [76], we call the vector space spanned by the basic set of binary alternatives

$$\begin{bmatrix} 1 \\ 0 \end{bmatrix} \quad \text{and} \quad \begin{bmatrix} 0 \\ 1 \end{bmatrix} \tag{8.27}$$

a *cebit* (or *c-bit*, where "c" stands for "classical") space, then the classic-optical representation of n cebits requires a vector space correspondingly spanned by $N = 2^n$ linearly independent basis states. The preparation of an average state (8.23) then requires the use of 2^n beam splitters. Similarly, 2^n detectors would be needed to individually measure (i.e., discern) all the possible alternatives. Such an exponential increase in resources is called *inefficient*. The unavoidably inefficient use of classical optics for the implementation of *logic circuits* (i.e., networks of reversible logic gates like the ones we have introduced so far, e.g., CNOT gates, Walsh–Hadamard gates, swap gates, etc.) will be called by us Linear Optics Classical Computing (LOCC). The current trend in LOCC research is to push the intrinsic inefficiency into a direction which makes it technically less cumbersome. Instead of coding cebits using R/L-like alternatives, a pair of parallely displaced light beams (e.g., an "upper" and a "lower" one) could be used as well. Ref. [76] describes several ways to build fundamental logic gates with such a coding. Making use of *Dove prisms* (optical devices which essentially rotate an incoming beam profile by $90°$ at their exit, thus transforming a pair of "upper and lower" beams into a pair of "left-handed and right-handed" ones), the authors of [16] have experimentally realized a CNOT gate. In [78], an $N = 4$ path and polarization encoding scheme is used to demonstrate the essence of the *Deutsch–Jozsa algorithm* [11]; cf. Box 2.1. Path number encoding lends itself to integrated optics approaches. Reference [26] shows instructively how a single (commercially available) "symmetric $N \times N$ fiber coupler" could be used to substitute N conventional beam splitters. Hence, the problem is changed from having an exponential increase in the number of optical elements to having multiport devices with an exponential number of ports [27]. The number of detectors can also be reduced by playing LOCC in the time-domain, i.e., by appropriately encoding cebits into time-delays ("time bins"). A single detector and a single optical fiber were used in [7] to implement both the Deutsch–Jozsa and the *Bernstein–Vazirani* [5] algorithms. In this case, it is the number of optical pulses (and therefore, the computational time) that grows exponentially like 2^n. Most recently, frequency-encoding was employed to demonstrate an alternative LOCC implementation of the Bernstein–Vazirani algorithm [44]; cf. Box 2.3 in Chap. 2.

The exponential growth of required resources as the number n of cebits increases is unavoidable both at the state preparation front and at the state measurement end of any LOCC circuit. Although it has been shown in general [69] that a universal n-cebit computation requires an exponentially large number of 2-cebit unitary transformations, the meaning of this (as well as any other similar "no-go"-type theorems) has to be understood and interpreted with caution. For instance, one may be led to think that one individual 2-cebit transformation corresponds to one optical component. But since (ordinarily) light does not interact with light so easily, one and the same optical component can actually be used to manipulate several light beams *simultaneously*. By carefully designing the circuit geometry, it becomes possible not only to "recycle" optical elements, thus saving components and simplifying their allignment, but to make the system inherently more stable against undesirable mechanical vibrations and other imperfections. How much can actually be saved this way is a mathematical open question. For example, extensive use of such *compilation* techniques essentially simplifies the optical circuitry necessary to realize Grover's algorithm with 2 and 3 cebits [42]. The possible argument that such a purpose-made device is not universal, but rather a "hard-wired" computer, is not relevant. First, because in the long term it may turn out that we are actually more interested in hard-wired LOCC than in programmable all-purpose devices. Second, because also programmable all-purpose LOCC devices can be optimally compiled. It would be not only theoretically interesting but also technically important to know how fast the minimum (i.e., optimized) number of required optical components grows as n becomes larger. It seems unlikely (although not totally clear) that this growth is sub-exponential. Whatever the result, we can expect that the relative simplicity of LOCC would make it the method of choice for the implementation of low-n light-computers, the question being only how low this n may be, before LOCC is effectively outperformed by its more subtle, quantum-optical rival (to be presented in the next section).

Concrete examples of $n = 3$ LOCC circuits have been theoretically studied in [9] ("teleportation" algorithm), [42] (Grover's algorithm), [26] ("classically chaotic baker's map") and [76] ("Greenberger-Horne-Zeilinger state" production and characterization, an alternative teleportation algorithm, CNOT and "Toffoli" gates, cebit-flip "error correction network"). On the experimental front, we are witnessing the first realization of such $n = 3$ LOCC circuits, including a block-coding scheme to demonstrate Schumacher's "noiseless coding theorem" [48] and the implementation of Deutsch–Jozsa's and Bernstein–Vazirani's algorithms using the time-bins method [7]. The Bernstein–Vazirani algorithm was also demonstrated in [44]. The "spectral encoding" used in this case seems to be particularly promising in increasing the number of cebits that can be manipulated.

So far we have been very careful in our formulation not to use any expressions ("photons", "qubits", etc.) or notational conventions ("Dirac's bras and kets", "Pauli operators", etc.) that may suggest anything quantum-mechanical. The physics involved in this kind of classical implementations of "quantum" algorithms would have been perfectly intelligible to any good physicist at the end of the nineteenth century. In fact, disregarding the enormous technical problems this would involve, even sound- or water-waves could be used for such simulations (necessarily, in these cases the missing polarization degree of freedom would have to be substituted by another directional one). Light computers of the kind heretofore discussed are but physical computers that make clever use of classical wave behavior. Mathematical features characteristic of linear wave propagation have been used for computational purposes for a long time, the most prominent example being perhaps the well established field of Fourier-optics. Lately it has been suggested [50] that the know-how developed in this field may be used to implement "quantum" algorithms as well. In [6] a (non-universal) Grover-search Fourier-optics light computer was demonstrated. A method to create programmable (universal) Fourier-optics computers is proposed in [62]. Although fascinating in their own right, we skip the presentation of those setups (as well as a long list of alternative ideas involving near- and far-field grating diffraction, interferometry and many other wave-manipulating tools), since they do not seem to be compatible with the production of information-carrying light beams that could be sent over long distances for the purpose of secure, i.e., eavesdropping-proof communication. The possibility to create correlations over long distances is a distinctive feature of light and one of the main motivations for the intense reseach in this direction. We close this section by remarking that, instead of using "good old Fourier optics" to implement cebit logic gates, it may prove fruitful to proceed the other way around, by designing practical optical devices using insights from an algorithmic point of view. For what is called a "quantum" algorithm in one context may well be a useful tool in some other classical optics applications. The Grover device described by (8.25), for instance, may be viewed as a phase-controlled optical switch that could have its own merits.

8.2 Quantum electrodynamics— Quantum computers

As light fields turn weaker and weaker, one would "classically" expect light detector signals to become correspondingly smaller. Experiment contradicts this prediction. Instead, one observes that a sensitive detector with a high enough temporal resolution starts to "click" intermitently when the light intensity is sufficiently reduced. Only the time-averaged number of these clicks does behave as expected from classical considerations. By analyzing higher-order statistical properties of the number of clickings, one concludes that different light sources may indeed behave quite differently. This clearly indicates that there is a whole new degree of freedom involved, a freedom not having been modeled by the classical theory of electromagnetism. The branch of contemporary physics dealing with this "quantum" electromagnetic degree of freedom is called *quantum optics*. Viewed in retrospective, it is impressive to notice for how long a time an entire degree of freedom can remain unnoticed by not possessing appropriate instrumentation or—perhaps—by not posing the appropriate set of questions. Extrapolating, we may wonder how many degrees of freedom still remain hidden for posterior discovery. By now it has become clear, however, that to any classical theory there seems to exist an underlying quantum theory, of which it somehow represents the "classical limit". Although it might be tempting to call a classical computer operating in the "quantum limit" a *quantum computer*, this concept is usually reserved for devices which explicitly make use of a quantum degree of freedom.

Going over the literature quoted in Section 8.1, the reader will notice that many experimental realizations of LOCC gates and circuits are actually made in the quantum limit. From a conceptual point of view, it is indeed quite interesting to observe that in some special cases (the simplest implementation of Grover's search algorithm described by (8.25) being an example), it becomes possible to make the computer work with a minimal amount of light. By this we mean that only so much light is used that one and only one of the involved detectors will click. Although we would like to say that one and only one of the involved detectors will click *with certainty*, no actual experiment is really perfect: light can be absorbed or the detector may fail to register it. In the classical limit, one click less does not spoil a computation. But once a single click contains all of the relevant information, its failed detection represents a failed computation, and the whole operation will have to be repeated. However, since most algorithms (including Grover's search algorithm as soon as more than 2 cebits are involved) do not produce in general as clear-cut "single-click" results as (8.25), they will anyhow need to be run several times if operated in the single-click regime, even under ideally lossless experimental conditions. Inasmuch as the use of classical light essentially corresponds to performing a huge number of independent single-click computer runs (i.e., individual *quantum measurements*) in one stroke, from a practical point of view there would be little sense in first reducing light output to a minimum in order to have then to repeat a measurement a large number of times: a classical light field does the same for free, since it gives rise to multiple measurements without further ado.

The actual reason why LOCC at the quantum-limit is being pursued goes beyond the scope of this chapter and is related to its potential use in secure communication between two distant partners. The underlying idea may be roughly described as follows. As long as two partners communicate via a classical light channel, a potential eavesdropper may choose to intercept a fraction of (information carrying) light as small as required in order not to be discovered. This is possible because—classically—there is no lower limit on light intensity in the above sense. This immediately suggests that if the light intensity used by the two communicating partners is chosen so small that its true quantum nature becomes important, it might be possible to conceive schemes in which the eavesdropper cannot *even in principle* reduce the intercepted fraction of light required to decipher a message to arbitrarily small values. Possible scenarios are actually much more subtle than this, inasmuch as the information content of the message may also and much more conveniently be stored in the quantum degrees of freedom and the eavesdropper may opt to resend so much light that on the average the communicating partners would not be able to detect any loss in the total *amount* of light transmitted. The eavesdropper's presence will in this case be still detectable due to an unavoidable change in *information content* caused by its undue interference. We encourage the mathematically interested reader to take a closer look at this fascinating field of research. A useful introduction can be found in [83].

Quantum mechanics associates a Hilbert space $\mathcal{H}$ with every degree of freedom available to a system. Any vector in such a complex vector space is identified with one possible *quantum state* of the physical system. The vector symbol $|\psi\rangle$ ($\in \mathcal{H}$) is most commonly used in physics literature. Vectors in the corresponding dual space are denoted by $\langle\psi|$. The scalar product $\langle\psi||\psi\rangle \equiv \langle\psi|\psi\rangle \in \mathbf{C}$ or "bra(c)ket" motivated Paul Dirac to coin the expressions *bra* (for left-handed vectors, $\langle\psi|$) and *ket* (for right-handed vectors, $|\psi\rangle$). Vectors representing physical states should be normalized in such a way that $|\langle\psi|\psi\rangle|^2 = 1$. Refer to Section 2.2 of Chap. 2 for a more careful discussion of the relevant mathematical properties.

8.2.1 Quantum optical states

As already discussed with all pertinent details in Remark 3.2.1, in the case of electromagnetic fields, every individual optical mode j is quantum-mechanically described not simply by a corresponding amplitude and phase, but by a whole vector space spanned by a set of countably infinite *number states*. These basis vectors, also known as *Fock states*, are conventionally denoted by $|n\rangle_j$, where $n = 0, 1, 2, 3, \ldots$ and satisfy the orthonormalization condition ${}_j\langle m|n\rangle_j = \delta_{mn}$. Number states are eigenstates of the *number operator* N_j of mode j, with eigenvalues defined by the identity $N_j|n\rangle_j = |n\rangle_j n$. The energy content of an optical mode j in the quantum state $|n\rangle_j$ is proportional to n. The *difference* in electromagnetic energy content of mode j between two consecutive states $|n + 1\rangle_j$ and $|n\rangle_j$ equals $h\nu_j$, where $h = 6.626 \times 10^{-34}$ Js is Planck's constant and ν_j is the frequency of light in mode j. Typical frequencies encountered in optical experiments are in the order of $\sim 10^{15}$ s^{-1}, so that a detector which is able to discern states that differ by only this minimum quantum of energy must have a sensitivity at the 10^{-19} J level. For the practical implementation of the ideas to be described later in this section, being able to clearly distinguish, for instance, state $|1\rangle_j$ from $|2\rangle_j$ is of crucial importance.

The linear dependence on n of the energy content of an optical mode is characteristic of the excitation spectrum of a simple (quantum-mechanical) harmonic oscillator. Instead of referring to the nth excitation of such an oscillator, it is commonplace to say that there are n *photons* in the mode. Using this terminology, the aforementioned detectors are also called photon-number *resolving* (or *discriminating*). An ideal (photon-number resolving) detector should give the correct photon number in an optical mode every time it is used. In particular, if performed immediately after a measurement, a second identical measurement should again give the same result. Such detectors do not yet exist, but approximations thereof are currently being investigated [41, 61]. Standard photodetectors, on the other hand, are highly destructive devices, since they determine the quantity of light impinging on them simply by absorbing it. In the most typical case, at the p-n junction of a semiconducting material, cf. Fig. 8.8(a), electron-hole pairs are generated via photon absorption. The ratio of the number of such "photocarriers" generated to the number of photons absorbed is called the *quantum efficiency* of the device. By introducing an intrinsic layer of material at the junction, the quantum efficiency can be tailored. Fig. 8.8(b) shows the basic scheme of such a "p-i-n (photo-)diode". Quite generally, thicker layers enhance the quantum efficiency but reduce the speed of response of the device. The p-i-n diode, which is usually operated under reverse bias, has no built-in amplification, i.e., it has unity "gain". The produced signal must therefore be amplified externally. In very low light level applications, however, since a large value feedback resistor is needed, there is a substantial Johnson current noise associated with it, which reduces the signal-to-noise ratio down to unacceptable levels. So called *avalanche photodiodes* make use of internal photocarrier multiplication instead to achieve intrinsic gains as large as 10000. The gain in this case is achieved by applying a high reverse bias of several hundred Volts. The resulting strong electric field accelerates the electrons so much that they can generate secondary electrons by collision ionization, thus starting a chain reaction or "avalanche" process. In addition to their high sensitivity, avalanche photo-detectors typically have very short response times. Their main drawback is that the stochastic nature of the avalanche breakdown leads to a much higher intrinsic noise than in p-i-n photodiodes. Conventional avalanche photodiodes cannot distinguish 2-photon events from 1-photon events [40]. Use of N such detectors after beam splitting and, therefore, evenly distributing the input to N branches would effectively lead to photon-number resolving detection, though [40] (see also [39] and the critical remarks in [21, p. 4]). A time-multiplexing analogue of this idea is discussed in [17]. Another solid-state device that can be used is the "visible light photon counter". It uses the avalanche multiplication effect of electrons in an impurity band in silicon and is essentially noise-free [33]. High-quantum-efficiency single-photon counting systems based on this principle have already been demonstrated [79].

Very difficult to build are also light sources which deliver a desired quantum state of mode j whenever required. In particular, a device which can be used to set the state of mode j to its corresponding 1-photon state $|1\rangle_j$ is called a *single-photon source*. Several schemes for such *on-demand* sources are being investigated, various QD approaches of this sort have been discussed in Subsection 7.1.2, and a recent review can be found in [73]. At this stage, however, most experiments relevant to this chapter are performed using either weakened *coherent states* or *heralded photon sources*. A coherent state is usually the most adequate quantum description of the output produced by a laser source in this kind of experiments. It is described by a complex number α, the modulus (argument) of which is proportional to the average amplitude (phase) of the emitted light. Its representation in terms of the number state eigenbasis is

$$|\alpha\rangle_j \equiv e^{-|\alpha|^2/2} \sum_{n=0}^{\infty} \frac{\alpha^n}{\sqrt{n!}} |n\rangle_j. \qquad (8.28)$$

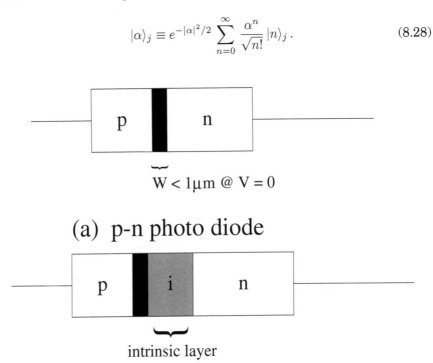

$W < 1\mu m \ @ \ V = 0$

(a) p-n photo diode

intrinsic layer

(b) p-i-n photo diode

Figure 8.8: Construction schematics for
(a) a *p-n*-diode, and
(b) a *p-i-n*-diode.
The letter W denotes the *depletion region* in which mobile charge carriers are removed by the high internal electric field arising at the junction between the positively (p) and negatively (n) doped regions.

According to the standard interpretation of quantum mechanics, the squared moduli of the expansion coefficients $c_n \equiv e^{-|\alpha|^2/2}\alpha^n/\sqrt{n!}$ are the probabilities of obtaining the result n when the "observable" N_j (which is proportional to the energy of mode j) is measured. The factors $|c_n|^2$ in this case describe a Poisson photon number distribution. By choosing $|\alpha|$ smaller than 1—a comparatively nonproblematic task—it is possible to reduce the probability of finding two or more photons in the mode to very small values (still, there always remains a finite chance to have more than just one photon excited; since most photodetectors are unable to discern this, a non-negligible amount of computational errors results). Better than such weakened coherent sources are heralded single-photon sources, in which a process called "spontaneous parametric down-conversion" is used to simultaneously produce two elementary excitations, one in each of two (independently accessible) optical modes. By detecting the photon in one of the modes ("trigger photon"), one can make sure that there is another photon in the second mode. This is important for proper timing, since the process being "spontaneous" implies that one cannot predict, with certainty, in which moment the pair of photons will be produced.

We have introduced the concept of a *cebit* as a means to describe information storage in a 2-dimensional (classical) complex vector space. The new degrees of freedom available through quantum mechanics allow one to store information in their associated Hilbert spaces. In order to emphasize the quantum character of such an encoding, the basic unit of information in the 2-dimensional case has long been called by us as a *qubit* (or *q-bit*, where "q" stands for "quantum"). Higher-dimensional Hilbert spaces are sometimes considered in the literature as well. The 3-dimensional case is commonly referred as a *qutrit*, the d-dimensional as a *qudit*. One viable implementation of qutrits makes use of a special kind of laser beams that possesses orbital angular momentum. The fundamentals and possible advantages of this rather young member in the family of encoding schemes are still being investigated. We won't consider this *ansatz* here; the interested reader may consult [1, 49] for an introduction to the subject.

An elementary representation of a qubit may be obtained by choosing a certain optical mode j and defining the qubit basis to be the pair of states $|0\rangle_j$ and $|1\rangle_j$. This, however, is not very advantageous for two basic reasons [77]. Firstly, any interaction coupling the $|0\rangle_j$ ("vacuum") state to the one photon state $|1\rangle_j$ easily couples this further to states with more photons. These are outside the space we utilize for information transfer and lead to losses. (Such higher order modes are sometimes referred to as *unmodeled* or *spurious* modes caused by "*leakage*"). The single mode Hilbert space is not a natural basis for 2-state coding. Secondly, the vacuum state is not easy to observe. Any measurement trying to detect its presence must be essentially a null-measurement, and as such it can never be distinguished from a failed observation [77]. The conventional way out of this second dilemma is to code information by means of a pair of *two* modes, labelled 1 and 2, either of which should contain precisely one photon. In analogy to the cebit case, also here the Hilbert space $\mathcal{H}$ associated with the two degrees of freedom is the product of the individual spaces $\mathcal{H}_1$ and $\mathcal{H}_2$ of the two modes, $\mathcal{H} = \mathcal{H}_1 \otimes \mathcal{H}_2$. Although this space is even less of a 2-dimensional one, the following definition of what is to be considered the qubit basis at least avoids problems involving vacuum detection. The identification

$$\begin{aligned} |\emptyset\rangle &\equiv |0\rangle_1 \otimes |1\rangle_2 \\ |1\!\!\!/\,\rangle &\equiv |1\rangle_1 \otimes |0\rangle_2 \end{aligned}, \qquad (8.29)$$

where we have introduced the dashed symbols $\emptyset$ and $1\!\!\!/$ for the sake of avoiding confusions with photon state numbering, is the standard convention of what is to be used as the qubit basis when photon occupation numbers are employed for (quantum) information coding. Following common usage, from now on we will drop the tensor-product symbol $\otimes$. Since it is in accord with the standard quantum-mechanical symbolism, it will be preferred in this chapter. Note, however, that the more particular notations $|a\rangle_1|b\rangle_2 \equiv |ab\rangle_{12} \equiv |a_1 b_2\rangle \equiv a_1 b_2 \equiv ab$ (where in the last expression a certain ordering of modes is presumed), preferred by some researchers, are also encountered in the literature.

Some other, less standard qubit encoding schemes make use of (properly normalized) linear superpositions $|\alpha\rangle_j \pm |-\alpha\rangle_j$ of coherent states in a given mode j (the so called "cat states") and "squeezed states" of light. Although the manipulation of such states presents a series of very convenient features, their highly "non-classical" character can make them quite hard to produce in the first place. Due to the potential advantages of such alternative coding schemes, however, much research is being done on them. Refs. [10, 22, 29, 30, 45, 64, 66, 81] contain a good amount of information on what has been published up to the moment. To the best of the authors' knowledge, no experiments using this kind of encoding have been performed so far.

It is in principle possible to excite only one optical mode in free space, which is done in practice by enforcing single-mode operation by transmitting light through physical media which due to their geometry simply do not allow for more than one spatial mode propagating in them. Such devices are called *single-mode optical fibers*. In order to avoid coupling losses, it is convenient to have all-fiber single-photon sources available. A heralded photon-pair fiber source for instance has been described in [15]. Since fibers are radially symmetric, all polarization states are physically equivalent as far as their propagation behavior is concerned. Since the spatial mode shape is the same one for all polarization directions, this degeneracy is not only permissible, but even extremely useful, for it allows to easily transfer quantum excitations—photons—from one mode (say, a vertically polarized one) to another (a horizontally polarized one, for instance). Instead of referring to two different single-mode fibers, the labels "1" and "2" in (8.29) could therefore also stand for "V" and "H" (the vertically and horizontally polarized modes of a single fiber). It is not uncommon in this case to introduce the notation $|V\rangle$ and $|H\rangle$, but one should always bear in mind that the symbols V and H actually label a mode and not the quantum state of a mode. For a thorough discussion of polarization coding, see [77].

8.2.2 Quantum operations and gates

The central advantage of using the quantum degree of freedom of an optical mode should already be evident at this point, since irrespective of the choice one makes as to how a qubit is encoded, the number of modes (say, fibers) necessary to represent n qubits grows proportionally to n. By using classical degrees of freedom instead, we find that this number grows exponentially like 2^n. In order to construct any possible algorithm also within the present (quantum) frame, however, it is important that all the 2-qubit analogs of the 2-cebit logic gates described in Section 8.1 are ready for experiments. But this precisely turns out to be the *crux* of the quantum approach. It is not *a priori* clear how the optical components described in Section 8.1 act on the quantum degrees of freedom, and this will be studied below. Even in the classical case, the action of an optics element depends on the degree of freedom considered: a mirror changes a beam from L- to R-travelling, but leaves its polarization unchanged. Wave plates, on the contrary, modifiy the polarization of a beam without altering its propagation direction. Polarizing beam splitters, on the other hand, play the role of an "optical language translator". In order to describe the quantum performance of these optical devices in a convenient way, we need to recall from Section 3.2 of Chap. 3 the concept of "raising" [or photon "creation", $a_j^\dagger$] and "lowering" [or (photon) "annihilation", a_j] operators of mode j, where as before the dagger ($\dagger$) symbol denotes Hermitian conjugation. The action of the linear operators $a_j^\dagger$ and a_j on a number state is defined by

$$
\begin{aligned}
a_j^\dagger |n\rangle_j &\equiv \sqrt{n+1} \;\; |n+1\rangle_j \\
a_j |n\rangle_j &\equiv \sqrt{n} \;\; |n-1\rangle_j
\end{aligned}. \tag{8.30}
$$

From this it follows that the commutator $[a_j, a_j^\dagger] \equiv a_j a_j^\dagger - a_j^\dagger a_j = 1$ and that the number operator $N_j = a_j^\dagger a_j$. (Note that the raising (lowering) operator actually lowers (raises) n when acting to the left, i.e., on a bra.) Since operators for different modes commute, $[a_j, a_k] = [a_j, a_k^\dagger] = 0$ whenever $j \neq k$, it does not lead to any confusion if, for the sake of notational simplicity, we prefer to write

$$
a_k |m\rangle_j |n\rangle_k = |m\rangle_j a_k |n\rangle_k = \sqrt{n} |m\rangle_j |n-1\rangle_k \tag{8.31}
$$

instead of using the more accurate $1_j \otimes a_k$ expression for the 2-mode operator.

Strictly speaking, a network of fibers and optical components that connect them defines the boundary conditions of an intricate Helmholtz equation, the eigensolutions of which are the (quantized) modes. It is indeed one of the advantages of the optical approach to quantum computation that such a cumbersome description is unnecessary. Instead, it is possible and highly accurate to consider a given network as consisting of interconnected, but otherwise independent modes. In such a modeling, optical elements play the role of mode junctions, and it is at these junctions where the essential physics takes place. Due to this modularity, it becomes unnecessary for the more mathematically inclined reader to understand the details of what is going on at the junctions, as long as we are able to specify how they redistribute mode excitations from the input- to the output-ports. Although it is this pragmatic approach that we plan to follow here, a few more physically motivated remarks may be helpful to understand why not all possible types of junction can be realized with current technology. Recall from Section 2.3 that in quantum theory, the time (t) evolution of a given state is generated by a unitary operator $U(t)$. Its mathematical form is dictated by a "Hamiltonian", an operator-valued function H that—in the present case—depends on the creation and annihilation operators of the modes participating in the interaction. When H is t-independent (which is the situation in passive optics), then $U(t) = \exp(i2\pi Ht/h)$. The unitary state evolution of a quantum mechanical system describes lossless and dissipation-free, *reversible* dynamics. The evolution in this case is also said to be *coherent*. (It should be clear that this definition is completely unrelated to our previous introduction of "coherent" states.) A phase shifter for mode j is described by a Hamiltonian H which is simply proportional to $a_j^\dagger a_j(= N_j)$. The unitary operator describing it is given by

$$[\mathbf{F}(\Theta)]_j = \exp(i\Theta a_j^\dagger a_j), \tag{8.32}$$

where the phase angle Θ is proportional to the propagation time t. A (symmetric) beam splitter with ports j and k, on the other hand, is described by the unitary operator

$$[\mathbf{B}]_{j,k} = \exp\left[\frac{\pi}{4}\left(a_j a_k^\dagger - a_j^\dagger a_k\right)\right]. \tag{8.33}$$

It is but a special case of the more general definition encountered in the literature, i.e.,

$$[\mathbf{B}(\vartheta, \varphi)]_{j,k} = \exp\left[-i\vartheta\left(e^{-i\varphi}a_j a_k^\dagger + e^{i\varphi}a_j^\dagger a_k\right)\right]. \tag{8.34}$$

All-optical networks consisting of only phase shifters $[\mathbf{F}(\Theta)]_j$ and beam splitters of the general type $[\mathbf{B}(\vartheta, \varphi)]_{j,k}$ preserve the total number of photons excited in the modes. These optical components are called "(passive) linear". Linear Optics Quantum Computing (LOQC) is by definition restricted to the use of this limited set of elements. The origin of this restriction is a physical one and is related to the extremely low light intensities that we are considering. It is indeed now possible to physically realize non-linear Hamiltonians in optics, a common one being the so called "Kerr-nonlinearity" [41, 47], described by the 2-mode function $H = \kappa a_j^\dagger a_j a_k^\dagger a_k$. An intuitive interpretation of the interaction so described is that the presence of light in one mode alters the properties of the propagation medium (typically a special kind of non-linear crystal) in such a way that light in another mode propagating through the same medium is affected. This effectively (though indirectly) causes light in one mode to interact with light in another. It is the usual size of the proportionality constant κ which is the problem [47]. Non-linear effects are easily observed with macroscopic light intensities, but no ordinary medium is non-linear to the extent that the presence of a single photon in one mode will have any noticeable effect on another mode. Therefore, for all practical purposes, optics of few-photons deals with linear optical elements *exclusively*.

How do the unitary operators $[\mathbf{F}]_j$ and $[\mathbf{B}]_{j,k}$ behave when acting on qubits in the standard form (8.29)? In the case of the beam shifter, its eigenvectors are Fock states and we immediately obtain

$$
\begin{aligned}
{[\mathbf{F}(\Theta_1)]_1\,[\mathbf{F}(\Theta_2)]_2\,|0\rangle_1|1\rangle_2} &= e^{i\Theta_2}|0\rangle_1|1\rangle_2, \\
{[\mathbf{F}(\Theta_1)]_1\,[\mathbf{F}(\Theta_2)]_2\,|1\rangle_1|0\rangle_2} &= e^{i\Theta_1}|1\rangle_1|0\rangle_2.
\end{aligned}
\tag{8.35}
$$

Such a one-by-one listing of all possible state transformation alternatives of the qubit basis is a typical representation of a quantum logic "truth table". The slightly more explicit (but since quantum mechanics is a linear theory, actually perfectly equivalent) formulation

$$
\mathbf{F}(\Theta_0, \Theta_1)\Big[c_0|0\rangle + c_1|1\rangle\Big] = c_0 e^{i\Theta_0}|0\rangle + c_1 e^{i\Theta_1}|1\rangle,
\tag{8.36}
$$

where for compactness we have introduced the definitions (8.29), is also frequently encountered. The complex-valued expansion coefficients c_0 and c_1 have to fulfill the normalization condition $|c_0|^2+|c_1|^2 = 1$ but are otherwise arbitrary. It is important for the reader to consciously realize that LOQC qubits are encoded with the help of *two* optical modes together containing exactly *one* photon. The action of a beam splitter using the truth-table representation is given by

$$
\begin{aligned}
{[\mathbf{B}(\vartheta, \varphi)]_{1,2}\,|0\rangle_1|1\rangle_2} &= \cos\vartheta|0\rangle_1|1\rangle_2 - e^{i\varphi}\sin\vartheta|1\rangle_1|0\rangle_2 \\
{[\mathbf{B}(\vartheta, \varphi)]_{1,2}\,|1\rangle_1|0\rangle_2} &= e^{-i\varphi}\sin\vartheta|0\rangle_1|1\rangle_2 + \cos\vartheta|1\rangle_1|0\rangle_2
\end{aligned}
\tag{8.37}
$$

Note that according to this result, beam splitters transform qubits into qubits. This idealized situation applies to well-aligned and manufactured devices. Any imperfections may lead to states outside of the qubit space and represent a potential source for losses or leakage. Representing optical devices in terms of creation and annihilation operators has the advantage of enabling one to understand their physical behavior from first principles and to study their response also outside of a chosen qubit space. Since losses may have many different origins, however, it is common to analyze their influence on a quantum computation in a more phenomenological way. The problems associated with losses and other sources of quantum computational errors (spurious phase shifts, in particular) shall not be considered here. The subject is of central importance, though, since it is a fact of life that any practical realization of a quantum computer will always have to cope with technical imperfections. Such imperfections are persistent, and so it does not come as a surprise that many types of quantum error correction schemes have been intensely studied since the earliest days of quantum information theory [52, 74] and that they were also considered from the very beginning in the context of LOQC [39].

Assuming *idealized* conditions, we can take things easy and avoid more rigorous expressions such as (8.32)–(8.34). Instead, a formalism tailored to the needs in 2-dimensional qubit-space is usually preferred. By identifying the qubit basis with the pair of column vectors

$$|0\rangle \quad \rightarrow \quad \begin{bmatrix} 1 \\ 0 \end{bmatrix}, \tag{8.38}$$

$$|1\rangle \quad \rightarrow \quad \begin{bmatrix} 0 \\ 1 \end{bmatrix}, \tag{8.39}$$

the linear transformation matrix describing the beam splitter (8.37) can be written

$$\begin{bmatrix} \cos\vartheta & e^{-i\varphi}\sin\vartheta \\ -e^{i\varphi}\sin\vartheta & \cos\vartheta \end{bmatrix}, \tag{8.40}$$

which is equivalent to our familiar Rabi rotation matrix $U_{\theta,\phi}$ in (2.46). (Incidentally, the matrix elements $\cos\vartheta$ and $e^{-i\varphi}\sin\vartheta$ are called the "transmission and reflection coefficients" of the beam splitter, respectively [72]. Perhaps more frequently, but at the cost of lost phase information, the words "transmitivity" and "reflectivity" are applied to the *squared norms* $T \equiv \cos^2\vartheta$ and $R \equiv \sin^2\vartheta$ thereof, such that $R + T = 1$ [25, 69]. Since the usage is not homogeneous among authors, we prefer to explicitly refer to *intensity* transmitivity/reflectivity in order to avoid possible confusion. Also the words "transmittance" and "reflectance" are oftentimes employed.) Similarly to (8.40), the phase shifter produces

$$\begin{bmatrix} \exp(i\Theta_0) & 0 \\ 0 & \exp(i\Theta_1) \end{bmatrix}. \tag{8.41}$$

The *Hadamard transform* of a single qubit, as seen in (2.49) is described by the matrix

$$\frac{1}{\sqrt{2}}\begin{bmatrix} 1 & 1 \\ 1 & -1 \end{bmatrix} \tag{8.42}$$

and can be obtained by placing a balanced beam splitter between two phase shifters acting on one of the modes, say, mode 1:

$$[F(-\pi/2)]_1 [B(\pi/4, -\pi/2)]_{1,2} [F(-\pi/2)]_1 .$$

It can be easily verified that by suitably combining a series of beam splitters and phase shifters, any possible matrix in SU(2) can be physically realized [14]. Since every matrix in SU(2) can be identified with a corresponding rotation matrix in O(3), it is customary to say that it produces a "qubit rotation". It is convenient then to use the fundamental set of *Pauli matrices* (introduced earlier in (2.42) and (2.43))

$$\sigma_x \equiv \begin{bmatrix} 0 & 1 \\ 1 & 0 \end{bmatrix} \qquad \sigma_y \equiv \begin{bmatrix} 0 & -i \\ i & 0 \end{bmatrix} \qquad \sigma_z \equiv \begin{bmatrix} 1 & 0 \\ 0 & -1 \end{bmatrix} \qquad (8.43)$$

to mathematically represent rotations about any axis $\vec{n}$, i.e.,

$$\exp\left(-i\vec{\sigma} \cdot \vec{n}\,\alpha/2\right) = 1\,\cos\frac{\alpha}{2} - i\vec{\sigma} \cdot \vec{n} \sin\frac{\alpha}{2}, \qquad (8.44)$$

with $n_x^2 + n_y^2 + n_z^2 = 1$. Again, different authors tend to use their own nomenclature and graphical representation for 1-qubit rotations (and their generalization in n-qubit spaces). Here we will try to use "standard" (text-book) notations for clarity. Note that even the terminology for these simple transformations is rather heterogeneous. Single-qubit rotations, "single- or 1-qubit (quantum) gates" or, alternatively, "Pauli gates" are all equivalent and frequently used expressions.

Two-qubit transformations (2-qubit "quantum gates") will be described in terms of the 4×4 generalization of the above matrices. We identify

$$|\emptyset\rangle_1|\emptyset\rangle_2 \quad \longleftrightarrow \quad \begin{bmatrix} 1 \\ 0 \\ 0 \\ 0 \end{bmatrix} \qquad (8.45)$$

$$|\emptyset\rangle_1|1\rangle_2 \quad \longleftrightarrow \quad \begin{bmatrix} 0 \\ 1 \\ 0 \\ 0 \end{bmatrix} \qquad (8.46)$$

$$|1\rangle_1|\emptyset\rangle_2 \quad \longleftrightarrow \quad \begin{bmatrix} 0 \\ 0 \\ 1 \\ 0 \end{bmatrix} \qquad (8.47)$$

$$|1\rangle_1|1\rangle_2 \quad \longleftrightarrow \quad \begin{bmatrix} 0 \\ 0 \\ 0 \\ 1 \end{bmatrix}, \qquad (8.48)$$

where the indices "1" and "2" refer to the qubit (not the mode!) number. This kind of qubit labelling will also be used to refer to single-qubit rotations. For instance, a beam splitter

$$\begin{bmatrix} \cos\vartheta & 0 & -e^{i\varphi}\sin\vartheta & 0 \\ 0 & \cos\vartheta & 0 & -e^{i\varphi}\sin\vartheta \\ e^{-i\varphi}\sin\vartheta & 0 & \cos\vartheta & 0 \\ 0 & e^{-i\varphi}\sin\vartheta & 0 & \cos\vartheta \end{bmatrix} \qquad (8.49)$$

affecting only qubit 1 will be denoted $\mathbf{B}_1(\vartheta, \varphi)$. Choosing $\vartheta = \pi/4$ and $\varphi = -\pi/2$ the matrix becomes identical to the 2-cebit balanced beam splitter matrix (8.7). In that case it was the spatial-cebit which was affected, while the polarization-cebit remains unaltered. (From now on, mathematically there is no need to formally distinguish between the two alternative ways of encoding a qubit, i.e., using a pair of different polarization modes in a single optical fiber or using two separate single-mode optical fibers with the same optical polarization. The indices in (8.29) may equally well be labelling polarization states or numbering fiber paths. Conversion of a polarization (-encoded) qubit into spatial encoding can be achieved experimentally by passing the photon through a polarizing beam splitter to spatially separate the optical modes, and then using a half-wave (retardation) plate to rotate one of the modes into the same polarization as the other. The reverse process can be used to return the encoding to the polarization degree of freedom [67]. Quite generally, single-fiber (polarization-encoded) qubits tend to be more resistant to certain kinds of experimental errors and are easier to manipulate than 2-fiber (path-encoded) qubits. For the realization of single-qubit gates, in particular, one only requires waveplates (and phase delays), instead of beam splitters as in spatial encoding. Ref. [12] contains a detailed listing of 1- and 2-qubit gate realizations using polarization encoding.) In similar terms, we define the phase shift operator $\mathbf{F}_\nu(\Theta_\emptyset, \Theta_\natural)$ for qubit number ν. The (quantum) CNOT gate, in which qubit number 1 (*control qubit*) defines the (controlled) action on qubit number 2 (*target qubit*), is described by the matrix $\Lambda_1(\sigma_x)$ in Section 2.4:

$$
\begin{bmatrix}
1 & 0 & 0 & 0 \\
0 & 1 & 0 & 0 \\
0 & 0 & 0 & 1 \\
0 & 0 & 1 & 0
\end{bmatrix}, \qquad \text{cf. (2.44),} \qquad (8.50)
$$

which, up to a phase factor in qubit number 2, corresponds to the LOCC gate (8.17). Unlike its classical counterpart, however, the quantum CNOT gate *cannot* be implemented with passive linear optics components alone. This is a direct consequence of our previous observation, that at low light intensities we are (currently) technically unable to influence and therefore control light in one mode by light in another one. The CNOT gate can be obtained by surrounding a *controlled phase (shift)* gate (or CPHASE, for short) (see (2.48))

$$
\mathbf{Q}_\eta \equiv
\begin{bmatrix}
1 & 0 & 0 & 0 \\
0 & 1 & 0 & 0 \\
0 & 0 & 1 & 0 \\
0 & 0 & 0 & e^{i\eta}
\end{bmatrix}
\qquad (8.51)
$$

with $\eta = \pi$ by a pair of rotations on qubit number 2 and correcting the phase shifts. This gate (8.51) was called by us the QPG in (2.47). Here, we could continue to use the very same nomenclature. However, due to the *optical nature* of the setup, we feel that a discerning name CPHASE (which is also a standard notation) has merit here. The particular phase gate with $\eta = \pi$ is also called a "controlled sign (flip)" gate (or "CSIGN" for short). Some authors prefer calling such gates "conditional" instead of "controlled". The CNOT gate (or equivalently, the CSIGN or the more general CPHASE gates) were noted earlier to be "universal" in the sense that any n-qubit gate can be obtained by composing 1-qubit gates (i.e., arbitrary rotations) and one of these controlled 2-qubit gates, following Theorem 2.6.2. Since the CNOT and CSIGN gates are related by Hadamard conjugation on the target qubit, in the context of LOQC, where 1-qubit gates are relatively straightforward, the two gates are practically equivalent and are used almost interchangeably. Gates like these, that cannot be constructed with linear optics alone, are also called *nonlinear gates*. For several years, the mathematically proven impossibility of constructing *reliable* (i.e., theoretically 100% successful) controlled gates using LOQC seemed to have prohibited the construction of universal quantum computers using light.

However, any such would-be "no-go theorem" does not prevent us from constructing *unreliable* universal quantum gates. By "unreliable" we mean that the gate does not always produce the desired output but, and this is the essential point, after the gate operation we will know with certainty (as always, modulo possible imperfections from experiments) if it did or did not perform correctly. Quantum mechanics even predicts the probability of the event that such a "nondeterministic" (or "probabilistic") gate does its intended job. As discussed above, for the purpose of universal quantum computation, it is in principle sufficient to demonstrate how any one of the various universal gates could be physically implemented. Here we will present a method conceived by Knill, Laflamme and Milburn [39] to build a probabilistic CSIGN gate which in the average succeeds in 1 out of 16 trials. It is not the most efficient construction, but it well illustrates the modular thinking typical in this field and allows us to introduce a number of important new concepts in a more pedagogical way.

8.2.3 The approach of Knill, Laflamme and Milburn

The first new building block required is the so called "non-linear sign shift gate" (NLS for brevity). It does not act on qubits but on single modes only. The action of NLS on mode number 1 is defined by NLS: $\alpha_0|0\rangle_1 + \alpha_1|1\rangle_1 + \alpha_2|2\rangle_1 \rightarrow \alpha_0|0\rangle_1 + \alpha_1|1\rangle_1 - \alpha_2|2\rangle_1$. This gate is a nonlinear one, but a probabilistic version of it using only linear optics and photon detectors can be realized by introducing two additional modes in a tricky way. Such modes, which must be added in order to achieve a certain quantum operation, are *ancilla modes* and their elementary excitations are referred to as "helper photons". Also the word "auxiliary" is often used with the same meaning. These two new modes 2 and 3 we feed into the optical circuit as depicted in Fig. 8.9(a). As already commented in Section 8.1, any such optical circuit can be described in terms of a corresponding unitary matrix and vice versa: any unitary matrix can be physically realized by an appropriate choice of

linear optical components [69]. In this case (see [69, Fig. 2] for a different example also involving three inputs and outputs), the unitary matrix is
given by

(a)

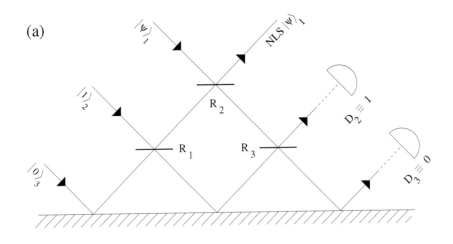

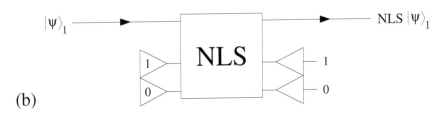

(b)

Figure 8.9: Non-linear sign shift (NLS) gate:
(a) optical realization using a mirror and three beam splitters with
reflectivities R_1, R_2, R_3, and
(b) its conventional symbol.

$$U \equiv \begin{bmatrix} 1-\sqrt{2} & 1/\sqrt{\sqrt{2}} & \sqrt{3/\sqrt{2}-2} \\ 1/\sqrt{\sqrt{2}} & 1/2 & 1/2-1/\sqrt{2} \\ \sqrt{3/\sqrt{2}-2} & 1/2-1/\sqrt{2} & \sqrt{2}-1/2 \end{bmatrix}, \qquad (8.52)$$

where the three beam splitter (intensity) reflectivities of the figure have
been chosen as follows: $R_1 = (4-2\sqrt{2})^{-1} = R_3$, $R_2 = (\sqrt{2}-1)^2$. If we
look at the device from a classical optics point of view, it would not seem
to produce anything particularly fascinating. (For instance, light entering
only through one of the three inputs 1, 2 or 3 will produce amplitudes proportional to the corresponding first, second or third columns of (8.52) at the
device's output.) Things become much more interesting when analyzed at

the quantum level. In order to do so, we use the operator formalism (sometimes also referred to as the "Heisenberg picture"; for a good treatment in the present context see, e.g., [22] or [43].) Passive linear optics components are photon-number-preserving. Let us first consider elementary mode excitations at the two entry ports of a single beam splitter (8.40) and see how they get redistributed at its exit:

$$
\begin{bmatrix} \cos\vartheta & e^{-i\varphi}\sin\vartheta \\ -e^{i\varphi}\sin\vartheta & \cos\vartheta \end{bmatrix} \begin{bmatrix} a_1^\dagger \\ a_2^\dagger \end{bmatrix} |0\rangle_1 |0\rangle_2 ,
\tag{8.53}
$$

where $|0\rangle_1|0\rangle_2$ is the vacuum state of the 2-mode system. In the case of a *symmetric* beam splitter (choose $\varphi = 0$, $\vartheta = \pi/4$, for instance), a single photon entering mode 1 will result in state

$$
\frac{|0\rangle_1|1\rangle_2 + |1\rangle_1|0\rangle_2}{\sqrt{2}}
\tag{8.54}
$$

at the exit. Similarly, a single photon introduced into mode 2 will produce

$$
\frac{|0\rangle_1|1\rangle_2 - |1\rangle_1|0\rangle_2}{\sqrt{2}}
\tag{8.55}
$$

at the output of the device. If one photon enters through each input port, the two photons exiting the beam splitter will be either detected together in mode 1 or together in mode 2,

$$
\frac{|0\rangle_1|2\rangle_2 - |2\rangle_1|0\rangle_2}{\sqrt{2}} ,
\tag{8.56}
$$

but never individually. This remarkable property, which will play an important role later on, can be verified by evaluating $Ua_1^\dagger a_2^\dagger|0\rangle_1|0\rangle_2 = [Ua_1^\dagger U^{-1}] Ua_2^\dagger|0\rangle_1|0\rangle_2$, or

$$
\frac{1}{\sqrt{2}}\left[a_1^\dagger + a_2^\dagger\right] \frac{|0\rangle_1|1\rangle_2 - |1\rangle_1|0\rangle_2}{\sqrt{2}} = \frac{|0\rangle_1|2\rangle_2 - |2\rangle_1|0\rangle_2}{\sqrt{2}} .
\tag{8.57}
$$

Unlike the input states $|1\rangle_1|0\rangle_2$, $|0\rangle_1|1\rangle_2$ and $|1\rangle_1|1\rangle_2$, the corresponding output states cannot be written in such a simple product form. One says that the initial *separable* state gets *entangled* by the optical device. Although the concept of "entanglement" is ordinarily used in the quantum context only, non-separability also arises very naturally in the case of cebits. For an insightful comparison of classical and quantum entanglement, see [75]. Much of the art of manipulating vector bits has to do with the engineering of entangled states. In the optics context, beam splitters represent the main tool used for this purpose. The particular entangler shown in Fig. 8.9(a), acting on a single photon entering through one of its inputs, produces

$$
U \begin{bmatrix} a_1^\dagger \\ a_2^\dagger \\ a_3^\dagger \end{bmatrix} |0\rangle_1 |0\rangle_2 |0\rangle_3 =
$$

$$
\begin{bmatrix}
(1-\sqrt{2})|1\rangle_1|0\rangle_2|0\rangle_3 + (2^{-1/4})|0\rangle_1|1\rangle_2|0\rangle_3 + \\
(3/\sqrt{2}-2)^{1/2}|0\rangle_1|0\rangle_2|1\rangle_3 \\
(2^{-1/4})|1\rangle_1|0\rangle_2|0\rangle_3 + (1/2)|0\rangle_1|1\rangle_2|0\rangle_3 + \\
(1/2-1/\sqrt{2})|0\rangle_1|0\rangle_2|1\rangle_3 \\
(3/\sqrt{2}-2)^{1/2}|1\rangle_1|0\rangle_2|0\rangle_3 + (1/2-1/\sqrt{2})|0\rangle_1|1\rangle_2|0\rangle_3 + \\
(\sqrt{2}-1/2)|0\rangle_1|0\rangle_2|1\rangle_3
\end{bmatrix} . \qquad (8.58)
$$

The squared moduli of the expansion coefficients in (8.58) are the probabilities of detecting the corresponding state when performing a photon-number sensitive measurement. (For this reason, the expansion coefficients themselves are also called *probability amplitudes*.) For example, if the device has an input state $a_2^\dagger|0\rangle_1|0\rangle_2|0\rangle_3$,

$$
U a_2^\dagger|0\rangle_1|0\rangle_2|0\rangle_3 = \left(U a_2^\dagger U^{-1}\right) U|0\rangle_1|0\rangle_2|0\rangle_3 = U|0\rangle_1|1\rangle_2|0\rangle_3 =
$$

$$
= \frac{1}{\sqrt{\sqrt{2}}}|1\rangle_1|0\rangle_2|0\rangle_3 + \frac{1}{2}|0\rangle_1|1\rangle_2|0\rangle_3 + \left(\frac{1}{2}-\frac{1}{\sqrt{2}}\right)|0\rangle_1|0\rangle_2|1\rangle_3 , \qquad (8.59)
$$

which is the second row in (8.58) (and where use is made of the fact that no light entering the circuit evidently produces no light at its exit: $U|0\rangle_1|0\rangle_2|0\rangle_3 = |0\rangle_1|0\rangle_2|0\rangle_3$), then the probability of detecting state $|0\rangle_1|1\rangle_2|0\rangle_3$ at the output equals $|1/2|^2 = 1/4$. In the same way, the probabilities for detecting $|1\rangle_1|0\rangle_2|0\rangle_3$ and $|0\rangle_1|0\rangle_2|1\rangle_3$ at the output ports equal $1/\sqrt{2}$ and $[1/2 - 1\sqrt{2}]^2$, respectively. (One confirms that the sum of these probabilities equals 1, as it should.)

Now, by construction, the entangler U will do its intended job only if fed with one (helper) photon in (ancilla) mode 2 and zero (helper) photons in (ancilla) mode 3. Of course, such a situation can always be prepared experimentally, at least in principle, and this is the situation where we will concentrate our attention now. We know already that under these circumstances the probability amplitude for output state $|0\rangle_1|1\rangle_2|0\rangle_3$ equals $1/2$ (input mode 1 fed with the vacuum state).

If the input mode is in state $|1\rangle_1$, there will be altogether two photons in the circuit, which U will redistribute and entangle according to

$$
U|1\rangle_1|1\rangle_2|0\rangle_3 = U a_1^\dagger a_2^\dagger|0\rangle_1|0\rangle_2|0\rangle_3 =
$$

$$
= U a_1^\dagger U^{-1}\left[\frac{1}{\sqrt{\sqrt{2}}}|1\rangle_1|0\rangle_2|0\rangle_3 + \frac{1}{2}|0\rangle_1|1\rangle_2|0\rangle_3 + \left(\frac{1}{2}-\frac{1}{\sqrt{2}}\right)|0\rangle_1|0\rangle_2|1\rangle_3\right] =
$$

$$
= (1-\sqrt{2})\left[\sqrt{\sqrt{2}}|2\rangle_1|0\rangle_2|0\rangle_3 + \frac{1}{2}|1\rangle_1|1\rangle_2|0\rangle_3 + \left(\frac{1}{2}-\frac{1}{\sqrt{2}}\right)|1\rangle_1|0\rangle_2|1\rangle_3\right] +
$$

$$
\frac{1}{\sqrt{\sqrt{2}}}\left[\frac{1}{\sqrt{\sqrt{2}}}|1\rangle_1|1\rangle_2|0\rangle_3 + \frac{1}{\sqrt{2}}|0\rangle_1|2\rangle_2|0\rangle_3 + \left(\frac{1}{2}-\frac{1}{\sqrt{2}}\right)|0\rangle_1|1\rangle_2|1\rangle_3\right] +
$$

$$\sqrt{3/\sqrt{2} - 2} \left[\frac{1}{\sqrt{\sqrt{2}}} |1\rangle_1 |0\rangle_2 |1\rangle_3 + \frac{1}{2} |0\rangle_1 |1\rangle_2 |1\rangle_3 + \left(\frac{1}{\sqrt{2}} - 1 \right) |0\rangle_1 |0\rangle_2 |2\rangle_3 \right].$$

$$(8.60)$$

Here we have made use of (8.30) and (8.59). One easily reads off the probability amplitudes for all possible output combinations involving two photons and, in particular, the one we will be interested in, i.e., the one with one photon in mode 1, one photon in mode 2 and no photons in mode 3. From (8.60) we infer that the probability amplitude for $|1\rangle_1 |1\rangle_2 |0\rangle_3$ equals $(1 - \sqrt{2})/2 + 1/\sqrt{2} = 1/2$. Thus, the probability for this measurement outcome is again $1/4$.

Finally, the last input state of interest is the one in which mode 1 contains two photons. Then there will be three photons all in all entering the device and we need to evaulate $U|2\rangle_1 |1\rangle_2 |0\rangle_3$ in order to learn how they will be distributed at the output ports. By noting that

$$U|2\rangle_1 |1\rangle_2 |0\rangle_3 = U \frac{1}{\sqrt{2}} a_1^\dagger a_1^\dagger a_2^\dagger |0\rangle_1 |0\rangle_2 |0\rangle_3 = \frac{1}{\sqrt{2}} \left[U a_1^\dagger U^{-1} \right] U a_1^\dagger a_2^\dagger |0\rangle_1 |0\rangle_2 |0\rangle_3,$$

$$(8.61)$$

one only needs to apply the operator

$$\left[(1 - \sqrt{2}) a_1^\dagger + 2^{-1/4} a_2^\dagger + (3/\sqrt{2} - 2)^{1/2} a_3^\dagger \right] / \sqrt{2}$$

to the right hand side of (8.60). This produces many combinations involving three photons at the output and the reader may easily calculate all the corresponding probability amplitudes. For our purposes here, however, we only need to know the probability amplitude for the event $|2\rangle_1 |1\rangle_2 |0\rangle_3$. We find

$$U|2\rangle_1 |1\rangle_2 |0\rangle_3 = \frac{1}{\sqrt{2}} |2\rangle_1 |1\rangle_2 |0\rangle_3 \left[(1 - \sqrt{2})^2 / \sqrt{2} + (1 - \sqrt{2}) + (1 - \sqrt{2}) \right] \quad (8.62)$$

$$(\quad + \quad \text{other distributions of 3 photons over 3 modes} \quad)$$

and evaluate the square brackets, giving $-1/2$. All in all, we conclude that the device in Fig. 8.9(a) will produce the desired NLS-mapping

$$\left. \begin{array}{ccc} |0\rangle_1 & \rightarrow & |0\rangle_1 \\ |1\rangle_1 & \rightarrow & |1\rangle_1 \\ |2\rangle_1 & \rightarrow & -|2\rangle_1 \end{array} \right\} \qquad (8.63)$$

as long as we accept only those events in which the ancilla modes remained in the state $|1\rangle_2 |0\rangle_3$ in which they were initially prepared. This will happen in 1 out of 4 attempts. In three out of four attempts, the output of the device is useless and must be discarded. Of central importance is that the destructive measurement is only performed on the ancilla modes: the actually relevant gate mode (number 1) is not subject to a measurement process and can be fed into another quantum gate, if so desired. The *a posteriori* decision to accept or discard a gate operation is called *postselection*.

It is this postselection process which effectively provides the nonlinearity required for the (quantum) logic operations, since the detection process is intrinsically nonlinear. The NLS gate, represented in Fig. 8.9(b) using a symbol introduced in [39], is universal in the sense that any non-linear gate useful for quantum information processing can be constructed with its help. Since its limited reliability directly translates to a correspondingly limited efficiency of the quantum gates based on it, two important questions arise: do there exist constructions different from the one shown in Fig. 8.9(a) which lead to higher success rates? And, does the addition of more ancilla modes improve the efficiency? In [36], an upper bound no larger than 0.5 was mathematically derived. Somewhat later, a semianalytical study [71] strongly suggests that the maximal success rate for the NLS gate is not larger than 0.25. Very recently, this value was rigurously confirmed by Eisert [13] using the methods of convex optimization and Lagrange duality. It is quite remarkable that the smallest known functioning scheme is already the optimal one and that it was even the first one to be discovered [39]. A somewhat simpler scheme, also involving two ancilla modes but only two beam splitters, was proposed in [84]. Due to this simplification, its efficiency is slightly reduced to 20%. A significant drawback, however, is that it requires a detector capable of distinguishing between no photon, one photon, or two photons. We will come to this point later when commenting about gates operating with respect to (or, colloquially, "in") the "coincidence basis"[1]

We have so far only *analyzed* the quantum behavior of two simple optical devices (a balanced beam splitter and an NLS gate). The reader interested in this area will need to learn how to *create* an optical device according to his/her needs. This is basically a 2-step process, the hardest part being to find the simplest unitary transformation best suited for the intended purpose. The second step, finding an optical arrangement realizing it, is more of a technical problem. Ref. [69] provides a general-purpose construction recipe for optical circuits. A more physical way of thinking is illustrated in [67] using the NLS gate as an example. Here we will demonstrate this more intuitive approach by describing how a probabilistic CSIGN gate may be implemented by combining two NLS gates via a pair of balanced beam splitters. The situation is depicted in Fig. 8.10. The idea is beautiful: as long as the NLS gates are fed with zero or one photon via the first beam splitter, the second beam splitter will simply revert the situation and reproduce at the output what came in at the input. Due to (8.54) and (8.55), this will be the case when one of the three states

$$|0\rangle_1|1\rangle_2|0\rangle_3|1\rangle_4$$

---

[1]In the "coincidence basis" one records only those events, in which two photons are detected in the same, narrow time window. Losses and detector inefficiency can be ignored because they take the system out of the coincidence basis and thus their only effect is to reduce the count rate [65]. Quite generally, coincidence basis operation is used to overcome the problems associated with the random nature of the photon sources, the various optical losses in experimental systems and the inability of the detectors to resolve photon-number [60].

$$|0\rangle_1|1\rangle_2|1\rangle_3|0\rangle_4$$
$$|1\rangle_1|0\rangle_2|0\rangle_3|1\rangle_4$$

is fed as input to the device. On the other hand, because of the special property (8.56), when the input state is

$$|1\rangle_1|0\rangle_2|1\rangle_3|0\rangle_4 \,,$$

there will be a vanishing probability amplitude to detect one single photon in any one of the two exit ports of the first beam splitter: the pair of NLS gates will do their duties (each with a probability $1/4$) and invert the sign of the superposition $|0\rangle_1|2\rangle_3 - |2\rangle_1|0\rangle_3$. The second beam splitter will again revert the initial situation, but leave the overall minus sign intact. By identifying the combination of modes 1 and 2 as one qubit and the combination of modes 3 and 4 as another qubit (using the standard LOQC definition (8.29)), the device is seen to behave like expected from a probabilistic CSIGN gate with an overall efficiency of $(1/4)^2 = 1/16$. The motivation for the invention of the NLS gate should at this point be clear to the reader who, in order to train this more pictorial way of thinking, may refer to [67, Fig. 3] and inspect how a probabilistic CNOT gate can be realized again with the help of two NLS gates and a few beam splitters. Similarly, the detailed analysis of concatenated CSIGN gates in [63] is very instructive.

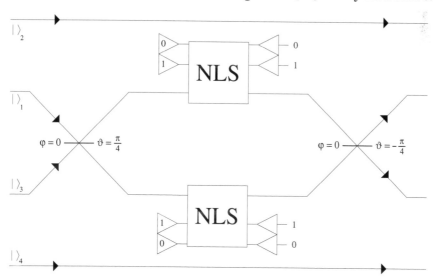

Figure 8.10: A possible realization of a CSIGN gate using a combination of two NLS gates (cf. Fig. 8.9(b)).

NLS gates allow one to conceive modular structures which can be conveniently represented with the graphical symbols introduced already in the seminal paper [39]. Experimentally, however, it is important to find the most economical representation of a quantum gate, and there is no *a priori* reason why the simplest solution should include any NLS gates. And, indeed, in general this is not the case. Knill [34, 35], for instance, discovered a simpler (though physically much less transparent) way of producing a nondeterministic CSIGN gate without NLS gates. In [67], an improved CNOT gate is discussed. Several probabilistic quantum logic operations using polarizing beam splitters were proposed in [56]. The authors of [72] went further and considered the following: instead of using only primitive quantum gates, they investigated what kind of operations can be constructed from linear optical elements, single-photon sources, and detectors. This paradigm shift is analogous to the shift in classical computing from a RISC (reduced instruction set computing) architecture to the CISC (complex instruction set computing) architecture [72]. As a rule of thumb, the fewer ancillas are required for a gate, the better!

Postselected gates of two types have been experimentally demonstrated so far. The ones discussed above, because of the possibility to feed their output into another gate, are called "(quantum) feedforwardable" or "scalable" [12]. (Note that the word "scalable" is sometimes also used in a different way, as a synonym of "computationally efficient".) As of this writing, only one experimental demonstration of a feedforwardable probabilistic linear optics quantum gate has been made [20]. The second type of postselected gates is sometimes called "destructive". They operate in the "coincidence basis", where a detector must be placed at the gate output in order to demonstrate that the gate has behaved as expected. The advantage of this is that it allows low-efficiency detectors and spontaneous single-photon sources to be used to demonstrate the basic operation of a gate. Some of these gates do not even require ancillas! Naturally, incorporating a quantum gate in a scalable system requires one to know that it has successfully operated *without* destroying its output [67]. The inherent drawback of destructive gates could in principle be overcome, if highly efficient, photon-number discriminating detectors became available. It has also been proposed to employ non-destructive ("quantum non-demolition", QND) photon detectors instead [41, 61]. At this writing, however, quantum gates operating in the coincidence basis are used basically for demonstration purposes [25, 53, 55, 58, 59, 60, 65, 70] only, even if their practical use under certain circumstances, particularly at the final stages of *quantum circuits*, could be conceived. (A sequence of state preparations, quantum gates, and measurements is called a "quantum circuit". In LOQC, the expressions "quantum circuit" and "quantum network" are used as synonyms. In the quantum communication context, on the other hand, the "network" concept typically implies the presence of several parties that are separated over somewhat larger distances.)

A direct concatenation of nondeterministic gates in a LOQC quantum network would be computationally inefficient, as the individual gate failure probabilities would accumulate exponentially. Every failed operation would have to be repeated, so that the inefficiency effectively translates into a computational time that grows exponentially with the number of gates in the circuit. At this stage, from the efficiency point of view, LOQC's is about equal to that of LOCC. A hybrid approach containing elements both from LOCC and LOQC could also be conceived [28]. The chief merit of [39] is not so much to have proposed a concrete way to construct probabilistic nonlinear gates (indeed, a CNOT and a CPHASE gate of this type have already been proposed before by Koashi *et al.* [40]), but to have shown how the success rate of these devices can in principle be improved to values arbitrarily close to 1 if appropriately linked to a "quantum teleportation" scheme. By doing so, they discovered not only the possibility to perform efficient quantum computations using (passive) linear optics components ("efficient-LOQC") but they provided, in addition, a constructive proof (i.e., a scheme) for its realization.

8.2.4 Quantum teleportation

The quantum teleportation concept was first introduced by Bennett *et al.* [4]. Their idea allows one to transfer a state from one mode to another *without* requiring a direct interaction between the two. We will briefly describe the method since, paraphrasing Englert *et al.* [14], "there is a lesson here about the wonderful things entanglement can do for you". An elementary implementation using two beam splitters and two photodetectors is sketched in Fig. 8.11(a). Let the arbitrary state $|\rangle_{0,in} = a_0|0\rangle_0 + a_1|1\rangle_0$ of the incoming mode (number 0) be transferred to what we call the outgoing mode number 2 (see Fig. 8.11(a)). Two ancilla modes (1 and 2) are required to be initially prepared in the states $|1\rangle_1$ and $|0\rangle_2$, respectively. From Eq. (8.54) we know that by applying to these ancilla modes the balanced beam splitter shown in Fig. 8.11(a), the outcome will be an entangled state, which will be denoted as

$$|t_1\rangle \equiv \frac{|0\rangle_1|1\rangle_2 + |1\rangle_1|0\rangle_2}{\sqrt{2}}.\tag{8.64}$$

By introducing a second beam splitter as shown in Fig. 8.11(a) and making use of identities (8.54), (8.55) and (8.56), the 3-mode state $|\rangle_{0,in}|t_1\rangle$ becomes

$$
\begin{aligned}
&\left.\frac{a_0}{\sqrt{2}}|0\rangle_0|0\rangle_1|1\rangle_2 \right\} &&\begin{pmatrix}\text{no photons in}\\\text{modes 0 and 1}\end{pmatrix}\\
&+\frac{a_0}{2}\left[|0\rangle_0|1\rangle_1|0\rangle_2 - |1\rangle_0|0\rangle_1|0\rangle_2\right] \\
&+\frac{a_1}{2}\left[|0\rangle_0|1\rangle_1|1\rangle_2 + |1\rangle_0|0\rangle_1|1\rangle_2\right] \Bigg\} &&\begin{pmatrix}\text{one photon in}\\\text{modes 0 and 1}\end{pmatrix}\\
&+\frac{a_1}{2}\left[|0\rangle_0|2\rangle_1|0\rangle_2 - |2\rangle_0|0\rangle_1|0\rangle_2\right] \Bigg\} &&\begin{pmatrix}\text{two photons in}\\\text{modes 0 and 1}\end{pmatrix}
\end{aligned}\tag{8.65}
$$

If two photodetectors D_0 and D_1 placed at the exit ports of this second beam splitter inform us during a particular measurement that no photon has been found by D_0 and one photon has been detected by D_1, then from (8.65) we conclude that mode number 2 will end up in the state

$$|\rangle_{2,out} = |\rangle_{0,in}.\tag{8.66}$$

According to (8.65), this will happen with a probability of $1/4$, irrespective of what the initial incoming state $|\rangle_{0,in}$ is.

(a)

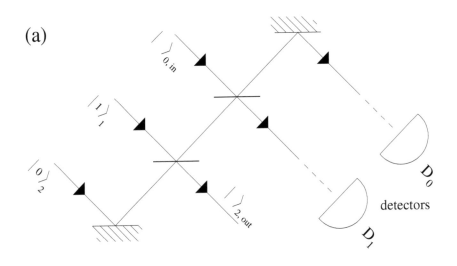

(b)

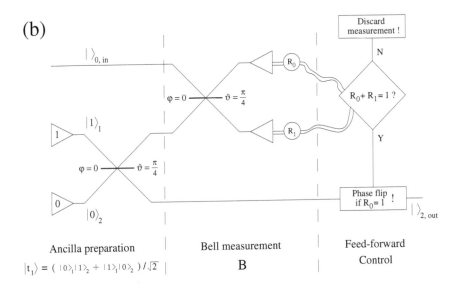

(c)

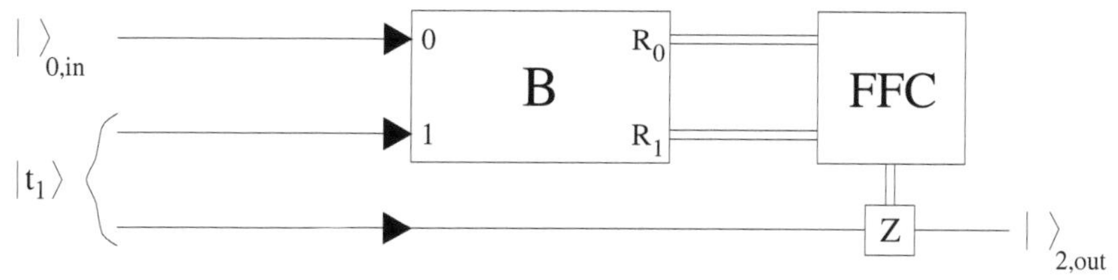

Figure 8.11: Probabilistic optical teleportation:
(a) optical circuit without feedforward-control,
(b) symbolic representation thereof including a feedforward-controlled phase-flipper, and
(c) compact block-diagram for such a partial Bell measurement (B) and feedforward-control (FFC).

The reader may look first at (8.66) and then at Fig. 8.11(a). Here we take note of the following fact. From what we have learned about a classical beam splitter in Section 8.1, it should be apparent that the light exiting the first beam splitter (port $|\rangle_{2,out}$) stems exclusively from the incoming modes 1 and 2, with no contribution from beam number 0. Yet, quantum mechanically we find that the *quantum state* of incoming mode 0 has been transferred to exiting mode 2. This is astonishing, since it means that according to the laws of quantum mechanics, it is possible to transfer information from one system to another without a direct interaction between the two (in the sense that, classically, playing around with the degrees of freedom of one system—the amplitude, phase, and polarization of incoming mode 0 in our case—does not produce any detectable effect on the second system— the amplitude, phase, and polarization of exiting mode 2 in our example). This means that quantum information *can indeed be teleported*. With the simple scheme of Fig. 8.11(a), this happens in one out of four attempts and we know from the readings of the detectors whether it has succeeded or not (according to our terminology above, this teleportation scheme thus works on the basis of post-selection).

One can easily improve the teleportation success rate by noticing that the complementary measurement result, in which one photon is detected by D_0 and none by D_1, implies that mode 2 will now be in the state $-a_0|0\rangle_2 + a_1|1\rangle_2$. Up to a phase flip (a wrong sign of the a_0 probability amplitude), this is the original state to be teleported. Such a phase error can be easily corrected by applying the σ_z Pauli operator (i.e., a phase shifter) on mode 2. With the help of a fast electronic circuit, which first verifies that all in all one photon has been detected (or, equivalently, that the total number of detected photons is odd) and then decides to send or not to send a phase-changing pulse to an actuator in path number 2 depending on which detector had signaled the photon, a (classical) "feed-forward control" mechanism can be implemented. By doing so, the teleportation success rate can effectively be doubled. The situation is depicted in Fig. 8.11(b) and symbolically compressed in Fig. 8.11(c). Note that double lines are ordinarily used to represent the flow of *classical* information (i.e., just an electronic signal), in contrast to single lines, which may transmit *quantum* information as well. An experimental demonstration of such a classical feed-forward mechanism is described in [57]. (In order to give the electronics time to do its job, in that experiment it became necessary to delay the light with a (single-mode) fiber-optic delay line. Since polarization encoding was used, the actuator in this case consisted of a Pockels cell.) We remark that from (8.65) it is also clear that the detection of no photons at all or of two photons (each of these possibilities again with probability $1/4$) in either detector cannot be used for teleportation, since the state produced in the output mode $|\rangle_{2,out}$ is in the general case (i.e., unless $a_1 = 0$) unrelated to the input state $|\rangle_{0,in}$.

Quantum teleportation is closely linked to "Bell (state) measurements". Bell states were introduced in the first section of this chapter in the context of LOCC as well as way back in Section 1.2 of Chap. 1. When talking about LOQC, the analogous states may be defined as follows:

$$|\Psi_{1,2}^-\rangle \equiv \frac{|0\rangle_1|1\rangle_2 - |1\rangle_1|0\rangle_2}{\sqrt{2}} \tag{8.67}$$

$$|\Psi_{1,2}^+\rangle \equiv \frac{|0\rangle_1|1\rangle_2 + |1\rangle_1|0\rangle_2}{\sqrt{2}} \tag{8.68}$$

$$|\Phi_{1,2}^-\rangle \equiv \frac{|0\rangle_1|0\rangle_2 - |1\rangle_1|1\rangle_2}{\sqrt{2}} \tag{8.69}$$

$$|\Phi_{1,2}^+\rangle \equiv \frac{|0\rangle_1|0\rangle_2 + |1\rangle_1|1\rangle_2}{\sqrt{2}} . \tag{8.70}$$

These quantum mechanical Bell states are mutually orthogonal and form a basis, the *Bell basis*. It is then possible to express the 3-mode state $|\rangle_{0,in}|t_1\rangle$ in terms of the Bell basis, i.e.,

$$\frac{1}{2} \left[(-a_0|0\rangle_0 - a_1|1\rangle_0) |\Psi_{1,2}^-\rangle + (-a_0|0\rangle_0 + a_1|1\rangle_0) |\Psi_{1,2}^+\rangle \right.$$
$$\left. + (\ a_0|1\rangle_0 + a_1|0\rangle_0) |\Phi_{1,2}^-\rangle + (\ a_0|1\rangle_0 - a_1|0\rangle_0) |\Phi_{1,2}^+\rangle \right] . \tag{8.71}$$

From expression (8.65) we then infer that the situation in which the detectors determined the presence of one photon (instead of zero or two) represents a projection on the Bell states $|\Psi_{1,2}^-\rangle$ (first case, no phase correction necessary) and $|\Psi_{1,2}^+\rangle$ (π phase-flip required). Due to the effective projection on Bell states, it is common to say that the measurement has been performed "in" the Bell basis (and not "in" the ordinary product basis $|0\rangle_0|0\rangle_1$, $|0\rangle_0|1\rangle_1$, $|1\rangle_0|0\rangle_1$, $|1\rangle_0|1\rangle_1$). Since the cases involving zero or two photons do *not* project the state on the remaining two Bell eigenstates, this is an example of a so-called "partial" or "incomplete" Bell measurement. It has indeed been rigorously shown [46, 80] that "complete" Bell-state measurements are inherently non-linear [32, 56] and cannot therefore be performed with certainty using linear optical devices alone. Such complete Bell measurements would be required for the purpose of determinstic (i.e., 100% efficient) quantum teleportation (in this case, the feedforward control mechanism would also be needed to eventually apply σ_x "bit flip" operations in order to swap states $|0\rangle_2$ and $|1\rangle_2$). For the time being, we must content ourselves with the partial (probabilistic) Bell measurement described above, if we want to apply teleportation in the context of LOQC. Teleportation is indeed a highly versatile tool! Among other existent LOQC-related proposals, we may quote its possible use for the implementation of a NLS gate [84], for the creation of a QND photon detection device [41], or for the construction of a heralded entangled photon-pair source [54]. But the by far most important application of quantum teleportation to LOQC was discovered by Gottesman and Chuang [23] and will be described next.

8.2.5 Application of quantum teleportation to LOQC

The essence of this is possibly best described graphically. Fig. 8.12(a) shows at first a seemingly quite uninteresting way of feeding a non-deterministic CSIGN gate with inputs coming in via teleportation. Inasmuch as the quantum gate is itself probabilistic and both teleportation steps (each with a success rate of $1/2$) together only do their job in one out of four attempts, a construction like this would be nonsensical. However, Gottesman and Chuang [23] noticed that a certain set of quantum gates commutes with the Bell measurement (teleportation) operation, except for trivial bit- and phase-flips that can be easily integrated into a modified feedforward-control scheme. Universal quantum gates are elements of this set, and Fig. 8.12(b) shows a quantum circuit which is mathematically equivalent to the one shown in Fig. 8.12(a). This can be easily verified by direct computation. For our present purposes, it is enough to look at the equivalence of the circuits shown in Figs. 12(a) and (b) as a fortuitous coincidence. Mathematical *aficionados* may alternatively like to look at it as a expression of the (not at all accidental) fact that such gates belong to the "Clifford group" that transforms Pauli gates into Pauli gates [23].

At first, one may have the impression that nothing is gained by swapping the order in which the nonlinear gate operation is performed. After all, we know by now that within the limitations of LOQC, a gate like CSIGN cannot be performed any better than probabilistically. Still there is an essential advantage in doing it "the other way around" by first applying the nonlinear gate to the pair of ancilla states: in general, constructing specific known states is easier than doing operations on unknown states. Since, if the construction of an ancilla fails, little is lost by discarding the ruined state and starting again. This option is not available when we try to perform logic gate operations on actual data [23]. If, for instance, a single (probabilistic) ancilla preparation device fails to create the required state in one out of two attempts, two such devices in parallel will fail to produce any useful ancilla in only one out of four trials. By adding more stages to

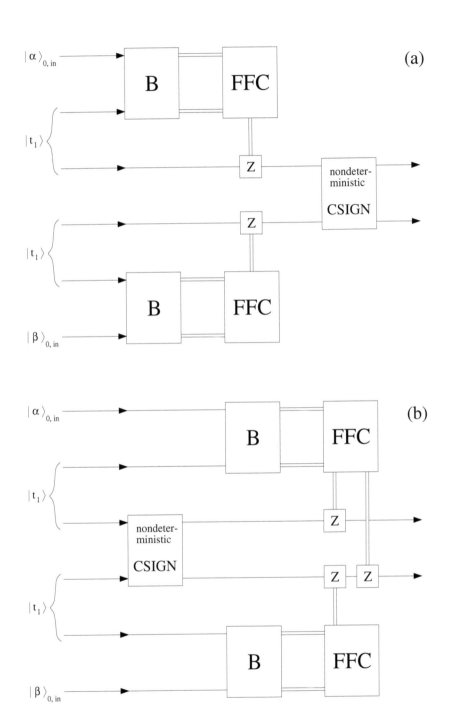

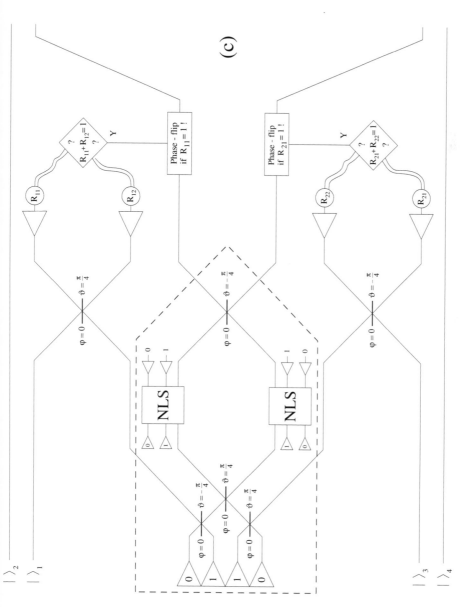

Figure 8.12: Construction schemes for a probabilistic CSIGN gate using either (a) teleported inputs, or (b) teleportation *through* the gate. A more detailed picture of the latter scheme is shown in (c).

such ancilla state "factory", the probability of a completely failed creation attempt falls off rapidly. It is the parallelism (i.e., the independence) of the individual creation processes which is the crucial point. On the contrary, due to the arbitrariness of the input (data) state, a failed gate operation cannot be undone just by discarding this one single operation, but the whole network operation will have to be repeated up to this point.

The Gottesman–Chuang technique is commonly referred to as "teleporting an incoming (unknown and arbitrary) state *through* a given quantum gate". This idea effectively allows one to decouple the problem of teleportation (and therefore, quantum gate) efficiency from the problem of how to produce the required ancilla states. Efficient on-demand ancilla state providers (highly parallel state factories that work off-line in a trial and error manner) can be conceived. It is then legitimate to measure the overall efficiency of, for instance, the CSIGN gate operation in Fig. 8.12(b) in terms of the Bell teleportation efficiency *exclusively*. For a thorough discussion of the resulting question, how the resources in such a split setup should be counted, see [39, p. 51].

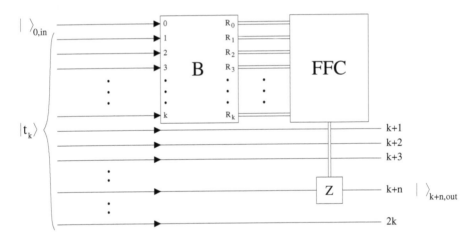

Figure 8.13: An enhanced (probabilistic) teleportation method requires additional ancilla modes, a generalized (partial) Bell measurement and a modified feedforward control algorithm. The graphic representation is a direct extension of the schematic in Fig. 8.11(c).

It was in this same reference that a novel method was proposed to boost teleportation efficiencies to values arbitrarily close to 1 by adding more ancilla, preparing them adequately, and performing a generalized (partial) Bell measurement B, as sketched in Fig. 8.13. Instead of one pair of two ancilla modes, the more efficient teleportation protocol requires two sets of k ancilla each. One set ("X register") is subsequently entangled with the state of the mode $|\rangle_{0,in} = a_0|0\rangle_0 + a_1|1\rangle_0$ to be teleported. The other one ("Y register") eventually carries the teleported quantum information. The proper generalization of $|t_1\rangle$ is

$$|t_k\rangle \equiv \frac{1}{\sqrt{k+1}} \sum_{j=0}^{k} \underbrace{|1\rangle_1 \ldots |1\rangle_j |0\rangle_{j+1} \ldots |0\rangle_k}_{\text{register } X} \underbrace{|0\rangle_{k+1} \ldots |0\rangle_{k+j} |1\rangle_{k+j+1} \ldots |1\rangle_{2k}}_{\text{register } Y} .$$

(8.72)

The so defined entangled state contains precisely k ancilla photons distributed over the two sets of k optical modes. We see that state (8.64) is but a special case of this broader definition (the X register in that case is also sometimes referred to as "half a Bell state") and that a Y register's distribution of elementary excitations is obtained by swapping ($0 \leftrightarrow 1$) the corresponding X register's contents mode by mode. The index j counts the number of photons in register X. (State $|t_k\rangle$ exists in the space of k qubits, where the nth qubit is encoded in modes $k+n$ and n, in that order. Using the qubit bases, the state $|t_k\rangle$ is given by $\sum_{j=0}^{j=k} |0\rangle_1 \ldots |0\rangle_j |1\rangle_{j+1} \ldots |1\rangle_k$.) Like $|t_1\rangle$, also state (8.72) can be prepared using a (linear size) beam splitter network applied to the initial state $|1\rangle_1 \ldots |1\rangle_k |0\rangle_{k+1} \ldots |0\rangle_{2k}$.

In the elementary teleportation scheme involving only two ancilla modes, a symmetric beam splitter is used to entangle $|\rangle_{0,in}$ with the X register. The matrix

$$F_1 \equiv \frac{1}{\sqrt{2}} \begin{bmatrix} 1 & 1 \\ 1 & -1 \end{bmatrix}$$

(8.73)

corresponding to this operation is a particular example of a *Fourier matrix* F_k, with components defined by $(F_k)_{m,n} \equiv w_k^{mn}/\sqrt{k+1}$, where $w_k \equiv \exp[i2\pi/(k+1)]$. Fourier matrices are unitary and therefore always implementable with passive linear optics. By applying F_{k+1} to modes 0 to k and then detecting the number R_j of photons at each output j (where $0 \leq j \leq k$), the information gained can be used to implement a phase-correcting feedforward mechanism that effectively (though non-deterministically) allows one to teleport state $|\rangle_{0,in}$ to mode $|\rangle_{k+n,out}$ (where $n \equiv \sum_{j=0}^{j=k} R_j$) with a success rate equal to $k/(k+1)$. In order to understand the basic trick, we introduce the mode-shifting

$$S_{k+1} \equiv \begin{bmatrix} 0 & 0 & \cdots & 0 & 1 \\ 1 & 0 & \cdots & \vdots & 0 \\ 0 & 1 & \cdots & \vdots & \vdots \\ \vdots & \vdots & \ddots & \vdots & \vdots \\ 0 & 0 & \cdots & 1 & 0 \end{bmatrix}$$

(8.74)

and phase-shifting

$$P_{k+1} \equiv \begin{bmatrix} \omega_k^0 & & & & \\ & \omega_k^1 & & \bigcirc & \\ & & \omega_k^2 & & \\ & & & \ddots & \\ & \bigcirc & & & \omega_k^k \end{bmatrix} \tag{8.75}$$

matrices and notice that

$$F_{k+1} S_{k+1} = P_{k+1} F_{k+1} . \tag{8.76}$$

The next observation is that, in case of $n = 0$, we have irreversibly lost information about the input state, since this measurement outcome necessarily implies that register Y has been projected on state $|1\rangle_{k+1} \ldots |1\rangle_{2k}$, with no reference whatsoever to coefficients a_0 and a_1. Similarly, the measurement outcome $n = k + 1$ implies a corresponding projection of the Y register on state $|0\rangle_{k+1} \ldots |0\rangle_{2k}$. In both cases, the teleportation attempt failed and must be discarded. If, on the contrary, $1 \le n \le k$, then (as long as we do not take into account the information contained in the individual measurement outcomes R_j) one can only infer that the state of the Y register after the photon detection must be in the subspace spanned by

$$|0\rangle_{k+1} \ldots |0\rangle_{k+n} |1\rangle_{k+n+1} \ldots |1\rangle_{2k}$$

and

$$|0\rangle_{k+1} \ldots |0\rangle_{k+n-1} |1\rangle_{k+n} \ldots |1\rangle_{2k} .$$

This is because, after Fourier-entangling modes 0 through k, there are always precisely two probability amplitudes that contribute to a measurement outcome with exactly n detected photons. By making use of our definition (8.74) above, expression

$$\alpha_0 F_{k+1}\Big(|0\rangle_0 |1\rangle_1 \ldots |1\rangle_n |0\rangle_{n+1} \ldots |0\rangle_k\Big)|0\rangle_{k+1} \ldots |0\rangle_{k+n} |1\rangle_{k+n+1} \ldots |1\rangle_{2k} \quad +$$

$$\alpha_1 F_{k+1}\Big(|1\rangle_0 |1\rangle_1 \ldots |1\rangle_{n-1} |0\rangle_n \ldots |0\rangle_k\Big)|0\rangle_{k+1} \ldots |0\rangle_{k+n-1} |1\rangle_{k+n} \ldots |1\rangle_{2k} \tag{8.77}$$

can be equivalently written as

$$\alpha_0 F_{k+1} S_{k+1}\Big(|1\rangle_0 \ldots |1\rangle_{n-1} |0\rangle_n \ldots |0\rangle_k\Big)|0\rangle_{k+1} \ldots |0\rangle_{k+n} |1\rangle_{k+n+1} \ldots |1\rangle_{2k} +$$

$$\alpha_1 F_{k+1}\Big(|1\rangle_0 \ldots |1\rangle_{n-1} |0\rangle_n \ldots |0\rangle_k\Big)|0\rangle_{k+1} \ldots |0\rangle_{k+n-1} |1\rangle_{k+n} \ldots |1\rangle_{2k} . \tag{8.78}$$

Now, due to the property (8.76), shifting modes $0 \ldots k$ circularly right before applying the Fourier transform F_{k+1} is equivalent to applying P_{k+1} after F_{k+1}, so that (8.78) equals

$$\alpha_0 P_{k+1} F_{k+1}\Big(|1\rangle_0 \ldots |1\rangle_{n-1} |0\rangle_n \ldots |0\rangle_k\Big)|0\rangle_{k+1} \ldots |0\rangle_{k+n} |1\rangle_{k+n+1} \ldots |1\rangle_{2k} +$$

$$\alpha_1 F_{k+1}\Big(|1\rangle_0 \ldots |1\rangle_{n-1}|0\rangle_n \ldots |0\rangle_k\Big)|0\rangle_{k+1} \ldots |0\rangle_{k+n-1}|1\rangle_{k+n} \ldots |1\rangle_{2k}. \quad (8.79)$$

And here comes the crucial point: since the vectors

$$|c\rangle \equiv F_{k+1}|1\rangle_0 \ldots |1\rangle_{n-1}|0\rangle_n \ldots |0\rangle_k$$

and $P_{k+1}|c\rangle$ only differ by phase factors (8.75) in the number state basis, they are (by construction) not linearly independent, and differ merely by a total phase

$$f \equiv \prod_{j=0}^{k} \omega_k^{jR_j}, \quad (8.80)$$

so that

$$P_{k+1}|c\rangle = f|c\rangle. \quad (8.81)$$

Thus, expression (8.79) finally equals

$$|c\rangle\Big[\alpha_0 f |0\rangle_{k+1} \ldots |0\rangle_{k+n}|1\rangle_{k+n+1} \ldots |1\rangle_{2k} + \alpha_1 |0\rangle_{k+1} \ldots |0\rangle_{k+n-1}|1\rangle_{k+n} \ldots |1\rangle_{2k}\Big]$$

$$= |c\rangle\Big[\underbrace{|0\rangle_{k+1} \ldots |0\rangle_{k+n-1}\,(fa_0|0\rangle_{k+n} + a_1|1\rangle_{k+n})\,|1\rangle_{k+n+1} \ldots |1\rangle_{2k}}_{\text{register } Y}\Big]. \quad (8.82)$$

This means that the observation of all in all n (with $1 \le n \le k$) photons by the detectors implies that the incoming state $|\rangle_{0,in}$ will have been teleported to mode number $k + n$, with the sole exception of a possibly wrong phase factor. According to (8.80), this phase can be calculated from our additional knowledge of the individual detector outcomes R_j. A feedforward control mechanism can correct this phase by means of a suitable actuator. We emphasize that both the necessary correction and which mode we teleported to are unknown until after the measurement [39].

An important and general lesson that can be learned from all this is that, before photons are destroyed in a detector, quantum information should be manipulated in such a way, that the amount of knowledge gained by the measurement process is not *too* large. As an example, the observation that the refined teleportation algorithm has an efficiency close but not equal to 1 signifies that the proposed scheme sometimes provides us with slightly too much information. (In order to get rid of even this small amount of extra information, non-linear interactions would be required.) The (coherent) removal of (eventually superfluous) quantum information before a measurement is performed is called *quantum erasure*; see Sections 3.5 and 3.6 of Chap. 3. The systematic use of quantum erasure techniques for the purpose of LOQC (and an alternative way of implementing the CNOT gate, in particular) has been proposed by Pittman, Jacobs, and Franson [56].

The next ingredient towards efficient-LOQC, introduced by Knill, Laflamme and Milburn in [39], is also the last one to be discussed in this chapter. It is a proposal for the implementation of an universal quantum gate that succeeds with a probability of $k^2/(k+1)^2$. The ancilla state appropriate for the enhanced-efficiency teleportation through a CSIGN gate, for instance, is the tensor product of two independent states of the form given by (8.72),

$$|\mathrm{cs}_k\rangle \equiv \sum_{j=0}^{k}\sum_{j'=0}^{k}(-1)^{jj'}|1\rangle_1 \ \cdots \ |1\rangle_j \ |0\rangle_{j+1} \ \cdots \ |0\rangle_{j+k}\,|1\rangle_{j+k+1} \ \cdots \ |1\rangle_{2k}$$

$$\times \ |1\rangle_{\substack{1\\+2k}} \ \cdots \ |1\rangle_{\substack{j'\\+2k}} \ |0\rangle_{\substack{j'+1\\+2k}} \ \cdots \ |0\rangle_{\substack{j'+k\\+2k}}\,|1\rangle_{\substack{j'+k+1\\+2k}} \ \cdots \ |1\rangle_{\substack{2k\\+2k}} \ , \tag{8.83}$$

aside from the factor $(-1)^{jj'}$ which entangles the two. Implementing this phase factor is a non-trivial problem. Some possible solutions have been discussed in [37] (see also the comment in [21, Ref. 7]) and [19]. It becomes evident from Eq. (8.83) that a notational simplification would at this stage be convenient in order to deal with increasingly lengthy and intricate expressions. For this purpose, it is common first to define a fixed order in which available modes are to be numbered, so that a corresponding indexing becomes unnecessary. It is then possible to introduce the compact notation

$$|j\rangle^k \equiv \underbrace{|j\rangle \ldots |j\rangle}_{k \text{ times}}, \tag{8.84}$$

where the kets on the right hand side are ordered according to the pre-established mode sequence. In terms of (8.84), state (8.83) then reads

$$|\mathrm{cs}_k\rangle = \sum_{j=0}^{k}\sum_{j'=0}^{k}(-1)^{jj'}|1\rangle^j|0\rangle^k|1\rangle^{k-j+j'}|0\rangle^k|1\rangle^{k-j'}. \tag{8.85}$$

Finding the optimal procedure for preparation of this state using only LOQC techniques is a useful and, to the best of the authors' knowledge, still open problem. (The possibility of using non-linear interactions for an intrinsically efficient creation of such states is discussed in [19].) The photonic qubit encoding introduced earlier for $|t_k\rangle$ is also adequate for $|\mathrm{cs}_k\rangle$, which can be equivalently written as

$$|\mathrm{cs}_k\rangle = \sum_{j,j'=0}^{k}(-1)^{jj'}|\emptyset\rangle^j|1\rangle^{k-j}|\emptyset\rangle^{j'}|1\rangle^{k-j'}. \tag{8.86}$$

The teleportation measurements involve the first modes of the two qubits to which CSIGN is to be applied, and modes $1\ldots k$ and $1+2k\ldots k+2k$ (left to right in order), respectively. An additional phase correction is needed after the measurement, depending on which modes the output appears in [39]. The probability that *both* teleportations succeed is $k^2/(k+1)^2$. In the literature, other quantum gates have been discussed along similar lines, also. A near-deterministic non-destructive parity measurement (or *quantum parity check*, see also [56, 57, 58]) would for instance be obtainable by applying the enhanced teleportation scheme of Fig. 8.13 to the state

$$|\mathbf{p}_k\rangle = \sum_{\substack{j,j'=0 \\ j+j' \text{ even}}}^{k} |\emptyset\rangle^j |1\rangle^{k-j} |\emptyset\rangle^{j'} |1\rangle^{k-j'} . \qquad (8.87)$$

For details, see [37].

Any realistic quantum computer will rely heavily on error-correcting codes. Recall from Section 2.9 that these are based on the encoding of a single logical qubit in the quantum states of multiple physical qubits. A simple example of such an encoding (involving j independent modes) is the transformation [59] (by majority voting)

$$\alpha|0\rangle + \beta|1\rangle \rightarrow \alpha|0\rangle^j + \beta|1\rangle^j . \qquad (8.88)$$

This kind of encoding, also known as *multiple-rail logic*, is a simple example of a *stabilizer code* (or, "quantum code") that can be used to provide a redundancy that protects a given quantum state against photon loss and other errors that occur in realistic environments. For instance, the standard LOQC qubit definition (8.29) can be viewed as an elementary example of *dual*-rail logic [12]. A *quantum encoder* is a device capable of encoding (copying) the value of a single qubit into a logic state represented by two or more photons [56, 59] (see also [20]). In the LOQC context, the inclusion of such redundancies has been considered from the very beginning for the purpose of quantum error detection and correction, but also to greatly increase the efficiency of implementing quantum information processing [37, 39]. These issues are well beyond the scope of this chapter. A quite detailed introduction into the subject from the LOQC point of view can be found in [38], where also a thorough resource analysis is given.

The efficient-LOQC scheme presented in this chapter appears exciting in principle but daunting in practice [67]. The required "over-head" resources would be enormous even under idealized conditions. If the effects caused by realistic, i.e., imperfect detectors are included in the picture, the technological requirements imposed by the method become prohibitive [21]. In its raw form, the original LOQC proposal by Knill, Laflamme and Milburn will probably never be realized and should be considered more of a theoretical proof-of-principle. A series of simplifications and improvements are possible, though, and were published soon after the seminal paper. Some have already been mentioned along this chapter, but a particularly important one, the "high-fidelity" approach [18], should also be cited. This proposal pragmatically starts from the assumption that real-world quantum computers will always require extensive quantum error-correcting systems. For this reason, those authors argue, always knowing with certainty if a quantum gate did or did not behave as expected is an unnecessary luxury, inasmuch as the built-in quantum error detection schemes will notice an eventual mistake anyway. The intuitive idea behind the "Hi-Fi" approach may be well illustrated by considering the enhanced-efficiency teleportation method based on the entangled ancilla state (8.72). Its failure rate of $1/k$ may be further improved by evenly distributing the probability amplitude for an error over all of the terms in the entangled ancilla state rather than concentrating it on just a few ($n = 0$ and $n = k + 1$). A generalization of (8.72),

$$
|t'_k\rangle \equiv \sum_{j=0}^{k} f_j \underbrace{|1\rangle_1 \ldots |1\rangle_j |0\rangle_{j+1} \ldots |0\rangle_k}_{\text{register } X} \underbrace{|0\rangle_{k+1} \ldots |0\rangle_{k+j} |1\rangle_{k+j+1} \ldots |1\rangle_{2k}}_{\text{register } Y} ,
$$

(8.89)

in which the coefficients f_j are numerically optimized so as to reduce the overall failure rate to a minimum, can be produced with the same methods discussed in [19] and is not more involved than generating state (8.72). A failure rate proportional to $1/k^2$ can be achieved this way. Similarly, a CSIGN gate with the same failure behavior can be obtained by generalizing state (8.83). This substantial improvement comes at a price, of course. Indeed, the resulting gates cease to be post-selected ones, inasmuch as we are unable to *a posteriori* tell if they have been successful or not: *all* of the logic operations are accepted after deterministic corrections have been applied. For this reason, it has been proposed that such gates be called "post-corrected" instead [18].

At present, a trend can be observed to pack more and more of the quantum circuitry that one wishes to construct into the preparation of the initial entangled ancilla state. The so-called "linked photon states" (and a "chain" state, in particular) were introduced by Yoran and Reznik [82]. Such highly entangled states can be prepared by employing post-selected gates of the type discussed so far. However, once they have been successfully created, the remaining computation process, which uses a teleportation scheme, can be efficiently completed. By doing so, the required number of elementary operations per logic gate can be dramatically reduced [82]. This approach is closely related to the more systematic proposal by Nielsen [51] to use "cluster states" [68] for the same purpose. In this spirit, Browne and Rudolph [8] have recently extended those ideas, resulting in several important relaxations of resource requirements, as well as providing an intuitively simple and clearly efficient scheme. It seems reasonable to assume that the further development of LOQC will follow closely along these new paths.

8.3 Summary and outlook

Two alternative and, in a sense, complementary approaches to pure-optical information processing using passive-linear elements have been introduced and described in this chapter. The purely classical approach is conceptually simpler and, due to the possibility to create *all* types of logic gates, experi-mentally easier to implement. On the other hand, the exponential growth of resources required to represent and manipulate classical vector bits (cebits) makes this *ansatz* inherently non-efficient from a computational point of view. The quantum approach has been shown to be potentially efficient, but the concepts and methods necessary for this efficiency are much less trivial than in the classical case and the experimental implementation becomes correspondingly much more involved. The required technical ingredients are: single photons sources, beam splitters, phase shifters, photo-detectors, and fast actuators for the controlled feedback from the photo-detector out-puts. The concept has several advantages, though, including the ability to teleport quantum vector bits (qubits) from one location to another using optical fibers, thus allowing quantum logic gates and quantum memory de-vices to be easily connected, in analogy with the wires of a conventional computer. This affords a type of modularity that is not readily available in other approches to quantum information processing. Other advantages include:

- the existence of well established experimental techniques for light ma-nipulation (and their relative simplicity: no cryogenics, no vacuum!)
- the ease of observing interference
- the long coherence times typical for photons
- the well understood sources of noise (and, in particular, the fact that they do not depend on difficult-to-predict or difficult-to-measure ther-mal interactions [39]),

etc. The linear optics quantum computational scheme that we decide to describe in this chapter is the seminal one introduced by Knill, Laflamme and Milburn [39]. Although computationally efficient by construction, it seems to be an unlikely candidate for the actual implementation of complex quantum networks, since it requires significant overhead resources and, in addition, is quite susceptible to detection losses [21]. More realistic schemes have been proposed in [82, 51, 8], but much work remains to be done. Quite generally, it is known by now that any nonlinear gate employing linear optics and photon counting *must* necessarily be nondeterministic. Besides photon counting, other processes may be employed to implement nonlinear transformations [2]. They remain to be investigated.

References

[1] L. Allen, M.W. Beijersbergen, R.J.C. Spreeuw, and J.P. Woerdman, *Phys. Rev. A* **45** (1992), 8185.

[2] S.D. Bartlett and B.C. Sanders, *Phys. Rev. Lett.* **89** (2002), 207903.

[3] A. Beige, B.-G. Englert, C. Kurtsiefer, and H. Weinfurter, *J. Phys. A* **35** (2002), L407.

[4] C.H. Bennett, G. Brassard, C. Crépeau, R. Jozsa, A. Peres, and W.K. Wootters, *Phys. Rev. Lett.* **70** (1993), 1895.

[5] E. Bernstein and U. Vazirani, *SIAM J. Comput.* **26** (1997), 1411.

[6] N. Bhattacharya, H.B. van Linden van den Heuvell, and R.J.C. Spreeuw, *Phys. Rev. Lett.* **88** (2002), 137901.

[7] E. Brainis, L.-P. Lamoureux, N.J. Cerf, Ph. Emplit, M. Haelterman, and S. Massar, *Phys. Rev. Lett.* **90** (2003), 157902.

[8] D.E. Browne and T. Rudolph, *Phys. Rev. Lett.* **95** (2005), 010501.

[9] N.J. Cerf, C. Adami, and P.G. Kwiat, *Phys. Rev. A* **57** (1998), R1477.

[10] P.T. Cochrane, G.J. Milburn, and W.J. Munro, *Phys. Rev. A* **59** (1999), 2631.

[11] D. Deutsch and R. Jozsa, *Proc. R. Soc. London, Ser. A* **439** (1992), 553.

[12] J.L. Dodd, T.C. Ralph, and G.J. Milburn, *Phys. Rev. A* **68** (2003), 042328.

[13] J. Eisert, *Phys. Rev. Lett.* **95** (2005), 040502.

[14] B.-G. Englert, C. Kurtsiefer, and H. Weinfurter, *Phys. Rev. A* **63** (2001), 032303.

[15] M. Fiorentino, P.L. Voss, J.E. Sharping, and P. Kumar, *IEEE Photonics Technol. Lett.* **14** (2002), 983.

[16] M. Fiorentino and F.N.C. Wong, *Phys. Rev. Lett.* **93** (2004), 070502.

[17] M.J. Fitch, B.C. Jacobs, T.B. Pittman, and J.D. Franson, *Phys. Rev. A* **68** (2003), 043814.

[18] J.D. Franson, M.M. Donegan, M.J. Fitch, B.C. Jacobs, and T.B. Pittman, *Phys. Rev. Lett.* **89** (2002), 137901.

[19] J.D. Franson, M.M. Donegan, and B.C. Jacobs, *Phys. Rev. A* **69** (2004), 052328.

[20] S. Gasparoni, J.-W. Pan, P. Walther, T. Rudolph, and A. Zeilinger, *Phys. Rev. Lett.* **93** (2004), 020504.

[21] S. Glancy, J.M. LoSecco, H.M. Vasconcelos, and C.E. Tanner, *Phys. Rev. A* **65** (2002), 062317.

[22] S. Glancy, H.M. Vasconcelos, and T.C. Ralph, *Phys. Rev. A* **70** (2004), 022317.

[23] D. Gottesman and I.L. Chuang, *Nature* **402** (1999), 390.

[24] L.K. Grover, *Phys. Rev. Lett.* **79** (1997), 325.

[25] H.F. Hofmann and S. Takeuchi, *Phys. Rev. A* **66** (2002), 024308.

[26] J.C. Howell and J.A. Yeazell, *Phys. Rev. A* **61** (1999), 012304.

[27] J.C. Howell and J.A. Yeazell, *Phys. Rev. A* **61** (2000), 052303.

[28] J.C. Howell and J.A. Yeazell, *Phys. Rev. Lett.* **85** (2000), 198.

[29] H. Jeong and M.S. Kim, *Phys. Rev. A* **65** (2002), 042305.

[30] H. Jeong, M.S. Kim, and J. Lee, *Phys. Rev. A* **64** (2001), 052308.

[31] Y.-H. Kim, *Phys. Rev. A* **67** (2003), 040301.

[32] Y.-H. Kim, S.P. Kulik, and Y. Shih, *Phys. Rev. Lett.* **86** (2001), 1370.

[33] J. Kim, S. Takeuchi, Y. Yamamoto, and H.H. Hogue, *Appl. Phys. Lett.* **74** (1999), 902.

[34] E. Knill, Techn. Rep. LAUR-01-5973 LANL [quant-ph/0110144] (2001).

[35] E. Knill, *Phys. Rev. A* **66** (2002), 052306.

[36] E. Knill, *Phys. Rev. A* **68** (2003), 064303.

[37] E. Knill, R. Laflamme, and G.J. Milburn, Supplementary Notes to Ref. [39].

[38] E. Knill, R. Laflamme, and G. Milburn, Techn. Rep. LAUR-00-3477 LANL [quant-ph/0006120] (2000).

[39] E. Knill, R. Laflamme, and G.J. Milburn, *Nature* **409** (2001), 46.

[40] M. Koashi, T. Yamamoto, and N. Imoto, *Phys. Rev. A* **63** (2001), 030301.

[41] P. Kok, H. Lee, and J.P. Dowling, *Phys. Rev. A* **66** (2002), 063814.

[42] P.G. Kwiat, J.R. Mitchell, P.D.D. Schwindt, and A.G. White, *J. Mod. Opt.* **47** (2000), 257.

[43] U. Leonhardt and A. Neumaier, *J. Opt. B: Quantum Semiclass. Opt.* **6** (2004), L1.

[44] P. Londero, C. Dorrer, M. Anderson, S. Wallentowitz, K. Banaszek, and I.A. Walmsley, *Phys. Rev. A* **69** (2004), 010302.

[45] A.P. Lund, H. Jeong, T.C. Ralph, and M.S. Kim, *Phys. Rev. A* **70** (2004), 020101.

[46] N. Lütkenhaus, J. Calsamiglia, and K.-A. Suominen, *Phys. Rev. A* **59** (1999), 3295.

[47] G.J. Milburn, *Phys. Rev. Lett.* **62** (1989), 2124.

[48] Y. Mitsumori, J.A. Vaccaro, S.M. Barnett, E. Andersson, A. Hasegawa, M. Takeoka, and M. Sasaki, *Phys. Rev. Lett.* **91** (2003), 217902.

[49] G. Molina-Terriza, J.P. Torres, and L. Torner, *Phys. Rev. Lett.* **88** (2002), 013601.

[50] J. Müller-Quade, H. Aagedal, T. Beth, and M. Schmidt, *Physica D* **120** (1998), 196.

[51] M.A. Nielsen, *Phys. Rev. Lett.* **93** (2004), 040503.

[52] M.A. Nielsen and I.L. Chuang, *Quantum Computing and Quantum Information* (Cambridge University Press, Cambridge, 2000).

[53] J.L. O'Brien, G.J. Pryde, A.G. White, T.C. Ralph, and D. Branning, *Nature* **426** (2003), 264.

[54] T.B. Pittman, M.M. Donegan, M.J. Fitch, B.C. Jacobs, J.D. Franson, P. Kok, H. Lee, and J.P. Dowling, *IEEE J. Sel. Top. Quantum Electron.* **9** (2003), 1478.

[55] T.B. Pittman, M.J. Fitch, B.C. Jacobs, and J.D. Franson, *Phys. Rev. A* **68** (2003), 032316.

[56] T.B. Pittman, B.C. Jacobs, and J.D. Franson, *Phys. Rev. A* **64** (2001), 062311.

[57] T.B. Pittman, B.C. Jacobs, and J.D. Franson, *Phys. Rev. A* **66** (2002), 052305.

[58] T.B. Pittman, B.C. Jacobs, and J.D. Franson, *Phys. Rev. Lett.* **88** (2002), 257902.

[59] T.B. Pittman, B.C. Jacobs, and J.D. Franson, *Phys. Rev. A* **69** (2004), 042306.

[60] T.B. Pittman, B.C. Jacobs, and J.D. Franson, *Phys. Rev. A* **71** (2005), 032307.

[61] G.J. Pryde, J.L. O'Brien, A.G. White, S.D. Bartlett, and T.C. Ralph, *Phys. Rev. Lett.* **92** (2004), 190402.

[62] G. Puentes, C. La Mela, S. Ledesma, C. Iemmi, J.P. Paz, and M. Saraceno, *Phys. Rev. A* **69** (2004), 042319.

[63] T.C. Ralph, *Phys. Rev. A* **70** (2004), 012312.

[64] T.C. Ralph, A. Gilchrist, G.J. Milburn, W.J. Munro, and S. Glancy, *Phys. Rev. A* **68** (2003), 042319.

[65] T.C. Ralph, N.K. Langford, T.B. Bell, and A.G. White, *Phys. Rev. A* **65** (2002), 062324.

[66] T.C. Ralph, W.J. Munro, and G.J. Milburn, *Proc. SPIE* **4917** (2002), 1.

[67] T.C. Ralph, A.G. White, W.J. Munro, and G.J. Milburn, *Phys. Rev. A* **65** (2001), 012314.

[68] R. Raussendorf and H.J. Briegel, *Phys. Rev. Lett.* **86** (2001), 5188.

[69] M. Reck, A. Zeilinger, H.J. Bernstein, and P. Bertani, *Phys. Rev. Lett.* **73** (1994), 58.

[70] K. Sanaka, K. Kawahara, and T. Kuga, *Phys. Rev. A* **66** (2002), 040301.

[71] S. Scheel and N. Lütkenhaus, *New J. Phys.* **6** (2004), 51.

[72] S. Scheel, K. Nemoto, W.J. Munro, P.L. Knight, *Phys. Rev. A* **68** (2003), 032310.

[73] See the special issue on Single Photons on Demand, *New J. Phys.* **6** (2004), which can be accessed on the webpage http://www.iop.org/EJ/abstract/1367-2630/6/1/E04.

[74] P.W. Shor, *Phys. Rev. A* **52** (1995), 2493.

[75] R.J.C. Spreeuw, *Found. Phys.* **28** (1997), 361.

[76] R.J.C. Spreeuw, *Phys. Rev. A* **63** (2001), 062302.

[77] S. Stenholm, *Opt. Commun.* **123** (1996), 287.

[78] S. Takeuchi, *Phys. Rev. A* **62** (2000), 032301.

[79] S. Takeuchi, J. Kim, Y. Yamamoto, and H.H. Hogue, *Appl. Phys. Lett.* **74** (1999), 1063.

[80] L. Vaidman and N. Yoran, *Phys. Rev. A* **59** (1999), 116.

[81] S.J. van Enk and O. Hirota, *Phys. Rev. A* **64** (2001), 022313.

[82] N. Yoran and B. Reznik, *Phys. Rev. Lett.* **91** (2003), 037903.

[83] H. Zbinden, H. Bechmann-Pasquinucci, N. Gisin, and G. Ribordy, *Appl. Phys. B* **67** (1998), 743.

[84] X. Zou, K. Pahlke, and W. Mathis, *Phys. Rev. A* **65** (2002), 064305.

Chapter 9

Superconducting Quantum Computing Devices

- Superconductivity history

- Components of superconducting circuits

- Quantization of superconducting circuits

- Types of qubits: charge, flux, phase and charge-flux

- Universality of quantum gates

- Measurements

Superconductivity manifests macroscopic quantum phenomena. A superconducting circuit composed of Josephson junctions, Cooper pair boxes, and rf-/dc-SQUID, properly miniaturized, becomes quantized and demonstrates Rabi oscillations and entanglement. In this chapter, we begin with an introduction of the history and elementary theory of superconductivity. Then we describe the building blocks of superconducting classical circuits and derive their canonical quantizations. The set up of qubits and superconducting quantum logic gates are then examined. Finally, ways to make measurements are discussed. We should remark that SQUID are widely regarded as the *most scalable* quantum computing device.

9.1 Introduction

Even though quantum effects are mostly observed in microscopic scales, they also manifest macroscopically. A particular case of such is *superconductivity*. Superconducting devices composed of Josephson junctions (JJ),

Cooper-pair boxes and SQUID (superconducting quantum interference devices) have been developed since the 1980s as magnetometers, gradiometers, gyroscopes, sensors, transistors, voltmeters, etc., to perform measurements on small magnetic fields, and to demonstrate the quantum effects of tunneling, resonance and coherence [6, 11, 18, 27, 30, 35]. Many industrial and medical applications have also resulted: maglev trains, superconducting power generator, cables and transformers, MRI and NMR for medical scans, to mention a few. With the advances in solid-state lithography and thin-film technology, superconducting devices have the great advantage of being easily scalable and engineering-designable. For a bulk superconductor, if its size is reduced smaller and smaller, then the quasi-continuous electron conduction band therein turns into discrete energy levels. In principle, such energy levels can be used to constitute a qubit. The first demonstration of quantum-coherent oscillations of a Josephson "charge qubit" in a superposition of eigenstates was made by Nakamura, et al. [20] in 1999. Ever since, theoretically and experimentally there has been steady progress. New proposals for qubits based on *charges, flux, phase* and *charge-flux* have been made, with observations of microwave-induced Rabi oscillations of 2-level populations in those qubit systems [8, 9, 10, 37, 38].

The organization of this chapter is made as follows: we will first introduce superconductivity in Section 9.2, the Josephson junction in Section 9.3, and the elementary superconducting circuits in Section 9.4. Superconducting quantum circuits and gates are studied in Sections 9.5 and 9.6, and conclude with measurements in Section 9.7.

9.2   Superconductivity

We begin by giving a brief historical account [13, 23, 33]. Superconductivity was discovered in 1911 by the Dutch physicist Heike Kamerlingh Onnes (1853–1926), who dedicated his career to the exploration of extremely cold refrigeration. In 1908, he successfully liquefied helium by cooling it to $-452°$ F (4 K). In 1911, he began the investigation of the electrical properties of metals in extremely cold temperatures, using liquid helium. He noticed that for solid mercury at cryogenic temperature of 4.2 K, its electric resistivity abruptly disappeared (as if there were a jump discontinuity). This is the discovery of superconductivity, and Onnes was awarded the Nobel Prize of Physics in 1913.

Subsequently, superconductivity was found in other materials. For example, lead was found to superconduct at 7 K, and (in 1941) niobium nitride was found to superconduct at 16 K.

Important understanding of superconductivity was made by Meissner and Ochsenfeld in 1933 who discovered that superconductors expelled applied magnetic fields, a phenomenon which has come to be known as the *Meissner effect*. In 1935, F. and H. London showed that the Meissner effect was a consequence of the minimization of the electromagnetic free energy carried by superconducting current. This causes the complete absence of electrical resistance and the exclusion of the interior magnetic field below some critical temperature T_c. As a consequence, the electric current density inside a superconductor must be zero. Shielding currents, which are confined on the surface of the superconducting body, are not damped and can circulate indefinitely.

In 1950, Russian scientists Ginzburg and Landau [12] developed a phenomenological theory of superconductivity which can successfully explain macroscopic properties of superconductors. From that theory, Abrikosov showed that it can predict the classification of superconductors into two types (see Type I and II superconductors below). Abrikosov and Ginzburg were awarded the Nobel prize for their contributions to superconductivity in 2003.

The theory of superconductivity was further advanced in an epoch in 1957 by three American physicists (then at the University of Illinois), J. Bardeen, L. Cooper, and J. Schrieffer, called the *BCS theory* [3]. The BCS theory explains superconductivity at temperatures close to absolute zero. Cooper theorized that atomic lattice vibrations were directly responsible for unifying and moderating the entire current. Such vibrations force the electrons to pair up into partners that enable them to pass all of the obstacles which cause resistance in the conductor. These partners of electrons are known as *Cooper pairs*. This electron coupling is viewed as an exchange of *phonons*, with phonons being the quanta of lattice vibration energy. The electron Cooper pairs are coupled over a range of hundreds of nanometers, three orders of magnitude larger than the lattice spacing. The effective net attraction between the normally repulsive electrons produces a pair binding energy on the order of milli-electron volts, enough to keep them paired at extremely low temperatures. Experimental corroboration of an interaction with the lattice was provided by the *isotope effect* on the superconducting transition temperature T_c. More on Cooper pairs in the next section.

Superconductivity phenomena have been found in metals, alloys, heavily doped semiconductor and ceramic materials at low temperatures. There are two types of superconductors, Type I and II. Thirty metals, see Table 9.1 below, together with their critical temperatures, are called Type I (or soft) superconductors. While superconductors made from alloys and ceramics of the high temperature kind are called Type II superconductors.

Among Type II, *cuprate* perovskite superconductors are certain ceramic compounds containing planes of copper and oxygen CuO_2 atoms. They can have much higher critical temperatures: $YBa_2Cu_3O_7$ (YBCO), one of the first cuprate superconductors to be discovered (by P.C.W. Chu and M.K. Wu in 1987), has a critical temperature of 93 K, and mercury-based cuprates have been found with critical temperatures in excess of 130 K. These are high temperature superconductors and so far, there is no explanation for their high critical temperatures. (The BCS theory explains superconductivity in conventional superconductors, but it does not explain supercon-

Mat.	$T_c(K)$
Be	0
Rh	0
W	0.015
Ir	0.1
Lu	0.1
Hf	0.1
Ru	0.5
Os	0.7
Mo	0.92
Zr	0.546
Cd	0.56
U	0.2
Ti	0.39
Zn	0.85
Ga	1.083

Mat.	$T_c(K)$
Al	1.2
Pa	1.4
Th	1.4
Re	1.4
Tl	2.39
In	3.408
Sn	3.722
Hg	4.153
Ta	4.47
V	5.38
La	6.00
Pb	7.193
Tc	7.77
Nb	9.46

Table 9.1: The above are thirty Type I superconductors and their critical temperature, excerpted from (14).

ductivity in the newer class of superconductors with high T_c.) Reports on materials with high T_c are constantly undergoing updating and verification. The highest figure of T_c known today is 130 K of a mercury compound, whose T_c under high pressure is 165 K.

Interested readers may find more information in superconductivity textbooks [15, 25, 26, 33], for example.

9.3   More on Cooper pairs and Josephson junctions

In the preceding section, we briefly introduced Cooper pairs. For electrons in a metal at low temperature, despite the fact that the electrons Coulomb force repel each other, the lattice of positive ions in the metal can have phonon vibration energy that mediates the coupling or pairing of eletrons to overcome the repelling force. It works as follows [23]. When one of the electrons that make up a Cooper pair and passes close to an ion in the crystal lattice, the attraction between the negative electron and the positive ion cause a vibration (i.e., phonon) to pass from ion to ion until the other electron of the pair absorbs the vibration. The net effect is that the electron has emitted a phonon and the other electron has absorbed the phonon. It is this exchange that keeps the Cooper pairs together. It is important to understand, however, that the pairs are constantly breaking and re-forming. Because electrons are indistinguishable particles, it is easier to think of

Mat.	$T_c(K)$
NbTi	10
PbMoS	14.4
V_3Ga	14.8
NbN	15.7
V_3Si	16.9
Nb_3Sn	18.0
Nb_3Al	18.7
Nb_3(AlGe)	20.7
Nb_3Ge	23.2
MgB_2	39

Mat.	$T_c(K)$
YBaCuO	93
BiSrCaCuO	110
TlBaCaCuO	125
HgBaCaCuO	135

Table 9.2: Some Type II superconductors and their critical temperature. (Data taken partly from (4).)

them as permanently paired. The composite entity, the Cooper pair, thus behaves as a single particle. These coupled electrons can take the character of a *boson* with charge twice that of an electron and zero spin. The first excited state of Cooper pairs has a minimum energy of 2Δ, where Δ is what we had referred to earlier as the *superconducting gap*. See also Δ in (9.1) and (9.2) below. Cooper pairs carry the current in a superconductor.

Now, consider two superconductors with currents. If they are kept apart and totally isolated from each other, then the phases of their wavefunctions will be independent. Bring them close together but separate by a thin non-conducting oxide barrier of tens of angströms thickness. Then Cooper pairs begin to tunnel stronger across the barrier as the separation decreases. This current is called the *Josephson current*. The "sandwich-like" arrangement is called the *Josephson junction*; see Fig. 9.1. Both were named after the British physicist B.D. Josephson (Nobel laureate in physics 1973).

The basic equations governing the dynamics of the Josephson tunneling are

$$V(t) = \frac{\hbar}{2e} \frac{\partial \phi(t)}{\partial t}, \qquad I(t) = I_c \sin(\phi(t)), \tag{9.1}$$

where $V(t)$ and $I(t)$ are, respectively, the voltage and current across the JJ, $\phi(t)$ is the phase difference of the superconductors across the JJ, and I_c, a constant, is the *critical current*. In the microscopic theory of superconductivity [33], it is known that

$$I_c = \frac{\pi\Delta}{2eR_N} \tanh \frac{\Delta}{2T}, \tag{9.2}$$

where Δ is the superconducting order parameter *energy gap*, T is the temperature, and R_N is a constant.

It follows from the equations in (9.1) that there are three major effects:

(1) The *DC Josephson effect*: This is the phenomenon of a direct current crossing the insulator in the absence of any external electromagnetic field due to tunneling. The second equation in (9.1) applies and the DC Josephson current is proportional to the sine of the phase difference across the insulator, and may take values between $-I_c$ and I_c.

(2) The *AC Josephson effect*: If the voltage U_{DC} is fixed across the junctions, the phase will vary linear with time and the current will be an AC current with amplitude Ic and frequency $\frac{2e}{h}U_{DC}$. Thus, a Josephson junction can act as a perfect *voltage-to-frequency converter*.

(3) The *inverse AC Josephson effect*: This works in a reverse way as (2) above, where for distinct DC voltages, the junction may carry a DC current and acts like a perfect *frequency-to-voltage converter*.

For example, one can apply (3) above to make the JJ a superfast voltage-switching device. JJ can perform voltage-switching functions approximately ten times faster than ordinary semiconducting circuits. This is a distinct and ideal advantage for building superfast electronic computers.

There are two general types of JJ: *overdamped* and *underdamped*, and they behave differently when $I(t) > I_c$ and $T \ll T_c$. For an overdamped junction, the junction's internal electrical resistance is small and its effect will be large. The time average voltage across the junction is defined uniquely by the going-through current [33] as

$$V = R(I^2 - I_c^2)^{1/2}.$$

The function changes smoothly from $V = 0$ when $I < I_c$ to $V = RI$ when $I \gg I_c$.

In contrast, for an underdamped junction, the barrier is an insulator. The junction's internal resistance (R in (9.21) below) will be maximum and the current-voltage curve is hysteretic near I_c.

A *superconducting quantum interference device* (SQUID) consists of two superconductors separated by thin insulating layers of JJ. SQUID are usually made of either a lead alloy (with 10% gold or indium) and/or niobium, often consisting of the tunnel barrier sandwiched between a base electrode of niobium and the top electrode of lead alloy.

There are two types of SQUID:

(1) dc-SQUID: It was invented by R. Jaklevic, J. Lambe, A. Silver, and J. Mercereau of Ford Research Labs in 1964. It consists of two JJ placed in parallel such that electrons tunneling through the junctions manifest quantum interference, depending upon the strength of the magnetic field within a loop.

(2) rf-(or ac-) SQUID: It was invented by J. E. Zimmerman and A. Silver at Ford in 1965. It is made up of one Josephson junction, which is mounted on a superconducting ring. An oscillating current is applied to an external circuit, whose voltage changes as an effect of the interaction between it and the ring. The magnetic flux can then be measured.

DC-SQUID are more difficult and expensive to fabricate, but they are much more sensitive. A SQUID can detect a change of energy as much as 100 billion times weaker than the electromagnetic energy that moves a compass needle. We will study dc- and rf-SQUID in more technical detail in the following sections.

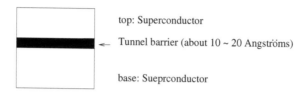

top: Superconductor

Tunnel barrier (about 10 ~ 20 Angströms)

base: Sueprconductor

Figure 9.1: Schematic of a simple Josephson junction. It has a "sandwich" structure. The base is an electrode made of a very thin niobium layer, formed by deposition. The midlayer, the tunnel barrier, is oxidized onto the niobium surface. The top layer, also an electrode, made of lead alloy (with about 10% gold or indium) is then deposited on top of the other two.

9.4 Superconducting circuits: classical

There are about a half dozen major proposals for superconducting qubits. We will introduce some of them in this section. First, *classical* superconducting circuits characterized by their *Lagrangians* will be presented. Then we advance to their quantum versions through the *canonical quantization* procedure when only a few electrons are present on such circuits. Our discussions mainly follow the tutorial paper by Wendin and Shumeiko [39].

In superconducting quantum computing applications, four basic types of circuits with JJ are commonly used as *building blocks*:

(1) single current-biased JJ;

(2) single Cooper-pair box (SCB);

(3) rf-SQUID;

(4) dc-SQUID.

We address each of them separately in the following subsection.

9.4.1 Current-biased JJ

This is the simplest superconducting circuit, consisting of a tunnel Josephson junction with superconducting electrodes connected to a current source. A schematic is given in Fig. 9.2.

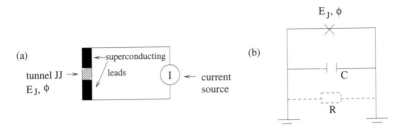

Figure 9.2: (a) A current biased Josephson junction;
(b) An equivalent lumped circuit, where × signifies the barrier of the JJ.
(Adapted from (39).)

Let $\phi(t)$ be the phase difference between the wavefunctions in the two superconductors across the junction. Let $V(t)$ denote the voltage difference across the junction. Then by the first equation in (9.1),

$$\phi(t) = \frac{2e}{\hbar} \int_{t_0}^{t} V(\tau)d\tau + \phi_0. \tag{9.3}$$

(The superconducting phase $\phi(t)$ is also related to magnetic flux $\Phi(t)$ as

$$\phi(t) = \frac{2e}{\hbar}\Phi(t) = 2\pi\frac{\Phi(t)}{\Phi_0}, \tag{9.4}$$

where $\Phi_0 = h/(2e)$ is the magnetic flux quantum.) As noted in the second equation of (9.1) in Section 9.3, the JJ current is proportional to the sine of $\phi(t)$ across the insulator:

$$I_J = I_c \sin\phi, \quad I_c \equiv \text{the critical Josephson current, (cf. (9.1)).} \tag{9.5}$$

Differentiating (9.3), we have

$$\dot{\phi}(t) = \frac{2e}{\hbar}V(t). \tag{9.6}$$

Refer to Fig. 9.2 (b). The current-voltage relations for the junction capacitance C and resistance R are given by the standard formulas

$$I_C = C\frac{dV}{dt}, \quad I_R = V/R. \tag{9.7}$$

From (9.6) and (9.7), by the Kirchhoff law of the circuit (see Fig. 9.2 (b)), we now have

$$\frac{\hbar}{2e}C\ddot{\phi} + \frac{\hbar}{2eR}\dot{\phi} + I_c \sin\phi = I_e, \tag{9.8}$$

where I_e is the *bias current*. Eq. (9.8) takes the form of a *damped forced pendulum*.

The damping term $\frac{\hbar}{2eR}\dot{\phi}$ in (9.8) determines the lifetime of the (future superconducting quantum circuit) qubit. Thus, the dissipation must be extremely small. Ideally, we assume that it is zero. So we consider an undamped Eq. (9.8):

$$\frac{\hbar}{2e}C\ddot{\phi} + I_c\sin\phi = I_e. \qquad (9.9)$$

Remark 9.4.1. It is necessary to emphasize that dropping the damping term $\hbar\dot{\phi}/(2eR)$ in (9.8) constitutes a reasonable approximation only under the following conditions of superconductivity:

(i) low temperature, i.e., T is small;

(ii) $|\dot{\phi}|$ is very small;

(iii) $T, \hbar\omega \ll \Delta$, where Δ is the energy gap in (9.2). □

For the undamped Eq. (9.9), Lagrangian and Hamiltonian variational forms can now be obtained by kinetic and potential energies:

$$\text{kinetic energy } K = K(\dot{\phi}) = \left(\frac{\hbar}{2e}\right)^2\frac{C}{2}\dot{\phi}^2, \qquad (9.10)$$

$$\text{potential energy } U = U(\phi) = \frac{\hbar}{2e}\int [I_c\sin\phi - I_e]d\phi$$

$$= \frac{\hbar}{2e}I_c(1 - \cos\phi) - \frac{\hbar}{2e}I_e\phi, \qquad (9.11)$$

where the kinetic energy is proportional to the electrostatic energy of the junction capacitor (corresponding to the first term in (9.9)), while the potential energy consists of the energy of the Josephson current and the magnetic energy of the bias current (corresponding to the last two terms in (9.9)).

For future quantum superconducting circuit applications, we introduce several useful constants. The first is the *charging energy* of the junction capacitor charged with a single Cooper pair (of electrons)

$$E_C \equiv \frac{(2e)^2}{2C}. \qquad (9.12)$$

The second,

$$E_J \equiv \frac{\hbar}{2e}I_c \qquad (9.13)$$

is called the *Josephson energy*. The third constant,

$$\omega_J \equiv \sqrt{\frac{2eI_c}{\hbar C}}, \qquad (9.14)$$

is called the *plasma frequency* of the JJ. This is the frequency of the small-amplitude oscillation of the unforced pendulum (i.e., Eq. (9.9) with $I_e = 0$). With (9.12) and (9.13), we can write (9.10) and (9.11) as

$$K = \frac{\hbar^2\dot{\phi}^2}{4E_C}, \quad U = E_J(1 - \cos\phi) - \frac{\hbar}{2e}I_e\phi.$$

Thus, we obtain the Lagrangian

$$L(\phi, \dot{\phi}) = K - U = \frac{\hbar^2 \dot{\phi}^2}{4E_C} - E_J(1 - \cos\phi) + \frac{\hbar}{2e}I_e\phi,$$

whose Lagrangian variational equation

$$\frac{d}{dt}\frac{\partial L}{\partial \dot{\phi}} - \frac{\partial L}{\partial \phi} = 0$$

is exactly (9.9).

The Hamiltonian H is related to the Lagrangian L through

$$H(p, \phi) = p\dot{\phi} - L, \quad \text{where} \quad p = \frac{\partial L}{\partial \dot{\phi}} = \frac{\hbar^2}{2E_C}\dot{\phi}, \qquad (9.15)$$

with p being the canonical momentum operator conjugate to ϕ. Then

$$H(p, \phi) = \frac{E_C}{\hbar^2}p^2 + E_J(1 - \cos\phi) - \frac{\hbar}{2e}I_e\phi, \qquad (9.16)$$

and the Hamiltonian equations of motion

$$\dot{\phi} = \frac{\partial H}{\partial p}, \quad \dot{p} = -\frac{\partial H}{\partial \phi} \qquad (9.17)$$

are again equivalent to (9.9).

9.4.2 Single Cooper-pair box (SCB)

An SCB is driven by an applied voltage Vg through capacitance Cg to induce an offset charge. The circuit consists of a small superconducting "island" connected via a Josephson tunnel junction to a large superconducting reservoir. See a schematic in Fig. 9.3.

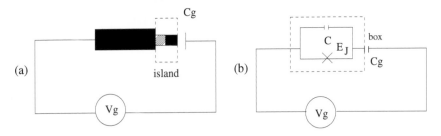

Figure 9.3: (a) A single Cooper-pair box.
(b) An equivalent lumped circuit, where × signifies the barrier of JJ.
(Adapted from (39) and (43).)

The electrostatic energy of the SCB is the sum

$$K = \frac{CV^2}{2} + \frac{Cg(Vg - V)^2}{2},$$

which, after using (9.6) and completing the square, gives

$$K = \frac{(C + Cg)}{2}\left(\frac{\hbar}{2e}\dot{\phi} - \frac{Cg}{C + Cg}Vg\right)^2 + \frac{1}{2}\left(Cg - \frac{Cg^2}{C + Cg}\right)Vg^2. \quad (9.18)$$

Dropping the (last) constant term in (9.18) and denoting $C_\Sigma \equiv C + Cg$, we have

$$K = K(\dot{\phi}) = \frac{C_\Sigma}{2}\left(\frac{\hbar}{2e}\dot{\phi} - \frac{Cg}{C_\Sigma}Vg\right)^2.$$

The potential energy U from (9.11) (by dropping the bias current I_e as it is no longer present) is

$$U = U(\phi) = E_J(1 - \cos\phi).$$

Therefore, we obtain the Lagrangian

$$L(\phi, \dot{\phi}) = \frac{C_\Sigma}{2}\left(\frac{\hbar}{2e}\dot{\phi} - \frac{Cg}{C_\Sigma}Vg\right)^2 - E_J(1 - \cos\phi). \quad (9.19)$$

The Hamiltonian, according to (9.15), is

$$H(\phi, p) = \frac{1}{2C_\Sigma}\left(\frac{2e}{\hbar}\right)^2 p^2 + E_J(1 - \cos\phi). \quad (9.20)$$

9.4.3 rf- or ac-SQUID

The rf-SQUID, also called an ac-SQUID or a magnetic-flux box, is depicted in Fig. 9.4. It is the magnetic analogue of the (electrostatic) SCB discussed in Subsection 9.4.2. It consists of a tunnel JJ inserted in a superconducting loop.

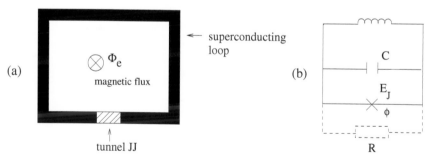

Figure 9.4: (a) An rf-SQUID.
(b) An equivalent lumped circuit.
(Adapted from (39, Fig. 8).)

Let I_L denote the current associated with the inductance L of the superconducting leads. Then by (9.4),

$$I_L = \frac{\hbar}{2eL}(\phi - \phi_e), \quad \phi_e = \frac{2e}{\hbar}\Phi_e,$$

where Φ_e is the external magnetic flux piercing the rf-SQUID loop. Using the same arguments as in Subsection 9.4.1, by the Kirchhoff circuit law we arrive at

$$\frac{\hbar}{2e}C\ddot{\phi} + \frac{\hbar}{2eR}\dot{\phi} + I_c \sin\phi + \frac{\hbar}{2eL}(\phi - \phi_e) = 0, \tag{9.21}$$

where in (9.8) the bias current I_e is replaced by $-I_L$.

If the damping is very small, then the term containing $\dot{\phi}$ can again be dropped and the Lagrangian of the rf-SQUID is

$$L(\phi, \dot{\phi}) = \frac{\hbar^2 \dot{\phi}^2}{4E_C} - E_J(1 - \cos\phi) - E_L\frac{(\phi - \phi_e)^2}{2}, \quad \left(E_L \equiv \frac{\hbar^2}{(2e)^2 L}\right). \tag{9.22}$$

The Hamiltonian is then obtained as

$$H(\phi, p) = \frac{E_C}{\hbar^2}p^2 + E_J(1 - \cos\phi) + E_L\frac{(\phi - \phi_e)^2}{2}. \tag{9.23}$$

9.4.4   dc-SQUID

A dc-SQUID consists of two JJ in parallel coupling to a current source. It has some similarity to the current-biased single junction (Fig. 9.2), except that there is an additional magnetic flux piercing the SQUID loop, which serves as a control on the effective Josephson energy of the double JJ. See Fig. 9.5 for a schematic of a dc-SQUID.

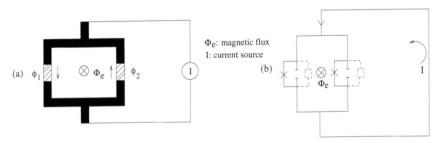

Figure 9.5: (a) Schematic of a dc-SQUID.
(b) An equivalent (nominal) lumped circuit.

Let ϕ_1 and ϕ_2 be superconducting phase differences across the JJ 1 and 2, respectively. Assume that the inductance of the SQUID loop is small so that the magnetic energy of the circulating currents can be neglected. Then the total voltage drop over the two JJ is zero:

$$V_1 + V_2 = 0.$$

From (9.6),

$$\dot{\phi}_1 + \dot{\phi}_2 = 0,$$

and, thus $\phi_1 + \phi_2$ is a constant, and

$$\phi_1 + \phi_2 = \phi_e, \tag{9.24}$$

where ϕ_e is the biasing superconducting phase related to the biasing magnetic flux. Define

$$\phi_\pm = \frac{\phi_1 \pm \phi_2}{2}.$$

Then

$$\phi_+ = \frac{\phi_1 + \phi_2}{2} = \frac{1}{2}\phi_e, \quad \phi_- = \frac{\phi_1 - \phi_2}{2},$$

which leads to

$$\phi_1 = \phi_- + \frac{\phi_e}{2}, \quad \phi_2 = \frac{\phi_e}{2} - \phi_-. \tag{9.25}$$

For the symmetric case, the two JJ have the same E_J, C and I_c. Assume that there is no dissipation, thus we can neglect the $\dot{\phi}$ term. The equation (9.24) can be rewritten using ϕ_+ as

$$2\phi_+ - \phi_e = 0.$$

The Kirchhoff circuit law requires

$$\frac{\hbar}{2e}C\ddot{\phi}_1 + I_c \sin\phi_1 - \frac{\hbar}{2e}C\ddot{\phi}_2 - I_c \sin\phi_2 - I_e = 0,$$

or

$$\frac{\hbar}{e}C\ddot{\phi}_- + 2I_c \cos\phi_+ \sin\phi_- - I_e = 0,$$

by using trigonometric identities. Thus the dynamic equation for the system for ϕ_+ and ϕ_- can be obtained as

$$\begin{cases} \dfrac{\hbar}{e}C\ddot{\phi}_- + 2I_c \cos\phi_+ \sin\phi_- - I_e = 0 \\ \dfrac{\hbar}{eL}(2\phi_+ - \phi_e) = 0. \end{cases} \tag{9.26}$$

The system has in fact only one degree of freedom since $2\phi_+ = \phi_e$. By substituting ϕ_+ with $\phi_e/2$ and comparing (9.26) with

$$\frac{d}{dt}\frac{\partial L}{\partial \dot{\phi}_-} - \frac{\partial L}{\partial \phi_-} = 0, \tag{9.27}$$

we can obtain the Lagrangian of the dc-SQUID as

$$L = \left(\frac{\hbar}{2e}\right)^2 C\dot{\phi}_-^2 + \frac{\hbar}{2e}2I_c \cos\phi_+ \cos\phi_- + \frac{\hbar}{2e}I_e\phi_-.$$

Its Hamiltonian, in turn, is

$$H = \left(\frac{\hbar}{2e}\right)^2 C\dot{\phi}_-^2 - \frac{\hbar}{2e}2I_c \cos\phi_+ \cos\phi_- - \frac{\hbar}{2e}I_e\phi_-.$$

The kinetic energy of the dc-SQUID can be obtained from the Lagrangian as

$$K(\phi_-) = \left(\frac{\hbar}{2e}\right)^2 2C\frac{\dot{\phi}_-^2}{2}. \tag{9.28}$$

It has a simple interpretation as the charging energy of the two junction capacitances (cf. Fig. 9.5 (b)) by looking at the following identity:

$$
\begin{aligned}
2C\frac{\hbar}{2e}\dot{\phi}_- &= C\frac{\hbar}{2e}(\dot{\phi}_1 - \dot{\phi}_2) \\
&= C(V_1 - V_2) \\
&= q.
\end{aligned}
$$

By setting $E_C \equiv \frac{(2e)^2}{2 \cdot 2C}$ and define E_J and E_L as before, we can rewrite the Hamiltonian as

$$H = \frac{\hbar^2}{4E_C}\dot{\phi}_-^2 - 2E_J \cos\frac{\phi_e}{2}\cos\phi_- - \frac{\hbar}{2e}I_e\phi_-,$$

and, alternatively, in terms of $p = \frac{\hbar^2}{2E_C}\dot{\phi}_-$, as

$$H = \frac{E_C}{\hbar^2}p^2 - 2E_J \cos\frac{\phi_e}{2}\cos\phi_- - \frac{\hbar}{2e}I_e\phi_-. \tag{9.29}$$

9.5 Superconducting circuits: quantum

We know that the quantization of the electromagnetic field gives a simple harmonic oscillator. A classical superconducting circuit may be viewed as an antenna. It can thus radiate electromagnetic waves. From this analogue, we see that superconducting circuits can be quantized as well, in a way analogous to Section 3.2 in Chap. 3, when the JJ becomes microscopically small, and the continuous electric current becomes discretely charged.

We now formalize the above argument by following the standard approach of *canonical quantization*. From the classical Lagrangian L, and then $p = \partial L/\partial\dot{\phi}$ we have the Hamiltonian H just as in (9.15). Now consider the simplest case of a single junction (Subsection 9.4.1, in particular Fig. 9.2). From (9.15),

$$p = \frac{\partial L}{\partial\dot{\phi}} = \frac{\hbar^2}{2E_C}\dot{\phi}, \quad \text{(cf. (9.12) for } E_C\text{)}. \tag{9.30}$$

From the first equation in (9.1),

$$V = \frac{1}{2e}\frac{h}{2\pi}\dot{\phi} = \frac{\hbar}{2e}\dot{\phi}. \tag{9.31}$$

Thus,

$$
\begin{aligned}
p &= \frac{\hbar^2}{2E_C}\dot{\phi} = \left(\frac{\hbar}{2e}\right)^2 C\dot{\phi} \\
&= \left(\frac{\hbar}{2e}\right)^2 C\left(\frac{2e}{\hbar}\right)V = \frac{\hbar}{2e}CV \\
&= \frac{\hbar}{2e}q \qquad (q = CV \text{ on the junction capacitor}) \\
&= \hbar\frac{q}{2e} = \hbar n,
\end{aligned}
\tag{9.32}
$$

where $q/(2e)$ is n, the number of Cooper pairs. Therefore, the momentum p has a simple interpretation that it is proportional to the number of Cooper pairs n on the junction capacitor. Substituting (9.32) into (9.16), we obtain the (quantum) Hamiltonian for the current-biased JJ:

$$
H = E_C n^2 - E_J \cos\phi - \frac{\hbar}{2e}I_e\phi,
\tag{9.33}
$$

where the constant E_J in (9.16) has been dropped.

For the SCB, from (9.19) we have the conjugated momentum

$$
p = \frac{\partial L}{\partial \dot{\phi}} = \frac{\hbar C_\Sigma}{2e}\left(\frac{\hbar}{2e}\dot{\phi} - \frac{C_g}{C_\Sigma}Vg\right),
\tag{9.34}
$$

and by using (9.32) and (9.34) in (9.20), we have

$$
H = \overline{E}_c(n - n_g)^2 - E_J \cos\phi,
\tag{9.35}
$$

where

$$
\overline{E}_c \equiv (2e)^2/(2C_\Sigma), \quad n_g = C_g V_g/(2e),
\tag{9.36}
$$

and n_g is the number of Cooper pairs on the gate capacitor. This n_g is tunable through different designs of Cg and Vg.

For the dc-SQUID, according to the derivations of (9.29), we obtain the Hamiltonian

$$
H = E_C n_-^2 - 2E_J \cos\frac{\phi_e}{2}\cos\phi_- - \frac{\hbar}{2e}I_e\phi_-,
\tag{9.37}
$$

where $E_C = \frac{(2e)^2}{4C}$ and $n_- = 2C\frac{\hbar}{(2e)^2}\dot{\phi}_-$.

In quantization, the classical momentum p in (9.30) becomes the *differential operator*

$$
\hat{p} = -i\hbar\frac{\partial}{\partial\phi},
\tag{9.38}
$$

where using ϕ we mean ϕ_- for the dc-SQUID. From (9.32), we thus also have the *operator of the pair number*

$$
\hat{n} = -i\frac{\partial}{\partial\phi},
\tag{9.39}
$$

and the commutator relation

$$[\phi, \hat{n}] = i. \tag{9.40}$$

The time evolution of the wave function $\psi = \psi(\phi, t)$ satisfies the Schrödinger equation

$$i\hbar \frac{\partial}{\partial t}\psi(\phi, t) = \hat{H}\psi(\phi, t) = H\left(\phi, \frac{\hbar}{i}\frac{\partial}{\partial \phi}\right)\psi(\phi, t), \tag{9.41}$$

where $H = H(\phi, p) = H(\phi, \hbar n)$ is the Hamiltonian derived in (9.33) through (9.37).

9.6 Quantum gates

We begin the discussion by using CPB as a major reference model for this section. Recall from (9.36), that the Hamiltonian for a CPB is given by

$$H = E_C(\hat{n} - n_g)^2 - E_J \cos\phi. \tag{9.42}$$

Here we assume that

$$E_C \gg E_J. \tag{9.43}$$

The pair-number operator $\hat{n}$ is defined by

$$\hat{n}|n\rangle = n|n\rangle, \qquad n = \text{an integer}, \tag{9.44}$$

where $|n\rangle$ is called the number state, an analogue of the number state $|n\rangle$ first introduced in Subsection 3.2.3 in Chap. 3. From (9.39), we see that the wave function $\psi = \psi(\phi)$ of $|n\rangle$ satisfies the differential equation

$$-i\frac{\partial}{\partial \phi}\psi = n\psi. \tag{9.45}$$

To allow only integer n in (9.45) for consideration in solving ψ, a periodic constraint must be imposed:

$$\psi(\phi + 2\pi) = \psi(\phi). \tag{9.46}$$

(Without such a constraint, the number of electrons on the island may be odd, or n could be a real value number. But here the electrode is miniaturized small enough that such cases would not happen due to the quantum effect that only a finite number of Cooper pairs can exist on the island.) Therefore, from (9.45) and (9.46), we obtain

$$\psi(\phi) = \frac{1}{\sqrt{2\pi}}e^{in\phi}, \quad \text{for} \quad n = 0, \pm 1, \pm 2, \cdots, \tag{9.47}$$

where $1/\sqrt{2\pi}$ is the normalization factor with respect to the $L^2(0, 2\pi)$-norm. From (9.42), we see that for the lowest energy eigenstate $|0\rangle$ and $|1\rangle$ of $\hat{n}$, when (9.43) holds, the states $|0\rangle$ and $|1\rangle$ are nearly degenerate when $n_g = 0.5$:

$$H|0\rangle = [E_C(0 - 0.5)^2 - E_J \cos\phi]|0\rangle \approx \frac{1}{4}E_C|0\rangle,$$
$$H|1\rangle = [E_C(1 - 0.5)^2 - E_J \cos\phi]|1\rangle \approx \frac{1}{4}E_C|1\rangle. \tag{9.48}$$

This is a favorable situation. (Normally, if two states $|0\rangle$ and $|1\rangle$ differ much in energy levels, then even though they discriminate better, the higher lying state $|1\rangle$ is *less* stable, and the system tends to decohere and lie more often in $|0\rangle$ than in $|1\rangle$, an unbalanced situation in quantum computing which is to be avoided.)

Similarly, if $n_g = n + 1/2$, then the two states $|n\rangle$ and $|n + 1\rangle$ are nearly degenerate for any integer n. For simplicity, let us just consider $n_g \approx 0.5$.

Theorem 9.6.1. *Assume that (9.43) holds, and that $n_g \approx 0.5$. Let*

$$V = span\{|0\rangle, |1\rangle\}. \tag{9.49}$$

Then the projection of the Hamiltonian H in (9.42) with respect to the ordered basis in (9.49) satisfies

$$P_H = \begin{bmatrix} E_C[\frac{1}{4} + (n_g - 0.5)] & -\frac{1}{2}E_J \\ -\frac{1}{2}E_J & E_C[\frac{1}{4} - (n_g - 0.5)] \end{bmatrix} + \mathcal{O}(|n_g - 0.5|^2). \tag{9.50}$$

Proof. The projection matrix P_H of H on V is easily evaluated as

$$P_H = \begin{bmatrix} a_0 & b \\ c & a_1 \end{bmatrix}, \tag{9.51}$$

where

$$a_j = \langle j|H|j\rangle \qquad \text{for } j = 0, 1, \tag{9.52}$$

and

$$b = \langle 0|H|1\rangle, c = \langle 1|H|0\rangle. \tag{9.53}$$

Using (9.47) for $|0\rangle$ and $|1\rangle$, we compute, e.g.,

$$a_1 = \langle 1|H|1\rangle$$
$$= \int_0^{2\pi} \left(\frac{1}{\sqrt{2\pi}}e^{-i\phi}\right)\left(E_C\left(-i\frac{\partial}{\partial\phi} - n_g\right)^2 - E_J \cos\phi\right)\left(\frac{1}{\sqrt{2\pi}}e^{i\phi}\right) d\phi$$
$$= \frac{1}{2\pi}\int_0^{2\pi} \{E_C(1 - n_g)^2 - E_J \cos\phi\}d\phi$$
$$= \frac{E_C}{2\pi} \cdot 2\pi\left[(1 - 0.5) + (0.5 - n_g)\right]^2$$
$$= E_C\left[0.5^2 + 2(0.5)(0.5 - n_g) + (0.5 - n_g)^2\right]$$
$$= E_C\left[\frac{1}{4} - (n_g - 0.5)\right] + \mathcal{O}(|n_g - 0.5|^2). \tag{9.54}$$

Similarly, the entries a_0, b and c can be computed. We obtain (9.50). $\square$

As

$$P_H = \frac{1}{4}E_C \begin{bmatrix} 1 & 0 \\ 0 & 1 \end{bmatrix} + \begin{bmatrix} E_C(n_g - 0.5) & -\frac{1}{2}E_J \\ -\frac{1}{2}E_J & -E_C(n_g - 0.5) \end{bmatrix}$$
$$+ \mathcal{O}(|n_g - 0.5|^2), \tag{9.55}$$

we can just use the effective Hamiltonian

$$\bar{P}_H = \begin{bmatrix} E_C(n_g - 0.5) & -\frac{1}{2}E_J \\ -\frac{1}{2}E_J & -E_C(n_g - 0.5) \end{bmatrix}$$
$$= E_C(n_g - 0.5)\sigma_z - \frac{1}{2}E_J\sigma_x, \tag{9.56}$$

as an approximate Hamiltonian in the subsequent discussion. The state $|0\rangle$ and $|1\rangle$ constitute a charge-qubit system. In addition, a probe gate may be coupled to the box through a junction to perform measurement, shown in Fig. 9.6.

9.6.1   One qubit operation: charge-qubit

There are various methods to manipulate the information encoded in the CPB system, and the essence is to know how to control the time-varying Hamiltonian. In the constraint linear subspace V (cf. (9.49)) spanned by number states $|0\rangle$ and $|1\rangle$, the system Hamiltonian has been obtained in Eq. (9.56). We assume that n_g is nearly equal to 0.5 and E_C is far smaller than the superconducting gap Δ. The evolution matrix of this system in time duration τ can be easily computed using results from Appendices D and E (originally developed from techniques in NMR, Chap. 10):

$$e^{-i\bar{P}_H\tau/\hbar} = e^{-i(E_C(n_g-0.5)\sigma_z - \frac{1}{2}E_J\sigma_x)\tau/\hbar}, \tag{9.57}$$

by noting that what we have is a rotation around the following axis:

$$\frac{1}{\sqrt{E_J^2/4 + E_C^2(n_g - 0.5)^2}} \left(-\frac{1}{2}E_J e_x + E_C(n_g - 0.5)e_z \right)$$

with angle $\tau\sqrt{E_J^2/4 + E_C^2(n_g - 0.5)^2}/\hbar$.

In this section, our main objective is to show that we can derive the Rabi (1-qubit) rotation gate $U_{\theta,\phi}$ in (2.46) by using the evolution matrix (9.57) with different choices of the parameter n_g and time duration τ. Note that the only tunable parameter is n_g. So we signify the dependence of $\bar{P}_H$ on n_g from (9.56) as

$$\bar{P}_H = \bar{P}_H(n_g). \tag{9.58}$$

Lemma 9.6.2. *We have the x-rotation matrix*

$$R_{x,\psi} = e^{-i\bar{P}_H(\bar{n}_g)\tau/\hbar} = \begin{bmatrix} \cos(\psi/2) & -i\sin(\psi/2) \\ -i\sin(\psi/2) & \cos(\psi/2) \end{bmatrix} \tag{9.59}$$

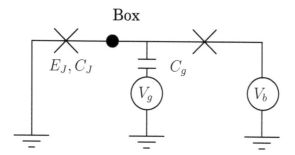

Figure 9.6: Schematic of a charge qubit constructed with a Cooper pair box. The box is denoted by a black dot and the two Josephson junctions are denoted by two crosses. The pulse gate voltage V_g can change the offset charge of the junction. The other junction is connected to a voltage V_b used for measurement, and the gate is called the probe gate.

where $\bar{n}_g = 0.5$ and $\psi = -E_J \tau / \hbar$. In particular,

$$R_{x, \pi} = \begin{bmatrix} 0 & -i \\ -i & 0 \end{bmatrix} = -i\sigma_x. \tag{9.60}$$

Proof. When $\bar{n}_g = 0.5$, we have from (9.56)

$$\bar{P}_H(\bar{n}_g) = \begin{bmatrix} 0 & -\frac{1}{2}E_J \\ -\frac{1}{2}E_J & 0 \end{bmatrix}. \tag{9.61}$$

The rest follow immediately from taking the exponential matrix $e^{-i\bar{P}_H(\bar{n}_g)\tau/\hbar}$. □

Remark 9.6.1. The operation in Lemma 9.6.2 is achieved through several steps. First, the offset charge $n_g = C_g V_g / (2e)$ as controlled by V_g is abruptly switched to the degeneration point $n_g = 0.5$, kept for duration τ, and then abruptly switched back. Time duration τ is in the order of 10^{-10} s, and the switching must be fast enough to avoid any adiabatic transition. □

Lemma 9.6.3. *Define*

$$R_{+,\theta} = e^{-i\bar{P}_H(\bar{n}_g^1)\tau/\hbar}, \quad R_{-,\phi} = e^{-i\bar{P}_H(\bar{n}_g^2)\tau/\hbar}, \tag{9.62}$$

where $\bar{n}_g^1$ and $\bar{n}_g^2$ satisfy, respectively,

$$E_C(\bar{n}_g^1 - 0.5) = -\frac{1}{2}E_J \equiv -\delta, \quad E_C(\bar{n}_g^2 - 0.5) = \frac{1}{2}E_J = \delta, \tag{9.63}$$

and

$$\theta = \phi = 2\sqrt{2}\tau\delta/\hbar. \tag{9.64}$$

Then we obtain the y-rotation and z-rotation matrices as

$$R_{y,\theta} \equiv \begin{bmatrix} \cos\frac{\theta}{2} & -\sin\frac{\theta}{2} \\ \sin\frac{\theta}{2} & \cos\frac{\theta}{2} \end{bmatrix} = -R_{-,3\pi/2}R_{+,\theta}R_{-,\pi/2}, \tag{9.65}$$

$$R_{z,\phi} \equiv \begin{bmatrix} e^{-i\phi/2} & 0 \\ 0 & e^{i\phi/2} \end{bmatrix} = -R_{x,\pi/2}R_{y,\phi}R_{x,3\pi/2}. \tag{9.66}$$

Proof. With the choice of $\bar{n}_g^1$, $\bar{n}_g^2$, δ and τ given in (9.63) and (9.64), we have

$$R_{+,\theta} = e^{i\frac{\theta}{2\sqrt{2}}(\sigma_x + \sigma_z)}, \quad R_{-,\phi} = e^{i\frac{\phi}{2\sqrt{2}}(\sigma_x - \sigma_z)}. \tag{9.67}$$

Note that $R_{+,\theta}$ and $R_{-,\theta}$ are rotations with respect to axes $-\frac{1}{\sqrt{2}}(e_x + e_z)$, $-\frac{1}{\sqrt{2}}(-e_x + e_z)$, respectively. According to the properties in Appendix D, we have

$$\begin{aligned} R_{-,\pi/2}R_{+,-\theta}R_{-,3\pi/2} &= -R_{y,\theta} \\ &= -\begin{bmatrix} \cos\frac{\theta}{2} & -\sin\frac{\theta}{2} \\ \sin\frac{\theta}{2} & \cos\frac{\theta}{2} \end{bmatrix}, \\ R_{x,\pi/2}R_{y,\phi}R_{x,3\pi/2} &= -R_{z,\phi} \\ &= -\begin{bmatrix} e^{-i\phi/2} & 0 \\ 0 & e^{i\phi/2} \end{bmatrix}. \end{aligned} \tag{9.68}$$

The negative sign comes from the fact that $R_{n,2\pi} = -I_2$ for any unit vector n. $\qquad\square$

Corollary 9.6.4. *We have the Rabi rotation gate*

$$\begin{aligned} U_{\theta/2,\alpha} &= e^{-i\frac{\theta}{2}(\cos\alpha\sigma_x + \sin\alpha\sigma_y)} \\ &= -R_{x,\pi/2}R_{y,-\alpha}R_{x,\theta}R_{y,\alpha}R_{x,3\pi/2}, \end{aligned} \tag{9.69}$$

through the cascading of quantum operations $e^{-i\bar{P}_H(n_g)\tau/\hbar}$ by tuning the parameter n_g and time duration τ. $\qquad\square$

Next, we construct the Rabi rotation gate $U_{\theta,\phi}$ in an alternative approach which is perhaps easier to implement. From (9.56), if we let the voltage V_g be oscillating (called a phase gate [20]) such that

$$E_C(n_g - 0.5) = \epsilon\cos(\omega t + \alpha), \tag{9.70}$$

where ϵ is the amplitude, then (9.56) gives (an approximate Hamiltonian)

$$H = \epsilon\cos(\omega t + \alpha)\sigma_z - \frac{1}{2}E_J\sigma_x. \tag{9.71}$$

The above Hamiltonian is with reference to the ordered basis $\{|0\rangle, |1\rangle\}$. Now define a new basis

$$|\uparrow\rangle \equiv \frac{1}{\sqrt{2}}(|0\rangle + |1\rangle), \quad |\downarrow\rangle \equiv \frac{1}{\sqrt{2}}(|0\rangle - |1\rangle). \tag{9.72}$$

Then, with respect to the above ordered basis, the Hamiltonian (9.6.1) becomes

$$\tilde{H} = \epsilon \cos(\omega t + \alpha)\sigma_x + \frac{1}{2}E_J\sigma_z, \qquad (9.73)$$

where we rename E_J to $-E_J$ just for some notational expedience.

We now utilize a procedure by transforming the system into a rotating frame. (This is actually a *standard* procedure in NMR, to be discussed in the next Chap. 10.) Namely, for the original wave function $|\chi(t)\rangle$ with Hamiltonian (9.73), let

$$|\psi(t)\rangle = e^{i\omega t\sigma_z/2}|\chi(t)\rangle. \qquad (9.74)$$

The $|\psi(t)\rangle$ satisfies the Schrödinger equation:

$$i\hbar\frac{d}{dt}|\psi(t)\rangle = (-\frac{\hbar\omega}{2}\sigma_z + e^{i\omega t\sigma_z/2}\tilde{H}e^{-i\omega t\sigma_z/2})|\psi(t)\rangle, \qquad (9.75)$$

which, by using

$$\begin{aligned}e^{i\omega t\sigma_z/2}\sigma_z e^{-i\omega t\sigma_z/2} &= \sigma_z, \\ e^{i\omega t\sigma_z/2}\sigma_x e^{-i\omega t\sigma_z/2} &= \sigma_x\cos(\omega t) - \sigma_y\sin(\omega t),\end{aligned} \qquad (9.76)$$

gives

$$\begin{aligned}i\hbar\frac{d}{dt}|\psi(t)\rangle = & (\epsilon(\sigma_x\cos(\omega t) - \sigma_y\sin(\omega t))\cos(\omega t + \alpha) \\ & +(\frac{1}{2}E_J - \frac{1}{2}\hbar\omega)\sigma_z)|\psi(t)\rangle.\end{aligned} \qquad (9.77)$$

We choose $\hbar\omega = E_J$, the *resonance* case, and obtain

$$\begin{aligned}i\hbar\frac{d}{dt}|\psi(t)\rangle &= (\sigma_x\epsilon\cos(\omega t)\cos(\omega t + \alpha) - \sigma_y\epsilon\sin(\omega t)\cos(\omega t + \alpha))|\psi(t)\rangle \\ &= (\sigma_x(\frac{\epsilon}{2}(\cos(2\omega t + \alpha) + \cos\alpha)) \\ &\quad - \sigma_y(\frac{\epsilon}{2}(\sin(2\omega t + \alpha) - \sin\alpha)))|\psi(t)\rangle.\end{aligned} \qquad (9.78)$$

We now invoke the *rotating-wave approximation* by dropping the high frequency terms $\cos(2\omega t + \alpha)$ and $\sin(2\omega t + \alpha)$. (Recall from Subsection 3.1.2 in Chap. 3 that these represent high frequency oscillations which either can not be observed in laboratory conditions or contribute little to the measurement data.) Then the Schrödinger equation (9.6.1) is further simplified to

$$i\hbar\frac{d}{dt}|\psi(t)\rangle = \frac{\epsilon}{2}(\cos\alpha\sigma_x + \sin\alpha\sigma_y)|\psi(t)\rangle,$$

whose evolution matrix is

$$\begin{aligned}U_{\theta/2,\alpha} &= e^{-i\frac{\epsilon t}{2\hbar}(\cos\alpha\sigma_x + \sin\alpha\sigma_y)} \\ &= \begin{bmatrix} \cos(\frac{\theta}{2}) & -i\sin(\frac{\theta}{2})e^{-i\alpha} \\ -i\sin(\frac{\theta}{2})e^{i\alpha} & \cos(\frac{\theta}{2}) \end{bmatrix},\end{aligned} \qquad (9.79)$$

where $\theta = \epsilon t/\hbar$. This is a Rabi rotation with respect to the ordered basis $\{|\uparrow\rangle, |\downarrow\rangle\}$. A Rabi rotation with respect to the ordered basis $\{|0\rangle, |1\rangle\}$ can be obtained by using a similarity transformation using the Walsh-Hadamard gate.

The density matrix of the system, according to the Boltzmann distribution, is given by

$$e^{\frac{-H}{k_B T}},$$

where $k_B = 1.381 \times 10^{-23}$ J/K is the Boltzmann constant and T is the absolute temperature. When $k_B T \ll E_C$, and $n_g \neq 0.5$, the Coulomb energy dominates the Hamiltonian and the system is initialized to its ground state, and this initializes the system.

9.6.2   Flux-qubit, charge-flux qubit and phase qubit

In this subsection, we briefly describe three other ways of setting up qubits in a superconducting circuit.

In an rf-SQUID, the magnetic flux Φ through the loop is quantized and must satisfy

$$(\Phi_0/2\pi)\phi + \Phi_{ext} + \Phi_{ind} = m\Phi_0, \tag{9.80}$$

where $\Phi_0 = 2.07 \times 10^{-15} Wb$; as before, m is an integer, Φ_{ext} is the external magnetic field and Φ_{ind} is induced by a current through the loop as in Fig. 9.7. That surface current through the loop is induced to compensate Φ_{ext} and its direction can be either clockwise or counterclockwise. If we denote the two surface current states as $|\uparrow\rangle$ and $|\downarrow\rangle$, they form a basis and the qubit is called a *flux qubit*. The main references for this qubit setup are [28, 36, 44]. When Φ_{ext} is near one half of Φ_0, the current can be either clockwise or counterclockwise and the system behaves like a Cooper pair box when n_g is near 0.5. Recall the Hamiltonian of an rf-SQUID in (9.23). When the self-inductance L is large enough such that $\beta_0 = E_J 4\pi^2 L/\Phi_0^2 = E_J/E_L > 1$ and Φ_{ext} is near $\Phi_0/2$ (this means $\phi_{ext} = 2\pi\Phi_{ext}/\Phi_0$ is near π), the Hamiltonian has a shape of a double-well near $\Phi = \Phi_0/2$ ($\phi = 2\pi\Phi/\Phi_0 = \pi$); see Fig. 9.8. The two lowest states at the bottom of each well are well separated from other excited levels in low temperature and suitable for quantum computation. When $\Phi_{ext} = \Phi_0/2$, the two states are degenerate and they are maximum superpositions of $|\uparrow\rangle$ and $|\downarrow\rangle$ [44]. When Φ_{ext} is away from $\Phi_0/2$, they approach $|\uparrow\rangle$ and $|\downarrow\rangle$. The Hamiltonian of this two level system has a simplified form as

$$H = -\frac{1}{2}B_z\sigma_z - \frac{1}{2}B_x\sigma_x,$$

where B_z can be tuned by Φ_{ext} and B_x is a function of E_J which is also tunable if the junction is replaced by a dc-SQUID. Thus, any 1-qubit operation can be realized through combinations of different choices of Φ_{ext} and E_J. When $\Phi_{ext} = \Phi_0/2$, $B_z = 0$.

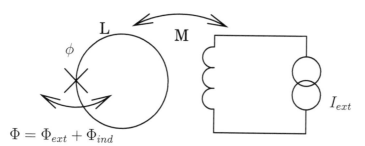

$$\Phi = \Phi_{ext} + \Phi_{ind}$$

Figure 9.7: Schematic of a flux qubit constructed with a Josephson junction in a loop.

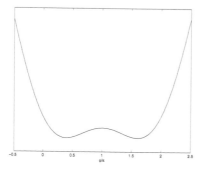

Figure 9.8: The double-well shape potential of a flux qubit with Hamiltonian (9.23). We take $\Phi_{ext} = \Phi_0/2$ and plot the potential curve near $\phi = \pi$.

A shortcoming of the simple rf-SQUID design is that its size is large in order to obtain high self-inductance and that makes it very susceptible to external noise. A better design uses more junctions in the loop and makes the size smaller. A three junction flux-qubit is shown in Fig. 9.9. Two of the junctions are designed to be the same while the third is different. The quantum constraint (9.80) applies and $\phi_1 + \phi_2 + \phi_3 = \phi_{ext} = 2\pi\Phi_{ext}/\Phi_0$. By neglecting the magnetic energy and the Coulomb term, we obtain the potential of the Hamiltonian which is dominated by the Josephson terms:

$$U(\phi_1, \phi_2) = -E_J \cos\phi_1 - E_J \cos\phi_2 - \alpha E_J \cos(\phi_{ext} - \phi_1 - \phi_2).$$

A similar 2-well potential appears in the 2-D plane of ϕ_1 and ϕ_2 when $\alpha > 0.5$. The setup has been used to observe the transitions between the two states when irradiated by an rf-photon field, and demonstrate superpositions of $|\uparrow\rangle$ and $|\downarrow\rangle$ in spectroscopic experiments.

For a flux qubit, E_J is much larger than E_C. When E_J is almost equal to E_C, both the Coulomb and JJ terms are important, and the qubit is called the *charge-flux qubit* [7, 34]. Neither ϕ nor n is a good quantum number and the lowest energy states are superpositions of several charge states. A typical design is shown in Fig. 9.10, which is developed from that of a Cooper pair box with a dc-SQUID. A larger junction is inserted in the loop for measurement, which is shunted by capacitors to reduce phase fluctuations. An external flux Φ_{ext} is also imposed as in the dc-SQUID case. Normally, the qubit works near $n_g = 1/2$, and the two lowest eigenstates are superpositions of number states $|0\rangle$ and $|1\rangle$. Denoted by $|+\rangle$ and $|-\rangle$, the two states have an energy difference E_J and the system Hamiltonian can be written as $H = \frac{1}{2}E_J\sigma_z$ when $n_g = 0.5$ exactly. Control signal with resonant frequency can be applied on the gate to manipulate the system. After putting the system in a "rotating frame" as before, the system Hamiltonian changes to

$$H = h\nu(\sigma_x \cos\alpha + \sigma_y \sin\alpha),$$

when the control signal is $\Delta n_g \cos(\omega t + \alpha)$, while $\nu = 2E_C\Delta n_g\langle +|\hat{n}|-\rangle/h$. The system behaves like an NMR spin (see Chap. 10) and all technologies, such as composite pulses can be used to increase the accuracy and robustness of the operation [7]. Charge-flux qubit shows better decoherence than charge- or flux- qubit in experiments.

Readout of the charge-flux qubit is realized through the current in the loop instead of the charge on the island. When a biased current I_b slightly below the critical current I_c of the large junction is applied, the large junction is switched into a finite voltage state depending on the qubit state. In theory, the measurement efficiency $p_+ - p_- = 0.95$ holds, where p_i is the probability to obtain a voltage in the read out when the qubit is in state $|i\rangle$.

Lastly, we address the *phase-qubit* setup, which is a current-biased Josephson junction. Its special feature is that the junction energy E_J is much larger than the Coulomb energy E_C. See Fig. 9.2. Here, our references are

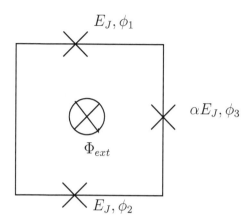

Figure 9.9: A three junction superconducting loop serving as a flux qubit. Compared with the simple design of rf-SQUID in Fig. 9.7, it has a smaller size and better coherence performance.

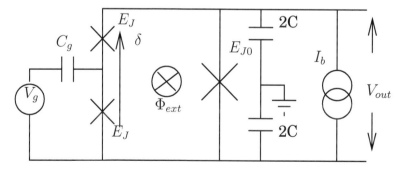

Figure 9.10: Circuit of a charge-flux qubit, with two small junctions and one large junction in a loop. An external flux Φ_{ext} penetrates the loop and a voltage V_g is applied through capacitor C_g to control the bias charge n_g. A bias current I_b is used for measurement.

[19, 31, 45]. For such, the Coulomb term is neglected, so its Hamiltonian can be obtained from (9.33) as

$$H = -E_J \cos\phi - \frac{\hbar}{2e} I_e \phi,$$

and the potential is a periodic function of ϕ offset by $I_e\phi$, with shape appearing like a "washboard", see Fig. 9.11. Normally, the JJ is undamped and we choose I_e not too large so that there are a series of wells on the potential curve. In every well formed by $\cos\phi$, it is well-known that the energy is quantized and has different levels. Besides the lowest two states serving as qubit states $|0\rangle$ and $|1\rangle$, sometimes there are one or more other states in the well. The extra level or levels may be used for measurement. Transitions between $|0\rangle$ and $|1\rangle$, in the form of Rabi rotation, are realized by applying a resonant electromagnetic field with $\omega = E_{10}/\hbar$, where E_{10} is the energy difference between $|0\rangle$ and $|1\rangle$. Measurement is accomplished by inspecting the tunneling probability of states through the well. There are two methods. One is to use a microwave field resonant with E_{21} (i.e., the energy difference between $|1\rangle$ and $|2\rangle$) to pump $|1\rangle$ to the second excited state $|2\rangle$, which has a higher tunneling probability. The other is to tilt the washboard by increasing I_e so that $|1\rangle$ can tunnel through the barrier with high probability.

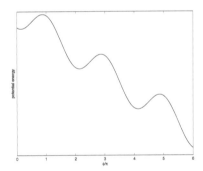

Figure 9.11: A "washboard" shape potential energy curve of a phase qubit. It is obtained by tilting the cosine function of ϕ by $-\frac{\hbar}{2e}I_e\phi$. When $I_e > \frac{2e}{\hbar}E_J$, there will be no well on the curve.

9.6.3   Two-qubit operations

Various proposals have been suggested to couple two qubits for different kinds of superconducting qubits. Capacitors, for example, can be used to couple two charge qubits. Experiments have shown 2-qubit oscillations using this scheme [24], and a conditional gate operation has also been demonstrated using the same device [40]. One disadvantage of the capacitor coupling is that it is not switchable, which makes the pulse design inflexible. It is also difficult to couple two qubits far away from each other because only the neighboring qubit coupling is convenient.

Inductance, instead, seems more promising. The simplest design is to construct a weak coupling between the qubits through the CL (capacitance-inductance) oscillation; see Fig. 9.12 [32]. But the coupling is still not switchable and thus lacks engineering flexibility. An improved design embeds a dc-SQUID into the qubit circuit with the advantage that the Josephson energy can be controlled [17]. The junction in Fig. 9.12 is replaced by a dc-SQUID. See Fig. 9.13. An external magnetic field Φ_e^i penetrates the SQUID and changes the term of the Josephson Hamiltonian to $-2E_J^0 \cos(\pi\Phi_e/\Phi_0)\cos\phi$, where the effective phase difference ϕ equals half of the difference of the two phase drops at the two junctions and $\frac{2\pi\Phi_e}{\Phi_0} = \phi_e$; cf. Section 9.4.4. This means that we replace E_J in equation (9.42) by a tunable $E_J(\Phi_e)$:

$$E_J(\Phi_e) = 2E_J^0 \cos(\pi\Phi_e/\Phi_0).$$

In this configuration, the additional effective interaction Hamiltonian induced by the oscillation in the LC-circuit can be given in the form of Pauli matrices as

$$H_{int} = -\sum_{i<j} \frac{E_J(\Phi_e^i)E_J(\Phi_e^j)}{E_L}\sigma_y^i\sigma_y^j,$$

where $E_L = [\Phi_0^2/(\pi^2 L)](C_J/C_{qb})^2$, while C_{qb} is the capacitor of the qubit defined by $C_{qb}^{-1} = C_J^{-1} + C^{-1}$.

Assume that we can still constrain every qubit in the projected subspace spanned by $|0\rangle$ and $|1\rangle$, see V in (9.49), and note that the whole Hamiltonian of the n-qubit system can be written as

$$H = \sum_{i=1}^{n}(\epsilon(V_g^i)\sigma_z^i - \frac{1}{2}E_J(\Phi_e^i)\sigma_x^i) - \sum_{i<j} \frac{E_J(\Phi_e^i)E_J(\Phi_e^j)}{E_L}\sigma_y^i\sigma_y^j, \qquad (9.81)$$

where we collect all parameters before σ_z in $\epsilon(V_g^i)$ for simplicity. If we let all

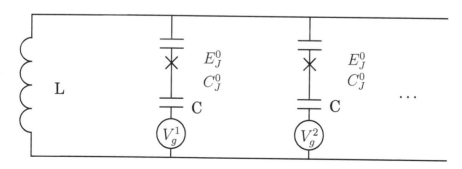

Figure 9.12: A simple design to couple charge qubits with inductance. The inductance and the effective capacitance of the charge qubits configured in parallel form a weak coupling among the qubits.

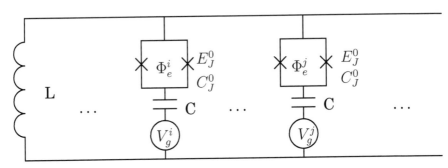

Figure 9.13: A design for coupling charge qubits with inductance where the junctions in the charge qubits are replaced by a dc-SQUID. All qubits are coupled through an inductor L, and an external field Φ_e^i penetrates every dc-SQUID. This changes the effective Josephson term in the Hamiltonian to $-2E_J^0 \cos(\pi\Phi_e/\Phi_0)\cos\phi$ and makes E_J tunable by Φ_e^i.

$\Phi_e^j = \Phi_0/2$ and $n_g^j = 0.5$ when $j \neq i$, the whole system Hamiltonian changes to

$$H = \epsilon(V_g^i)\sigma_z^i - \frac{1}{2}E_J(\Phi_e^i)\sigma_x^i,$$

and all the other terms are turned off. We can perform any single qubit operation through the approximation offered by (9.56).

Similarly, a two qubit operation between qubits i and j can be performed by turning off all $E_J^k(\Phi_e^k)$ and n_g^k except qubits i and j. By doing this, now the Hamiltonian becomes

$$H = \epsilon(V_g^i)\sigma_z^i + \epsilon(V_g^j)\sigma_z^j - \frac{1}{2}E_J(\Phi_e^i)\sigma_x^i - \frac{1}{2}E_J(\Phi_e^j)\sigma_x^j + \Pi_{ij}\sigma_y^i\sigma_y^j.$$

If we also move the two qubits to their degenerate state, i.e., $n_g^i = n_g^j = 0.5$, the Hamiltonian is simplified to

$$H = -\frac{1}{2}E_J(\Phi_e^i)\sigma_x^i - \frac{1}{2}E_J(\Phi_e^j)\sigma_x^j + \Pi_{ij}\sigma_y^i\sigma_y^j.$$

Because σ_x does not commute with σ_y, the computation of the evolution matrix is tedious and the design of the CNOT gate and conditional phase change gate is complicated. Although it provides a mechanism to realize any qubit gates in combination with one qubit gates, more simplification will be helpful.

You, et al. [41] improved this design further and obtained a simpler pulse sequence for two qubit operations. In fact, the conditional phase gate can be achieved with just one 2-qubit pulse combined with several 1-qubit operations, leading to a much more efficient scheme. The improved design has two dc-SQUID instead of one; see Fig. 9.14. Similar to the previous design, the JJ term is tunable through the magnetic field:

$$H_J^i = -E_J^i(\Phi_e^i)(\cos(\phi_A^i) + \cos(\phi_B^i)),$$

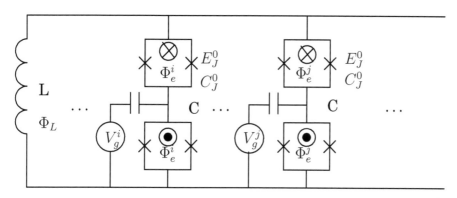

Figure 9.14: An improved design to couple charge qubits with inductance. The top and bottom magnetic fluxes piercing through each of the two SQUID are designed to have the same amplitude but different directions. Similar to the design in Fig.9.13, the JJ term is tunable through the magnetic fluxes, and the interaction term now has the form of $\sigma_x^i \sigma_x^j$, which is more preferable.

where ϕ_A^i and ϕ_B^i are the effective phase drops of the top and bottom SQUID, respectively, in Fig. 9.14. The new effective junction energy is given by

$$E_J^i(\Phi_e^i) = 2E_J^0 \cos(\pi\Phi_e^i/\Phi_0)$$

as previously. The inductance couples all qubits and the whole system Hamiltonian of n qubits now is

$$H = \sum_{k=1}^{n} H_k + \frac{1}{2}LI^2,$$

where $H_k = E_C^k(\hat{n}_k - n_{gk})^2 - E_J^k(\Phi_e^k)(\cos(\phi_A^i) + \cos(\phi_B^i))$, E_C^k is the Coulomb energy of qubit k, and I is the persistent current through the superconducting inductance. Written in Pauli matrices form, the new overall Hamiltonian is

$$H = \sum_{k=i,j} [\epsilon_k(V_g^k)\sigma_z^k - \bar{E}_J^k(\Phi_e^k, \Phi_L, L)\sigma_x^k] + \Pi_{ij}\sigma_x^i\sigma_x^j. \tag{9.82}$$

The $\sigma_x^i\sigma_x^j$ forms the interaction term which brings the advantage that it commutes with the Josephson term, and we will show later that it make the 2-qubit gate design much more straightforward and simple. Also, note that the effective junction energy $\bar{E}_J^k$ in (9.82) is not the same as the E_J in (9.81) and also depends on the inductance L and its magnetic flux Φ_L, but it is still tunable through Φ_e^k. Similarly, the interaction coefficients Π_{ij} are also functions of Φ_L, Φ_e^i and Φ_e^j. Thus all terms are *switchable*.

By setting $\Phi_e^k = \frac{1}{2}\Phi_0$ and $n_g^k = 0.5$ for all qubits, we can let all terms vanish and obtain $H = 0$. The system state will not change. If we need to perform an operation on qubit i, we change the corresponding Φ_e^i from $\frac{1}{2}\Phi_0$ and n_g^i from 0.5, and then the Hamiltonian becomes

$$H = \epsilon_i(V_g^i)\sigma_z^i - \bar{E}_J^i(\Phi_e^i, \Phi_L, L)\sigma_x^i.$$

Because both n_g^i and Φ_e^i can be tuned separately, the 1-qubit operators $e^{i\alpha\sigma_z^i}$ and $e^{i\beta\sigma_x^i}$ can be obtained easily by choosing $\bar{E}_J^i = 0$ or $\epsilon_i(V_g^i) = 0$, with an appropriate time duration. Any other 1-qubit operations can be constructed by combining these two operators.

Two-qubit operations can now be performed by tuning Φ_e^i and Φ_e^j away from $\Phi_0/2$. Then the Hamiltonian becomes

$$H = -\bar{E}_J^i\sigma_x^i - \bar{E}_J^j\sigma_x^j + \Pi_{ij}\sigma_x^i\sigma_x^j. \tag{9.83}$$

Theorem 9.6.5. *For the Hamiltonian (9.83) with tunable coefficients $\bar{E}_J^i$, $\bar{E}_J^j$ and Π_{ij}, we can construct the 2-bit quantum phase gate Q_π and the CNOT gate in conjunction with 1-bit Rabi gate $U_{\theta,\phi}$ (cf. (2.46), as warranted by Corollary 9.6.4), where*

$$Q_\pi = \begin{bmatrix} 1 & 0 & 0 & 0 \\ 0 & 1 & 0 & 0 \\ 0 & 0 & 1 & 0 \\ 0 & 0 & 0 & -1 \end{bmatrix}, \quad CNOT = \begin{bmatrix} 1 & 0 & 0 & 0 \\ 0 & 1 & 0 & 0 \\ 0 & 0 & 0 & 1 \\ 0 & 0 & 1 & 0 \end{bmatrix}. \tag{9.84}$$

Proof. We choose the control parameters such that $\bar{E}_J^i = \bar{E}_J^j = \Pi_{ij} = \delta$. Then the evolution matrix for the Hamiltonian (9.83) becomes

$$U = e^{-iH\tau/\hbar} = e^{-(i\delta\tau/\hbar)(-\sigma_x^i - \sigma_x^j + \sigma_x^i\sigma_x^j)}. \tag{9.85}$$

It is easy to check that the eigenvalue equations for H now are:

$$\begin{aligned} H|++\rangle = -\delta|++\rangle, \quad H|+-\rangle = -\delta|+-\rangle, \\ H|-+\rangle = -\delta|-+\rangle, \quad H|--\rangle = 3\delta|--\rangle, \\ (|\pm\rangle = \tfrac{1}{\sqrt{2}}(|0\rangle \pm |1\rangle)) \end{aligned} \tag{9.86}$$

By choosing $\delta\tau/\hbar = \pi/4$ in (9.85), we see that (9.85) gives the evolution matrix

$$\tilde{U} = e^{i\pi/4}\begin{bmatrix} 1 & & & \\ & 1 & & \\ & & 1 & \\ & & & -1 \end{bmatrix} \tag{9.87}$$

with respect to the ordered basis $\{|++\rangle, |+-\rangle, |-+\rangle, |--\rangle\}$. We can convert the matrix representation (9.87) to a representation with respect to the standard ordered basis $\{|00\rangle, |01\rangle, |10\rangle, |11\rangle\}$ by

$$Q_\pi = H_i^\dagger H_j^\dagger \tilde{U} H_i H_j,$$

where H_i and H_j are, respectively, the Walsh-Hadamard gate for the i-th and j-th qubit. Since the Walsh-Hadamard gate satisfies

$$H_i = H_j = \frac{1}{\sqrt{2}} \begin{bmatrix} 1 & 1 \\ 1 & -1 \end{bmatrix} = e^{-i\pi/2} R_{y,\pi/2} R_{z,\pi}, \tag{9.88}$$

we have obtained Q_π as promised.

From Q_π, we have

$$\text{CNOT} = U_{\pi/4,\pi/2}^2 \tilde{U} U_{\pi/4,-\pi/2}^2, \tag{9.89}$$

and, thus, we also have the CNOT-gate. □

Corollary 6.5 Superconducting 1-bit gates $U_{\theta,\phi}$ obtained in Corollary 9.6.4 together with 2-bit gates Q_π or U_{CNOT} obtained in Theorem 9.6.5 are universal.

Proof. This follows from Theorem 2.6.2. □

9.7 Measurement of charge qubit

The energy level of the first excited state $|1\rangle$ changes with the offset charge; when it is higher than the superconducting gap, the Cooper pair is broken apart into two quasi-particles. In Fig. 9.6, a read pulse applied on the probe gate will break the pair and let them tunnel through the junction. Repeating the experiment and measurement at frequency ν and assuming that the probability of observing the qubit at state $|1\rangle$ is P_1, we can obtain a classical current through the probe gate which is proportional to P_1:

$$I = 2eP_1\nu.$$

This measurement is *destructive*. Although state $|0\rangle$ is kept unchanged, state $|1\rangle$ is destroyed after measurement. Nakamura has used this method to observe the coherence in an SCB and quantum oscillation in two coupled charge qubits [21, 22, 24].

The above method is easy to apply, but it requires many repeated experiments and measurements. A single shot measurement requires only one measurement and would save much time. One example is realized by a group in Japan [2] using single electron transistor (SET), a sensitive electrometer, and similar setups is also investigated by other groups. See Fig. 9.15. When an appropriate pulse V_p is applied to the probe gate, such as mentioned in the preceding paragraph, the extra Cooper pair in the box is broken into two quasi-particles and tunnels into the trap. If the box is originally in state $|0\rangle$, no electron will tunnel through the junction. Then the extra charge in the trap may be detected by the SET. This completes the measurement. During normal operations, the trap junction is kept unbiased and the charge qubit is isolated from the trap and SET.

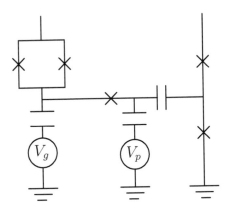

Figure 9.15: Schematic of a circuit for measuring a charge qubit using low frequency SET. The charge qubit is coupled capacitively to an SET through a charge trap which is connected to the Cooper pair box with a tunnel junction. To reduce dissipation, the junction has high resistance. The SET is in Coulomb blockade state and there is no current through the junctions when there is no charge in the trap. When a read pulse moves extra charges from the charge qubit to the trap, the SET is biased and a current is observed through the SET.

The qubit may be coupled to the SET directly through a capacitor without the trap and junction, but this may induce more decoherence to the qubit. The above low frequency SET can be replaced by an rf-SET [1, 29], a more sensitive and fast electrometer, see Fig. 9.16. Different from the low frequency SET where it is the current from the source to the drain to be measured, the rf-SET measures the conductance.

There is a worrisome aspect of measurement due to the effects of noise in hight T_c superconductors as Kish and Svedlindh [16] and others have reported excessively strong magnetic and conductance noise on such superconductors.

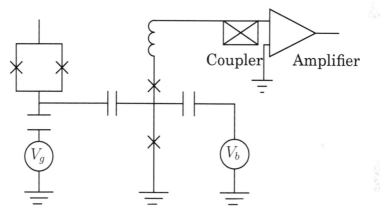

Figure 9.16: Schematic of circuit for measuring a charge qubit using rf-SET. Different from the low frequency SET where it is the current from the source to the drain to be measured, the rf-SET measures the conductance and this makes it faster and more sensitive. A radio frequency (rf) signal resonant to the SET, referred to as "carrier" but not shown in this figure , is launched toward the SET though the coupler. Then a conductance change of the SET due to the extra charge in the charge qubit results in the change of the damping of the SET circuit, and it is reflected in the output of the amplifier.

References

[1] A. Aassime, G. Johansson, G. Wendin, R.J. Schoelkopf, and P. Delsing, Radio-frequency single-electron transistor as readout device for qubits: charge sensitivity and backaction, *Phys. Rev. Lett.* **86** (2001), 3376.

[2] O. Astafiev, Y.A. Pashkin, T. Yamamoto, Y. Nakamura, and J.S. Tsai, Single-shot measurement of Josephson charge qubit, *Phys. Rev. B* **69** (2004), 180507.

[3] J. Bardeen, L.N. Cooper and J.R. Schrieffer, Theory of Superconductivity, *Phys. Rev.* **108**(1957), 1175.

[4] F.J. Blatt, *Modern Physics*, McGraw–Hill, New York, 1992.

[5] V. Bouchiat, D. Vion, P. Joyez, D. Esteve and M.H. Devoret, Quantum coherence with a single Cooper pair, *Physica Scripta* **T76** (1998), 165–170.

[6] C. Cosmelli, P. Carelli, M.G. Castellano, F. Chiarello, Palazzi G. Diambrini, R. Leoni, and G. Torrioli, *Phys. Rev. Lett.* **82** (1999), 5357.

[7] E. Collin, G. Ithier, A. Aassime, P. Joyez, D. Vion, and D. Esteve, NMR-like control of a quantum bit superconducting circuit, *Phys. Rev. Lett.* **93** (2004), 157005.

[8] M.H. Devoret and J.M. Martinis, Implementing qubits with superconducting circuits, *Quantum Information Processing* **3** (2004), in press.

[9] M.H. Devoret, A. Wallraff, and J.M. Martinis, Superconducting qubits: a short review, (2004); cond-mat/0411174.

[10] D. Esteve and D. Vion, Solid state quantum bit circuits, *Les Houches Summer School-Session LXXXI on Nanoscopic Quantum Physics*(2004), Elsevier Science, Amsterdam, 2005.

[11] J.R. Friedman, V. Patel, W. Chen, S.K. Tolpygo, and J.E. Lukens, *Nature* **406** (2000) , 43.

[12] V.L. Ginzburg and L.D. Landau, On the theory of superconductivity, *Zh. Eksperim. i. Teor. Fiz.* **20**(1950), 1064.

[13] http://en.wikipedia.org/wiki/Superconductivity.

[14] http://hyperphysics.phy-astr.gsu.edu/hbase/solids/scond.html#c4.

[15] J.B. Ketterson and S.N. Song, *Superconductivity*, Cambridge University Press, Cambridge, U.K., 1999.

[16] L.B. Kish and P. Svedlindh, Noise in hight T_c superconductors, *IEEE Trans. Electron Devices* **41**(1994), 2112–2122.

[17] Y. Makhin, G. Schön, and A. Shnirman, Josephson-junction qubits with controlled coupling, *Nature* **398** (Mar. 1999), 305–307.

[18] J.M. Martinis, M.H. Devoret, and J. Clark, *Phys. Rev. B* **35** (1987), 4682.

[19] J.M. Martinis, S. Nam, and J. Aumentado, Rabi oscillations in a large Josephson-junction qubit, *Phys. Rev. Lett.* **89**(2002), 117901.

[20] Y. Nakamura, Y.A. Pashkin, and J.S. Tsai, Coherent control of macroscopic quantum states in a single-Cooper-pair box, *Nature* **398** (1999), 786–788.

[21] Y. Nakamura, Y.A. Pashkin, and J.S. Tsai, Quantum coherence in a single-Cooper-pair box: experiments in the frequency and time domains, *Physica B* **280** (2000), 405–409.

[22] Y. Nakamura and J.S. Tsai, Quantum-state control with a single-Cooper-pair box, *J. of Low Temperature Physics* **118** (2000), 765–779.

[23] http://www.ornl.gov/info/reports/m/ornlm3063r1/pt2.html.

[24] Y. Pashkin, T. Yamamoto, O. Astaflev, Y. Nakamura, D.V. Averin, and J.S. Tsai, Quantum oscillations in two coupled charge qubits, *Nature* **421** (Feb. 2003), 823–826.

[25] C.P. Poole, H.A. Farach, and R.J. Creswick, *Superconductivity*, Academic Press, New York, 1996.

[26] A.C. Rose-Innes and E.H. Rhoderick, *Introduction to Superconductivity*, Second Edition. Pergamon Press, Oxford, 1978.

[27] R. Rouse, S. Han, and J.S. Lukens, *Phys. Rev. Lett.* **75** (1995), 514.

[28] S. Saito, M. Thorwart, H. Tanaka, M. Ueda, H. Nakano, K. Semba, and H. Takayan, Multiphoton transitions in a macroscopic quantum 2-state system, *Phys. Rev. Lett.* **93** (2004), 037001.

[29] R.J. Schoelkof, P. Wahlgren, A.A. Kozhevnikov, P. Delsing, and D.E. Prober, The radio-frequency single-electron transistor(RF-SET): a fast and ultrasensitive electrometer, *Science* **280** (May 1998), 1238–1242.

[30] P. Silvestrini, V.G. Palmieri, B. Ruggiero, and M. Russo, *Phys. Rev. Lett.* **79** (1997), 3046.

[31] R.W. Simmonds, K.M. Lang, D.A. Hite, S. Nam, D.P. Pappas, and J.M. Martinis, Decoherence in Josephson junction phase qubits from junction resonators, *Phys. Rev. Lett.* **93**(2004), 077003.

[32] A. Shnirman, G. Schön, and Z. Hermon, Quantum manipulation of small Josephson junctions, *Phys. Rev. Lett* **79** (1997), 2317.

[33] M. Tinkham, *Introduction To Superconductivity*, McGraw-Hill, Singapore, 1996.

[34] D. Vion, A. Aassime, A. Cottet, P. Joyez, H. Pothier, C. Urbina, D. Esteve, and M.H. Devoret, Manipulating the quantum state of an electrical circuit, *Science* **296** (2002), 886–889.

[35] R.F. Voss and R.A. Webb, *Phys. Rev. Lett.* **47** (1981), 265.

[36] Caspar H. van der Wal, A.C.J. ter Haar, F.K. Wilhelm, R.N. Schouten, C.J.P.M. Harmans, T.P. Orlando, Seth Lloyd, and J.E. Mooij, Quantum superposition of macroscopic persistent-current states, *Science* **290** (2000), 773–776.

[37] G. Wendin, Scalable solid state qubits: challenging decoherence and readout, *Phil. Trans. R. Soc. Lond. A* **361** (2003), 1323.

[38] G. Wendin, Superconducting quantum computing, *Physics World*, May 2003.

[39] G. Wendin and V.S. Shmeiko, Superconducting quantum circuits, qubits and computing, arXiv:cond-mat/0508729v1, Aug. 30, 2005.

[40] T. Yamamoto, Y.A. Pashkin, O. Astaflev, Y. Nakamura, and J.S. Tsai, Demonstration of conditional gate operation using superconducting charge qubits, *Nature* **425** (Oct. 2003), 941–944.

[41] J.Q. You, J.S. Tsai, and F. Nori, Scalable quantum computing with Josephson charge qubits, *Phys. Rev. Lett.* **89** (2002), 197902.

[42] J.Q. You, J.S. Tsai, and F. Nori, Controllable manipulation and entanglement of macroscopic quantum states in coupled charge qubits, *Phys. Rev. B* **68** (2003), 024510.

[43] J.Q. You and F. Nori, Quantum information processing with superconducting qubits in a microwave field, *Phys. Rev. B* **68** (2003), 064509.

[44] J.Q. You and F. Nori, Superconducting circuits and quantum information, *Physics Today* (Nov. 2005), 42–47.

[45] Y. Yu, S. Han, X. Chu, S. Chu, and Z. Wang, Coherent temporal oscillations of macroscopic quantum states in a Josephson junction, *Science* **296** (2002), 889–892.

Chapter 10

NMR Quantum Computing

- NMR history

- Qubit setup and gating operation

- The refocusing technique

- New solid state NMR technology

- Measurements

- Applications of NMR quantum computing to Shor's algorithm and a lattice-gas algorithm

In this chapter, we explain the basic ideas and principles of nuclear spins and resonance, NMR quantum gates and operations. Liquid NMR quantum computing may be regarded as the progenitor of the general quantum computing technology. Its techniques have deeply influenced the development of today's QC and, thus, in chronological order, we should have placed this chapter in front of most of the chapters on QC devices. However, the liquid NMR technology is faced with the very uncertain future of getting phased out (due to its lack of scalability). Presently, a revival of NMR quantum computing is taking place, making new progress in optically addressed solid state NMR, which is expounded. Examples of the Shor's algorithm for factorization of composite integers and quantum lattice-gas algorithm for the diffusion partial differential equation are also illustrated.

This chapter is mostly based on the work of Zhang, Chen, Diao and Hemmer [94]. We gratefully acknowledge the **copyright permission from Springer-Verlag, New York** for the reproduction of text and figures throughout this chapter.

10.1   Nuclear magnetic resonance

We first note that this area of NMR quantum computing is the most mature and established, and there already exist many papers on this topic, see, e.g., [5, 7, 8, 37, 39, 80], written by physicists and computer scientists.

10.1.1   Introduction

In NMR quantum computing, we use molecules as a small computer. The logic bits are the nuclear spins of atoms in custom designed molecules. Spin flips are achieved through the application of radio-frequency (RF) fields on resonance at the nuclear spin frequencies. The system can be initialized by cooling the system down to the ground state or known low-entropy state, or using a special technology called averaging, especially for liquid NMR working in room temperature. Measurement or readout is carried out by measuring the magnetic induction signal generated by the precessing spin on the receiver coil. Numerous experiments have been successfully tried for different algorithms, mostly using liquid NMR technology. The algorithms tested include Grover's search algorithm [27, 39, 82, 95], other generalized search algorithms [56], quantum Fourier transforms [12, 87], Shor's algorithm [83], Deutsch-Jozsa algorithm [6, 10, 11, 54, 63], order finding [75, 81], error correcting code [46], and dense coding [17]. There are also other implementations reported, such as cat-code benchmark [45], information teleportation [66] and quantum system simulation [73].

NMR is an important tool in chemistry which has been in use for the determination of molecular structure and composition of solids, liquid and gases since the mid 1940s, by research groups in Stanford and MIT independently, led by F. Bloch and E.M. Purcell, both of whom shared the Nobel prize in physics in 1952 for the discovery.

There are many excellent monographs on NMR [15, 62, 69]. There are also many other nice Internet website resources offering concise but highly useful information about NMR; cf., e.g., [13, 32, 88]. Let us briefly explain the physics of NMR by following Edwards [13]. The NMR phenomenon is based on the fact that the spin of nuclei of atoms have magnetic properties that can be utilized to yield chemical, physical, and biological information. Through the famous Stern–Gerlach experiment (cf. Chap. 1) in the earlier development of quantum mechanics, it is known that subatomic particles (protons, neutrons and electrons) have spins. Nuclei with spins behave like a bar magnet in a magnetic field. In some atoms, e.g., ^{12}C (carbon-12), ^{16}O (oxygen-16), ^{32}S (sulphur-32), these spins are paired and cancel each other out so that the nucleus of the atom has no overall spin. However, in many atoms (^1H, ^{13}C, ^{31}P, ^{15}N, ^{19}F etc.) the nucleus does possess an overall spin. To determine the spin of a given nucleus one can use the following rules:

(1) If the number of neutrons and the number of protons are both even, the nucleus has no spin.

(2) If the number of neutrons plus the number of protons is odd, then the nucleus has a half-integer spin (i.e., $1/2, 3/2, 5/2$).

(3) If the number of neutrons and the number of protons are both odd, then the nucleus has an integer spin (i.e., $1, 2, 3$).

In quantum mechanical terms, the nuclear magnetic moment of a nucleus can align with an externally applied magnetic field of strength B_0 in only $2I + 1$ ways, either with or against the applied field B_0, where I is the nuclear spin given in (1), (2) and (3) above. For example, for a single nucleus with $I = 1/2$, only one transition is possible between the two energy levels. The energetically preferred orientation has the magnetic moment aligned parallel with the applied field (spin $m = +1/2$) and is often denoted as α, whereas the higher energy anti-parallel orientation (spin $m = -1/2$) is denoted as β. See Fig. 10.1. In NMR quantum computing, these spin-up and spin-down quantum states resemble the two binary states 0 and 1 in a classical computer. Such a nuclear spin can serve as a *qubit*. The rotational axis of the spinning nucleus cannot be orientated exactly parallel (or anti-parallel) with the direction of the applied field B_0 (aligned along the z axis) but must precess (motion similar to a gyroscope) about this field at an angle, with an angular velocity, ω_0, given by the expression $\omega_0 = \gamma B_0$. The precession rate ω_0 is called the Larmor frequency; cf. Fig. 10.2. See more discussion of ω_0 below. The constant γ is called the magnetogyric ratio. This precession process generates an magnetic field with frequency ω_0. If we irradiate the sample with radio waves (MHz), then the proton can absorb the energy and be promoted to the higher energy state. This absorption is called resonance because the frequencies of the applied radiation and the precession coincide at that frequency, leading to resonance.

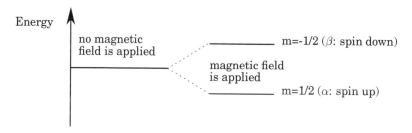

Figure 10.1: Splitting of energy levels of a nucleus with spin quantum number 1/2.

There is another technique related to NMR, called *electron spin resonance (ESR)*, that deals with the spins of electrons instead of those of the nuclei. The principles for ESR are nevertheless similar.

Quantum entanglement is accomplished through spin-spin coupling from the electronic bonds between the nuclei within the molecule and special RF pulse manipulations.

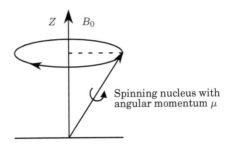

Figure 10.2: A magnetic field B_0 is applied along the z-axis, causing the spinning nucleus to precess around the applied magnetic field.

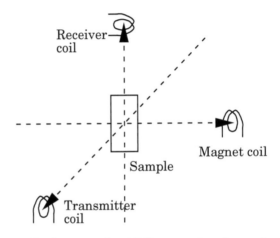

Figure 10.3: Schematic diagram of an NMR apparatus. A sample which has non-zero spin nuclei is put in a static magnetic field regulated by the current through the magnet coil. A transmitter coil provides the perpendicular field and a receiver coil picks up the signal. We can change the current through the magnet coil or change the frequency of the current in the transmitter coil to reach resonance.

10.1.2   More about the Hamiltonian of NMR

A classical way to explain NMR is to regard it as a rotating charged particle that acts like a current circulating in a loop [3, 15], which creates a magnet with magnetic moment μ, $\mu = qvr/2$, where q is the electronic charge. The particle is rotating at $v/(2\pi r)$ revolutions per second.

Converting μ to electromagnetic units by dividing it by the velocity of light, and using angular momentum of the particle rather than the velocity of the particle, we obtain

$$\boldsymbol{\mu} = (q/2Mc)\boldsymbol{p},$$

where $\boldsymbol{p}$ is the angular momentum oriented along the rotating axis. The ratio μ/p is called the magnetogyric ratio, denoted by γ. A static magnetic field with strength B will apply a torque, which is equal to $\boldsymbol{\mu} \times \boldsymbol{B}$, on this particle. Newton's law states that the angular momentum will change according to a differential equation

$$\frac{d\mathbf{p}}{dt} = \boldsymbol{\mu} \times \mathbf{B} = \frac{q}{2Mc}\mathbf{p} \times \mathbf{B}.$$

Computation shows that $\mathbf{p}$ will rotate around the direction of $\mathbf{B}$ with frequency ω_0 defined by

$$\omega_0 = \frac{q}{2Mc}B.$$

The above is called the *Larmor equation*, and the frequency ω_0 is called the Larmor frequency, the precession frequency, or the resonance frequency as mentioned previously in Fig. 10.2.

The above classical considerations are now modified by *quantization* to incorporate the quantum-mechanical behaviors of the nuclear spin. The vector variable $\mathbf{p}$ is quantized with quantum number $(I(I+1))^{1/2}$ (cf. Box 1.1 in Chap. 1), and its projection to z-axis (the direction of the magnetic field) is $m\hbar$. In total, there are $2I+1$ valid values of m evenly distributed from $-I$ to I, i.e., $m = -I, -I+1, \cdots, I-1, I$. A factor g is introduced to include both the spin and orbital motion in the total angular momentum, called the Landé or spectroscopic splitting factor. For a free electron and proton, the magnetic momenta can be given as

$$\mu_e = \frac{g_e}{2}\left(\frac{he}{4\pi M_e c}\right) = \frac{g_e \beta}{2},$$

$$\mu_n = g_n I\left(\frac{he}{4\pi M_N c}\right) = g_n I \beta_N,$$

where $g_e = 2.0023$, $g_n = 5.58490$. Numbers β and β_N are called, respectively, the Bohr and the nucleus magneton where $\beta = 9.27 \times 10^{-21}$ *erg gauss*$^{-1}$ and $\beta_N = 5.09 \times 10^{-24}$ *erg gauss*$^{-1}$. These values vary for different particles. In NMR, it is convenient to use the resonance frequency ω_0:

$$\hbar\omega_0 = g_e \beta B_0,$$
$$\hbar\omega_0 = g_N I \beta_N B_0.$$

Now we can write the Hamiltonian of a free nucleus as

$$H = -\boldsymbol{\mu} \cdot \mathbf{B} = -\hbar\gamma \mathbf{I} \cdot \mathbf{B}, \tag{10.1}$$

where γ is the magnetogyric ratio defined by $\gamma = \frac{\mu}{I\hbar}$ just as in the classical case. It is a characteristic constant for every type of nuclei; different nuclei have different magnetogyric ratios. Vector $\mathbf{I}$ after quantization, becomes the operator of angular momentum. The eigenvalues of this system, or the energy levels are

$$E = \gamma\hbar m B, \quad m = -I, -I+1, \cdots, I-1, I. \tag{10.2}$$

The difference between two neighboring energy levels is $\gamma \hbar B$, which defines the resonance frequency depending on the magnetic field B and the particle.

There are other factors to be considered. The resonance frequency changes with the chemical environment of the nucleus. An example is the fluorine resonance spectrum of perfluorioisopropyl iodide. Two resonance lines of fluorine are observed in the spectrum, and the intensities ratio 6:1 agrees with the population ratio of the two groups of fluorine atoms. This phenomenon, called the *chemical shift*, is proportional to the strength of the magnetic field applied. This effect comes up because electrons close to the nucleus change the magnetic field around it; in other words, they create a diamagnetic shielding surrounding the nucleus. If the static field applied is B_0, then the electrons precessing around the magnetic field direction produce an induced magnetic field opposing B_0. The total effective magnetic field around the nucleus is then

$$\boldsymbol{B} = \boldsymbol{B}_0 - \boldsymbol{B}' = (1 - \sigma)\boldsymbol{B}_0,$$

where the parameter σ is called *shielding coefficient*. In some cases σ is dependent on the temperature.

High resolution NMR spectroscopy has found that the chemical shifted peaks are also composed of several lines, a result of the spin-spin coupling, which is the second term in the NMR Hamiltonian:

$$H_{II} = \sum_{i>j} \boldsymbol{I_i} \cdot J_{ij} \cdot \boldsymbol{I_j};$$

see (2.31) in Chap. 2.

10.1.3 Organization of the chapter

Section 10.1 so far has introduced some basic facts of nuclear spins and atomic physics.

In Section 10.2, we will give a motivation of what quantum computing is about, and introduce universal quantum gates based on liquid NMR.

Section 10.3 describes the most recent progress in solid state NMR quantum gate controls and designs.

Section 10.4 and 10.5 explain applications of the NMR quantum computer to the Shor's algorithm and a lattice gas algorithm.

10.2 Basic technology used in quantum computation with NMR

10.2.1 Realization of a qubit

A molecule with several nuclear spins may work as a quantum computer where each spin constitutes a *qubit*. In fact, NMR has a long history in information science. Back in the 1950s, nuclear spins were already proposed for the purpose of information storage in computers.

Liquid NMR receives more interest due to its mature technology and readiness for application. For now, spin-$\frac{1}{2}$ nuclei such as proton and ^{13}C are preferred because they naturally represent a qubit, but multi-level qubits formed by spin-n nuclei, $n = 1, 2, \cdots$, may provide more freedom in the future. Through careful design, the potential qubits or nuclei are configured with different resonance frequencies and can be distinguished from each other. In a low viscosity liquid, dipolar coupling between nuclei is averaged away by the random motion of the molecules. The J-coupling (scalar coupling) dominates the spin-spin interaction, which is an indirect through-bond electronic interaction. Previously, a very difficult part of the system operation was to set the quantum system to a special state (or to initialize it). Now a very complicated technology has been developed to solve this problem.

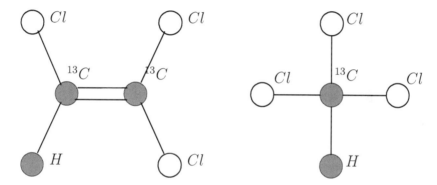

Figure 10.4: The molecule structure of a candidate 3-qubit quantum system, trichloroethylene (left), and a candidate 2-qubit quantum system, chloroform. The trichloroethylene molecule has two labelled ^{13}C and a proton, all having 1-half-spin nuclei. By considering the static magnetic field and spin-spin interaction, its Hamiltonian can be written as $H = -\sum_{i=1}^{3} g_{ni}\beta_{ni}\boldsymbol{I_i} \cdot \boldsymbol{B} + \sum_{i=1}^{2}\sum_{j=i+1}^{3} \boldsymbol{I_i} \cdot J_{i,j} \cdot \boldsymbol{I_j}$. The chloroform has one labelled ^{13}C and one proton.

Fig. 10.4 shows the structure of a trichloroethylene (TCE) molecule and a chloroform molecule used in NMR quantum computers. The hydrogen nucleus (proton) and two ^{13}C nuclei in a TCE molecule form three qubits which can be manipulated, while the chloroform molecule provides two qubits. The sample used by an NMR quantum computer has a large number ($\sim 10^{23}$) of such molecules. This is also called a *bulk* quantum computer. Although most molecules are in a totally random state at room temperature, there are still a small amount of spins standing out and serving our purpose. Theoretically, we use a statistical spin state called a pseudo-pure state, which has the same transformation property as that of a pure quantum state.

Let $|\phi\rangle = a|0\rangle + b|1\rangle$ be the state of a single qubit, $|0\rangle$ for spin-up and $|1\rangle$ for spin-down. We also assume that a is real since only the relative phase is important. Thus this state can be represented using two angles θ and ψ:

$$|\phi\rangle = \cos\frac{\theta}{2}|0\rangle + e^{i\psi}\sin\frac{\theta}{2}|1\rangle, \tag{10.3}$$

where $\theta \in [0, \pi]$ and $\psi \in [0, 2\pi)$. If we think $|0\rangle$ and $|1\rangle$ as the standard basis in $\mathbf{C}^2$, the quantum state corresponds to a unit vector in $\mathbf{C}^2$.

For the study of NMR spectroscopy with many nuclei, density matrices are preferred and are often written as the linear combination of *product operators* [64]:

$$\begin{aligned}
\rho &= |\phi\rangle\langle\phi| \\
&= \begin{bmatrix} \cos^2\frac{\theta}{2} & e^{-i\psi}\frac{\sin\theta}{2} \\ e^{i\psi}\frac{\sin\theta}{2} & \sin^2\frac{\theta}{2} \end{bmatrix} \\
&= I_0 + \sin\theta\cos\psi I_x + \sin\theta\sin\psi I_y + \cos\theta I_z, \tag{10.4}
\end{aligned}$$

where

$$I_0 \equiv \frac{1}{2}\begin{bmatrix} 1 & 0 \\ 0 & 1 \end{bmatrix}, \quad I_x \equiv \frac{1}{2}\begin{bmatrix} 0 & 1 \\ 1 & 0 \end{bmatrix}, \quad I_y \equiv \frac{1}{2}\begin{bmatrix} 0 & -i \\ i & 0 \end{bmatrix}, \quad I_z \equiv \frac{1}{2}\begin{bmatrix} 1 & 0 \\ 0 & -1 \end{bmatrix}. \tag{10.5}$$

They are different from the Pauli matrices only by a constant factor $\frac{1}{2}$ and share the similar commutative law. Upon collecting all the coefficients of I_x, I_y, and I_z together, we obtain a vector

$$\boldsymbol{v} = [\sin\theta\cos\psi \quad \sin\theta\sin\psi \quad \cos\theta]^T, \tag{10.6}$$

which is called a *Bloch vector*; cf. Subsection 5.6.1. In essence, we have defined a mapping from the set of unit vectors $|\phi\rangle \in \mathbf{C}^2$ to the set of unit vectors $\boldsymbol{v} \in \mathbf{R}^3$. We have good reasons to ignore the coefficient of I_0, since it has no effect on the spectroscopy and remains unchanged under any unitary transformation. Each Bloch vector determines a point on the unit sphere, called the *Bloch sphere*, which is displayed in Fig. 10.5 [50, 64]. Bloch vectors have proven to be a very good tool for NMR quantum operations.

The mapping defined above is surjective, because every point on the Bloch sphere gives rise to a unit vector $\boldsymbol{v} = [\sin\theta\cos\psi \quad \sin\theta\sin\psi \quad \cos\theta]^T$ for some pair of (θ, ψ). Conversely, if $\boldsymbol{v}(\theta', \psi') = \boldsymbol{v}(\theta, \psi)$, we get

$$\begin{cases} \cos\theta &= \cos\theta', \\ \sin\theta\cos\psi &= \sin\theta'\cos\psi', \\ \sin\theta\sin\psi &= \sin\theta'\cos\psi', \end{cases} \tag{10.7}$$

which can be used to show that the mapping is also injective except when $\theta = 0$ or $\theta = \pi$. But we can just identify all pairs of $(0, \psi)$ with the north pole of the Bloch sphere and all pairs of (π, ψ) with the south pole. In fact, these two sets correspond to two states $|0\rangle$ and $|1\rangle$, respectively.

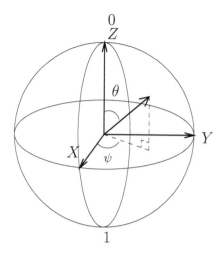

Figure 10.5: The Bloch sphere representation of a quantum state.

10.2.2 Construction of quantum gates

From Appendix E, we know that the Hadamard gate H can be decomposed as $H = e^{i\frac{\pi}{2}} Y_{\pi/2} Z_\pi$. Clearly, the $x/y/z$ rotation gates provide building blocks sufficient to construct any one qubit unitary gate. In this subsection, we will show how to realize these 1-qubit rotation gates and the 2-qubit CNOT gate using NMR. We will also show how to decouple the interaction between two spins, a process called *refocusing* [65].

One-qubit gates

A single spin system has Hamiltonian $H = -\boldsymbol{\mu} \cdot \boldsymbol{B}$, where $\boldsymbol{\mu}$ is the magnetic moment, and

$$\boldsymbol{B} = B_0 \boldsymbol{e}_z + B_1(\boldsymbol{e}_x \cos(\omega t) + \boldsymbol{e}_y \sin(\omega t)) \qquad (10.8)$$

is the magnetic field applied. B_0, a large constant, is the amplitude of the static magnetic field, and B_1 is the amplitude of the oscillating magnetic field in the x-y plane. When $B_1 = 0$, the Hamiltonian and Schrödinger equation can be obtained as [65]

$$H = \frac{\omega_0}{2} \sigma_z \qquad (10.9)$$

and

$$i\partial_t |\psi(t)\rangle = H|\psi(t)\rangle, \qquad (10.10)$$

respectively, where $\hbar$ has been divided from both sides in the second equation and we take $\hbar$ away from H in the first one just for simplicity. The Larmor frequency $\omega_0 = -B_0\gamma$ is defined by the nuclei and the magnetic field, see (10.2). Assume that the initial state is $|\psi_0\rangle = a_0|0\rangle + b_0|1\rangle$. Then

the evolution of the quantum state of the spin and the density matrix can
be solved directly and given as

$$|\psi(t)\rangle = e^{-i\omega_0\sigma_z t/2}|\psi_0\rangle$$

$$= \begin{bmatrix} e^{-i\omega_0 t/2} & 0 \\ 0 & e^{i\omega_0 t/2} \end{bmatrix} \begin{bmatrix} a_0 \\ b_0 \end{bmatrix}$$

$$= e^{-i\omega_0 t/2} \begin{bmatrix} 1 & 0 \\ 0 & e^{i\omega_0 t} \end{bmatrix} |\psi_0\rangle,$$

$$\rho(t) = e^{-itH}\rho(0)e^{itH}.$$

This evolution is also called a *chemical shift evolution*, resembling the
precessing of a magnet in a static field. Recall the Bloch vector on the Bloch
sphere. It is exactly Z_θ, the rotation operator around the z axis with $\theta = \omega_0 t$.

To achieve an x-rotation operator, we need a small magnetic field transverse to the z direction to control the evolution of the quantum state. The
Hamiltonian is given as in (10.8) by choosing B_1 different from zero:

$$H = -\mu \cdot B = \frac{\omega_0}{2}\sigma_z + \frac{\omega_1}{2}\left(\sigma_x \cos(\omega t) + \sigma_y \sin(\omega t)\right),$$

where ω_1 depends on the x-y plane component B_1 of the magnetic field,
$\omega_1 = -B_1\gamma$. To solve the Schrödinger equation, we put $|\psi(t)\rangle$ in a "frame"
rotating with the magnetic field around the z axis at frequency ω, $|\phi(t)\rangle = e^{i\omega t\sigma_z/2}|\psi(t)\rangle$. With this substitution, the Schrödinger equation (10.10) be-
comes

$$i\partial_t|\phi(t)\rangle = (e^{i\omega\sigma_z t/2}He^{-i\omega\sigma_z t/2} - \frac{\omega}{2}\sigma_z)|\phi(t)\rangle. \qquad (10.11)$$

Using properties

$$\begin{array}{rcl} e^{i\omega\sigma_z t/2}\sigma_z e^{-i\omega\sigma_z t/2} & = & \sigma_z, \\ e^{i\omega\sigma_z t/2}\sigma_x e^{-i\omega\sigma_z t/2} & = & \sigma_x \cos(\omega t) - \sigma_y \sin(\omega t), \\ e^{i\omega\sigma_z t/2}\sigma_y e^{-i\omega\sigma_z t/2} & = & \sigma_x \sin(\omega t) + \sigma_y \cos(\omega t), \end{array} \qquad (10.12)$$

we obtain

$$i\partial_t|\phi(t)\rangle = \left(\frac{\omega_0 - \omega}{2}\sigma_z + \frac{\omega_1}{2}\sigma_x\right)|\phi(t)\rangle,$$

$$|\phi(t)\rangle = e^{-i((\omega_0-\omega)\sigma_z/2+\omega_1\sigma_x/2)t}|\phi(0)\rangle. \qquad (10.13)$$

We know from (D.2) in Appendix D that this is a rotation around the axis

$$n = \frac{1}{\sqrt{1 + (\frac{\omega_1}{\omega_0-\omega})^2}}\left(z + \frac{\omega_1}{\omega_0 - \omega}x\right). \qquad (10.14)$$

An important case is $\omega_0 = \omega$, also called the *resonance* case where its
name came from the zero denomination in (10.14). By (10.13), we see that

a relatively weak transverse magnetic field causes a rotation around the x axis:

$$|\psi(t)\rangle = e^{-i\omega_0\sigma_z t/2}|\phi(t)\rangle = e^{-i\omega_0 t\sigma_z/2}e^{-i\omega_1 t\sigma_x/2}|\phi(0)\rangle = Z_\theta X_\beta|\psi(0)\rangle, \quad (10.15)$$

where $X_\beta = e^{-i\omega_1 t\sigma_x/2}$, $\beta = \omega_1 t$. By applying another $Z_{-\theta}$, we obtain a rotation X_β as desired. Since the frequency of the precession is in radio frequency band, the field applied is called an *RF pulse*.

When $|\omega_0 - \omega| \gg \omega_1$, the rotation axis direction is almost along z and the RF pulse has no effect on it:

$$|\psi(t)\rangle = e^{-i\omega\sigma_z t/2}|\phi(t)\rangle \approx e^{-i\omega_0 t\sigma_z/2}|\psi(0)\rangle = Z_{\omega_0 t}|\psi(0)\rangle,$$

thus we can tell one qubit from another because their resonance frequencies are designed to be different. There are still cases where the difference of resonance frequencies between spins is not large enough. The RF pulse may cause similar rotations on all those spins. To avoid or at least minimize it, a *soft* pulse is applied instead of the so called *hard* pulse. It is a pulse with longer time span and weaker magnetic field, in other word, a smaller ω_1. This strategy makes these "close" qubits fall into the $|\omega_0 - \omega| \gg \omega_1$ case.

If we change the magnetic field to

$$\boldsymbol{B} = B_0\boldsymbol{e}_z + B_1(\boldsymbol{e}_x \cos(\omega_0 t + \alpha) + \boldsymbol{e}_y \sin(\omega_0 t + \alpha)), \quad (10.16)$$

the Hamiltonian will become

$$H = \frac{\omega_0}{2}\sigma_z + \frac{\omega_1}{2}(\sigma_x \cos(\omega_0 t + \alpha) + \sigma_y \sin(\omega_0 + \alpha)) \quad (10.17)$$

where ω_1 is defined as before. The RF field is almost the same as (10.8) in the resonance case except a phase shift. Using the same rotation frame as before with $\omega = \omega_0$, we obtain

$$i\partial_t|\phi(t)\rangle = \frac{\omega_1}{2}(\sigma_x \cos(\alpha) + \sigma_y \sin(\alpha))|\phi(t)\rangle, \quad (10.18)$$

after simplification. After time duration t, the new system state is given as

$$|\phi(t)\rangle = e^{-i\frac{\omega_1}{2}(\sigma_x \cos(\alpha) + \sigma_y \sin(\alpha))t}|\phi(0)\rangle, \quad (10.19)$$

and the evolution operator can be computed using (D.2) as

$$\begin{aligned} U_{\theta/2,\alpha} &= e^{-i\frac{\omega_1}{2}(\sigma_x \cos(\alpha) + \sigma_y \sin(\alpha))t} \\ &= \begin{bmatrix} \cos(\frac{\theta}{2}) & -i\sin(\frac{\theta}{2})e^{-i\alpha} \\ -i\sin(\frac{\theta}{2})e^{i\alpha} & \cos(\frac{\theta}{2}) \end{bmatrix}, \end{aligned} \quad (10.20)$$

where $\theta = \omega_1 t$. This is a 1-qubit rotation operator (a Rabi rotation gate). When $\alpha = \pi/2$,

$$\begin{aligned} U_{\theta/2,\pi/2} &= \begin{bmatrix} \cos(\frac{\theta}{2}) & -\sin(\frac{\theta}{2}) \\ \sin(\frac{\theta}{2}) & \cos(\frac{\theta}{2}) \end{bmatrix} \\ &= Y_\theta. \end{aligned} \quad (10.21)$$

We have achieved a y-rotation operator just by adding a phase shift to the RF field.

Two-qubit gates

The construction of a 2-qubit gate requires the coupling of two spins. In a liquid sample of NMR, *J-coupling* is the dominating coupling between spins. Under the assumption that the resonance frequency difference between the coupled spins is much larger than the strength of the coupling (a so-called weak coupling regime), the total Hamiltonian of a two-spin system without transverse field may be given as

$$H = \frac{1}{2}w_1\sigma_z^1 + \frac{1}{2}w_2\sigma_z^2 + \frac{1}{2}J\sigma_z^1\sigma_z^2, \tag{10.22}$$

where w_i is the frequency corresponding to spin i, σ_z^i is the z projection operator of spin i, for $i = 1, 2$, and J is the coupling coefficient. Take the chloroform in Fig. 10.4 for example [5, 50]. In a 11.7T magnetic field, the precession frequency of ^{13}C is about $2\pi \times 500$ MHz and the precession frequency of proton is about $2\pi \times 125$ MHz. The coupling constant J is about $2\pi \times 100$ Hz. Here we set $B_1 = 0$, which means no transverse magnetic field is applied and those terms such as σ_x, σ_y do not appear. The remaining terms in the Hamiltonian only contains operators σ_z^1 or σ_z^2, which are commutative. Thus, we can obtain the eigenstates and eigenvalues of this 2-spin system and we map the set of eigenstates to the standard basis of $\mathbb{C}^4$, as follows:

$$|00\rangle = \begin{bmatrix} 1 \\ 0 \\ 0 \\ 0 \end{bmatrix}, \ |01\rangle = \begin{bmatrix} 0 \\ 1 \\ 0 \\ 0 \end{bmatrix}, \ |10\rangle = \begin{bmatrix} 0 \\ 0 \\ 1 \\ 0 \end{bmatrix}, \ |00\rangle = \begin{bmatrix} 0 \\ 0 \\ 0 \\ 1 \end{bmatrix}; \tag{10.23}$$

$$\begin{aligned} H|00\rangle &= k_{00}|00\rangle, & k_{00} &= \tfrac{1}{2}w_1 + \tfrac{1}{2}w_2 + \tfrac{1}{2}J; \\ H|01\rangle &= k_{01}|01\rangle, & k_{01} &= \tfrac{1}{2}w_1 - \tfrac{1}{2}w_2 - \tfrac{1}{2}J; \\ H|10\rangle &= k_{10}|10\rangle, & k_{10} &= -\tfrac{1}{2}w_1 + \tfrac{1}{2}w_2 - \tfrac{1}{2}J; \\ H|11\rangle &= k_{11}|11\rangle, & k_{11} &= -\tfrac{1}{2}w_1 - \tfrac{1}{2}w_2 + \tfrac{1}{2}J. \end{aligned} \tag{10.24}$$

Since the matrix is diagonal, the evolution of this 2-spin system can be easily derived as

$$|\psi(t)\rangle = e^{-iHt}|\psi(0)\rangle = \begin{bmatrix} e^{-ik_{00}t} & & & \\ & e^{-ik_{01}t} & & \\ & & e^{-ik_{10}t} & \\ & & & e^{-ik_{11}t} \end{bmatrix} |\psi(0)\rangle. \tag{10.25}$$

We can also rewrite the one qubit rotation operators for this two $\frac{1}{2}$-spin system in matrix form with respect to the same basis:

$$Z^1_{\pi/2} = \begin{bmatrix} e^{-i\pi/4} & & & \\ & e^{-i\pi/4} & & \\ & & e^{i\pi/4} & \\ & & & e^{i\pi/4} \end{bmatrix}, \tag{10.26}$$

$$Z^2_{-\pi/2} = \begin{bmatrix} e^{i\pi/4} & & & \\ & e^{-i\pi/4} & & \\ & & e^{i\pi/4} & \\ & & & e^{-i\pi/4} \end{bmatrix}, \tag{10.27}$$

$$Y^2_{\pi/2} = \frac{\sqrt{2}}{2} \begin{bmatrix} 1 & -1 & & \\ 1 & 1 & & \\ & & 1 & -1 \\ & & 1 & 1 \end{bmatrix}, \tag{10.28}$$

$$Y^2_{-\pi/2} = \frac{\sqrt{2}}{2} \begin{bmatrix} 1 & 1 & & \\ -1 & 1 & & \\ & & 1 & 1 \\ & & -1 & 1 \end{bmatrix}, \tag{10.29}$$

where Z^i_θ is the rotation operator for spin i with angle θ around the z axis while keeping another spin unchanged, and all Y^i_θ are similarly defined operators about the y axis; see (D.5). A careful reader may raise issues about the 1-qubit gate we have obtained in Subsection 10.2.2 because the coupling between two qubits always exists and has not been considered. We need to turn off the coupling when we only want to operate one spin but the coupling is non-negligible. This is in fact one of the major characteristic difficulties associated with the NMR quantum computing technology. A special technology called *refocusing* is useful. It works as follows. We apply a soft π pulse on the spare spin that we don't want to change at the middle point of the operation time duration while we are working on the target spin. The effect is that the coupling before the pulse cancels the one after the pulse, so the result of no-coupling is achieved. Another π pulse will be needed to turn the spin back. All pulses are soft.

This technology is so important that we now state it here as a theorem.

Theorem 10.2.1. *Let $H = \frac{\omega_1}{2}\sigma^1_z + \frac{J}{2}\sigma^1_z\sigma^2_z + A$ be a given Hamiltonian, where A is a Hamiltonian that does not act on spin 1 and commutes with σ^2_z. Then the evolution operators of A and H satisfy*

$$e^{-iAt} = -X^1_\pi e^{-iHt/2} X^1_\pi e^{-iHt/2}, \tag{10.30}$$

i.e., the collective evolution of the quantum system with Hamiltonian H and additional two X^1_π-pulses at the middle and the end of the time duration, equals that of a system with Hamiltonian A (up to a global phase shift π, or a factor -1). □

Proof. Assume that the time duration is t and denote U for

$$U = X_\pi^1 e^{-iHt/2} X_\pi^1 e^{-iHt/2}. \tag{10.31}$$

Note that $X_\pi^1 = e^{-i\frac{\pi}{2}\sigma_x^1}$ and it commutes with A which contains no operators acting on spin 1, thus

$$U = X_\pi^1 e^{-i(\frac{\omega_1}{2}\sigma_z^1 + \frac{J}{2}\sigma_z^1\sigma_z^2)\frac{t}{2}} X_\pi^1 e^{-i(\frac{\omega_1}{2}\sigma_z^1 + \frac{J}{2}\sigma_z^1\sigma_z^2)\frac{t}{2}} e^{-iAt}. \tag{10.32}$$

It suffices to prove that the part before e^{-iAt} satisfies

$$B = X_\pi^1 e^{-i(\frac{\omega_1}{2}\sigma_z^1 + \frac{J}{2}\sigma_z^1\sigma_z^2)\frac{t}{2}} X_\pi^1 e^{-i(\frac{\omega_1}{2}\sigma_z^1 + \frac{J}{2}\sigma_z^1\sigma_z^2)\frac{t}{2}} = -I. \tag{10.33}$$

We first check the effect of B on the four basis vector. We have

$$
\begin{aligned}
B|11\rangle &= X_\pi^1 e^{-i(\frac{\omega_1}{2}\sigma_z^1 + \frac{J}{2}\sigma_z^1\sigma_z^2)\frac{t}{2}} X_\pi^1 e^{-i(\frac{\omega_1}{2}\sigma_z^1 + \frac{J}{2}\sigma_z^1\sigma_z^2)\frac{t}{2}}|11\rangle \\
&= e^{-i\frac{-\omega_1 + J}{4}t}(-i)X_\pi^1 e^{-i(\frac{\omega_1}{2}\sigma_z^1 + \frac{J}{2}\sigma_z^1\sigma_z^2)\frac{t}{2}}|01\rangle \\
&= (-i)e^{-i\frac{-\omega_1+J}{4}t} X_\pi^1 e^{-i\frac{\omega_1 - J}{4}t}|01\rangle \\
&= (-i)^2|11\rangle \\
&= -|11\rangle,
\end{aligned} \tag{10.34}
$$

$$
\begin{aligned}
B|01\rangle &= X_\pi^1 e^{-i(\frac{\omega_1}{2}\sigma_z^1 + \frac{J}{2}\sigma_z^1\sigma_z^2)\frac{t}{2}} X_\pi^1 e^{-i(\frac{\omega_1}{2}\sigma_z^1 + \frac{J}{2}\sigma_z^1\sigma_z^2)\frac{t}{2}}|01\rangle \\
&= e^{-i\frac{\omega_1 - J}{4}t}(-i)X_\pi^1 e^{-i(\frac{\omega_1}{2}\sigma_z^1 + \frac{J}{2}\sigma_z^1\sigma_z^2)\frac{t}{2}}|11\rangle \\
&= (-i)e^{-i\frac{-\omega_1+J}{4}t} X_\pi^1 e^{-i\frac{-\omega_1+J}{4}t}|11\rangle \\
&= (-i)^2|01\rangle \\
&= -|01\rangle,
\end{aligned} \tag{10.35}
$$

and similarly,

$$
\begin{aligned}
B|10\rangle &= -|10\rangle, \\
B|00\rangle &= -|00\rangle.
\end{aligned} \tag{10.36}
$$

In the computation above, we have used the fact that X_π^1 has no effect on the second spin and the four basis vectors $|00\rangle$, $|01\rangle$, $|10\rangle$ and $|11\rangle$ are the eigenstates of the operator $\frac{\omega_1}{2}\sigma_z^1 + \frac{J}{2}\sigma_z^2\sigma_z^1$. The result shows that $B = -I$, and we are done. $\square$

When the Hamiltonian is given in the form as (10.22), the above theorem tells us that both the chemical shift evolution (precession) and the J-coupling effect on spin 1 are removed and only the term $\frac{\omega_2}{2}\sigma_z^2$ remains. We obtain a z-rotation of spin 2 while freezing spin 1. By combining it with several hard pulses, we can also achieve any arbitrary rotation on spin 2 with the motion of spin 1 frozen [55]. Similar computation shows that a hard π pulse applied at the middle point of the time duration cancels the chemical shift evolution of both spins. This can be seen by checking the identity

$$
e^{-iHt/2} X_\pi^1 X_\pi^2 e^{-iHt/2} =
\begin{bmatrix}
 & & & e^{-iJt/2} \\
 & & e^{iJt/2} & \\
 & e^{iJt/2} & & \\
e^{-iJt/2} & & &
\end{bmatrix}. \tag{10.37}
$$

Another hard π pulse can rotate two spins back, so we have achieved an evolution which has only the J-coupling effect, denoted by $\mathbb{Z}_\theta$:

$$
\mathbb{Z}_\theta = \begin{bmatrix} e^{-i\theta/2} & & & \\ & e^{i\theta/2} & & \\ & & e^{i\theta/2} & \\ & & & e^{-i\theta/2} \end{bmatrix},
$$

and when $\theta = \pi/2$,

$$
\mathbb{Z}_{\pi/2} = \begin{bmatrix} e^{-i\pi/4} & & & \\ & e^{i\pi/4} & & \\ & & e^{i\pi/4} & \\ & & & e^{-i\pi/4} \end{bmatrix}. \tag{10.38}
$$

Although we give only an example of the 2-qubit system in the above, the reader should note that a general method is available to reserve only the couplings wanted while keeping all the others canceled for multi-qubit systems [38, 52, 55]. Combining operators in (10.26) through (10.29) and (10.38), we can now construct a CNOT gate as in Fig. 10.6 which includes four 1-qubit $\pi/2$ rotations around y or z axes and one 2-qubit $\pi/2$ rotation. The total operator, denoted by CN, can be computed as

$$
CN = Z^1_{\pi/2} Y^2_{-\pi/2} Z^2_{-\pi/2} \mathbb{Z}_{\pi/2} Y^2_{\pi/2} = e^{-\frac{\pi}{4}i} \begin{bmatrix} 1 & & & \\ & 1 & & \\ & & 0 & 1 \\ & & 1 & 0 \end{bmatrix}, \tag{10.39}
$$

which is a CNOT gate up to a phase of $-\pi/4$ [50].

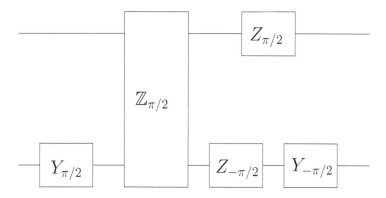

Figure 10.6: The quantum circuit used to realize a quantum controlled-not gate.

We have shown how to construct 1-qubit gates and the 2-qubit CNOT gate using the NMR technology. The simple pulse design works fine in ideal situations. In practice, errors arise from various factors. Decoherence causes the lost of quantum information with time. Thus, all operations should be completed within a short time, roughly constrained by the energy relaxation time T_1 and the phase randomization time T_2. Again, take the chloroform for an example. For protons, $T_1 \approx 7$ sec and $T_2 \approx 2$ sec; for carbons, $T_1 \approx 16$ sec and $T_2 \approx 0.2$ sec [5, 50]. The pulses have to be short enough so that all the pulses can be jammed in the time window. Ideally, a pulse can be completed quite fast, but this may incur undesirable rotations in other qubits because the frequency band width is inversely proportional to the time length of the pulse. A shorter and stronger pulse will have a wider frequency band that may cover the resonance frequency of another spin, called *cross-talking*. It should also be noted that both T_1 and T_2 are defined and measured in simplified situation, and they can only be used as an approximation of the decoherence rate for the quantum computation. Coupling is also a problem which makes the pulse design much more complicated. Finally, any experimental facility is not perfect, which may introduce more errors. Typical error resources include *inhomogeneities in the static and RF field, pulse length calibration errors, frequency offsets*, and *pulse timing/phase imperfections*.

If the quantum circuit can be simplified and the number of gates needed is reduced, the requirements on the pulses can be alleviated. Mathematicians are looking for methods to find time-optimal pulse sequences [25, 41, 42, 74], with the goal of finding the shortest path between the identity and a point in the space of $\mathbf{SU}(n)$ allowed by the system and the control Hamiltonians. Besides that, NMR spectroscopists have already developed advanced pulse techniques to deal with system errors such as cross-talking and coupling. They turn out to work well and are now widely used in NMR quantum computation. Such techniques include composite pulses [9, 19, 35, 36, 53, 82] and pulse shaping. The latter consists mainly of two methods: phase profiles [67] and amplitude profiles [22, 47].

10.2.3   Initialization

An NMR sample eventually will go into its equilibrium state when no RF pulse is applied for a long time. Then the density matrix is proportional to $e^{-H/kT}$, according to the Boltzmann distribution, where $k = 1.381 \times 10^{-23}$ J/K and T is the absolute temperature. Normally, the environment temperature is far larger than the energy difference between the up and down states of the spin, and H/kT is very small, about 10^{-4}. We also make the assumption that the coupling terms are small enough compared with the resonant frequency; thus we can make a reasonable approximation of the equilibrium state density matrix of a system with n spins:

$$\rho_{eq} = \frac{e^{-H/kT}}{tr(e^{-H/kT})} \approx I - \frac{1}{kT}(\epsilon_1 \sigma_z^1 + \epsilon_2 \sigma_z^2 + \cdots + \epsilon_n \sigma_z^n). \qquad (10.40)$$

In the four operators appearing in the density matrix (10.4), only those with zero traces can be observed in NMR. The operator I_0 is invisible, and moreover, it remains invariant under any unitary similarity transformation. Therefor, we only need to take care of the zero-trace part of the initial density matrix, noting that only that part (called deviation) is effective. Most algorithms prefer an initial state such as

$$\rho_0 = \frac{1-\epsilon}{2^n} I + \epsilon |00 \cdots 0\rangle \langle 0 \cdots 00|,$$

which is an example of the so called *pseudo-pure* states, corresponding to the pure state $|00 \cdots 0\rangle$.

To initialize the system to a pseudo-pure state as above, we may use a scheme called *averaging*. Let us explain this for a 2-spin system. Suppose we have three 2-spin subsystems with density matrices

$$\rho_1 = \begin{bmatrix} a & 0 & 0 & 0 \\ 0 & b & 0 & 0 \\ 0 & 0 & c & 0 \\ 0 & 0 & 0 & d \end{bmatrix}, \quad \rho_2 = \begin{bmatrix} a & 0 & 0 & 0 \\ 0 & c & 0 & 0 \\ 0 & 0 & d & 0 \\ 0 & 0 & 0 & b \end{bmatrix}, \quad \rho_3 = \begin{bmatrix} a & 0 & 0 & 0 \\ 0 & d & 0 & 0 \\ 0 & 0 & b & 0 \\ 0 & 0 & 0 & c \end{bmatrix},$$

$$(10.41)$$

respectively, where a, b, c, and d are nonnegative, and $a + b + c + d = 1$. These are three diagonal matrices with three of their diagonal elements in cyclic permutation.

Now, we mix these three subsystems together (for n-qubit system, we may have $2^n - 1$ subsystems) and assume that the three subsystems have the same signal scale. Because the readout is linear with respect to the initial state, we are in fact working on a system with an effective initial density matrix

$$\frac{1}{3} \sum_{i=1}^{3} \rho_i = \frac{1}{3} \begin{bmatrix} 3a & & & \\ & b+c+d & & \\ & & b+c+d & \\ & & & b+c+d \end{bmatrix}$$

$$= \frac{b+c+d}{3} I + \frac{1}{3} \begin{bmatrix} 4a-1 & 0 & 0 & 0 \\ 0 & 0 & 0 & 0 \\ 0 & 0 & 0 & 0 \\ 0 & 0 & 0 & 0 \end{bmatrix}, \quad (10.42)$$

which is a pseudo-pure state corresponding to $|00 \cdots 0\rangle$.

Various methods have been developed to achieve this effect of averaging. Because ρ_1, ρ_2, and ρ_3 differ only by a permutation of the diagonal elements, a sequence of CNOT pulses can be used to transform one to another. In most cases, we only have one sample, the same algorithm can be repeated on the very sample three times but with different initial states ρ_1, ρ_2, and ρ_3, respectively. At last, after all the three outputs are obtained and added together (average), we achieve the same result as what we will get when the algorithm is employed on a system with the expected initial state $|00 \cdots 0\rangle$. This is called "temporal averaging" [44]. Gradient fields can also

be used to divide the sample into different slices in space which are prepared into different initial states, and the averaging is realized spatially,
called "spatial averaging" [7]. The number of the experiments and pulses
needed grows very large when the number of qubits increases. For example, 9 experiments are combined in order to prepare one pseudo-pure state
for a 5-qubit system and 48 pulses are used to form one pseudo-pure state
in a 7-qubit system [23] after modifications such as logical labeling [24, 84]
and selective saturation [43].

10.2.4 Measurement

An NMR computer differs from other quantum computers in that it works
on an ensemble of spins instead of just a single one. It produces an observable macroscopic signal which can be picked up by a set of coils positioned
on the x-y plane, as shown in Fig. 10.3. The signal measures the change
rate of the magnetic field created by a large number of spins in the sample
rotating around the z-axis, called *free induction decay* (FID). Due to relaxation, peaks of the Fourier transform of the signal, or spectra, have width.
However, we do not need to worry about that since it will not make any
substantial difference in our discussion here. One disadvantage is that the
readout from NMR is an average of all the possible states, in contrast to
most existing quantum algorithms that ask for the occurrence of only a single state. But it is possible for one to modify ordinary quantum algorithms
to make NMR results usable.

The magnetization detected by the coil in Fig. 10.3 is proportional to the
trace of the product of the density matrix with $\sigma_+ = \sigma_x + i\sigma_y$:

$$M_x + iM_y = nV\langle\mu_x + i\mu_y\rangle = nV\gamma\hbar Tr(\rho(\sigma_x + i\sigma_y)), \qquad (10.43)$$

where γ is the magnetogyric ratio as in (10.2) and ρ is the density matrix.
When the external RF magnetic field is removed, the density matrix will
change according to the system's Hamiltonian as we discussed earlier. If we
decompose the density matrix into a sum of product operators as in (10.4),
only I_x and I_y contribute to the readout. We can not "see" the coefficients
of I_0 and I_z. Recall (10.12): if a 1-spin system begins from density matrix
$\rho_0 = I_0 + \sin\theta\cos\psi I_x + \sin\theta\sin\psi I_y + \cos\theta I_z$, the magnetization will rotate
with the resonant frequency as

$$
\begin{aligned}
M_z + iM_y &= C\,Tr(e^{-iHt}\rho_0 e^{iHt}\sigma_+) \\
&= C\,Tr(e^{-iHt}(I_0 + \sin(\theta)\cos(\psi)I_x + \\
&\quad \sin(\theta)\sin(\psi)I_y + \cos(\theta)I_z)e^{iHt}\sigma_+) \\
&= C\,Tr((\sin\theta\cos\psi(\cos(\omega t)I_x + \sin(\omega t)I_y) + \\
&\quad sin\theta\sin\psi(\cos(\omega t)I_y - \sin(\omega t)I_x))\sigma_+) \\
&= C\sin\theta\,e^{i(\omega t+\psi)},
\end{aligned}
\qquad (10.44)
$$

where $C = nV\gamma\hbar$. This rotating magnetization will introduce an oscillating
electric potential in the receiver coils, which will be processed by a computer

to generate the spectra. Note that the signal is proportional to $\sin\theta$. If an x-rotation with angle $\pi/2$ is applied on the spin before the measurement, the magnetization will become

$$M_z + iM_y = \frac{\sqrt{2}}{2}C(\sin\theta - i\cos\theta)e^{i\omega t}.$$

For simplicity, we have chosen $\psi = 0$. The imaginary part is proportional to the population difference:

$$\cos\theta = \cos^2\frac{\theta}{2} - \sin^2\frac{\theta}{2}.$$

Computation of a 2-spin system is complicated, so we will only give some partial results here. The purpose is to point out what methodology is used. We will still use the basis given by (10.23) and the Hamiltonian in (10.22). The system begins from a density matrix as

$$\rho_0 = \begin{bmatrix} \rho_{11} & \rho_{12} & \rho_{13} & \rho_{14} \\ \rho_{21} & \rho_{22} & \rho_{23} & \rho_{24} \\ \rho_{31} & \rho_{32} & \rho_{33} & \rho_{34} \\ \rho_{41} & \rho_{42} & \rho_{43} & \rho_{44} \end{bmatrix}. \tag{10.45}$$

The operator σ_+ is a summation of operators from the two subsystems:

$$\begin{aligned} \sigma_+ &= \sigma_+^1 + \sigma_+^2 \\ &= \begin{bmatrix} 0 & 2 & 2 & 0 \\ 0 & 0 & 0 & 2 \\ 0 & 0 & 0 & 2 \\ 0 & 0 & 0 & 0 \end{bmatrix}. \end{aligned} \tag{10.46}$$

The magnetization in the x-y plane is composed of four frequencies:

$$\begin{aligned} M_x + iM_y &= C\, Tr(e^{-iHt}\rho_0 e^{iHt}\sigma_+) \\ &= C\left(\rho_{31}e^{i(\omega_1+J)t} + \rho_{42}e^{i(\omega_1-J)t} + \rho_{43}e^{i(\omega_2-J)t} + \rho_{21}e^{i(\omega_2+J)t}\right). \end{aligned} \tag{10.47}$$

The spectrum has two pairs of peaks, one pair around the precession frequency ω_1, another pair around ω_2. See Fig. 10.7. The splitting is a result of coupling. If the system have more than two spins, the coupling will split up a peak into up to 2^{n-1} peaks where n is the number of spins. We also combine all the constants in C to make the formula concise. Only four of the elements out of the density matrix appear in this spectrum, so we need to design certain control pulses to move the expected information to these four positions where numbers can be shown via free induction signal. If multi-tests are allowed, theoretically, all the elements of the density matrix can be retrieved [4, 6]. It is also possible to transport the desired information (computational results) to the four positions where the observer can see.

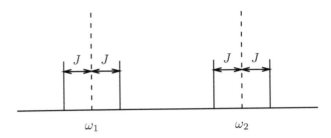

Figure 10.7: Simplified stick spectra of a 2-qubit molecule. The two dotted lines show two peaks at ω_1 and ω_2, respectively, when no coupling is applied ($J = 0$). After coupling, every peak is split into two small peaks with the intensities reduced to half.

A typical pulse used in reading out is a hard $X_{\pi/2}$ pulse which rotates all the spins about the x-axis with angle $\pi/2$. Let us still use 2-spin systems as an example. The operation is the tensor product of two x-rotation operators, i.e., $X_{\pi/2} = X^1_{\pi/2}X^2_{\pi/2}$. The imaginary part of the four effective elements of the density matrix ρ' after the operation, utilizing the fact that the density matrix is Hermitian, are

$$
\begin{aligned}
Im(\rho'_{31}) &= \tfrac{1}{4}(\rho_{33} + \rho_{44} - \rho_{11} - \rho_{22} - 2Im(\rho_{21}) - 2Im(\rho_{34})), \\
Im(\rho'_{42}) &= \tfrac{1}{4}(\rho_{33} + \rho_{44} - \rho_{11} - \rho_{22} + 2Im(\rho_{21}) + 2Im(\rho_{34})), \\
Im(\rho'_{43}) &= \tfrac{1}{4}(\rho_{22} + \rho_{44} - \rho_{11} - \rho_{33} + 2Im(\rho_{31}) + 2Im(\rho_{24})), \\
Im(\rho'_{21}) &= \tfrac{1}{4}(\rho_{22} + \rho_{44} - \rho_{11} - \rho_{33} - 2Im(\rho_{31}) - 2Im(\rho_{24})).
\end{aligned} \quad (10.48)
$$

Find the sum of $Im(\rho'_{31})$ and $Im(\rho'_{42})$ and that of $Im(\rho'_{43})$ and $Im(\rho'_{21})$:

$$
\begin{aligned}
Im(\rho'_{31} + \rho'_{42}) &= -\tfrac{1}{2}(\rho_{11} + \rho_{22} - \rho_{33} - \rho_{44}), \\
Im(\rho'_{43} + \rho'_{21}) &= -\tfrac{1}{2}(\rho_{11} - \rho_{22} + \rho_{33} - \rho_{44}).
\end{aligned} \quad (10.49)
$$

Because what the coils pick up is the change rate of the magnetic field, rather than the magnetic field itself, the imaginary part we have listed above is reflected in the real part of the spectra. The computation above shows that the sum of the real parts of each pair of peaks in the spectra is proportional to the population difference between the spin-up and the spin-down states of the corresponding spin.

10.3 Solid state NMR

Liquid NMR, discussed in Section 10.2, has several constrains that make a liquid NMR quantum computer *not scalable*. At first, as the result of the pseudo-pure state preparation, the signal-noise ratio decreases exponentially when the number of qubits increases, limiting its ability to realize more qubits. Another difficulty arises when we want to control the system as accurately as desired. Because the range of the chemical shift is limited by nature, the number of qubits represented by the same type of nuclei, such as carbon, is constrained as the resonance frequency gaps between any two qubits must be large enough so that we can distinguish the qubits easily and control them with great precision. It is estimated that a quantum computer realized by liquid state NMR can have at most 10 to 20 qubits.

Solid state NMR has the potential to overcome many of the problems of its liquid state counterpart as listed in the preceding paragraph. These advantages are derived partly from the lack of motion of the molecules and partly from the ability to cool to low temperatures. As with many potential solutions, there are tradeoffs to consider. Here we summarize:

(1) At low temperatures, near or below that of liquid helium, it is possible to initialize electron spins using the thermal Boltzmann distribution. Nuclear spins do not become significantly oriented until much lower temperatures because of their 1000 times lower energies, but there are existing pulse RF sequences that can transfer an electron spin orientation to nearby nuclear spins using their mutual spin-spin interaction. In principle, this solves the problem of qubit initialization. In practice, the thermal initialization process can be slow since it depends on the electron spin population lifetime. It is possible to find systems with short electron spin lifetimes, but this will tend to result in faster decoherence of the nuclear spins, since they must be coupled to the electron in order to initialize in the first place.

(2) Because the molecules in a solid are usually not tumbling, the dipole coupling between nearby spins does not average out. This has the advantage of making multi-qubit gates faster, since the dipole coupling is much larger than the scalar coupling. The orientation dependent chemical shifts also do not average out, in principle making individual qubits easier to address so that more qubits can be used. Here, it should be noted that custom molecules containing electron spins [83] can be used to enhance this effect. There is a tradeoff to consider, in that the faster interaction with nearby spins provided by dipole coupling can also lead to faster decoherence times.

(3) Spin lifetimes in solids can be much longer than in liquids. Lack of molecular motion eliminates the spatial diffusion of spins, which is a problem in liquid NMR for times in the range of milliseconds or longer [18]. Phonons can cause decoherence in solids at room temperature, but this can be strongly suppressed at temperatures achievable in liquid helium. It is not unusual to see spin population lifetimes of minutes in solids, especially at low temperatures. Unfortunately, spin coherence times are usually somewhat shorter due to dephasing caused by mutual spin flips through the strong dipole coupling. To eliminate this decoherence mechanism, there are two main approaches. One is to disperse the active molecule, as a dopant in a spin-free host. Actually the host does not need to be completely spin-free, provided its spins are far enough off resonance with those of the active molecule. Another technique is to use stoichiometric materials consisting of relatively large unit cells containing many spin-free atoms. The idea for both these approaches is to keep the active nuclei relatively far apart, except for nearest neighbors.

Beside above differences, nuclei with non-zero spin in *solid state* can also be used for quantum computation [49] and manipulated similar to the liquid state NMR. Because all the nuclei are fixed in space, a static magnetic field with strong gradient in one direction separates the nuclei into different layers along the direction. Every layer of nuclei can be regarded as a qubit and the qubits have different resonance frequencies as the magnetic field is different from one layer to another. Readout also can be made to take advantage of the bulk quantum computer much like the liquid NMR. Signal is picked up using methods like magnetic resonance force microscopy.

There are two types of methods to make such nuclei arrangement. Crystal, such as cerium-monophosphide (CeP), is a natural choice, where the 1/2 spin ^{31}P nuclei form periodical layers in the crystal with inter-layer distance about 12 Å [26, 89]. Another method is to grow a chain of ^{29}Si that has 1/2 spin along the static field direction on a base of pure ^{28}Si or ^{30}Si which are both 0 spin nuclei [1, 48]. The last one combines the mature crystal growth and processing technology for silicon from the semiconductor industry. Liquid crystal [90] or solid-state sample [51] are also candidates for realizing NMR quantum computer.

Recently, there has been considerable progress made in the area of *optically addressed spins in solids*. As a result, some highly scalable designs have recently come forward that have the potential to eliminate all of the limitations of NMR. Aside from potentially solving NMR's problems, optical addressing has the important advantage that it would provide an interface between spin qubits and optical qubits, which is essential to interface with existing quantum communication systems, and for quantum networking in general.

Optically addressed spins are better known in the literature as *spectral hole burning* (SHB) materials [61]. Most of these are dopant-host systems that exhibit strong zero-phonon optical absorption lines at low temperature. Due to the inherent inhomogeneity of dopant-host systems it is often found that this optical zero-phonon linewidth is much larger than that of the individual atoms. Furthermore, when these transitions are excited with a narrowband laser, the resonant atoms can be optically pumped into a different ground state, making the material more transparent at the laser frequency. This is known as burning a spectral hole, hence the name *SHB*. In many SHB materials, the optical pumping is into different ground state spin sublevels, and hence the hole burning process can initialize spin qubits, as illustrated in Fig. 10.8. This type of spin qubit initialization can be much faster than Boltzmann initialization, especially in spin systems with long spin population lifetimes, since the tradeoff between spin lifetime and initialization speed is removed.

In addition to optical initialization of spins, the hole burning process can also be used to readout the spin state of the qubits. This happens when the quantum algorithm returns some of the spin qubits to the state that was initially emptied by the laser, resulting in a temporary increase in laser

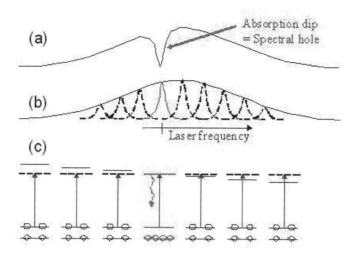

Figure 10.8: (a) The signature of spectral hole burning is a narrowband dip in the optical absorption spectrum. (b) This dip occurs when an optical laser bleaches out an ensemble of atoms at a particular transition frequency. (c) In the case when bleaching is due to spin sublevel optical pumping, it can be used to initialize qubits.

absorption and/or fluorescence that is proportional to the final population of this spin state. Of course, the readout process also re-initializes, so one must take care to work with a large enough ensemble to achieve the desired readout fidelity. In general, optical readout is orders of magnitude more sensitive than the typical NMR coil, and so it is possible to work with small ensembles consisting of very dilute dopant-host systems that can have very long spin coherence lifetimes.

Spin-qubit coherence lifetime in SHB materials can be lengthened by a variety of techniques. In dilute dopant-host systems, the choice of a spin-free or low-spin host has the largest benefit. Examples include praseodymium [31] or europium [16] doped in a yttrium-silicate host ($Pr : YSO$ or $Eu : YSO$) and nitrogen-vacancy [85] (NV) color centers doped in diamond; see Fig. 10.9. In $Pr : YSO$ only the yttrium host nuclei have spin but the magnetic moment is very weak. In NV diamond, the only host spins are 1% abundant ^{13}C which can be virtually eliminated with isotopically pure material. In dopant-host systems, dephasing due to host spins is reduced by the so-called *frozen core effect* [76], wherein the magnetic field generated by the active (qubit) spin system tunes nearby host nuclei out or resonance with the rest of the crystal up to a distance which defines the frozen core radius. This suppresses the energy conserving mutual spin flips that are the main source of spin decoherence.

In $Pr : YSO$ the spin Hamiltonian is given by [58]:

$$H = \boldsymbol{B} \cdot \left(g_J^2 \mu_B^2 \boldsymbol{\Lambda}\right) \cdot \boldsymbol{B} + \boldsymbol{B} \cdot \left(\gamma_N I_3 + 2A_J g_J \mu_B \boldsymbol{\Lambda}\right) \cdot \boldsymbol{I} + \boldsymbol{I} \cdot \left(A_J^2 \boldsymbol{\Lambda} + \boldsymbol{T}_Q\right) \cdot \boldsymbol{I}, \quad (10.50)$$

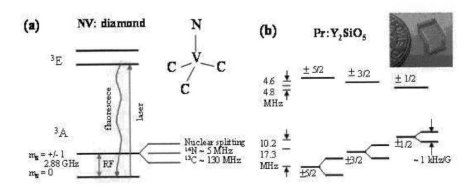

Figure 10.9: (a) Spin sublevels of nitrogen-vacancy (NV) color center in diamond. (b) Spin sublevels of Pr:YSO.

where the tensor Λ is given by

$$\Lambda_{\alpha\beta} = \sum_{n=1}^{2J+1} \frac{\langle 0|J_\alpha|n\rangle\langle n|J_\beta|0\rangle}{\Delta E_{n,0}}, \qquad (10.51)$$

I_3 is the 3×3 identity matrix, B is the magnetic field, and I is the nuclear spin vector, g_J is the Lande g, γ_N is the nuclear magnetogyric ratio, A_J is the hyperfine interaction. The term $I \cdot T_Q \cdot I$ described the nuclear electric quadrupole interaction and $A_J^2 I \cdot \Lambda \cdot I$ is the second order magnetic hyperfine or pseudoquadrupole interaction.

Recently, a spin coherence lifetime of 1/2 minute has been observed in $Pr : YSO$ [21]. This impressive result is made possible by combining two techniques. The first technique involves magnetically tuning the qubit spin to a level anti-crossing [20]. These are common in systems with spin 1 or larger. Near such an anti-crossing there is no first order magnetic Zeeman shift. Consequently, spin flips of nearby host and active spins, which ordinarily introduce coherence by perturbing the local magnetic field of the qubit, no longer have a first order effect. The complication is that the magnetic field is a vector so that the level anti-crossing must exist in all three directions. Nonetheless, such global level-crossings were found in $Pr : YSO$ and were used to lengthen the qubit spin coherence lifetime by orders of magnitude. More importantly, the residual spin decoherence was found to decay as a quadratic exponential in time, meaning it decays as $e^{-(t/\tau)^2}$. This is critical because most quantum error correction schemes require the short-time decay to be slower than the usual linear exponential decay. Since this condition was satisfied in $Pr : YSO$, a version of bang-bang error correction was successfully applied to give the observed half minute coherence times.

Manipulation of spin-qubits in SHB materials is generally done using RF coils similar to those used in liquid NMR. Recently, optical Raman transitions have been explored as an alternative to this, in which case the spin-qubits are manipulated by lasers instead of an RF coil. The advantages of this are 2-fold. First, the gate time can be made faster because it depends on the optical Rabi frequency, rather than that of the spin transition. One reason for this is that spin transitions are generally magnetic dipole allowed transitions, whereas optical transitions are often electric dipole allowed. Another reason is that it is often easier to insert strong laser fields into a cryostat than strong RF fields. Second, the selectivity of qubit excitation can be improved considerably because only spins with the correct optical transition frequency and spin transition frequency are manipulated. Additional spatial selectivity exists because the optical laser beams can be focused down to microns, and only the illuminated part undergoes qubit manipulations. This is especially important for algorithms like those designed for a Type II quantum computer; see Section 10.5.

The real power of optically addressed NMR lies in multi-qubit manipulations. The optical "handle" allows several options to increase scalability. First, the relatively long range of optical interactions frees NMR from the restrictions imposed by near-neighbor interactions. An example of this is an ensemble-based two qubit gate demonstration in $Eu : YSO$ involving ions separated by $100\,nm$ [59], which is orders of magnitude larger than distances required by conventional NMR. In this demonstration, a series of optical pulses refocuses a "target" optical qubit with a different phase depending on whether the "control" qubit is excited or not. Since these qubits are defined only by their transition frequency, neither the exact location nor number of spins located in between is unimportant. This demonstration also illustrates the interesting fact that optical transitions in some SHB materials have a coherence lifetime that is similar to that of many room temperature NMR transitions.

In principle, long range optical interactions such as in the $Eu : YSO$ example are scalable. In practice, however, the $Eu : YSO$ demonstration experiment is not very scalable because well-defined pairs of qubits must be distilled out of a random ensemble [72], and this incurs an exponential penalty with number of qubits. To make this technique more scalable one approach would be to apply it to special solid state pseudo-molecules. These pseudo-molecules exist in a number of stoichiometric crystals, the most interesting of which are those containing europium, for example $EuVO_4$ or Eu_2O_3 because of their narrow optical transitions at low temperatures [29, 57]. In these pseudo-molecules, localized defects have a large effect on the optical transition frequency of the Eu ions. Up to 50 optical transitions can be easily resolved in these materials. Assuming that all the defects are identical, each optical transition would correspond to a Eu spin system in a well-defined location near the defect center, thereby producing a pseudo-molecule. By using the long range optical coupling demonstrated in $Eu : YSO$, one could in principle construct up to a 50 qubit quantum computer without most of the usual scaling limitations of NMR.

To achieve scalability beyond 100 qubits, single spin manipulation is preferred. The excitation and especially detection of single spins in a solid is a very active area of research. Much of this research is based on a proposal to build a quantum computer using qubits consisting of nuclear spins of a phosphorous atom implanted in a silicon host [40], see Fig. 10.10. ^{31}P nuclei in spin-free ^{28}Si have a spin population lifetime on the order of hours at ultra-cold temperatures, with coherence lifetimes on the scale of 10s of ms so far. Single qubit manipulations would be done with the usual NMR pulse sequences. However, to avoid driving all the qubits at once, an off resonant RF field is applied and the active qubit is tuned into resonance when desired by using the interaction with its electron spin. The electron spin in turn is controlled by distorting the P electron cloud with a voltage applied to a nearby gate, called the A-electrode. To achieve 2-qubit logic, the electron clouds of two neighboring P atoms are overlapped using a J-electrode. The resulting exchange interaction between the two electrons can then be transferred to the P nuclei using RF and/or gate pulse sequences. Since the P atom has electron spin, Boltzmann initialization can be used, though a number of faster alternatives such as spin injection are being explored. For readout, the nuclear spin state is transferred to the electron spin which in turn is converted into a charge state via spin exchange interactions with a nearby readout atom. The charge state is then detected with a single electron transistor.

The Hamiltonian for P in Si is given by [40]:

$$H = \mu_B B \sigma_z^e - g_n \mu_n B \sigma_z^n + A \sigma^e \cdot \sigma^n \tag{10.52}$$

where μ_B is the Bohr magneton, μ_n is the nuclear magneton, g_n is the nuclear g-factor, B is the applied magnetic field (assumed parallel to z) and σ are the Pauli matrices, with superscripts e and n for electron and nuclear. For two coupled qubits the Hamiltonian becomes:

$$H = H(B) + A_1 \sigma^{1n} \cdot \sigma^{1e} + A_2 \sigma^{2n} \cdot \sigma^{2e} + J \sigma^{1n} \cdot \sigma^{2e} \tag{10.53}$$

where $H(B)$ is the magnetic field part and the superscripts 1 and 2 refer to the two spins,

$$4J \cong 1.6 \frac{e^2}{\varepsilon a_B} \left(\frac{r}{a_B} \right)^{5/2} e^{\frac{-2r}{a_B}} \tag{10.54}$$

where r is the distance between P atoms, ε is the dielectric constant of silicon, and a_B is the effective Bohr radius.

Many of the more challenging operations proposed for the P in Si quantum computer have recently be demonstrated in quantum wells/dots using individual electrons and/or excitons [77]. In most of these experiments electron spins, rather than nuclear spins, are the qubits. Unfortunately, these experiments are usually done in GaAs which is not a spin-free host and so decoherence is a problem. Spin-free semiconductor hosts exist, but fabrication in these systems is not yet as mature (except for silicon).

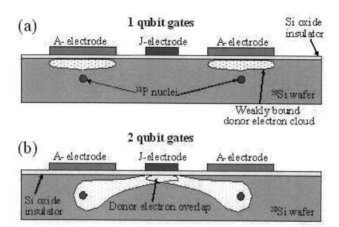

Figure 10.10: Operation of the P doped Si quantum computer. (a) Single qubits are tuned into resonance with applied microwave field using voltages on A-electrodes to distort weakly bound electron cloud. (b) Two qubit gates are enabled by using voltages on J-electrodes to overlap neighboring electron clouds, thereby switching on spin exchange coupling.

While the exciting prospect of single electron and/or nuclear spin detection is currently a technical challenge for most solids, it was done long ago in nitrogen-vacancy (NV) diamond [28]. In this optically active SHB material, single electron spin qubits are routinely initialized and read out with high fidelity at liquid helium temperature. In fact, the fidelity is so high that it begins to compare to trapped ions [34]. The electron spin coherence has also been transferred to nearby nuclear spins to perform (non-scalable) two qubit logic [33]. Single qubit logic is usually done with RF pulses, but optical Raman transitions between spin sublevels have been observed in NV diamond under certain experimental conditions [30]. Two qubit gates can be performed using the electron spin coupling between adjacent qubits. Initialization of such a 2-qubit system consisting of a NV and nearby N atom has recently been demonstrated using optical pumping. Scalability of this system can be achieved using an electron spin resonance (ESR) version of a Raman transition to transfer this spin coupling to the nuclear spins [86]. This ESR Raman has already been demonstrated for single NV spins.

The spin Hamiltonian for NV diamond is given by [70]:

$$H = D\left(S_z^2 - \frac{1}{3}S^2\right) + \beta_e \boldsymbol{B}g_e\boldsymbol{S} + \boldsymbol{SAI} + P\left(I_z^2 - \frac{1}{3}I^2\right) \tag{10.55}$$

where D is the zero-field splitting, B is the applied magnetic field, $\boldsymbol{A}$ is the electron-nuclear hyperfine coupling tensor, $\boldsymbol{S}$ and $\boldsymbol{I}$ are the electron and nuclear spin vectors, and P is the nuclear quadrupole contribution.

With optical Raman, it should be possible to use long-range optical dipole-dipole coupling [60], or eventually cavity-based Quantum ElectroDynamic
(QED) coupling, to perform two qubit spin logic, see Fig. 10.11. If successful, this will be highly scalable because the optical transition frequency can be tuned where desired using dc electric Stark shifts created by gate electrodes (A-electrodes). By requiring both optical laser frequency and electrode voltages to be correctly tuned, qubit excitation becomes very selective and 2-qubit coupling only exists when needed, in contrast to the usual case in NMR.

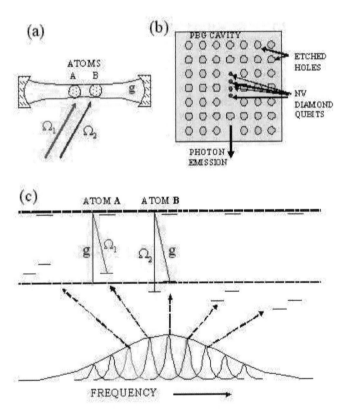

Figure 10.11: (a) Two distant qubits coupled by vacuum mode of cavity using cavity QED. (b) Possible implementation of multi-qubit bus using photon band gap cavity in NV diamond. (c) Illustration of the use of optical Raman with cavity QED to couple spectrally adjacent qubits.

Electron spin population lifetimes up to minutes have so far been observed at low temperature, with coherence lifetimes up to 0.3 milliseconds at room temperature [34]. Interestingly, the optical initialization and read-out still works at room temperature (though with less fidelity), as well as the ESR Raman transitions. This raises the intriguing question of whether or not a room-temperature solid-state NMR quantum computer can eventually be built using NV diamond.

10.4 Shor's algorithm and its experimental realization

In the rest of the chapter, we will describe two applications of the NMR quantum computer: Shor's algorithm and a lattice algorithm. We return to Subsection 2.7.5. A successful experiment has shown the potential capability of the implementation of Shor's algorithm, although it is still very simple and tentative. In [83], Vandersypen et al. factor 15 into 3 times 5. That work has demonstrated the liquid NMR quantum computer to be the most successful quantum computer so far.

10.4.1 Shor's algorithm ("hardwired" NMR experiment)

Let n be an odd integer to be factored, and choose another random integer x less than n. We require x to be coprime with n; otherwise, we find a factor of n immediately by the Euclidean method. It is then known that function $f(s) = x^s \mod n$ is periodic. The period of f (and also of x) is the smallest integer r such that $x^r = 1 \mod n$. For example, when $n = 15$ and $x = 3$, the moduli of x^s, with s being 1, 2, 3, ..., are 3, 9, 12, 6, 3, 9, 12, 6, ..., and the period is 4.

Now we check the period r. If r is even, $r = 2t$, then $x^{2t} - 1 = (x^t + 1)(x^t - 1) = 0 \mod n$, so either $x^t - 1$ or $x^t + 1$ has a common factor with n. A classical computer can use the Euclidean algorithm to compute the greatest common divisors, denoted as $\gcd(x^t + 1, n)$ and $\gcd(x^t - 1, n)$, in polynomial time. It is possible that we only obtain the trivial factors 1 or n using the x we choose. This happens only when $x^t = -1 \mod n$, since $x^t - 1 = 0 \mod n$ can not happen with r being already the smallest integer such that $x^r = 1 \mod n$. Fortunately it has been proved that the probability to meet such a bad x is at most $1/2^k$, where k is the number of distinct prime factors of n. Since k is at least 2, the probability is still large enough for us to find a good x, which has an even period r and $x^t \neq -1 \mod n$.

We now utilize quantum Fourier transform (QFT). As indicated in Box 2.5 in Chap. 2, we write $L = b$, $n = N$ and $2^L (= 2^b) = S$ therein, and obtain

$$|\psi_1\rangle = \frac{1}{\sqrt{S}} \sum_{k=0}^{S-1} |k\rangle|0\rangle, \tag{10.56}$$

where $S = 2^b$, the number of the total b-qubit states of the first register, with b large enough such that $2n^2 > S > n^2$.

We now design a certain series of pulses to compute $f(k) = x^k \bmod n$, and change the quantum state to

$$|\psi_2\rangle = \frac{1}{\sqrt{S}} \sum_{k=0}^{S-1} |k\rangle |f(k)\rangle. \qquad (10.57)$$

Now, apply QFT [68] to the first register in (10.57), which is a unitary transform mapping every $|k\rangle$ to another state:

$$|k\rangle \rightarrow \frac{1}{\sqrt{S}} \sum_{k=0}^{S-1} e^{2\pi i u k / S} |u\rangle. \qquad (10.58)$$

Then the quantum state of the system changes to

$$|\psi_3\rangle = \frac{1}{S} \sum_{u=0}^{S-1} |u\rangle \sum_{k=0}^{S-1} e^{2\pi i u k / S} |f(k)\rangle. \qquad (10.59)$$

Assume that $f(k)$ has period r, and we write $k = d + jr$ such that $0 \leq d < r$, where d is the remainder of k after it is divided by r and j ranges from 0 to A, the largest integer such that $Ar < S$. This way, we can write $|\psi_3\rangle$ as

$$|\psi_3\rangle = \frac{1}{S} \sum_{u=0}^{S-1} |u\rangle \sum_{d=0}^{r-1} |f(d)\rangle e^{2\pi i u d / S} \sum_{j=0}^{A} e^{2\pi i u r j / S} I_{(d+rj<S)},$$

where $I_{(d+rj<S)} = 1$ when $d + rj < S$, and 0 otherwise. If $S = (A+1)r$, $I_{(d+rj<S)} = 1$ for every d and j. If $S \neq (A+1)r$, it is still reasonable to ignore the difference and let $I_{(d+rj<S)} = 1$ everywhere because we have chosen S large enough. In this case, we let

$$b_u = \frac{1}{S} \sum_{j=0}^{A} e^{2\pi i u r j / S} = \frac{1}{S} \left(\frac{1 - e^{2\pi i u r (A+1)/S}}{1 - e^{2\pi i u r / S}} \right), \qquad (10.60)$$

thus our quantum state is now

$$|\psi_3\rangle = \frac{1}{S} \sum_{u=0}^{S-1} \sum_{d=0}^{r-1} b_u e^{2\pi i u d / S} |u\rangle |f(d)\rangle.$$

We can now measure the first register, and we want to find such a u, for which there is an l satisfying

$$\left| \frac{u}{S} - \frac{l}{r} \right| \leq \frac{1}{2S}. \qquad (10.61)$$

There are about r such u's, and it has been estimated that the probability to find such a u is at least 0.4 [68]. Because $\frac{1}{2S} < \frac{1}{2n^2}$, and we know that

$r < n$, there is at most one fraction $\frac{k}{r}$ satisfying the condition and we can use *continued fraction expansions* to find the fraction. If k and r are coprime, we obtain r as the denominator of the fraction. If not, we only find a factor of r. If r is odd or $x^{r/2}$ does not give us a useful result, choose another x and try again. It may be necessary to try several (of the order $\mathcal{O}(\log\log n)$) times until r is successfully found, but the overall running time is still reasonable.

10.4.2 Circuit design for Shor's algorithm

Before we introduce the experiment by Vandersypen, et al. [83], we extend the above discussion a little further to the case when r divides S. Now S/r becomes an integer and (10.60) always holds so that S doesn't have to be very large. Moreover, (10.61) becomes an identity

$$u = \frac{l \cdot S}{r}, \qquad (10.62)$$

i.e., r is the denominator of the fraction $\frac{u}{S}$ after canceling the common factor between u and S if l and r are coprime. The integer 15 falls into this situation. The possible x can be 2, 4, 7, 8, 11, or 13. When we choose x to be 2, 7, 8 or 13, the period r is 4. In other cases, r is 2. The period r divides $S = 2^b$ in both cases. Only 2 qubits at most are required to compute one period of f. In the experiment, 3 qubits are used to obtain more periods.

Vandersypen, et al. [83] used liquid NMR to realize Shor's algorithm in factorizing 15. The sample in the experiment is a custom-synthesized material whose molecules have five ^{19}F and two ^{13}C, so it has seven qubits ready for use. Those seven qubits are divided into two registers, 3 to store the number k (the first register, represented by $|k_2k_1k_0\rangle$) and 4 to store the modular exponentiation y (the second register, represented by $|y_3y_2y_1y_0\rangle$); see Figs. 10.12 and 10.13. The total Hamiltonian is

$$H = \sum_{i=1}^{7} \frac{1}{2}\omega_i\sigma_z^i + \sum_{i<j} 2\pi J_{ij}\sigma_z^i\sigma_z^j.$$

Each run of the experiment consists of 4 steps. In the first step, the sample is initialized to a certain pseudo-pure state; in the second step, a series of specially designed pulses are applied to realize the computation of modular exponentiation; in the third step, QFT is applied to the first register; finally, the period was obtained through the reading of the spectrum. The system begins from thermal equilibrium, where the density matrix is given by $\rho_0 = e^{-H/kT} \approx I - \frac{H}{kT}$. A suitable initial pseudo-pure state $|\psi_1\rangle = |0000001\rangle$ is obtained by the temporal averaging method.

Although it is difficult to design a general circuit for the modular exponentiation, it is easy to "hard-wire" for this special case in consideration. As the exponent k can be written as $k = k_0 + 2k_1 + 4k_2$, we can change the modular exponential x^k mod 15 into successive operations of modular multiplications by $x^{2^i k_i}$, with $i = 0, 1, 2$, applied to the second register y beginning from $y = 1$.

When $i = 0$, $y \cdot x = x = 1 + (x - 1)$, so the multiplication is actually a controlled-addition with $(x - 1)$ in case $k_0 = 1$. For $x = 7 = (0111)_2$, it is equal to flip the state of y_1 and y_2 ($y = (0001)_2$ before the multiplication). For $x = 11 = (1011)_2$, the same reasoning shows that we only have to flip the state of y_3 and y_1, depending on k_0. Gates A and B in Figs. 10.12 and 10.13 accomplish the modular multiplication x^{k_0}.

The situation is a little more complicated for $i = 1$. We only discuss the situation when $k_1 = 1$, since y will not change when $k_1 = 0$. Different strategies are needed for $x = 7$ and $x = 11$. When $x = 11$, since $11^2 = 121 = 15 \times 8 + 1$, $y \times 11^2 = y \pmod{15}$. We need to do nothing and the same result holds for the third qubit k_2. When $x = 7$, we can design the circuit by first investigating the following identity

$$
\begin{aligned}
y \cdot 7^2 &= y \cdot 4 \quad \mod 15 \\
&= (y_0 + 2y_1 + 4y_2 + 8y_3) \cdot 4 \quad \mod 15 \\
&= (4y_0 + 8y_1 + 16y_2 + 32y_3) \quad \mod 15 \\
&= (y_2 + 2y_3 + 4y_0 + 8y_1) \quad \mod 15 \\
&= (y_2 \cdot 2^0 + y_3 \cdot 2^1 + y_0 \cdot 2^2 + y_1 \cdot 2^3) \quad \mod 15.
\end{aligned}
$$

It shows that the modular multiplication can be achieved by exchanging the first qubit y_0 with the third qubit y_2, and the second qubit y_1 with the fourth qubit y_3. In Fig. 10.12, gates C, D, and E are used to accomplish the former, and gates F, G, and H the latter. Further simplification of the circuit can be made. Since the control bit y_3 is $|0\rangle$ before gate C, that gate can just be omitted. Gates H and E have no effect on the period; they can be omitted, too.

The circuit design for the quantum Fourier transform is just a standard design; see, e.g., [14, Fig. 5]. It has 3 Hadamard gates and 3 controlled-phase gates. Figs. 10.12 and 10.13 show the circuit designs for $y = 7$ and $y = 11$. Totally about 300 pulses are used in the experiment and it takes about 700ms to accomplish all steps in the case of $x = 7$.

10.4.3 Experimental result

Readout of the experiment needs a careful interpretation of the data. Because an NMR sample consists of many molecules, the readout is the average value of u from all molecules instead of the reading from a single molecule.

Both qubits k_0 and k_1 are found to be in state $|0\rangle$ after the extraction of the spectra [83] for the easy case of $x = 11$, while qubit k_2 is in a equally mixture state of $|0\rangle$ and $|1\rangle$. Thus the possible u can be 0 and 4, i.e., 000 and 100 in binary form. From (10.62), r can be obtained as $r = 8/4 = 2$, and the greatest common divisors are computed as $\gcd(11^{2/2} + 1, 15) = 3$ and $\gcd(11^{2/2} - 1, 15) = 5$.

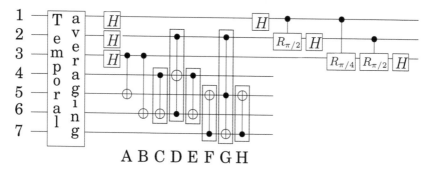

Figure 10.12: The quantum circuit for the (hard) case for the realization of Shor's algorithm ($x = 7$). From top to bottom, the qubits are k_2, k_1, k_0, y_3, y_2, y_1 and y_0, respectively, in sequential order.

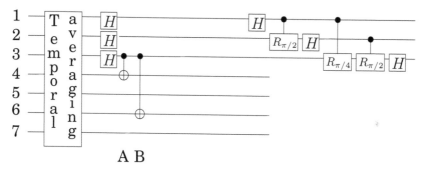

Figure 10.13: The quantum circuit for the (easy) case for realization of Shor's algorithm ($x = 11$). From top to bottom, the qubits are k_2, k_1, k_0, y_3, y_2, y_1 and y_0, respectively, in sequential order.

In the case of $x = 7$, the spectra in [83] indicate that only qubit k_0 is in state $|0\rangle$, and both qubits k_1 and k_2 are in equal mixture of states $|0\rangle$ and $|1\rangle$. Thus u is in a mixture of states $|0\rangle$, $|2\rangle$, $|4\rangle$ and $|6\rangle$. We can see that the period of u is 2, thus the period of the modular exponent r is $8/2 = 4$. The factors of 15 can finally be obtained as $\gcd(7^{4/2} - 1, 15) = 3$ and $\gcd(7^{4/2} + 1, 15) = 5$.

10.5 Quantum algorithm for lattice-gas systems

In the previous sections, we have explained how to construct a quantum computer using liquid NMR and illustrated a successful experiment. We have taken it for granted that the coherence can be maintained long enough and different qubits can be entangled even they are separated far apart in space. Unfortunately, these assumptions are not always practical and in fact they constitute great obstacles to overcome. The problem becomes more serious when more qubits are involved. *Type-II quantum computers* are proposed to alleviate this problem. A Type-II quantum computer is composed of a network or array of small quantum computers interconnected by classical communication channels [92]. Instead of the global coherence and entanglement, only local coherence and entanglement within every small quantum computer, called a *node*, are required, and the difficulty faced by the centralized quantum computer is dramatically eased.

The wave function of the whole Type-II quantum computer system is a tensor product over all the nodes:

$$|\psi(t)\rangle = |\psi(x_1, t)\rangle \otimes \quad \cdots \quad \otimes |\psi(x_N, t)\rangle, \tag{10.63}$$

where N is the number of nodes. The lattice gas algorithm (LGA) is specially suited for this structure. Every computation cycle can be broken up into three steps with two intermediate states $|\psi'\rangle$ and $|\psi''\rangle$:

$$\begin{aligned}
|\psi'\rangle &= \hat{C}|\psi(t)\rangle, \\
|\psi''\rangle &= \Gamma|\psi'\rangle, \\
|\psi(t+1)\rangle &= T|\psi''\rangle,
\end{aligned} \tag{10.64}$$

where $\hat{C}$ is a unitary operator acting locally on every node, while Γ is a projection operator, such as a measurement, and T is the streaming operator which exchanges information among nodes. The Type-II quantum computer takes advantage of parallelism in two ways: one classical, all the nodes work simultaneously; the other quantum, quantum entanglement is still kept inside every node. Because measurement is applied and the system is reset at the end of every computation cycle, the coherence only needs to be maintained for a short time.

10.5.1 Quantum algorithm for a lattice-gas model

Consider a 1-dimensional diffusion equation without boundary condition

$$\frac{\partial \rho}{\partial t} = \frac{\partial^2 \rho}{\partial x^2}, \tag{10.65}$$

where ρ is the mass density or temperature function along the x-axis. Using the finite difference method, we can write a finite difference approximation to solve the above partial differential equation numerically:

$$\frac{\rho(x, t+\tau) - \rho(x, t)}{\tau} = \frac{\rho(x+l, t) - 2\rho(x, t) + \rho(x-l, t)}{l^2}, \tag{10.66}$$

where τ is the time step size and l is the space step size. From physics, we know that the above equation may be studied by a lattice gas algorithm. Without loss of generality, we assume that τ and l are normalized so that the difference equation can be written as

$$\rho(x_i, k+1) - \rho(x_i, k) = \frac{1}{2}(\rho(x_{i+1}, k) - 2\rho(x_i, k) + \rho(x_{i-1}, k)).$$

Points x_i are evenly distributed along the x-axis, also called *nodes*. To study the above equation, two functions, $f_1(x_i, k)$ and $f_2(x_i, k)$, called *channels*, are defined for each node x_i at time k. The set of values of these two functions are called the *state* of node x_i. Any physical observable, such as the density function $\rho(x_i, k)$, is a function of the state at the node. The evolution of the lattice, or the state of all nodes, consists of two operations: *collision* and *propagation*. A collision is a local operator only defined by the state of the node itself. The propagation operator transfers information from one node to another and the state at one node changes according to the state of other nodes. This is completed by defining a velocity vector for every channel which gives the information flow a direction. In our special example here, information in the two channels flows in opposite directions. After propagation, one channel gets its new value from its left neighbor, while the other from its right neighbor. This LGA is completed with a Type II quantum computer by J. Yepez of the Air Force Research Laboratory and M.A. Pravia, et al. of the Department of Nuclear Engineering at MIT [2, 71, 91, 92]. The actual result is not as good as desired, but improvement is still possible.

To store a floating point number, a classical computer uses a register with 32 or 64 bits, depending on the machine. Quantum computers presently have difficulty in doing it the same way as classical computers because there is not yet the technology for 32 or 64 qubits. In this quantum lattice-gas algorithm, a two qubit system is proposed for every node. The two qubits are represented by $|q_1(x_i, k)\rangle$ and $|q_2(x_i, k)\rangle$, respectively, and

$$|q_1(x_i, k)\rangle = \sqrt{f_1(x_i, k)}|0\rangle + \sqrt{1 - f_1(x_i, k)}|1\rangle,$$

$$|q_2(x_i, k)\rangle = \sqrt{f_2(x_i, k)}|0\rangle + \sqrt{1 - f_2(x_i, k)}|1\rangle.$$

The state of the whole system $|\psi(x_i, k)\rangle$ at node x_i and time k is a tensor product:

$$
\begin{aligned}
|\psi(x_i, k)\rangle &= |q_1(x_i, k)\rangle|q_2(x_i, k)\rangle \\
&= \sqrt{f_1(x_i, k)f_2(x_i, k)}|00\rangle + \sqrt{(1 - f_1(x_i, k))f_2(x_i, k)}|10\rangle \\
&\quad + \sqrt{f_1(x_i, k)(1 - f_2(x_i, k))}|01\rangle \\
&\quad + \sqrt{(1 - f_1(x_i, k))(1 - f_2(x_i, k))}|11\rangle.
\end{aligned}
$$

Quantities $f_1(x_i, k)$ and $f_2(x_i, k)$ are the probabilities of occurrence of the state $|0\rangle$ for qubit 1 and 2, respectively, corresponding to the two channels, and $1 - f_{1,2}(x_i, k)$ are the occurrence probabilities of the state $|1\rangle$. Since the states are normalized, $0 \leq f_{1,2}(x_i, k) \leq 1$, and we let $\rho(x_i, k) = f_1(x_i, k) + f_2(x_i, k)$. It is noted that our Type-II quantum computer assigns ρ a continuous value (a function of the occurrence probabilities) rather than a discrete value as a digital computer does. An array of two qubit systems are used in the computation, corresponding to a series of nodes.

The quantum LGA here has three steps in every cycle that complete a step of the finite difference algorithm computation: *collision, measurement,* and *re-initialization*. The last two composed are equal to one propagation operation in a normal lattice gas algorithm. Because the propagation needs information exchange among different nodes, measurement and classical communication are needed to accomplish one operation. We map the quantum state to a vector in $\mathbf{C}^4$ as that given in (10.23).

In the collision step, a unitary operator is applied simultaneously to all nodes:

$$|\bar{\psi}(x_i, k)\rangle = U|\psi(x_i, k)\rangle,$$

where

$$U = \begin{bmatrix} 1 & 0 & 0 & 0 \\ 0 & \frac{1}{2} - \frac{i}{2} & \frac{1}{2} + \frac{i}{2} & 0 \\ 0 & \frac{1}{2} + \frac{i}{2} & \frac{1}{2} - \frac{i}{2} & 0 \\ 0 & 0 & 0 & 1 \end{bmatrix}. \tag{10.67}$$

The new occurrence probabilities of the state $|0\rangle$ of the two qubits after the operation, $\bar{f}_1$ and $\bar{f}_2$, can be computed using

$$\bar{f}_1 = \langle\bar{\psi}|n_1|\bar{\psi}\rangle, \qquad n_1 = \begin{bmatrix} 1 & 0 & 0 & 0 \\ 0 & 1 & 0 & 0 \\ 0 & 0 & 0 & 0 \\ 0 & 0 & 0 & 0 \end{bmatrix},$$

$$\tag{10.68}$$

$$\bar{f}_2 = \langle\bar{\psi}|n_2|\bar{\psi}\rangle, \qquad n_2 = \begin{bmatrix} 1 & 0 & 0 & 0 \\ 0 & 0 & 0 & 0 \\ 0 & 0 & 1 & 0 \\ 0 & 0 & 0 & 0 \end{bmatrix},$$

leading to

$$\bar{f}_1(x_i, k) = \tfrac{1}{2}(f_1(x_i, k) + f_2(x_i, k)),$$
$$\bar{f}_2(x_i, k) = \tfrac{1}{2}(f_1(x_i, k) + f_2(x_i, k)).$$

The collision operator is actually doing a job of averaging. The state after the collision is also called the local equilibrium.

In the second step, a measurement is applied at every node and $\bar{f}_{1,2}(x_i, k)$ of all the nodes are retrieved for future use.

In the third step, using information from the measurement from the previous step, the state of all the nodes are re-initialized to a separable state

$$|q_1(x_i, k+1)\rangle = \sqrt{\bar{f}_1(x_{i+1}, k)}|0\rangle + \sqrt{1 - \bar{f}_1(x_{i+1}, k)}|1\rangle,$$
$$|q_2(x_i, k+1)\rangle = \sqrt{\bar{f}_2(x_{i-1}, k)}|0\rangle + \sqrt{1 - \bar{f}_2(x_{i-1}, k)}|1\rangle. \tag{10.69}$$

It can be seen that the second and third steps have accomplished the propagation operation. At node x_i, the new state of channel one is acquired from the same channel of its right neighbor node x_{i+1}, and channel two acquires its state from its left neighbor. It is complicated here only because the communication between two quantum systems is difficult.

To see how this LGA works, let us begin from a local equilibrium state, $f_1(x_i, k) = f_2(x_i, k) = \rho(x_i, k)/2$, where the states come off a collision operation (step 2). We list the $f_{1,2}$ around position x_i before the third step in two rows

$$f_1: \quad \ldots \quad \frac{\rho(x_{i-2},k)}{2} \quad \frac{\rho(x_{i-1},k)}{2} \quad \frac{\rho(x_i,k)}{2} \quad \frac{\rho(x_{i+1},k)}{2} \quad \ldots$$
$$f_2: \quad \ldots \quad \frac{\rho(x_{i-2},k)}{2} \quad \frac{\rho(x_{i-1},k)}{2} \quad \frac{\rho(x_i,k)}{2} \quad \frac{\rho(x_{i+1},k)}{2} \quad \ldots$$

and after the third step

$$f_1: \quad \ldots \quad \frac{\rho(x_{i-1},k)}{2} \quad \frac{\rho(x_i,k)}{2} \quad \frac{\rho(x_{i+1},k)}{2} \quad \frac{\rho(x_{i+2},k)}{2} \quad \ldots$$
$$f_2: \quad \ldots \quad \frac{\rho(x_{i-3},k)}{2} \quad \frac{\rho(x_{i-2},k)}{2} \quad \frac{\rho(x_{i-1},k)}{2} \quad \frac{\rho(x_i,k)}{2} \quad \ldots$$

We can see that the row of f_1 (channel one) moves left and the row of f_2 (channel two) moves right. According to our definition, the new value of ρ is the sum of $f_1(x_i, k+1)$ and $f_2(x_i, k+1)$, i.e., $\rho(x_i, k+1) = \frac{1}{2}(\rho(x_{i+1}, k) + \rho(x_{i-1}, k))$. It is easy to check that

$$\rho(x_i, k+1) - \rho(x_i, k) = \frac{1}{2}(\rho(x_{i+1}, k) - 2\rho(x_i, k) + \rho(x_{i-1}, k)),$$

as desired.

Applications of the Type-II quantum computer with quantum LGA also have been reported in the simulation of the time-dependent evolution of a many-body quantum mechanical system [93], solution of a 1-dimensional magnetohydrodynamic turbulence [78], representation of solitons [79] and other equations.

10.5.2 Physical realization and result

The experiment in Subsection 10.5.1 uses a 2-qubit molecule, chloroform, whose structure is shown in Fig. 10.4. The hydrogen and carbon nuclei serve as qubit 1 and 2, respectively.

The actual results obtained from the experiment are compared with simulation results [71]. After 12 steps, the error becomes very large. Imperfection in the decoupling sequences is blamed and it is believed that the problem can be mitigated when the technology is improved in the future. Extreme requirement of high accuracy in the control pulse and readout is a disadvantage of this Type-II quantum computer, because it uses a continuous representation (the probability of occurrence) instead of a discrete one. Thus, it is more vulnerable to the inaccuracy in the NMR operation. Small errors in every step accumulate and finally become intolerable. Repeated measurement and re-initialization ease the requirement for coherence time, but place a high requirement on the fidelity at the same time.

10.6 Conclusion

In this chapter, we present the basic technology used to construct a quantum computer with liquid state NMR. The successful experiments for many algorithms have shown that liquid state NMR is capable of simulating a quantum computer and forms a test bed for quantum algorithms. It is so far the only technology available to realize a 7 qubit algorithm in laboratory. One reason for its success is the robustness of the spin system which only interacts with the external magnetic field, and it is possible to maintain the coherence for a long time (from seconds to hours). Besides, over the 60 years, history of NMR spectroscopy, analytic tools have been developed for the purpose of chemical and medical applications, and exact description and dedicated coherence control of the dynamics of the quantum spin system is now available to achieve high accuracy in the pulse design and application. In fact, the experimental techniques established in NMR, especially the coherence control technology, can be easily transferred to other quantum systems if they have a similar Hamiltonian, and the research in NMR is therefor helpful for the development of other more complicated and powerful quantum computers.

Liquid NMR has played a pioneering role in the quantum computer technology development. But its lack of scalability has constituted a severe obstacle to its future applicability. However, in Section 10.3, we show new technology of *solid state NMR* which has the potential to overcome liquid NMR's difficulties. For solid state NMR, under low temperature, the relaxation times of spins are typically very long, and the coupling between qubits is strong so that the control can be fast and easy. The small ratio of the gate time and the decoherence time makes more gates available, and more complicated algorithms can be tested. The nuclei can be cooled down easily and the spin system is highly polarized. The signal is much stronger so that fewer nuclei are needed. Even without the help of a gradient field and silicon technology, as we have mentioned, a quantum computer with 30 to 40 qubits is envisioned with designed molecules similar to that of the liquid state NMR computer except that the ensemble is in a solid crystal state. This is already a quantum system that reaches the limit a classical computer can simulate. Although it is still not scalable and not a standard quantum computer, these small and medium scale quantum computers will help in the building of a scalable and working quantum computer.

References

[1] E. Abe, K.M. Itoh, T.D. Ladd, J.R. Goldman, F. Yamaguchi, and Y. Yamamoto, Solid-state silicon NMR quantum computer, *Journal of Superconductivity: Incorporating Novel Magnetism* **16** (2003), 1, 175–178.

[2] G.P. Berman, A.A. Ezhov, D.I. Kamenev, and J. Yepez, Simulation of the diffusion equation on a type-II quantum computer, *Phys. Rev. A* **66** (2002), 012310.

[3] F. Bloch, Nuclear induction, *Phys. Rev.* **70** (1946), 460–474.

[4] A.M. Childs, I.L. Chuang, and D.W. Leung, Realization of quantum process tomography in NMR, *Phys. Rev. A* **64** (2001), 012314.

[5] I.L. Chuang, Quantum computation with nuclear magnetic resonance, in *Introduction to Quantum Computation and Information*, edited by H. Lo, S. Popescu, and T. Spiller, World Scientific, Singapore, 1998.

[6] I.L. Chuang, L.M.K. Vandersypen, X. Zhou, D.W. Leung, and S. Lloyd, Experimental realization of a quantum algorithm, *Nature* **393** (1998), 143–146.

[7] D.G. Cory, A.F. Fahmy, and T.F. Havel, Ensemble quantum computing by NMR spectroscopy, *Proc. Natl. Acad. Sci. USA* **94** (1997), 1634–1639.

[8] D.G. Cory, M.D. Price, and T.F. Havel, Nuclear magnetic resonance spectroscopy: An experimentally accessible paradigm for quantum computing, *Physica D* **120** (1998), 82–101.

[9] H.K. Cummings and J.A. Jones, Use of composite rotations to correct systematic errors in NMR quantum computation, *New Journal of Physics* **2** (2000), 6.1–6.12.

[10] D. Deutsch and R. Jozsa, Rapid solution of problems by quantum computer, *Proc. R. Soc. London. A* **439** (1992), 553–558.

[11] K. Dorai, Arvind, and A. Kumar, Implementing quantum-logic operations, pseudopure states, and the Deutsh-Jozsa algorithm using non-commuting selective pulses in NMR, *Phys. Rev. A* **61** (2000), 042306.

[12] K. Dorai, and D. Suter, Efficient implementations of the quantum Fourier transform: an experimental perspective, *International Journal of Quantum Information* **3** (2005), 2, 413–424.

[13] J.C. Edwards, http://www.process-nmr.com/nmr.htm.

[14] A. Ekert and R. Jozsa, Quantum computation and Shor's factoring algorithm, *Reviews of Modern Physics* **68** (1996), 3, 733–753.

[15] J.W. Emsley, J. Feeney, and L.H. Sutcliffe, *High Resolution Nuclear Magnetic Resonance Spectroscopy*, Pergamon Press, Oxford, 1965.

[16] R.W. Equall, Y. Sun, and R.M. Macfarlane, Ultraslow Optical Dephasing In Eu-3+-Y2SiO5, *Phys. Rev. Lett.* **72** (1994), 2179–2181.

[17] X. Fang, X. Zhu, M. Feng, X. Mao, and F. Du, Experimental implementation of dense coding using nuclear magnetic resonance, *Phys. Rev. A* **61** (2000), 022307.

[18] S. Fernbach and W.G. Proctor, Spin-Echo memory device, *J. of Applied Physics* **26** (1955), 2, 170–181.

[19] E.M. Fortunato, M.A. Pravia, and N. Boulant et al., Design of strongly modulating pulses to implement precise effective Hamiltonians for quantum information processing, *J. of Chemical Physics* **116** (2002), Issue 17, 7599–7606.

[20] E. Fraval, M.J. Sellars, and J.J. Longdell, Method of extending hyperfine coherence times in Pr3+: Y2SiO5, *Phys. Rev. Lett.* **92** (2004), 077601.

[21] E. Fraval, M.J. Sellars, and J.J. Longdell, Dynamic decoherence control of a solid-state nuclear-quadrupole qubit, *Phys. Rev. Lett.* **95** (2005), 030506.

[22] R. Freeman, Shaped radiofrequency pulses in high resolution NMR, *J. of Progress in Nuclear Magnetic Resonance Spectroscopy* **32** (1998), 59–106.

[23] B. Fung and V.L. Ermakov, A simple method for the preparation of pseudopure states in nuclear magnetic resonance quantum information processing, *Journal of Chemical Physics* **121** (2004), 17, 8410–8414.

[24] N. Gershenfeld and I.L. Chuang, Bulk spin-resonance quantum computation, *Science* **275** (1997), 350–356.

[25] S.J. Glaser, T. Schulte-Herbruggen, M. Sieveking, O. Schedletzky, N.C. Nielsen, O.W. Sorensen, and C. Griesinger, Unitary control in quantum ensembles: maximizing signal intensity in coherent spectroscopy, *Science* **280** (1998), Issue 5362, 421–424.

[26] J.R. Goldman, T.D. Ladd, F. Yamaguchi, and Y. Yamamoto, Magnet designs for a crystal-lattice quantum computer, *Appl. Phys. A* **71** (2000), 11–17.

[27] L.G. Grover, Quantum mechanics helps in searching for a needle in a haystack, *Phys. Rev. Lett.* **79** (1997), 325.

[28] A. Gruber, A. Drabenstedt, C. Tietz, L. Fleury, J. Wrachtrup, and C. vonBorczyskowski, Scanning confocal optical microscopy and magnetic resonance on single defect centers, *Science* **276** (1997), Issue 5321, 2012–2014.

[29] P.C. Hansen, M.J.M. Leask, B.M. Wanklyn, Y. Sun, R.L. Cone, and M.M. Abraham, Spectral hole burning and optically detected nuclear quadrupole resonance in flux-grown stoichiometric europium vanadate crystals, *Phys. Rev. B* **56** (1997), 7918–7929.

[30] P.R. Hemmer, A.V. Turukhin, M.S. Shahriar, and J.A. Musser, Raman excited spin coherences in N-V diamond, *Optics Letters* **26** (2001), 361–363.

[31] K. Holliday, M. Croci, E. Vauthey, and U.P. Wild, Spectral hole-burning and holography in an Y2SiO5Pr3+ crystal, *Phys. Rev. B* **47** (1993), 14741–14752.

[32] J.P. Hornak, http://www.cis.rit.edu/htbooks/nmr/inside.htm.

[33] F. Jelezko, T. Gaebel, I. Popa, M. Domhan, A. Gruber, and J. Wrachtrup, Observation of coherent oscillation of a single nuclear spin and realization of a 2-qubit conditional quantum gate, *Phys. Rev. Lett.* **93** (2004), 130501.

[34] F. Jelezko and J. Wrachtrup, Quantum information processing in diamond, quant-ph/0510152.

[35] J.A. Jones, Robust quantum information processing with techniques from liquid–state NMR, *Phil. Trans. Roy. Soc. Lond. A* **361** (2003), 1429–1440.

[36] J.A. Jones, Robust Ising gates for practical quantum computation, *Phys. Rev. A* **67** (2003), 012317.

[37] J.A. Jones, NMR quantum computation: a critical evaluation, *Fortschr. Phys.* **48** (2000), 909–924.

[38] J.A. Jones and E. Knill, Efficient refocusing of 1-spin and 2-spin interaction for NMR quantum computation, *J. of Magnetic Resonance* **141** (1999), 322–325.

[39] J.A. Jones and M. Mosca, Implementation of a quantum algorithm on a nuclear magnetic resonance quantum computer, *Journal of Chemical Physics* **109** (1998), 5, 1648–1653. Another short version can be found at *Nature* **393** (1998), 344–346.

[40] B.E. Kane, A silicon-based nuclear spin quantum computer, *Nature* **393** (1998), 133–137.

[41] N. Khaneja, R. Brockett, and S.J. Glaser, Time optimal control in spin system, *Phys. Rev. A* **63** (2001), 032308.

[42] N. Khaneja, S.J. Glaser, and R. Brockett, Sub-Riemannian geometry and time optimal control of three spin systems: quantum gates and coherent transfer, *Phys. Rev. A* **65** (2002), 032301.

[43] A.K. Khitrin, H. Sun, and B.M. Fung, Method of multifrequency excitation for creating pseudopure states for NMR quantum computing, *Phys. Rev. A* **63** (2001), 020301.

[44] E. Knill, I. Chuang, and R. Laflamme, Effective pure states for bulk quantum computation, *Phys. Rev. A* **57** (1998), 3348.

[45] E. Knill, R. Laflamme, R. Martinez, and C.H. Tseng, An algorithm benchmark for quantum information processing, *Nature* **404** (2000), 368–370.

[46] E. Knill, R. Laflamme, R. Martinez, and C. Negrevergne, Benchmarking quantum computers: the five-qubit error correcting code, *Phys. Rev. Lett.* **86** (2001), 5811–5814.

[47] E. Kupce and R. Freeman, Close encounters between soft pulses, *J. of Magnetic Resonance Series A* **112** (1995), 261–264.

[48] T.D. Ladd, J.R. Goldman, F. Yamaguchi, and Y. Yamamoto, All-silicon quantum computer, *Phys. Rev. Lett.* **89** (2002), No. 1, 017901.

[49] T.D. Ladd, D. Maryenko, Y. Yamamoto, E. Abe, and K.M. Itoh, Coherence time of decoupled nuclear spins in silicon, *Phys. Rev. B* **71** (2005), 014401.

[50] R. Laflamme, E. Knill, C. Negrevergne, R. Martinez, S. Sinha, and D.G. Cory, Introduction to NMR quantum information processing, in *Experimental Quantum Computation and Information, Proceeding of the International School of Physics "Enrico Fermi"*, edited by F. De Martini and C. Monroe, IOS Press, Amsterdam, Netherlands, 2002.

[51] G.M. Leskowitz, N. Ghaderi, R.A. Olsen, and L.J. Mueller, Three-qubit nuclear magnetic resonance quantum information processing with a single-crystal solid, *J. of Chemical Physics* **119** (2003), 1643–1649.

[52] D.W. Leung, I.L. Chuang, F. Yamaguchi, and Y. Yamamoto, Efficient implementation of coupled logic gates for quantum computation, *Phys. Rev. A* **61** (2000), 042310.

[53] M.H. Levitt, Composite pulses, *Progress in Nuclear Magnetic Resonance Spectroscopy* **18** (1986), 61–122.

[54] N. Linden, H. Barjat, and R. Freeman, An implementation of the Deutsch-Jozsa algorithm on a 3-qubit NMR quantum computer, *Chemical Physics Letters* **296** (1998), 61–67.

[55] N. Linden, B. Herve, R.J. Carbajo, and R. Freeman, Pulse sequences for NMR quantum computers: how to manipulate nuclear spins while freezing the motion of coupled neighbors, *Chemical Physics Letters* **305** (1999), 28–34.

[56] G. Long, H. Yan, Y. Li, C. Lu, J. Tao, and H. Chen et al., Experimental NMR realization of a generalized quantum search algorithm, *Physics Letters A* **286** (2001), 121–126.

[57] J.J. Longdell and M.J. Sellars, Selecting ensembles for rare earth quantum computation, quant-ph/0310105, 2003.

[58] J.J. Longdell, M.J. Sellars, and N.B. Manson, Hyperfine interaction in ground and excited states of praseodymium-doped yttrium orthosilicate, *Phys. Rev. B* **66** (2002), 035101.

[59] J.J. Longdell, M.J. Sellars, and N.B. Manson, Demonstration of conditional quantum phase shift between ions in a solid, *Phys. Rev. Lett.* **93** (2004), 130503.

[60] M.D. Lukin and P.R. Hemmer, Quantum Entanglement via Optical Control of Atom-Atom Interactions, *Phys. Rev. Lett.* **84** (2000), 2818–2821.

[61] R.M. Macfarlane, High-resolution laser spectroscopy of rare-earth doped insulators: a personal perspective, *J. of Luminescence* **100** (2002), 1-4, 1–20.

[62] M.L. Martin and G.J. Martin, *Practical NMR Spectroscopy*, Heyden, London, U.K., 1980.

[63] R. Marx, A.F. Fahmy, J.M. Myers, W. Bermel, and S.J. Glaser, Approaching five-bit NMR quantum computing, *Phys. Rev. A* **62** (2000), 012310.

[64] G.D. Mateescu and A. Valeriu, *2D NMR : Density Matrix and Product Operator Treatment*, PTR Prentice-Hall, New Jersey, 1993.

[65] M.A. Nielsen and I.L. Chuang, *Quantum Computation and Quantum Information*, Cambridge University Press, Cambridge, U.K., 2000.

[66] M.A. Nielson, E. Knill, and R. Laflamme, Complete quantum teleportation using NMR, *Nature* **396** (1995), 52–55.

[67] S.L. Patt, Single- and multiple-frequency-shifted Laminar pulses, *J. of Magnetic Resonance* **96** (1992), 1, 94–102.

[68] A.O. Pittenger, *An Introduction to Quantum Computing Algorithms*, Birkhäuser, Boston, 2000.

[69] C.P. Poole, and H.A. Farach, *Theory of Magnetic Resonance*, Wiley, New York, 1987.

[70] I. Popa, T. Gaebel, M. Domhan, C. Wittmann, F. Jelezko, and J. Wrachtrup, Energy levels and decoherence properties of single electron and nuclear spins in a defect center in diamond, quant-ph/0409067, 2004.

[71] M.A. Pravia, Z. Chen, J. Yepez, and D.G. Cory, Towards a NMR implementation of a quantum lattice gas algorithm, *Computer Physics Communications* **146** (2002), Issue 3, 339–344.

[72] G.J. Pryde, M.J. Sellars, and N.B. Manson, Solid state coherent transient measurements using hard optical pulses, *Phys. Rev. Lett.* **84** (2000), 1152–1155.

[73] S. Somaroo, C.H. Tseng, T.F. Havel, R. Laflamme, and D.G. Cory, Quantum simulation on a quantum computer, *Phys. Rev. Lett.* **82** (1999), 5381–5384.

[74] O.W. Sorenson, Polorization transfer experiments in high-resolution NMR spectroscopy, *Progress of Nuclear Magnetic Resonance Spectroscopy* **21** (1989), Issue 6, 503–569.

[75] M. Steffen, L.M.K. Vandersypen, and I.L. Chuang, Toward quantum computation: a five-qubit quantum processor, *IEEE Micro*, 24–34, 2001.

[76] A. Szabo, Spin Dependence Of Optical Dephasing In Ruby – The Frozen Core, *Optics Letters* **8** (1983), 9, 486–487.

[77] J.M. Taylor, H.A. Engel, W. Dur, A. Yacoby, C.M. Marcus, P. Zoller, and M.D. Lukin, Fault-tolerant architecture for quantum computation using electrically controlled semiconductor spins, *Nature Physics* **1** (2005), 3, 177–183.

[78] L. Vahala, G. Vahala, and J. Yepez, Lattice Boltzmann and quantum lattice gas representations of 1-dimensional magnetohydrodynamic turbulence, *Physics Letters A* **306** (2003), 227–234.

[79] G. Vahala, J. Yepez, and L. Vahala, Quantum lattice gas representation of some classical solitons, *Physics Letters A* **310** (2003), 187–196.

[80] L.M.K. Vandersypen and I.L. Chuang, NMR techniques for quantum control and computation, *Reviews of Modern Physics* **76** (2004), 4, 1037–1069.

[81] L.M.K. Vandersypen, M. Steffen, G. Breyta, C. Yannoni, R. Cleve, and I.L. Chuang, Experimental realization of an order-finding algorithm with an NMR quantum computer, *Phys. Rev. Lett.* **85** (2000), 5452.

[82] L.M.K. Vandersypen, M. Steffen, M.H. Sherwood, C. Yannoni, G. Breyta, and I.L. Chuang, Implementation of a 3-quantum-bit search algorithm, *Applied Physics Letters* **76** (2000), 5, 646–648.

[83] L.M.K. Vandersypen, M. Steffen, G. Breyta, et al., Experimental realization of Shor's quantum factoring algorithm using nuclear magnetic resonance, *Nature* **414** (2001), 883–887.

[84] L.M.K. Vandersypen, C.S. Yannoni, M.H. Sherwood, and I.L. Chuang, Realization of logically labeled effective pure states for bulk quantum computation, *Phys. Rev. Lett.* **83** (1999), 3085.

[85] E. VanOort, N.B. Manson, and M. Glasbeek, Optically Detected Spin Coherence Of The Diamond N-V Center In Its Triplet Ground-State , *J. of Physics C-Solid State Physics* **21** (1988), 4385–4391.

[86] C. Wei and N.B. Manson, Observation of electromagnetically induced transparency within an electron spin resonance transition, *J. of Optics B-Quantum and Semiclassical Optics* **1** (1999), 464–468.

[87] Y.S. Weinstein, M.A. Pravia, E.M. Fortunato, S. Lloyd, and D.G. Cory, Implementation of the quantum Fourier transform, *Phys. Rev. Lett.* **86** (2001), 1889–1891.

[88] Wikipedia, http://en.wikipedia.org/wiki/Nuclear_magnetic_resonance.

[89] F. Yamaguchi and Y. Yamamoto, Crystal lattice quantum computer, *Microelectronic Engineering* **47** (1999), 273–275.

[90] C. Yannoni, M.H. Sherwood, D.C. Miller, I.L. Chuang, L.M.K. Vandersypen, and M.G. Kubines, Nuclear magnetic resonance quantum computing using liquid crystal solvents, *Applied Physics Letter* **75** (1999), No. 22, 3563–3562.

[91] J. Yepez, Quantum lattice-gas model for the diffusion equation, *International Journal of Modern Physics C* **12** (2001), 9, 1285–1303.

[92] J. Yepez, Type-II quantum computers, *International Journal of Modern Physics C* **12** (2001), 9, 1273–1284.

[93] J. Yepez and B. Boghosian, An efficient and accurate quantum lattice-gas model for the many-body Schrödinger wave equation, *Computer Physics Communications* **146** (2002), Issue 3, 280–294.

[94] Z. Zhang, G. Chen, Z. Diao, and P.R. Hemmer, NMR quantum computing, Chap. 4 in *Advances in Mechanics and Mathematics (AMMA)*, Vol. III, D.Y. Gao and R.W. Ogden (eds.), Springer, New York, 2006, pp. 135–189.

[95] J. Zhang, Z. Lu, L. Shan, and Z. Deng, Synthesizing NMR analogs of Einstein-Podolsky-Rosen state using the generalized Grover's algorithm, *Phys. Rev. A* **66** (2002), 044308.

Appendices

There are five appendices. The first, Appendix A, supplements the discussion of Section 1.2 of Chap. 1. The next two appendices, B and C, contain technical details needed for Chap. 7 in regard to laterally coupled quantum dots. The last two appendices, D and E, provide some fundamental information related to the Bloch sphere, rotations, unitary groups, evolution matrices and applications to SQUID (Chap. 9) and NMR (Chap. 10).

Appendix A

EPR, "locality" and "reality": the Bell inequalities à la Wigner

Consider the EPR gedanken experiment illustrated in Fig. A.1. A spin-zero system 'split' into two spin-1/2 particles. For the purpose of deriving Bell's theorem we consider the probability that particle 1 will pass through an SGA in Fig. A.1 oriented at an angle θ_a with the vertical $(+z)$ direction and that particle 2 will pass through an SGA oriented at an angle θ_b to the vertical. We call this joint passage probability $P(\theta_a, \theta_b) \equiv P_{ab}$. Let us first establish our notation by defining

$$P_{ab} = P(\overset{\text{particle 1}}{\underset{a\ \ b\ \ c}{+\ -\ \bigcirc}}\ |\ \overset{\text{particle 2}}{\underset{a\ \ b\ \ c}{-\ +\ \bigcirc}}\ ). \tag{A.1}$$

In Eq. (A.1) there are three "slots" on each side of the partition in which we have put either a plus sign, a minus sign, or a circle. The first, second, and third slots are reserved for information concerning passage through an SGA oriented at the angles θ_a, θ_b, and θ_c, respectively. A plus sign refers to passage and a minus sign to blockage. A circle means that the particular joint probability in question does not contain information about passage at that angle. We then write

$$P_{bc} = P(\bigcirc\ +\ -\ |\ \bigcirc\ -\ +),$$
$$P_{ac} = P(+\ \bigcirc\ -\ |\ -\ \bigcirc\ +). \tag{A.2}$$

In any classical probability theory, the remaining undetermined components, $\bigcirc$, will have specific values. That is, in any fixed direction, the spin component of every particle will be either $+1/2$ or $-1/2$; and we can write

493

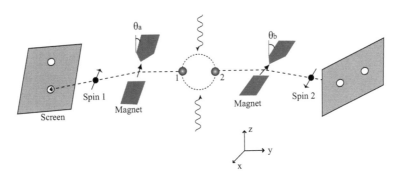

Figure A.1: Schematic of the EPR gedanken experiment. A spin-zero system such as orthohydrogen is split by an external field. The two spin-1/2 particles (protons) proceed in the opposite directions where they pass through the Stern-Gerlach apparati oriented at an angle θ_a with the vertical ($+z$) direction in the case of particle 1 and at an angle θ_b in the case of particle 2.

each of Eqs. (A.2) as a sum of probabilities for these two possibilities

$$
\begin{aligned}
P_{ab} &= P(+-+) + P(+--) &\text{(a)}; \\
P_{bc} &= P(++-) + P(-+-) &\text{(b)}; \\
P_{ac} &= P(++-) + P(+--) &\text{(c)}.
\end{aligned}
\tag{A.3}
$$

Where we have dropped the redundant information involving the second particle by writing $P = P(+-+)$ for $P(+-+|-+-)$, etc. The quantities $P(+-+)$, etc., describe measurements which, contrary to the actual experimental, were never carried out and are called "counter-factual". That is P_{ab}, etc. are measurable probabilities, but we have no procedure for measuring the separate classical probabilities $P(+-+)$, $P(+--)$, etc., on the right hand sides of Eq. (A.3). Adding Eqs. (A.3)(a) and (A.3)(b) and comparing with Eq. (A.3)(c), we see that [31]

$$
P_{ab} + P_{bc} = P_{ac} + P(+-+) + P(-+-).
\tag{A.4}
$$

Assuming the "probabilities" $P(+-+)$ and $P(-+-)$ are positive, we can then obtain the Bell–Wigner inequality,

$$
P_{ab} + P_{bc} \geq P_{ac}.
\tag{A.5}
$$

As has been indicated in Section 1.2.3, and shown experimentally [13], the seemingly obvious expression, Eq. (A.5), and thus the argument leading to it, is wrong. What insights into quantum mechanics can we glean from this? To this question we now turn to the question: What must we give up – "locality" or "reality" in order to have a hidden variable theory?

We present a simple, quantum mechanical argument along the lines of the preceding derivation of Bell inequality [8]; this argument encourages us to relinquish certain elements of "physical reality" (rather than locality) in quantum mechanics. We will see that certain "elements of physical reality" should be sacrificed, not "locality".

To set the stage for our quantum mechanical considerations, we recall that the probability of observing spin-up with angle θ to the z-axis is

$$|\langle\theta|\Psi\rangle|^2 = \langle\Psi|\theta\rangle\langle\theta|\Psi\rangle = \langle\Psi|\hat{\pi}_\theta|\Psi\rangle, \tag{A.6}$$

where $\hat{\pi}_\theta = |\theta\rangle\langle\theta|$ is a projection operator, and $|\theta\rangle = \exp[-(i/2)\sigma_y\theta]|\uparrow\rangle$. Thus, the projection operator for spin-up θ to the z-axis is

$$\hat{\pi}_\theta = |\theta\rangle\langle\theta| = \frac{1}{2}(I + \sigma_z\cos\theta + \sigma_x\sin\theta) = \frac{1}{2}(I + \boldsymbol{\sigma}\cdot\boldsymbol{r}) \tag{A.7}$$

where $\boldsymbol{\sigma} = (\sigma_x, \sigma_y, \sigma_z)$ (for those who don't yet know what they are, see (2.42) and (2.43) in Chap. 2) are Pauli matrices, $\boldsymbol{r} = (\sin\theta, 0, \cos\theta)$ and I is the identity matrix. For the state given by

$$|\Psi\rangle = \frac{1}{\sqrt{2}}\{|\uparrow\rangle_1|\downarrow\rangle_2 - |\downarrow\rangle_1|\uparrow\rangle_2\}, \tag{A.8}$$

the quantum mechanical probabilities $\mathcal{P}_{jk}$ that particles 1 and 2 will both have spin up in the directions j and k, respectively, are then

$$\mathcal{P}_{ab} = \langle\Psi|\hat{\pi}_{\boldsymbol{a}}^{(1)}\hat{\pi}_{\boldsymbol{b}}^{(2)}|\Psi\rangle \text{ (a)}, \quad \mathcal{P}_{bc} = \langle\Psi|\hat{\pi}_{\boldsymbol{b}}^{(1)}\hat{\pi}_{\boldsymbol{c}}^{(2)}|\Psi\rangle \text{ (b)},$$
$$\mathcal{P}_{ac} = \langle\Psi|\hat{\pi}_{\boldsymbol{a}}^{(1)}\hat{\pi}_{\boldsymbol{c}}^{(2)}|\Psi\rangle \text{ (c)} \tag{A.9}$$

where each projection operator has a superscript indicating the particle number and a vector subscript defining the direction in the x-z plane. We write that Eq. (A.9)(a) using Eqs. (A.8) and (A.7) gives, $\mathcal{P}_{ab} = \frac{1}{4}\{1 - \cos(\theta_a - \theta_b)\} = \frac{1}{4}\{1 - \boldsymbol{a}\cdot\boldsymbol{b}\}$ where $\theta_a(\theta_b)$ is the angle of vector $\boldsymbol{a}(\boldsymbol{b})$ with respect to the z-axis. This is the usual result for the quantum mechanical prediction; it provides a maximum violation of the Bell inequality, Eq. (A.5), when $\theta_a = 0$, $\theta_b = \pi/3$, and $\theta_c = 2\pi/3$.

Let us now write the quantum mechanical joint passage probabilities $\mathcal{P}_{ab}$, $\mathcal{P}_{ac}$ and $\mathcal{P}_{bc}$ into a form that is in direct correspondence with the classical probability expressions Eqs. (A.3). First, we note that the completeness expression can, for example, be written $I = \hat{\pi}_{\boldsymbol{a}}^{(1)} + \hat{\pi}_{-\boldsymbol{a}}^{(1)}$. Note also that if the angle between $\boldsymbol{a}$ and the $+z$-axis is θ_a, then for $-\boldsymbol{a}$ this angle is $\theta_a + \pi$. Inserting the identity operator I into Eq. (A.9), we can obtain the quantum analog of Eq. (A.3),

$$\mathcal{P}_{ab} = \langle\Psi|\hat{\pi}_{+\boldsymbol{a}}^{(1)}\hat{\pi}_{+\boldsymbol{b}}^{(2)}I^{(1)}|\Psi\rangle = \langle\Psi|\hat{\pi}_{+\boldsymbol{a}}^{(1)}\hat{\pi}_{+\boldsymbol{b}}^{(2)}\hat{\pi}_{+\boldsymbol{c}}^{(1)}|\Psi\rangle + \langle\Psi|\hat{\pi}_{+\boldsymbol{a}}^{(1)}\hat{\pi}_{+\boldsymbol{b}}^{(2)}\hat{\pi}_{+\boldsymbol{c}}^{(1)}|\Psi\rangle$$
$$\equiv \quad \mathcal{P}(+-+) \quad + \quad \mathcal{P}(+--), \tag{A.10}$$

$$\mathcal{P}_{bc} = \langle\psi|I^{(1)}\hat{\pi}_{+\boldsymbol{b}}^{(1)}\hat{\pi}_{+\boldsymbol{c}}^{(2)}|\Psi\rangle = \langle\Psi|\hat{\pi}_{+\boldsymbol{a}}^{(1)}\hat{\pi}_{+\boldsymbol{b}}^{(1)}\hat{\pi}_{+\boldsymbol{c}}^{(2)}|\Psi\rangle + \langle\Psi|\hat{\pi}_{-\boldsymbol{a}}^{(1)}\hat{\pi}_{+\boldsymbol{b}}^{(1)}\hat{\pi}_{+\boldsymbol{c}}^{(2)}|\Psi\rangle$$
$$\equiv \quad \mathcal{P}(++-) \quad + \quad \mathcal{P}(-++), \tag{A.11}$$

$$\mathcal{P}_{ac} = \langle\Psi|\hat{\pi}_{+\boldsymbol{a}}^{(1)}I^{(1)}\hat{\pi}_{+\boldsymbol{c}}^{(2)}|\psi\rangle = \langle\Psi|\hat{\pi}_{+\boldsymbol{a}}^{(1)}\hat{\pi}_{+\boldsymbol{b}}^{(1)}\hat{\pi}_{+\boldsymbol{c}}^{(2)}|\Psi\rangle + \langle\Psi|\hat{\pi}_{+\boldsymbol{a}}^{(1)}\hat{\pi}_{-\boldsymbol{b}}^{(1)}\hat{\pi}_{+\boldsymbol{c}}^{(2)}|\Psi\rangle$$
$$\equiv \quad \mathcal{P}(++-) \quad + \quad \mathcal{P}(+--). \tag{A.12}$$

We proceed by adding Eqs. (A.10) and (A.11) and pairing off terms via Eq. (A.12) to obtain

$$\mathcal{P}_{ab} + \mathcal{P}_{bc} = \mathcal{P}_{ac} + \mathcal{P}(+-+) + \mathcal{P}(-+-). \tag{A.13}$$

Comparing Eqs. (A.4) and (A.13), we see that they are in a one-to-one correspondence in which $P(+-+) \leftrightarrow \mathcal{P}(+-+)$ and $P(-+-) \leftrightarrow \mathcal{P}(-+-)$. Thus, we have regained a generalized Bell–Wigner [10] expression, but now via purely quantum mechanical calculations. However, we now have explicit expressions for the $\mathcal{P}$'s and can evaluate them. For example, at the angles $\theta_a = 0$, $\theta_b = \pi/3$, and $\theta_c = 2\pi/3$ which give a maximum violation of the Bell–Wigner inequality, Eq. (A.5), we find $\mathcal{P}(+-+) = \mathcal{P}(-+-) = -0.0625$. These quantities are not positive as required for the derivation of a Bell inequality. They can, in fact, be negative, i.e., they are not physical probabilities. Thus, the preceding quantum treatment of EPR-Bell involves operator expectation values such as $\mathcal{P}(---) \equiv \langle \Psi | \hat{\pi}_{+a}^{(1)} \hat{\pi}_{-b}^{(1)} \hat{\pi}_{+c}^{(2)} | \Psi \rangle$ which are not probabilities. But the Wigner (counter-factual) arguments assume that such probabilities do exist. Nowhere is it asserted that the Bell inequalities are the result of quantum nonlocality; instead we see that in quantum mechanics, quantities such as $P(+-+)$ are not elements of (quantum) physical reality.

Thus, the Bell inequalities are violated by quantum mechanics because the quantities $P(+-+)$ and $P(-+-)$ in Eq. (A.4) have been treated as positive probabilities. But, in the real (quantum) world, the correct quantities $\mathcal{P}(+-+)$ and $\mathcal{P}(-+-)$ are not physical observables. Specifically, they are not elements of physical reality!

Appendix B

The Fock–Darwin States

The mathematical derivations of (7.17) in Chap. 7 rely heavily on the Fock–Darwin Hamiltonian, which models the motion of a conduction-band electron confined in a 2-dimensional parabolic potential well in an external magnetic field perpendicular to the 2-dimensional plane:

$$H_{FD} = \frac{1}{2m} \left| \boldsymbol{p} - \frac{e}{c}\boldsymbol{A} \right|^2 + \frac{1}{2M}\omega_0^2 r^2 \qquad (r = (x^2 + y^2)^{1/2}); \qquad (\text{B.1})$$

where the notation follows that introduced in Section 7.3.3. The Fock–Darwin Hamiltonian H_{FD} and its eigenstates have pleasant mathematical properties (Fock [4], Darwin [3]) and may be viewed as a 2-dimensional analog of the simple harmonic oscillator.

From (B.1) and (7.18), we have

$$
\begin{aligned}
H &= \frac{1}{2m}\left(|\boldsymbol{p}|^2 - 2\frac{e}{c}\boldsymbol{p} \cdot \boldsymbol{A} + \frac{e^2}{c^2}|\boldsymbol{A}|^2 \right) + \frac{1}{2}m\omega_0^2(x^2 + y^2) \\
&= \frac{|\boldsymbol{p}|^2}{2m} - \frac{1}{2}\frac{eB}{mc}(-p_x y + p_y x) + \frac{e^2 B^2}{8mc^2}(x^2 + y^2) + \frac{1}{2}m\omega_0^2(x^2 + y^2) \\
&= \frac{|\boldsymbol{p}|^2}{2m} + \frac{1}{2}m\left(\omega_0^2 + \frac{\omega_c^2}{4} \right)(x^2 + y^2) + \frac{1}{2}\omega_c L_z,
\end{aligned} \qquad (\text{B.2})
$$

where

$$\frac{eB}{mc} \equiv \omega_c = \text{ the cyclotron frequency;}$$

$$L_z = xp_y - yp_x = \text{the } z\text{-component of the angular momentum, } \boldsymbol{L} = \boldsymbol{r} \times \boldsymbol{p}.$$

Next, from the four independent operators x, y, p_x and p_y, we define four new operators:

$$
\left.
\begin{aligned}
a &= \varepsilon(x - iy) + \eta(ip_x + p_y); \\
a^+ &= \varepsilon(x + iy) + \eta(-ip_x + p_y); \\
b &= \varepsilon(x + iy) + \eta(ip_x - p_y); \\
b^+ &= \varepsilon(x - iy) + \eta(-ip_x - p_y),
\end{aligned}
\right\} \qquad (\text{B.3})
$$

where ε and η are real numbers. Using (the Poisson brackets)

$$[x, p_x] = [y, p_y] = i\hbar, \tag{B.4}$$

$$[x, y] = [x, p_y] = [y, x] = [y, p_x] = 0, \tag{B.5}$$

we can easily show that

$$[a^+, b] = [a, b^+] = [a, b] = [a^+, b^+] = 0, \tag{B.6}$$

and

$$[a^+, a] = \varepsilon\eta[x + iy, ip_x + p_y] + \varepsilon\eta[-ip_x + p_y, x - iy]$$
$$= -4\varepsilon\eta\hbar, \tag{B.7}$$

$$[b^+, b] = \varepsilon\eta[x - iy, ip_x - p_y] + [-ip_x - p_y, x + iy]$$
$$= -4\varepsilon\eta\hbar. \tag{B.8}$$

Thus, if we choose

$$\eta = 1/(4\varepsilon\hbar), \tag{B.9}$$

then

$$\begin{cases} [a, a^+] = [b, b^+] = 1; \\ \text{all other commutators are zero.} \end{cases} \tag{B.10}$$

We obtain

$$p_x^2 + p_y^2 = (ip_x + p_y)(-ip_x + p_y)$$
$$= \left(\frac{a - b^+}{2\eta}\right)\left(\frac{a^+ - b}{2\eta}\right); \tag{B.11}$$

$$x^2 + y^2 = (x + iy)(x - iy)$$
$$= \left(\frac{a^+ + b}{2\varepsilon}\right)\left(\frac{a + b^+}{2\varepsilon}\right), \tag{B.12}$$

and

$$L_z = xp_y - yp_x = \frac{1}{2}[(x + iy)(ip_x + p_y) - (x - iy)(ip_x - ip_y)]$$
$$= \frac{1}{2}\left[\left(\frac{a^+ + b}{2\varepsilon}\right)\left(\frac{a - b^+}{2\eta}\right) - \left(\frac{a + b^+}{2\varepsilon}\right)\left(\frac{b - a^+}{2\eta}\right)\right]. \tag{B.13}$$

Define

$$\Omega^2 = \omega_0^2 + \frac{\omega_c^2}{4}. \tag{B.14}$$

Then

$$H = \frac{1}{2m}\frac{1}{4\eta^2}(a - b^+)(a^+ - b) + \frac{1}{2}m\Omega^2 \cdot \frac{1}{4\varepsilon^2}(a^+ + b)(a + b^+)$$
$$+ \frac{1}{2}\omega_c \cdot \frac{1}{8\varepsilon\eta}[(a^+ + b)(a - b^-) - (a + b^+)(b - a^+)]$$
$$= \frac{1}{8m\eta^2}(aa^+ - ab - b^+a^+ + b^+b) + \frac{m\Omega^2}{8\varepsilon^2}(a^+a + a^+b^+ + ba + bb^+)$$
$$+ \frac{\omega_c}{16\varepsilon\eta}(a^+a + \cancel{ba} - \cancel{a^+b^+} - bb^+ - \cancel{ab} + aa^+ - b^+b + \cancel{b^+a^+}). \tag{B.15}$$

Recall from (B.9) that $1/\eta = 4\varepsilon\hbar$. If we further require that

$$\frac{1}{8m\eta^2} = \frac{m\Omega^2}{8\varepsilon^2} = \frac{m\Omega^2}{8}(4\eta\hbar)^2 = 2m\Omega^2\eta^2\hbar^2,$$

i.e.,

$$\eta = \frac{1}{2\sqrt{\hbar m\Omega}}, \tag{B.16}$$

then from (B.15) we see that cross-terms ab, ab^+, a^+b, a^+b^+, etc., cancel out:

$$H = \frac{\hbar\Omega}{2}\{[aa^+ - \cancel{ab} - b^+\cancel{a^+} + b^+b] + [a^+a + a^+\cancel{b^+} + \cancel{ba} + bb^+]\}$$
$$+ \frac{\hbar\omega_c}{4}[\underbrace{a^+a + aa^+}_{2a^+a+1} \underbrace{-bb^+ - b^+b}_{-2b^+b-1}]$$
$$= \frac{\hbar\Omega}{2}[2a^+a + 1 + 2b^+b + 1] + \frac{\hbar\omega_c}{2}[a^+a - b^+b]$$
$$= \hbar\left(\Omega + \frac{\omega_c}{2}\right)\left(a^+a + \frac{1}{2}\right) + \hbar\left(\Omega - \frac{\omega_c}{2}\right)\left(b^+b + \frac{1}{2}\right)$$
$$= \hbar\omega_+\left[a^+a + \frac{1}{2}\right] + \hbar\omega_-\left[b^+b + \frac{1}{2}\right], \tag{B.17}$$

where

$$\omega_\pm \equiv \Omega \pm \frac{\omega_c}{2}. \tag{B.18}$$

We can now define the Fock–Darwin states

$$|n_+, n_-\rangle = \frac{1}{[(n_+!)(n_-!)]^{1/2}}(a^+)^{n_+}(b^+)^{n_-}|0, 0\rangle, \tag{B.19}$$

for any integers n_+ and n_-, $n_+ \geq 0$, $n_- \geq 0$, where

$$\begin{cases} a = \varepsilon(x - iy) + \eta(ip_x + p_y), \\ b = \varepsilon(x + iy) + \eta(ip_x - p_y), \end{cases} \tag{B.20}$$

with

$$\eta = \frac{1}{2\sqrt{\hbar m\Omega}}, \quad \varepsilon = \frac{1}{4\hbar\eta} = \frac{1}{4\hbar} \cdot 2\sqrt{\hbar m\Omega} = \frac{1}{2}\sqrt{\frac{m\Omega}{\hbar}}. \tag{B.21}$$

From (B.17), (B.18) and (B.19), we have

$$H|n_+, n_-\rangle = \left[\hbar\omega_+ \left(n_+ + \frac{1}{2} \right) + \hbar\omega_- \left(n_- + \frac{1}{2} \right) \right] |n_+, n_-\rangle, \qquad \text{(B.22)}$$

for integers $n_+ \geq 0, n_- \geq 0$.

Instead of using x and y, we can also use the complex variable z and its conjugate $\bar{z}$:

$$z = x + iy, \quad \bar{z} = x - iy. \qquad \text{(B.23)}$$

Then

$$\begin{cases} \partial_x = \dfrac{\partial}{\partial x} = \dfrac{\partial z}{\partial x}\dfrac{\partial}{\partial z} + \dfrac{\partial \bar{z}}{\partial x}\dfrac{\partial}{\partial \bar{z}} = \partial_z + \bar{\partial}_z, \\[2mm] \partial_y = \dfrac{\partial}{\partial y} = \dfrac{\partial z}{\partial y}\dfrac{\partial}{\partial z} + \dfrac{\partial \bar{z}}{\partial y}\dfrac{\partial}{\partial \bar{z}} = i\partial_z - i\bar{\partial}_z, \end{cases} \qquad \text{(B.24)}$$

from where we obtain in turn

$$\partial_z = \frac{1}{2}(\partial_x - i\partial_y), \quad \bar{\partial}_z = \frac{1}{2}(\partial_x + i\partial_y). \qquad \text{(B.25)}$$

From (B.19)–(B.25), we thus have

$$\begin{cases} a = \dfrac{1}{2}\sqrt{\dfrac{m\Omega}{\hbar}}\, \bar{z} + \dfrac{1}{2\sqrt{\hbar m\Omega}} 2\hbar\partial_z = \dfrac{1}{\sqrt{2}} \left[\dfrac{\bar{z}}{2\ell_0} + 2\ell_0\partial_z \right], \\[4mm] b = \dfrac{1}{2}\sqrt{\dfrac{m\Omega}{\hbar}}\, z + \dfrac{1}{2\sqrt{\hbar m\Omega}} 2\hbar\bar{\partial}_z = \dfrac{1}{\sqrt{2}} \left[\dfrac{z}{2\ell_0} + 2\ell_0\bar{\partial}_z \right], \end{cases} \qquad \text{(B.26)}$$

where $\ell_0 \equiv [\hbar/(2m\Omega)]^{1/2}$.

Theorem B.1. *The ground state of the Fock–Darwin states are given by*

$$|0,0\rangle = \sqrt{\frac{m\Omega}{\pi\hbar}}\, e^{-\frac{m\Omega}{2\hbar}(x^2 + y^2)}. \qquad \text{(B.27)}$$

Proof. Since

$$a|0,0\rangle = 0,$$

we have

$$|0,0\rangle = \tilde{c}\, e^{-\frac{m\Omega}{2\hbar}\bar{z}z}, \qquad \text{(B.28)}$$

where c is a normalization constant. We also see that (B.28) satisfies

$$b|0,0\rangle = 0.$$

Thus

$$|0,0\rangle = \tilde{c}\, e^{-\frac{m\Omega}{2\hbar}(x^2 + y^2)}.$$

The constant of normalization is easily computed to be $\tilde{c} = [(m\Omega)/(\pi\hbar)]^{1/2}$. The rest can also be easily verified. $\qquad \square$

Appendix C

Evaluation of the exchange energy for laterally coupled quantum dots

The point of view taken by Burkard, Loss and DiVincenzo [2] is to regard the coupled two quantum dots as a "molecule" obtained by combining two quantum dots through perturbation with a Fock–Darwin-like ground state as the ground state of the single electron spin on each dot.

Let us rewrite the overall Hamiltonian in (7.23) of Chap. 7 of the coupled system as

$$H_{\rm orb} = H_1(\boldsymbol{p}_1, \boldsymbol{r}_1) + H_2(\boldsymbol{p}_2, \boldsymbol{r}_2) + C(\boldsymbol{r}_1, \boldsymbol{r}_2) + W(\boldsymbol{r}_1, \boldsymbol{r}_2), \tag{C.1}$$

$$(\boldsymbol{p}_1 = (p_{x_1}, p_{y_1}, 0), \boldsymbol{r}_1 = (x_1, y_1), \boldsymbol{p}_2 = (p_{x_2}, p_{y_2}, 0), \boldsymbol{r}_2 = (x_2, y_2))$$

$$H_1(\boldsymbol{p}_1, \boldsymbol{r}_1) = \frac{1}{2m} \left| \boldsymbol{p}_1 - \frac{e}{c} \boldsymbol{A}(x_1, y_1, 0) \right|^2 + eEx_1 + \frac{m\omega_0^2}{2}[(x_1 + a)^2 + y_1^2], \tag{C.2}$$

$$H_2(\boldsymbol{p}_2, \boldsymbol{r}_2) = \frac{1}{2m} \left| \boldsymbol{p}_2 - \frac{e}{c} \boldsymbol{A}(x_2, y_2, 0) \right|^2 + eEx_2 + \frac{m\omega_0^2}{2}[(x_2 - a)^2 + y_2^2], \tag{C.3}$$

$$C(\boldsymbol{r}_1, \boldsymbol{r}_2) = \frac{e^2}{\kappa|\boldsymbol{r}_1 - \boldsymbol{r}_2|}, \text{ same as (7.21)}$$

$$W(\boldsymbol{r}_1, \boldsymbol{r}_2) = W_1(x_1) + W_2(x_2), \text{ with} \tag{C.4}$$

$$W_j(x_j) \equiv \frac{m\omega_0^2}{2} \left[\frac{1}{4a^2}(x_j^2 - a^2)^2 - (x_j - a)^2 \right], \text{ for } j = 1, 2. \tag{C.5}$$

The H_1 and H_2 given above are not Fock–Darwin Hamiltonians. However, after simple similarity transformations, they become Fock–Darwin plus a constant.

Lemma C.1. *Given H_1 and H_2 as in (C.1) and (C.3), define*

$$\left. \begin{array}{l} \tilde{H}_1 = e^{\frac{i}{\hbar}\left(\frac{e^2BE}{2m\omega_0^2c}+\frac{eBa}{2c}\right)y_1} H_1 e^{-\frac{i}{\hbar}\left(\frac{e^2BE}{2m\omega_0^2c}+\frac{eBa}{2c}\right)y_1}, \\[2mm] \tilde{H}_2 = e^{\frac{i}{\hbar}\left(\frac{e^2BE}{2m\omega_0^2c}-\frac{eBa}{2c}\right)y_2} H_2 e^{-\frac{i}{\hbar}\left(\frac{e^2BE}{2m\omega_0^2c}-\frac{eBa}{2c}\right)y_2}. \end{array} \right\} \qquad \text{(C.6)}$$

Then

$$\tilde{H}_j = H_{j,FD} - \left(-\frac{e^2E^2}{2m\omega_0^2} \mp eEa \right); \text{``$-$'' for $j=1$, ``$+$'' for } j = 2, \qquad \text{(C.7)}$$

where $H_{j,FD}$ is a Fock–Darwin Hamiltonian for $j = 1, 2$ defined by

$$H_{j,FD} = \frac{1}{2m} \left| \mathbf{p}_j - \frac{e}{c}\mathbf{A}(x_{j\mp}, y_j, 0) \right|^2 + \frac{m\omega_0^2}{2}(x_{j\mp}^2 + y_j^2); \qquad \text{(C.8)}$$

where

$$x_{j\mp} \equiv x_j - (-1)^j a + \frac{eE}{m\omega_0^2}, \qquad \text{(C.9)}$$

and "$-$" for $j = 1$ and "$+$" for $j = 2$.

Proof. For $j = 1$, the similarity transformation (C.6)$_1$ effects a translation of p_{y_1}, the y-component of $\mathbf{p}_1$, as follows

$$p_{y_1} \longrightarrow p_{y_1} + \frac{e^2BE}{2m\omega_0^2c} + \frac{eBa}{2c}, \qquad \text{(C.10)}$$

while the remaining variables x_1, y_1 and p_{x_1} are left unchanged. Thus from (C.6)

$$\tilde{H}_1 = \frac{1}{2m} \left[\left(p_{x_1} + \frac{eBy_1}{2c} \right)^2 + \left(p_{y_1} - \frac{e^2BE}{2m\omega_0^2c} + \frac{eBa}{2c} - \frac{eBx_1}{2c} \right)^2 \right]$$

$$+ \frac{m\omega_0^2}{2} \left[\left(x_1 - a + \frac{eE}{m\omega_0^2} \right)^2 + y_1^2 \right] - eE \left(\frac{eE}{2m\omega_0^2} - a \right). \qquad \text{(C.11)}$$

Define

$$x_{1-} = x_1 + a + \frac{eE}{m\omega_0^2} \qquad \text{(C.12)}$$

as in (C.9). Then

$$\tilde{H}_1 = \frac{1}{2m} \left| \mathbf{p}_1 - \frac{e}{c}\mathbf{A}(x_{1-}, y_1, 0) \right|^2 + \frac{m\omega_0^2}{2}(x_{1-}^2 + y_1^2) - \frac{e^2E^2}{2m\omega_0^2} + eEa \qquad \text{(C.13)}$$

$$\equiv H_{1,FD} + \left(eEa - \frac{e^2E^2}{2m\omega_0^2} \right), \qquad \text{(C.14)}$$

where $H_{1,FD}$ is a Fock–Darwin Hamiltonian (of variables $\mathbf{p}_1, x_{1-}$ and y_1). H_2 and $\tilde{H}_2$ can be similarly treated. $\qquad \square$

We thus have

$$H_1 = e^{-\frac{i}{\hbar}\left(\frac{e^2 BE}{2m\omega_0^2 c} - \frac{eBa}{2c}\right)y_1} H_{1,FD} e^{\frac{i}{\hbar}\left(\frac{e^2 BE}{2m\omega_0^2 c} - \frac{eBa}{2c}\right)y_1} + \left(eEa - \frac{e^2 E^2}{2m\omega_0^2}\right), \quad \text{(C.15)}$$

whose eigenstates are

$$e^{-\frac{i}{\hbar}\left(\frac{e^2 BE}{2m\omega_0^2 c} - \frac{eBa}{2c}\right)y_1} |n_+^{(1)}, n_-^{(1)}\rangle, \quad \text{cf. (B.22) in Appendix B,} \quad \text{(C.16)}$$

with eigenvalues

$$\varepsilon(n_+^{(1)} n_-^{(1)}) \equiv \hbar\omega_+ \left(n_+^{(1)} + \frac{1}{2}\right) + \hbar\omega_- \left(n_-^{(1)} + \frac{1}{2}\right) + \left(eEa - \frac{eE^2}{2m\omega_0^2}\right), \quad \text{(C.17)}$$

$$\left(\omega_\omega \equiv \sqrt{\omega_0^2 + \left(\frac{eB}{2mc}\right)^2} \pm \frac{eB}{2mc}\right). \quad \text{(C.18)}$$

Similarly, H_2 and $H_{2,FD}$ can be obtained from (C.3), (C.14) and (C.15) by simply replacing the index 1 by 2 and x_- by

$$x_+ \equiv x - a + \frac{eE}{m\omega_0^2}. \quad \text{(C.19)}$$

Since the ground state $|0,0\rangle$ of the Fock–Darwin Hamiltonian, H_{FD}, is (cf. (B.27))

$$\Phi_0(x, y) = \sqrt{\frac{m\Omega}{\pi\hbar}} e^{-\frac{m\Omega}{2\hbar}(x^2 + y^2)}, \quad \left(\Omega \equiv \sqrt{\omega_0^2 + \left(\frac{eB}{2mc}\right)^2}\right),$$

therefore, the ground state of the Hamiltonians H_1 and H_2 are, respectively,

$$\Phi_0^{(1)}(x, y) = e^{-\frac{i}{\hbar}\left(\frac{e^2 BE}{2m\omega_0^2 c} + \frac{eBa}{2c}\right)y} \sqrt{\frac{m\Omega}{\pi\hbar}} e^{-\frac{m\Omega}{2\hbar}(x^2 + y^2)}, \quad \text{(C.20)}$$

$$\Phi_0^{(2)}(x, y) = e^{-\frac{i}{\hbar}\left(\frac{e^2 BE}{2m\omega_0^2 c} - \frac{eBa}{2c}\right)y} \sqrt{\frac{m\Omega}{\pi\hbar}} e^{-\frac{m\Omega}{2\hbar}(x_+^2 + y^2)}. \quad \text{(C.21)}$$

We are now in a position to apply the well known Heitler–London method in quantum molecular chemistry to model the coupled system. The method utilizes "quantum dot" orbitals:

$$\left.\begin{array}{l} a(j) \equiv \Phi_0^{(1)}(x_j, y_j), \quad j = 1, 2, \\ b(j) \equiv \Phi_0^{(2)}(x_j, y_j), \quad j = 1, 2, \end{array}\right\} \quad \text{(C.22)}$$

from which, further define

$$|\Psi_\pm\rangle = \nu[a(1)b(2) \pm a(2)b(1)] \quad \text{(C.23)}$$

where ν is the normalization factor. Note that $|\Psi_+\rangle$ is the singlet state, while $|\Psi_-\rangle$ is the triplet state. Note that our notation in (C.21) and (C.23) follows from the convention used by Slater [9, Chap. 3].

Lemma C.2. *We have the overlap integral*

$$S \equiv \langle \Phi_0^{(2)} | \Phi_0^{(1)} \rangle = e^{-bd^2 - d^2(b - \frac{1}{b})}, \tag{C.24}$$

where

$$b \equiv \frac{\Omega}{\omega_0}, \quad d = (m\omega_0/\hbar)^{1/2} a. \tag{C.25}$$

Consequently, the normalized singlet and triplet states are

$$|\Psi_\pm\rangle = \frac{1}{\sqrt{2(1 \pm S^2)}} [a(1)b(2) \pm a(2)b(1)], \tag{C.26}$$

satisfying

$$\langle \Psi_+ | \Psi_+ \rangle = 1, \quad \langle \Psi_- | \Psi_- \rangle = 1$$

and

$$\langle \Psi_+ | \Psi_- \rangle = 0.$$

Proof. We evaluate (C.24):

$$S \equiv \int\limits_{-\infty}^{\infty} \int\limits_{-\infty}^{\infty} \bar{\Phi}_0^{(2)}(x, y) \Phi_0^{(1)}(x, y) dx dy$$

$$= \int\limits_{-\infty}^{\infty} \int\limits_{-\infty}^{\infty} e^{-\frac{i}{\hbar}(\frac{e^2 BE}{2m\omega_0^2 c} + \frac{eBa}{2C} - \frac{e^2 BE}{2m\omega_0^2 c} + \frac{eBa}{2c})y} \frac{m\Omega}{\pi\hbar} \cdot e^{-\frac{m\Omega}{\hbar}[(x + \frac{eE}{m\omega_0^2})^2 + a^2 + y^2]} dx dy$$

$$= e^{-\frac{m\Omega}{\hbar} a^2 - \frac{e^2 B^2 a^2}{4\hbar m\Omega c^2}} \left(\frac{m\Omega}{\pi\hbar}\right) \underbrace{\int\limits_{-\infty}^{\infty} e^{-\frac{m\Omega}{\hbar}(x + \frac{eE}{m\omega_0^2})^2} dx}_{\sqrt{\frac{\pi}{(m\Omega/\hbar)}}} \cdot \underbrace{\int\limits_{-\infty}^{\infty} e^{-\frac{m\Omega}{\hbar}(y + i\frac{eBa}{2m\Omega c})^2} dy}_{\sqrt{\frac{\pi}{(m\Omega/\hbar)}}}$$

$$= e^{-\frac{m\Omega}{\hbar} a^2 - \frac{e^2 B^2 a^2}{4\hbar m\Omega c^2}} = e^{-bd^2 - d^2(b - \frac{1}{2})}.$$

The rest follows from straightforward calculations. □

The exchange energy, by (7.25), now can be written as

$$J \equiv \langle \Psi_- | H_{\text{orb}} | \Psi_- \rangle - \langle \Psi_+ | H_{\text{orb}} | \Psi_+ \rangle$$

$$= \frac{1}{2(1 - S^2)} \Big\{ \langle a(1)b(2) | H_{\text{orb}} | a(1)b(2) \rangle + \langle a(2)b(1) | H_{\text{orb}} | a(2)b(1) \rangle$$

$$- \langle a(1)b(2) | H_{\text{orb}} | a(2)b(1) \rangle - \langle a(2)b(1) | H_{\text{orb}} | a(1)b(2) \rangle \Big\}$$

$$- \frac{1}{2(1 + S^2)} \Big\{ \langle a(1)b(2) | H_{\text{orb}} | a(1)b(2) \rangle + \langle a(2)b(1) | H_{\text{orb}} | a(2)b(1) \rangle$$

$$+ \langle a(1)b(2) | H_{\text{orb}} | a(2)b(1) \rangle + \langle a(2)b(1) | H_{\text{orb}} | a(1)b(2) \rangle \Big\}$$

$$= \cdots \text{(combining the two parentheses, using (C.1) and expanding)}$$

$$= \frac{S^2}{1-S^4}\left\{\left[\langle a(1)|H_1|a(1)\rangle + \langle a(2)|H_2|a(2)\rangle + \langle b(1)|H_1|b(1)\rangle\right.\right.$$

$$+ \langle a(2)|H_2|a(2)\rangle\Big]$$

$$- \frac{1}{S^2}\Big[\langle a(1)|H_1|b(1)\rangle\langle b(2)|a(2)\rangle + \langle b(2)|H_2|a(2)\rangle\langle a(1)|b(1)\rangle$$

$$+ \langle b(1)|H_1|a(1)\rangle\langle a(2)|b(2)\rangle + \langle a(2)|H_2|b(2)\rangle\langle b(1)|a(1)\rangle\Big]$$

$$+ \big[\langle a(1)b(2)|C|a(1)b(2)\rangle + \langle a(2)b(1)|C|a(2)b(1)\rangle\big]$$

$$- \frac{1}{S^2}\Big[\langle a(1)b(2)|C|a(2)b(1)\rangle + \langle a(2)b(1)|C|a(1)b(2)\rangle\Big]$$

$$+ \Big[\langle a(1)b(2)|W|a(1)b(2)\rangle + \langle a(2)b(1)|W|a(2)b(1)\rangle$$

$$\left.\left.- \frac{1}{S^2}\Big(\langle a(1)b(2)|W|a(2)b(1)\rangle + \langle a(2)b(1)|W|a(1)b(2)\rangle\Big)\right]\right\} \qquad \text{(C.27)}$$

$$\equiv \frac{S^2}{1-S^4}\left\{\mathcal{B}_1 - \frac{1}{S^2}\mathcal{B}_2 + \mathcal{B}_3 - \frac{1}{S^2}\mathcal{B}_4 + \mathcal{B}_5\right\}, \qquad \text{(C.28)}$$

where each $\mathcal{B}_j, j = 1, 2, 3, 4$ and 5, represents a square bracket inside the curly parentheses in (C.27) in the correct sequential order. We evaluate these $\mathcal{B}_j$ one by one below.

Lemma C.3. *We have*

$$\mathcal{B}_1 - \frac{1}{S^2}\mathcal{B}_2 = 4ma^2\omega_0^2. \qquad \text{(C.29)}$$

Proof. Note the following pairs of cancellations

$$\langle a(1)|H_1|a(1)\rangle - \frac{\langle b(1)|H_1|a(1)\rangle\langle b(2)|a(2)\rangle}{S^2} = 0, \qquad \text{(C.30)}$$

$$\langle b(2)|H_2|b(2)\rangle - \frac{\langle a(2)|H_2|b(2)\rangle\langle b(1)|a(1)\rangle}{S^2} = 0, \qquad \text{(C.31)}$$

because

$$H_1|a(1)\rangle = E_0|a(1)\rangle$$

as $|a(1)\rangle$ is the ground state of H_1 and E_0 is the ground state energy (cf. (C.17) with $n_+^{(1)} = n_-^{(1)} = 0$ therein) and so

$$\text{Left Hand Side of (C.30)} = E_0\langle a(1)|a(1)\rangle - \frac{E_0\langle b(1)|a(1)\rangle\langle b(2)|a(2)\rangle}{S^2}$$

$$= E_0 - \frac{E_0 \cdot S \cdot S}{S^2} = 0.$$

Similarly,

$$H_2|b(2)\rangle = E_0|b(2)\rangle,$$

so (C.31) also holds.

For the two remaining terms in $\mathfrak{B}_1$, we have

$$
\langle b(1)|H_1|b(1)\rangle + \langle a(2)|H_2|a(2)\rangle
$$
$$
= 2\langle b(1)|H_1|b(1)\rangle \tag{C.32}
$$

and by translation along the x_2-axis.

$$
2\langle b(1)|H_1|b(1)\rangle = 2\langle b(2)|e^{\frac{i}{\hbar}[(\frac{eBa}{c})y_2 - 2ap_{x_2})]}H_2e^{-\frac{i}{\hbar}[(\frac{eBa}{c})y_2 - 2ap_{x_2}]}|b(2)\rangle.
$$

For $H_2(\boldsymbol{p}_2, \boldsymbol{r}_2)$ in (C.3), we have

$$
H_2(p_{x_2}, p_{y_2}, x_2, y_2) = \frac{1}{2m}\left[\left(p_{x_2} + \frac{eB}{2c}y_2\right)^2 + \right.
$$
$$
\left. \left(p_{y_2} - \frac{eBa}{2c} + \frac{e^2BE}{2mc\omega_0^2} - \frac{eB}{2c}x_{2+}\right)^2\right]
$$
$$
+ \frac{m\omega_0^2}{2}(x_{2+}^2 + y_2^2) - eE\left(\frac{eE}{2m\omega_0^2} - a\right),
$$

so

$$
e^{\frac{i}{\hbar}[(\frac{eBa}{c})y_2 - 2ap_{x_2}]}H_2(p_{x_2}, p_{y_2}, x_2, y_2)e^{-\frac{i}{\hbar}[(\frac{eBa}{c})y_2 - 2ap_{x_2}]}
$$
$$
= H_2\left(p_{x_2}, p_{y_2} - \frac{eBa}{c}, x_2 - 2a, y_2\right)
$$
$$
= \cdots \text{(substituting and simplifying)}
$$
$$
= H_2(p_{x_2}, p_{y_2}, x_2, y_2) + \frac{m\omega_0^2}{2}(4a^2 - 4ax_{2+}).
$$

Therefore

$$
(C.32) = 2\langle b(1)|H_2|b(1)\rangle
$$

$$
= 2[\langle b(2)|H_2|b(2)\rangle - \underbrace{2am\omega_0^2\langle b(2)|x_{2+}|b(2)\rangle}_{\substack{=0 \text{ because the} \\ \text{integrand is an odd} \\ \text{function of } x_{2+}}} + 2ma^2\omega_0^2]
$$

$$
= 2(E_0 + 2ma^2\omega_0^2). \tag{C.33}
$$

The remaining terms in $-\frac{1}{S^2}\mathfrak{B}_2$ are

$$
-\frac{1}{S^2}[\langle a(1)|H_1|b(1)\rangle S + \langle b(2)|H_2|a(2)\rangle S]
$$
$$
= -\frac{S}{S^2}[\langle E_0a(1)|b(1)\rangle + \langle E_0b(2)|a(2)\rangle]
$$
$$
= -\frac{S}{S^2}\cdot 2E_0S = -2E_0. \tag{C.34}
$$

By adding (C.33) and (C.34), we obtain (C.29). $\qquad\square$

Lemma C.4. *We have*

$$\mathcal{B}_3 - \frac{1}{S^2}\mathcal{B}_4 = 2\hbar\omega_0\left[c\sqrt{b}\,e^{-bd^2}\,I_0(bd^2) - c\sqrt{b}\,e^{d^2(b-\frac{1}{b})}I_0\left(d^2\left(b-\frac{1}{b}\right)\right)\right].$$
(C.35)

Proof. Note that by the symmetry $C(\mathbf{r}_1,\mathbf{r}_2) = C(\mathbf{r}_2,\mathbf{r}_1)$, we have

$$\mathcal{B}_3 = 2\langle a(1)b(2)|C|a(1)b(2)\rangle$$

$$= \int_{\mathbb{R}^2}\int_{\mathbb{R}^2} \overline{\Phi_0^{(1)}(x_1,y_1)}\,\overline{\Phi_0^{(2)}(x_2,y_2)}\frac{e^2}{\kappa|\mathbf{r}_1-\mathbf{r}_2|}\Phi_0^{(1)}(x_1,y_1)\Phi_0^{(2)}(x_2,y_2)$$

$$dx_1dy_1dx_2dy_2$$

$$= \left(\frac{m\Omega}{\pi\hbar}\right)^2\frac{e^2}{\kappa}\int_{\mathbb{R}^2}\int_{\mathbb{R}^2}\frac{1}{|\mathbf{r}_1-\mathbf{r}_2|}e^{-\frac{m\Omega}{\hbar}[(x_1+a+\frac{eE}{m\omega_0^2})^2+y_1^2+(x_2-a+\frac{eE}{m\omega_0^2})^2+y_2^2]}$$
(C.36)

$$dx_1dy_1dx_2dy_2.$$

Introduce the *center of mass coordinates*:

$$\begin{cases} \mathbf{R} = \dfrac{1}{2}(\mathbf{r}_1+\mathbf{r}_2) & \text{(center of mass)} \\[2mm] \mathbf{r} = \mathbf{r}_1-\mathbf{r}_2 & \text{(relative coordinates)} \end{cases}$$
(C.37)

$$\begin{cases} X = \dfrac{1}{2}(x_1+x_2), Y = \dfrac{1}{2}(y_1+y_2); & X = R\cos\Phi, Y = R\sin\Phi, \\[2mm] x = \dfrac{1}{2}(x_1-x_2), y = \dfrac{1}{2}(y_1-y_2); & x = r\cos\phi, y = r\sin\phi. \end{cases}$$
(C.38)

This change of coordinates has Jacobian equal to 1. Then the integral in (C.36) becomes

$$\text{(C.36)} = \left(\frac{m\Omega e}{\pi\hbar}\right)^2\frac{1}{\kappa}\int_{\mathbb{R}^2}\int_{\mathbb{R}^2}\frac{1}{r}$$

$$e^{-\frac{m\Omega}{\hbar}[X^2+xX+\frac{x^2}{4}+(a+\frac{eE}{m\omega_0^2})(2X-x)+(a+\frac{eE}{m\omega_0^2})^2}$$

$$^{+X^2-xX+\frac{x^2}{4}+(\frac{eE}{m\omega_0^2}-a)(2X-x)+(\frac{eE}{m\omega_0^2}-a)^2+Y^2+yY+\frac{y^2}{4}+Y^2-yY+\frac{y^2}{4}]}$$

$$rdrd\phi RdRd\Phi$$

$$= \left(\frac{m\Omega e}{\pi\hbar}\right)^2\frac{1}{\kappa}\int_0^{2\pi}\int_0^\infty\int_0^{2\pi}\int_0^\infty (drd\phi)(RdRd\Phi)$$

$$\left\{e^{-\frac{m\Omega}{\hbar}[2R^2+\frac{r^2}{2}+2(a^2+(\frac{eE}{m\omega_0^2})^2)+2ax+\frac{4eE}{m\omega_0^2}X^2]}\right\}$$

$$= \left(\frac{m\Omega e}{\pi\hbar}\right)^2 \frac{1}{\kappa} e^{-\frac{2m\Omega}{\hbar}(a^2+\frac{e^2E^2}{m^2\omega_0^4})} \int_0^{2\pi}\int_0^\infty RdRd\Phi \cdot e^{-\frac{m\Omega}{\hbar}[2R^2+\frac{4cE}{m\omega_0^2}X]}$$

$$\cdot \int_0^{2\pi}\int_0^\infty drd\phi \cdot e^{-\frac{m\Omega}{\hbar}[\frac{r^2}{2}+2ar\cos\phi]}$$

$$= \left(\frac{m\Omega e}{\pi\hbar}\right)^2 \frac{1}{\kappa} e^{-\frac{2m\Omega}{\hbar}(a^2+\frac{e^2E^2}{m^2\omega_0^4})}$$

$$\left\{ \int_{-\infty}^\infty e^{-\frac{m\Omega}{\hbar}[2X^2+\frac{4eE}{m\omega_0^2}X]} \left[\int_{-\infty}^\infty e^{-\frac{2m\Omega}{\hbar}Y^2}dY\right]dX \right\} \cdot$$

$$\cdot \left\{ \int_0^\infty e^{-\frac{m\Omega}{2\hbar}r^2}\underbrace{\left[\int_0^{2\pi} e^{-\frac{2m\Omega a}{\hbar}r\cos\phi}d\phi\right]}_{(\mathcal{J}_1)}dr \right\}. \tag{C.39}$$

We evaluate the integral $(\mathcal{J}_1)$ above by using the expansion

$$e^{-(\frac{2m\Omega a}{\hbar}r)\cos\phi} = \sum_{m=-\infty}^\infty (-1)^m I_m\left(\frac{2m\Omega a}{\hbar}r\right)e^{im\phi}$$

$$= I_0\left(\frac{2m\Omega a}{\hbar}r\right) + 2\sum_{m=1}^\infty I_m\left(\frac{2m\Omega a}{\hbar}r\right)\cos(m\phi)$$

(cf. Abramowitz and Stegun [1, p. 376, Formula 9.6.34])

$$(\mathcal{J}_1) = \int_0^{2\pi}\left\{ I_0\left(\frac{2m\Omega a}{\hbar}r\right) + 2\sum_{m=1}^\infty I_m\left(\frac{2m\Omega a}{\hbar}r\right)\cos m\phi \right\}d\phi$$

$$= 2\pi I_0\left(\frac{2m\Omega a}{\hbar}r\right).$$

Substituting $(\mathcal{J}_1)$ into (C.39) above and continuing, we obtain

$$\text{(C.36)} = \left(\frac{m\Omega e}{\pi\hbar}\right)^2 \frac{1}{\kappa} e^{-\frac{2m\Omega}{\hbar}(a^2+\frac{e^2E^2}{m^2\omega_0^4})} \cdot e^{\frac{2m\Omega}{\hbar}\frac{e^2E^2}{m^2\omega_0^4}} \cdot \underbrace{\int_{-\infty}^\infty e^{-\frac{2m\Omega}{\hbar}(X+\frac{eE}{m\omega_0^2})^2}dX}_{(\frac{\pi\hbar}{2m\Omega})^{1/2}} \cdot$$

$$\cdot \underbrace{\int_{-\infty}^\infty e^{-\frac{2m\Omega}{\hbar}Y^2}dY}_{(\frac{\pi\hbar}{2m\Omega})^{1/2}} \cdot 2\pi \underbrace{\int_0^\infty e^{-\frac{m\Omega}{2\hbar}r^2}I_0\left(\frac{2m\Omega a}{\hbar}r\right)dr}_{(\mathcal{J}_2)}.$$

To evaluate the integral $(\mathcal{J}_2)$, we use

$$\int_0^\infty e^{-ax^2} I_\nu(bx)\,dx = \frac{1}{2}\sqrt{\frac{\pi}{a}}\, e^{\frac{b^2}{8a}} I_{\frac{1}{2}\nu}\left(\frac{b^2}{8a}\right) \qquad \text{(for Re } \nu > -1,\ \text{Re } a > 0)$$

(cf. Abramowitz and Stegun [1, p. 487, Formula 11.4.31]).

Then

$$(\mathcal{J}_2) = \frac{1}{2}\sqrt{\frac{2\hbar\pi}{m\Omega}}\, e^{\frac{m\Omega a^2}{\hbar}} I_0\left(\frac{m\Omega a^2}{\hbar}\right).$$

Therefore, we have arrived at

$$\mathfrak{B}_3 = 2\langle a(1)b(2)|C|a(1)b(2)\rangle = 2\left(\frac{\pi m\Omega}{2\hbar}\right)^{1/2} \frac{e^2}{\kappa} e^{-\frac{m\Omega a^2}{\hbar}} I_0\left(\frac{m\Omega a^2}{\hbar}\right)$$

$$= 2\hbar\omega_0 c\sqrt{b}\, e^{-bd^2} I_0(bd^2); \quad \left(\text{with } c = \frac{e^2}{\kappa}\frac{1}{\hbar\omega_0}\sqrt{\frac{\pi m\omega_0}{2\hbar}},\ \text{cf. (7.22)}\right).$$

$$(C.40)$$

Next, we proceed to evaluate integral in $\mathfrak{B}_4$:

$$\mathfrak{B}_4 = \langle a(1)b(2)|C|a(2)b(1)\rangle + \langle a(2)b(1)|C|a(1)b(2)\rangle$$
$$= 2\,\mathrm{Re}\langle a(1)b(2)|C|a(2)b(1)\rangle$$
$$= \cdots \text{ (similar to (C.36)–(C.39), using the center of mass coordinates}$$
$$\text{(C.37) and (C.38))}$$

$$= 2\,\mathrm{Re}\left(\frac{m\Omega e}{\pi\hbar}\right)^2 \frac{1}{\kappa} \int_0^{2\pi}\int_0^{2\pi}\int_0^\infty\int_0^\infty \frac{1}{r} \exp\left\{i\frac{eBa}{\hbar c}y - \frac{m\Omega}{\hbar}\left[\left(\frac{2X+x}{2}\right)^2\right.\right.$$
$$\left.\left.+\left(\frac{2X-x}{2}\right)^2 + 2a^2 + \left(\frac{2Y+y}{2}\right)^2 + \left(\frac{2Y-y}{2}\right)^2\right]\right\} r\,dr\,R\,dR\cdot d\phi\,d\Phi$$

$$= 2\,\mathrm{Re}\left(\frac{m\Omega e}{\pi\hbar}\right)^2 \frac{1}{\kappa} e^{-\frac{2m\Omega a^2}{\hbar}} \int_0^{2\pi}\int_0^{2\pi}\int_0^\infty\int_0^\infty$$

$$\exp\left\{\frac{ieBa}{\hbar c}y - \frac{m\Omega}{\hbar}\left[2X^2 + 2Y^2 + \frac{x^2}{2} + \frac{y^2}{2}\right]\right\} \cdot dr\,R\,dR\,d\phi\,d\Phi$$

$$= 2\,\mathrm{Re}\left(\frac{m\Omega e}{\phi\hbar}\right)^2 \frac{1}{\kappa} e^{-\frac{2m\Omega a^2}{\hbar}} \int_{-\infty}^\infty e^{-\frac{2m\Omega}{\hbar}X^2}dX \cdot \int_{-\infty}^\infty e^{-\frac{2m\Omega}{\hbar}Y^2}dY$$

$$\cdot \int_0^\infty e^{-\frac{m\Omega}{2\hbar}r^2} \underbrace{\left[\int_0^{2\pi} e^{\frac{ieBa}{\hbar c}r\sin\phi}\,d\phi\right]}_{(\mathcal{J}_3)} dr$$

$$= 2\,\mathrm{Re}\left(\frac{m\Omega e}{\pi\hbar}\right)^2 \frac{1}{\kappa} e^{-\frac{2m\Omega a^2}{\hbar}} \cdot \left(\frac{\pi\hbar}{2m\Omega}\right)^{1/2} \cdot \left(\frac{\pi\hbar}{2m\Omega}\right)^{1/2} \underbrace{\int_0^\infty e^{-\frac{m\Omega}{2\hbar}r^2}(\mathcal{J}_3)\,dr,}_{(\mathcal{J}_4)}$$

where

$$(\mathcal{J}_3) = \int_0^{2\pi} e^{\frac{ieBa}{\hbar c} r \sin\phi} d\phi = J_0\left(\frac{eBa}{\hbar c} r\right) \cdot 2\pi$$

(cf. Abramowitz and Stegun [1, p. 360, Formula (9.1.18)])

and

$$(\mathcal{J}_4) = 2\pi \int_0^\infty e^{-\frac{m\Omega}{2\hbar} r^2} J_0\left(\frac{eBa}{\hbar c} r\right) dr$$

$$= 2\pi \sqrt{\frac{\hbar\pi}{2m\Omega}} e^{-\left(\frac{eBa}{\hbar c}\right)^2 \frac{\hbar}{4m\Omega}} \cdot I_0\left(\left(\frac{eBa}{\hbar c}\right)^2 \frac{\hbar}{4m\Omega}\right)$$

(cf. Gradshteyn and Ryzhik [5, p. 732, Formula (6.618(1))]).

Therefore, we have arrived at

$$\mathcal{B}_4 = 2\,\mathrm{Re}\langle a(1)b(2)|C|a(2)b(1)\rangle$$

$$= 2\,\mathrm{Re}\left(\frac{m\Omega}{\hbar}\right)^{1/2} \sqrt{\frac{\pi}{2}} \frac{e^2}{\kappa} e^{-\frac{2m\Omega}{\hbar} a^2 - \frac{e^2 B^2 a^2}{4\hbar c^2 m\Omega}} I_0\left(\frac{e^2 B^2 a^2}{4\hbar c^2 m\Omega}\right)$$

$$= 2\hbar\omega_0 c\sqrt{b}\, e^{-2bd^2} e^{-d^2(b-\frac{1}{b})} I_0\left(d^2\left(b-\frac{1}{b}\right)\right). \tag{C.41}$$

Using S in (C.24), we obtain from (C.40) and (C.41) that $\mathcal{B}_3 - (1/S^2)\mathcal{B}_4$ is indeed equal to (C.35). □

Finally, we evaluate $\mathcal{B}_5$.

Lemma C.5. *We have*

$$\mathcal{B}_5 = -4m\omega_0^2 a^2 + 2 \cdot \left(\frac{m\omega_0^2}{2}\right)\left[\frac{3\hbar}{2m\Omega} + 3\left(\frac{eE}{m\omega_0^2}\right)^2 + \frac{3}{2}a^2\right]. \tag{C.42}$$

Proof. First, we want to show that

$$\langle a(1)b(2)|W|a(1)b(2)\rangle + \langle a(2)b(1)|W|a(2)b(1)\rangle$$

$$= 2\langle a(1)b(2)|W|a(1)b(2)\rangle - 4m\omega_0^2 a^2. \tag{C.43}$$

The first term in $\mathcal{B}_5$ (and on the left hand side of (C.43)) satisfies

$$\langle a(1)b(2)|W|a(1)b(2)\rangle = \langle a(1)|W_1|a(1)\rangle + \langle b(2)|W_2|b(2)\rangle \qquad \text{(see (C.4), (C.5))}$$

$$= \langle b(1)|e^{\frac{i}{\hbar}\left(\frac{eBa}{c} y_1 - 2ap_{x_1}\right)} W_1 e^{-\frac{i}{\hbar}\left(\frac{eBa}{c} y_1 - 2ap_{x_1}\right)}|b(1)\rangle$$

$$+ \langle a(2)|e^{-\frac{i}{\hbar}\left(\frac{eBa}{c} y_2 - 2ap_{x_2}\right)} W_2 e^{\frac{i}{\hbar}\left(\frac{eBa}{c} y_2 - 2ap_{x_2}\right)}|a(2)\rangle. \tag{C.44}$$

But

$$W_1(x) = \frac{m\omega_0^2}{2}\left[\frac{1}{4a^2}(x^2 - a^2)^2 - (x+a)^2\right],$$

so

$$e^{\frac{i}{\hbar}(\frac{eBa}{c}y_1 - 2ap_{x_1})} W_1(x_1) e^{-\frac{i}{\hbar}(\frac{eBa}{c}y_1 - 2ap_{x_1})} = W_1(x_1 - 2a)$$

$$= \frac{m\omega_0^2}{2}\left[\frac{1}{4a^2}((x_1 - 2a)^2 - a^2)^2 - (x_1 - 2a + a)^2\right]$$

$$= \cdots(\text{expanding and regrouping terms})$$

$$= W_1(x_1) + \frac{m\omega_0^2}{2}\left[4ax_1 - \frac{2}{a}(x_1 - a)^3\right]. \tag{C.45}$$

Similarly, for W_2 given in (C.5),

$$e^{-\frac{i}{\hbar}(\frac{eBa}{c}y_2 - 2ap_{x_2})} W_2(x_2) e^{\frac{i}{\hbar}(\frac{eBa}{c}y_2 - 2ap_{x_2})} = W_2(x_2 + 2a)$$

$$= W_2(x_2) + \frac{m\omega_0^2}{2}\left[-4ax_2 + \frac{2}{a}(x_2 + a)^3\right]. \tag{C.46}$$

Thus, continuing from (C.44) using (C.45) and (C.46), we have

$$(\text{C.44}) = \langle b(1)|W_1|b(1)\rangle + \langle b(1)|\frac{m\omega_0^2}{2}\left[4ax_1 - \frac{2}{a}(x_1 - a)^3\right]|b(1)\rangle$$

$$+ \langle a(2)|W_2|a(2)\rangle + \langle a(2)|\frac{m\omega_0^2}{2}\left[-4ax_2 + \frac{2}{a}(x_2 + a)^3\right]|a(2)\rangle$$

$$= \langle b(1)a(2)|W|b(1)a(2)\rangle + 2m\omega_0^2 a[\langle b(1)|x_1|b(1)\rangle - \langle a(2)|x_2|a(2)\rangle]$$

$$+ \frac{m\omega_0^2}{a}[\langle a(2)|(x_2 + a)^3|a(2)\rangle - \langle b(1)|(x_1 - a)^3|b(1)\rangle]$$

$$= \langle b(1)a(2)|W|b(1)a(2)\rangle + 2m\omega_0^2 a\left[\langle b(1)|(x_1)_+ + a - \frac{eE}{m\omega_0^2}|b(1)\rangle\right.$$

$$\left. - \langle a(2)|(x_2)_- - a - \frac{eE}{m\omega_0^2}|a(2)\rangle\right]$$

$$+ \frac{m\omega_0^2}{a}\left[\langle a(2)|\left((x_2)_- - \frac{eE}{m\omega_0^2}\right)^3|a(2)\rangle\right.$$

$$\left. - \langle b(1)|\left((x_1)_+ - \frac{eE}{m\omega_0^2}\right)^3|b(1)\rangle\right]$$

$$(\text{where, recall that } x_+ = x - a + \frac{eE}{m\omega_0^2} \text{ and } x_- = x + a + \frac{eE}{m\omega_0^2})$$

$$= \langle b(1)a(2)|W|b(1)a(2)\rangle + 4m\omega_0^2 a^2$$

$$+ \frac{m\omega_0^2}{a}\left[\langle a(2)|(x_2)_-^3 - \frac{3eE}{m\omega_0^2}(x_2)_-^2 + 3\left(\frac{eE}{m\omega_0^2}\right)^2(x_2)_-\right.$$

$$\left. -\left(\frac{eE}{m\omega_0}\right)^3|a(2)\rangle - \langle b(1)|(x_1)_+^3 - 3\frac{eE}{m\omega_0^2}(x_1)_+^2 + 3\left(\frac{eE}{m\omega_0^2}\right)^2(x_1)_+\right.$$

$$\left. -\left(\frac{eE}{m\omega_0}\right)^3|b(1)\rangle\right]$$

$$= \langle b(1)a(2)|W|b(1)a(2)\rangle + 4m\omega_0^2 a^2. \tag{C.47}$$

By (C.44) and (C.47), we have confirmed (C.43). So our objective now is to evaluate $\langle a(1)b(2)|W|a(1)b(2)\rangle$:

$$\langle a(1)b(2)|W|a(1)b(2)\rangle = \langle a(1)|W_1|a(1)\rangle + \langle b(2)|W_2|b(2)\rangle; \tag{C.48}$$

$$\langle a(1)|W_1|a(1)\rangle = \frac{m\Omega}{\pi\hbar}\int_{-\infty}^{\infty}\int_{-\infty}^{\infty} e^{-\frac{m\Omega}{\hbar}[(x_1+a+\frac{eE}{m\omega_0^2})^2+y_1^2]}.$$

$$\cdot \frac{m\omega_0^2}{2}\left[\frac{1}{4a^2}(x_1^2-a^2)^2 - (x_1+a)^2\right]\,dx_1dy_1 \equiv f(a); \tag{C.49}$$

$$\langle b(2)|W_2|b(2)\rangle = \frac{m\Omega}{\pi\hbar}\int_{-\infty}^{\infty}\int_{-\infty}^{\infty} e^{-\frac{m\Omega}{\hbar}[(x_2-a+\frac{eE}{m\omega_0^2})^2+y_2^2]}.$$

$$\cdot \frac{m\omega_0^2}{2}\left[\frac{1}{4a^2}(x_2^2-a^2)^2 - (x_2-a)^2\right]\,dx_2dy_2. \tag{C.50}$$

By comparing (C.49) and (C.50), we see that if the outcome of (C.49) is $f(a)$ (with all the parameters other than a being fixed), then the outcome of (C.50) will be $f(-a)$.

Similarly,

$$\langle a(1)|W_1|b(1)\rangle = \frac{m\Omega}{\pi\hbar}\int_{-\infty}^{\infty}\int_{-\infty}^{\infty} e^{\frac{i}{\hbar}\frac{eBa}{c}y_1}e^{-\frac{m\Omega}{\hbar}[(x_1+\frac{eE}{m\omega_0^2})^2+a^2+y_1^2]}.$$

$$\cdot \frac{m\omega_0^2}{2}\left[\frac{1}{4a^2}(x_1^2-a^2)^2 - (x_1+a)^2\right]\,dx_1dy_1 \equiv g(a), \tag{C.51}$$

then

$$\langle b(2)|W_2|a(2)\rangle = \frac{m\Omega}{\pi\hbar}\int_{-\infty}^{\infty}\int_{-\infty}^{\infty} e^{-\frac{i}{\hbar}\frac{eBa}{c}y_2}e^{-\frac{m\Omega}{\hbar}[(x_2+\frac{eE}{m\omega_0^2})^2+a^2+y_2^2]}.$$

$$\cdot \frac{m\omega_0^2}{2}\left[\frac{1}{4a^2}(x_2^2-a^2)^2 - (x_2-a)^2\right]\,dx_2dy_2 = g(-a). \tag{C.52}$$

By translating $x \mapsto x - a - \frac{eE}{m\omega_0^2}$, we have

$$\langle a(1)|W_1|a(1)\rangle = (\text{C.49}) = \frac{m\Omega}{\pi\hbar} \left[\int_{-\infty}^{\infty} e^{-\frac{m\Omega}{\hbar}y^2} dy \right]$$

$$\left[\int_{-\infty}^{\infty} e^{-\frac{m\Omega}{\hbar}x^2} W_1\left(x - a - \frac{eE}{m\omega_0^2}\right) dx \right]. \qquad (\text{C.53})$$

Here

$$W_1\left(x - a - \frac{eE}{m\omega_0^2}\right) = W_1(x - a - \beta) \quad \left(\beta \equiv \frac{eE}{m\omega_0^2}\right) \qquad (\text{C.54})$$

$$= \frac{m\omega_0^2}{2} \left\{ \frac{1}{4a^2}[(x - a - \beta)^2 - a^2]^2 - (x - \beta)^2 \right\}$$

$$= \frac{m\omega_0^2}{2} \left[\frac{1}{4a^2}(x - \beta)^4 - \frac{1}{a}(x - \beta)^3 \right]$$

$$= \frac{m\omega_0^2}{2} \left[\frac{1}{4a^2}x^4 - \left(\frac{\beta}{a^2} + \frac{1}{a}\right)x^3 + \left(\frac{3}{2}\frac{\beta^2}{a^2} + \frac{3\beta}{a}\right)x^2 \right.$$

$$\left. - \left(\frac{\beta^3}{a^2} + \frac{3\beta^2}{a}\right)x + \left(\frac{\beta^4}{4a^2} + \frac{\beta^3}{a}\right) \right]. \qquad (\text{C.55})$$

Recall the formulas for Gaussian integrals

$$\int_{-\infty}^{\infty} x^{2n} e^{-\alpha x^2} dx = -\left(\frac{\partial}{\partial\alpha}\right)^n \int_{-\infty}^{\infty} e^{-\alpha x^2} dx = \left(-\frac{\partial}{\partial\alpha}\right)^n \sqrt{\frac{\pi}{\alpha}}$$

$$= \left[\prod_{k=1}^{n}\left(\frac{1}{2} + k - 1\right) \right]\left(\frac{1}{\alpha^n}\sqrt{\frac{\pi}{\alpha}}\right), \quad \text{for} \quad n = 1, 2, \ldots, \qquad (\text{C.56})$$

$$\int_{-\infty}^{\infty} x^{2n+1} e^{-\alpha x^2} dx = 0, \quad \text{for} \quad n = 0, 1, 2, \ldots, \qquad (\text{C.57})$$

from (C.53), (C.55)–(C.57) we obtain

$$\langle a(1)|W_1|a(1)\rangle = \frac{m\Omega}{\pi\hbar}\left(\frac{\pi\hbar}{m\Omega}\right)^{1/2}\left(\frac{m\omega_0^2}{2}\right)$$

$$\cdot \left[\frac{1}{4a^2}\frac{\partial^2}{\partial\alpha^2} - \left(\frac{3\beta^2}{2a^2} + \frac{3\beta}{a}\right)\frac{\partial}{\partial\alpha} + \left(\frac{\beta^4}{4a^2} + \frac{\beta^3}{a}\right) \right]\left(\sqrt{\frac{\pi}{\alpha}}\right)$$

$$(\text{where } \alpha \equiv m\Omega/\hbar)$$

$$= \frac{m\Omega}{\pi\hbar}\left(\frac{\pi\hbar}{m\Omega}\right)^{1/2}\frac{m\omega_0^2}{2}\left(\frac{\pi\hbar}{m\Omega}\right)^{1/2}\left[\frac{1}{4a^2}\frac{3\hbar^2}{4m^2\Omega^2} \right.$$

$$+\frac{1}{2}\left(\frac{3\beta^2}{2a^2}+\frac{3\beta}{a}\right)\frac{\hbar}{m\Omega}+\frac{\beta^4}{4a^2}+\frac{\beta^3}{a}\Bigg]$$

$$=\frac{m\omega_0^2}{2}\left[\frac{3\hbar^2}{16m^2\Omega^2a^2}+\left(\frac{3\beta^2}{4a^2}+\frac{3\beta}{2a}\right)\frac{\hbar}{m\Omega}+\frac{\beta^4}{4a^2}+\frac{\beta^3}{a}\right]$$

$$=f(a);\qquad\text{cf. (C.49).}$$

Then $\langle b(2)|W_2|b(2)\rangle = f(-a)$ and so from (C.48), we obtain

$$\langle a(1)b(2)|W|a(2)b(1)\rangle = \frac{m\omega_0^2}{2}\left[\frac{3\hbar^2}{8m^2\Omega^2a^2}+\frac{3\beta^2\hbar}{2a^2m\Omega}+\frac{\beta^4}{2a^2}\right].\qquad\text{(C.58)}$$

Next, repeat similar procedures,

$$\text{(C.51)}=\frac{m\Omega}{\pi\hbar}e^{-\frac{m\Omega}{\hbar}a^2}\left[\int_{-\infty}^{\infty}e^{\frac{i}{\hbar}\frac{eBa}{c}y-\frac{m\Omega}{\hbar}y^2}\,dy\right]\left[\int_{-\infty}^{\infty}e^{-\frac{m\Omega}{\hbar}x^2}W_1(x-\beta)dx\right]$$

$$=\frac{m\Omega}{\pi\hbar}e^{-\frac{m\Omega}{\hbar}a^2}\left[\int_{-\infty}^{\infty}e^{-\frac{m\Omega}{\hbar}(y-i\frac{eBa}{2m\Omega c})^2-\frac{m\Omega}{\hbar}\frac{e^2B^2a^2}{4m^2\Omega^2c^2}}\,dy\right]$$

$$\cdot\int_{-\infty}^{\infty}e^{-\frac{m\Omega}{\hbar}x^2}\cdot\frac{m\omega_0^2}{2}\left[\frac{1}{4a^2}(x-\beta+a)^4-\frac{1}{a}(x-\beta+a)^3\right]dx$$

$$=\left(\frac{m\Omega}{\pi\hbar}\right)e^{-\frac{m\Omega}{\hbar}a^2-\frac{e^2B^2a^2}{4\hbar m\Omega c^2}}\left(\frac{\pi\hbar}{m\Omega}\right)^{1/2}\cdot\left(\frac{\pi\hbar}{m\Omega}\right)^{1/2}\cdot\frac{m\omega_0^2}{2}\cdot$$

$$\left\{\frac{3\hbar^2}{16m^2\Omega^2a^2}+\left[\frac{3(\beta-a)^2}{4a^2}+\frac{3(\beta-a)}{2a}\right]\frac{\hbar}{m\Omega}+\frac{(\beta-a)^4}{4a^2}+\frac{(\beta-a)^3}{a}\right\}$$

$$=g(a).\qquad\text{(C.59)}$$

Therefore,

$$\langle a(1)b(2)|W|a(2)b(1)\rangle = \langle a(1)|W_1|b(1)\rangle\langle b(2)|a(2)\rangle + \langle a(1)|b(1)\rangle$$
$$\cdot\langle b(2)|W_2|a(2)\rangle$$
$$= S[\langle a(1)|W_1|b(1)\rangle + \langle b(2)|W_2|a(2)\rangle]$$

$$= S[g(a)+g(-a)]\qquad\text{(by (C.52))}$$

$$= S\cdot e^{-\frac{a^2}{\hbar}(m\Omega+\frac{e^2B^2}{4m\Omega c^2})}\frac{m\omega_0^2}{2}\left[\frac{3\hbar^2}{8m^2\Omega^2a^2}\right.$$

$$\left.+\left(\frac{3\beta^2}{2a^2}+\frac{3}{2}-3\right)\frac{\hbar}{m\Omega}+\frac{\beta^4}{2a^2}+3\beta^2+\frac{a^2}{2}-6\beta^2-2a^2\right]$$

(by (C.58) and (C.52)).

But the factor $e^{-\frac{a^2}{\hbar}\left(m\Omega + \frac{e^2 B^2}{4m\Omega c^2}\right)}$ behind S is just S itself by (C.24). Thus

$$\frac{1}{S^2}\langle a(1)b(2)|W|a(2)b(1)\rangle$$
$$= \frac{m\omega_0^2}{2}\left[\frac{3\hbar^2}{8m^2\Omega^2 a^2} + \frac{3}{2}\left(\frac{\beta^2}{a^2} - 1\right)\frac{\hbar}{m\Omega} + \frac{\beta^4}{2a^2} - 3\beta^2 - \frac{3}{2}a^2\right]. \qquad (C.60)$$

Summarizing (C.43), (C.58) and (C.60), we have

$$\mathfrak{B}_5 = 2\langle a(1)b(2)|W|a(1)b(2)\rangle - 4m\omega_0^2 a^2$$
$$- 2\,\mathrm{Re}\cdot\frac{1}{S^2}\langle a(1)b(2)|W|a(2)b(1)\rangle$$
$$= -4m\omega_0^2 a^2 + 2\cdot\left(\frac{m\omega_0^2}{2}\right)\left(\frac{3\hbar}{2m\Omega} + 3\beta^2 + \frac{3}{2}a^2\right).$$

This is (C.42). $\qquad\qquad\qquad\square$

We can now combine all the preceding lemmas and finally obtain the following.

Theorem C.6. *The exchange energy is given by*

$$J = \langle \Psi_-|H_{orb}|\Psi_-\rangle - \langle \Psi_+|H_{orb}|\Psi_+\rangle$$
$$= (7.17).$$

Proof. We only need note that with S given in (C.24), we have

$$\frac{S^2}{1 - S^4} = \frac{1}{S^{-2} - S^2} = \frac{1}{2\sinh(2d^2(2b - \frac{1}{b}))}.$$

Thus, the coefficient outside the parentheses in (C.28) is determined as above. We now collect all the terms in (C.29), (C.35), and (C.42), noting the cancellation of the terms $4ma^2\omega_0^2$ and $-4ma^2\omega_0^2$ in (C.28) and (C.42), and then simplify (just a little). We then obtain (7.17). $\qquad\square$

Appendix D

Transformation of quantum states: SU(2) and SO(3)

When a quantum operation is applied to a quantum system, it may change the quantum state of the system from one to another. The representation of the operation depends on how the quantum state is represented. For example, (10.3) leads to an operator or matrix U which connects the new and old states of a single spin quantum system:

$$|\phi'\rangle = U|\phi\rangle,$$

where $|\phi'\rangle$ and $|\phi\rangle$ are the quantum state after and before the operation, respectively. The fact that both states are unit vectors implies that U is a 2×2 unimodular complex matrix. Moreover, U is also unitary, i.e., $U \in$ $\mathbf{SU}(2)$[1], a Lie group endowed with a certain topology.

If the quantum state is instead represented by a 3-dimensional Bloch vector (cf. (10.6)), the effect of a unitary operation can be viewed as that of a rotation which rotates the Bloch sphere, and the operator is represented by a 3×3 real matrix S. If the quantum system has states v and v' in Bloch vector form before and after the operation, respectively, then

$$v' = Sv'.$$

The matrix S is a proper rotation matrix, i.e., $S \in \mathbf{SO}(3)$[2]. It is isometric and preserves the 3-fold product.

---

[1] $\mathbf{SU}(n)$ is the special unitary group of $n \times n$ matrices. An $n \times n$ matrix $A \in \mathbf{SU}(n)$ if and only if A is unitary, i.e., $A \cdot A^\dagger = I_n$, where $A^\dagger$ is the Hermitian adjoint of A, and $\det A = 1$.
[2] $\mathbf{SO}(3)$ denotes the special orthogonal group of 3×3 matrices. An $n \times n$ matrix $A \in \mathbf{SO}(n)$ if and only if A is real, $AA^T = I_n$ and $\det A = 1$.

If both S and U represent "the same physical" operation, such as a transformation induced by a series of pulses in NMR, what is the "canonical" relation between S and U? One can show that there is a mapping R from $\mathbf{SU}(2)$ to $\mathbf{SO}(3)$ such that $S = R(U)$, for any $U \in \mathbf{SU}(2)$ and its corresponding Bloch-sphere representation S [7]. Straightforward computation shows that the entry of matrix $S = R(U)$ at the kth row and i^{th} column is given as

$$S_{ki} = Tr(\sigma_k \, U \, I_i \, U^\dagger), \tag{D.1}$$

where σ_k are the Pauli matrices, I_i are the operators in (10.5), with the subscripts x, y, z for k and i such that x, y, z correspond, respectively, to 1, 2 and 3, and Tr is the trace operator. It can also be shown that R is a two-to-one homomorphism between $\mathbf{SU}(2)$ and $\mathbf{SO}(3)$ with kernel $\ker(R) = \{I, -I\}$. It coincides with the fact that U and $-U$ in $\mathbf{SU}(2)$ represent the same operation because only the relative phase matters. This mapping is also surjective, so it defines an isomorphism from the quotient group $\mathbf{SU}(2)/\ker(R)$ to $\mathbf{SO}(3)$. We provide a more detailed discussion about this isomorphism in Appendix E.

It is known that any $U \in \mathbf{SU}(2)$ can be written into an exponential form parameterized by an angle $\theta \in [0, 2\pi)$ and a unit vector $\boldsymbol{n} = (n_1, n_2, n_3) \in \mathbb{R}^3$ such that

$$
\begin{aligned}
U = U(\theta, \boldsymbol{n}) \quad &= \quad e^{-i\frac{\theta}{2}\boldsymbol{n}\cdot\boldsymbol{\sigma}} \\
&= \quad \begin{bmatrix} \cos\frac{\theta}{2} - in_3\sin\frac{\theta}{2} & -\sin\frac{\theta}{2}(n_2 + in_1) \\ \sin\frac{\theta}{2}(n_2 - in_1) & \cos\frac{\theta}{2} + in_3\sin\frac{\theta}{2} \end{bmatrix} \\
&= \quad \cos\frac{\theta}{2}I - i\sin\frac{\theta}{2}\boldsymbol{n}\cdot\boldsymbol{\sigma},
\end{aligned} \tag{D.2}
$$

where $\boldsymbol{\sigma} = [\sigma_x, \sigma_y, \sigma_z]$. With this parameterization of $\mathbf{SU}(2)$, entries of $S = R(U)$ can be computed using (D.1) as

$$S_{ij} = R(U)_{ij} = \cos\theta \, \delta_{ij} + (1 - \cos\theta)n_i n_j + \sum_{k=1}^{3} \sin\theta \, \epsilon_{ikj} n_k; \tag{D.3}$$

see the details of the derivations in (E.9)–(E.12). It should be noted now that S coincides with a rotation about the axis along n with an angle θ in the three dimensional Euclidean space after comparing S_{ij} with the standard formula of a rotation matrix. This interpretation is important in understanding the terminologies used in NMR. For example, the rotations around x, y, and z axes ($x/y/z$-rotations) with an arbitrary angle θ define the following three unitary operators in $\mathbf{SU}(2)$, respectively:

$$X_\theta = e^{-i\theta\sigma_x/2} = \begin{bmatrix} \cos\frac{\theta}{2} & -i\sin\frac{\theta}{2} \\ -i\sin\frac{\theta}{2} & \cos\frac{\theta}{2} \end{bmatrix}, \tag{D.4}$$

$$Y_\theta = e^{-i\theta\sigma_y/2} = \begin{bmatrix} \cos\frac{\theta}{2} & -\sin\frac{\theta}{2} \\ \sin\frac{\theta}{2} & \cos\frac{\theta}{2} \end{bmatrix}, \tag{D.5}$$

$$Z_\theta = e^{-i\theta\sigma_z/2} = \begin{bmatrix} e^{-i\theta/2} & 0 \\ 0 & e^{i\theta/2} \end{bmatrix}. \tag{D.6}$$

From Theorem 2.6.1, we know that the collection of all the 1-qubit gates and the 2-qubit CNOT gate are universal. In addition, the following fact [6, p. 175] holds for 1-qubit quantum gates:

Theorem D.1. *Suppose U is a unitary operation on a single qubit. Then there exist real numbers α, β, γ, and δ such that*

$$U = e^{i\alpha} Z_\beta Y_\gamma Z_\delta.$$

$\square$

Every $R \in \mathbf{SO}(3)$ is a rotation with axis n and angle θ, where n is a unit vector in $\mathbf{R}^3$ and $\theta \in [0, 2\pi)$. So we denote

$$R = R(\theta, n). \tag{D.7}$$

We have the following [7]:

- Let $\rho \in \mathbf{SO}(3)$ be any rotation. Then

$$\rho R(\theta, n)\rho^{-1} = R(\theta, \rho n). \tag{D.8}$$

- The map

$$\begin{aligned} \mathcal{R} : \mathbf{SU}(2) &\to \mathbf{SO}(3) \\ \mathcal{R}(U(\theta, n)) &\equiv R(\theta, n) \end{aligned} \tag{D.9}$$

  is a 2-to-1 homomorphism, where

$$\begin{aligned} \mathcal{R}(U(\theta_1, n_1)U(\theta_2, n_2)) &= \mathcal{R}(U(\theta_1, n_1))\mathcal{R}(U(\theta_2, n_2)), \\ \mathcal{R}(I_2) &= \mathcal{R}(-I_2) = I_3. \end{aligned} \tag{D.10}$$

  Specifically, for $U \in \mathbf{SU}(2)$ and $R \in \mathbf{SO}(3)$ and

$$U = (u_{ij})_{2\times 2}, \quad R = (r_{ij})_{3\times 3}, \quad R = \mathcal{R}(U), \tag{D.11}$$

  then

$$r_{ij} = Tr(\frac{1}{2}\sigma_i U \sigma_j U^\dagger); \quad i, j = 1, 2, 3, \tag{D.12}$$

  where σ_1, σ_2 and σ_3 are, respectively, the Pauli matrices σ_x, σ_y and σ_z. The proof that the above is indeed a homomorphism will be given in Appendix E next.

Appendix E

The Homomorphism from SU(2) to SO(3)

We use the same notation as in Appendix D. Recall that U is a complex matrix in $\mathbf{SU}(2)$, and we want to find a mapping $\mathcal{R}$ from $\mathbf{SU}(2)$ to the space of 3×3 real matrices, such that $\mathcal{R}(U)$ and U represent the same physical operation. Let v be a Bloch vector corresponding to the old state $|\phi\rangle$ and v' to the new one. Then the following identity must be satisfied:

$$v' = \mathcal{R}(U)v, \tag{E.1}$$

where $U \in \mathbf{SU}(2)$. Using the definition of Bloch vector and density matrix, we obtain

$$\begin{aligned}
v' \cdot A &= U|\psi\rangle\langle\psi|U^\dagger - I_0 \\
&= Uv \cdot AU^\dagger,
\end{aligned} \tag{E.2}$$

where $A = [I_x, I_y, I_z]$ (cf. (10.5)) and we use the dot to denote the inner product between vectors.

The product of I_i and σ_j satisfies

$$Tr(I_i\sigma_j) = \delta_{ij}, \qquad i, j \in \{x, y, z\},$$

where σ_j are the Pauli matrices. By multiplying both sides of (E.2) with σ_k and taking the trace of both sides, we obtain

$$\begin{aligned}
v'_k &= Tr(v' \cdot A\sigma_k) \\
&= Tr(\sigma_k U I_i U^\dagger)v_i,
\end{aligned} \tag{E.3}$$

where v'_k and v_i are, respectively, the k-th an i-th entry of the corresponding vectors. We apply the summation convention in this section, where summing over repeated indices is implied unless otherwise stated. Comparing the above result with equation

$$v'_k = \mathcal{R}(U)_{ki}v_i,$$

we obtain the desirable matrix $\mathcal{R}(U)$:

$$\mathcal{R}(U)_{ki} = Tr(\sigma_k\, U\, I_i\, U^\dagger). \tag{E.4}$$

Thus we have constructed a mapping from $\mathbf{SU}(2)$ to the set of 3×3 matrices. Let us now tentatively accept the fact the target matrix is real which we will prove later. We first show that $\mathcal{R}(U)$ is in fact a proper rotation matrix, i.e., $\mathcal{R}(U) \in \mathbf{SO}(3)$ by proving that it is isometric and preserves the 3-fold product; cf. (E.7).

Let $\boldsymbol{v} = [a \quad b \quad c]^T \in \mathbb{R}^3$. First note that we can find its norm by computing the determinant of a special matrix:

$$
\begin{aligned}
\det(\boldsymbol{v} \cdot \boldsymbol{A}) &= \frac{1}{4}\begin{vmatrix} c & a - bi \\ a + bi & -c \end{vmatrix} \\
&= -\tfrac{1}{4}(c^2 + a^2 + b^2) \\
&= -\tfrac{1}{4}\|\boldsymbol{v}\|^2.
\end{aligned}
\tag{E.5}
$$

Together with (E.2), we see that the transformation is isometric:

$$
\begin{aligned}
\|\boldsymbol{v}'\|^2 &= -4\det(\boldsymbol{v}' \cdot \boldsymbol{A}) \\
&= -4\det(U\,\boldsymbol{v} \cdot \boldsymbol{A}U^\dagger) \\
&= -4\det(\boldsymbol{v} \cdot \boldsymbol{A}) \\
&= \|\boldsymbol{v}\|^2.
\end{aligned}
\tag{E.6}
$$

Direct computation also shows the preservation of the 3-fold product, as follows. Let $\boldsymbol{v}^l$, $l = 1, 2, 3$ be three vectors on the Bloch sphere, $\boldsymbol{v}'^l = \mathcal{R}(u)\boldsymbol{v}^l$. We have

$$
\begin{aligned}
Tr\left((\boldsymbol{v}^1 \cdot \boldsymbol{A})(\boldsymbol{v}^2 \cdot \boldsymbol{A})(\boldsymbol{v}^3 \cdot \boldsymbol{A})\right) &= Tr(v_i^1 v_j^2 v_k^3 I_i I_j I_k) \\
&= \tfrac{i}{4} v_i^1 v_j^2 v_k^3 \epsilon_{ijk} \\
&= \tfrac{i}{4}\boldsymbol{v}^1 \cdot (\boldsymbol{v}^2 \times \boldsymbol{v}^3).
\end{aligned}
\tag{E.7}
$$

The above identity can be used to show the preservation of the *3-fold product*:

$$
\begin{aligned}
\boldsymbol{v}'^1 \cdot (\boldsymbol{v}'^2 \times \boldsymbol{v}'^3) &= \tfrac{4}{i}Tr\left((\boldsymbol{v}'^1 \cdot \boldsymbol{A})(\boldsymbol{v}'^2 \cdot \boldsymbol{A})(\boldsymbol{v}'^3 \cdot \boldsymbol{A})\right) \\
&= \tfrac{4}{i}Tr\left((U\boldsymbol{v}^1 \cdot \boldsymbol{A}U^\dagger)(U\boldsymbol{v}^2 \cdot \boldsymbol{A}U^\dagger)(U\boldsymbol{v}^3 \cdot \boldsymbol{A}U^\dagger)\right) \\
&= \tfrac{4}{i}Tr\left((\boldsymbol{v}^1 \cdot \boldsymbol{A})(\boldsymbol{v}^2 \cdot \boldsymbol{A})(\boldsymbol{v}^3 \cdot \boldsymbol{A})\right) \\
&= \boldsymbol{v}^1 \cdot (\boldsymbol{v}^2 \times \boldsymbol{v}^3),
\end{aligned}
\tag{E.8}
$$

and we see that $\mathcal{R}(U) \in \mathbf{SO}(3)$. To prove that the mapping is surjective, we introduce a parameterization of $\mathbf{SU}(2)$ in some exponential form. For any $U \in \mathbf{SU}(2)$, we can find a $\theta \in [0, 2\pi)$ and a unit vector $\boldsymbol{n} \in \mathbb{R}^3$ such that (D.2) holds (in Appendix D). Every combination of θ and $\boldsymbol{n} \in \mathbb{R}^3$ also corresponds to a complex matrix in $\mathbf{SU}(2)$. By using (D.1) and the equality $U(\theta, \boldsymbol{n})^{-1} = U(\theta, -\boldsymbol{n})$, the element of $\mathcal{R}(U)$ can be computed as

$$
\begin{aligned}
\mathcal{R}(U)_{ij} &= Tr(\sigma_i\, U\, I_j\, U^\dagger) \\
&= Tr\left(\sigma_i(\cos\tfrac{\theta}{2}I - i\sin\tfrac{\theta}{2}\boldsymbol{n}\cdot\boldsymbol{\sigma})I_j(\cos\tfrac{\theta}{2}I + i\sin\tfrac{\theta}{2}\boldsymbol{n}\cdot\boldsymbol{\sigma})\right) \\
&= Tr(\sigma_i I_j \cos^2\tfrac{\theta}{2} + i\sigma_i I_j\boldsymbol{n}\cdot\boldsymbol{\sigma}\sin\tfrac{\theta}{2}\cos\tfrac{\theta}{2} \\
&\quad - i\sigma_i\boldsymbol{n}\cdot\boldsymbol{\sigma}I_j\sin\tfrac{\theta}{2}\cos\tfrac{\theta}{2} + \sigma_i\boldsymbol{n}\cdot\boldsymbol{\sigma}I_j\boldsymbol{n}\cdot\boldsymbol{\sigma}\sin^2\tfrac{\theta}{2}).
\end{aligned}
\tag{E.9}
$$

We divide the trace into four parts and compute them separately:

$$Tr(\sigma_i I_j) = \delta_{ij}, \quad Tr(\sigma_i I_j \boldsymbol{n} \cdot \boldsymbol{\sigma}) = i\epsilon_{ijk} n_k, \quad Tr(\sigma_i \boldsymbol{n} \cdot \boldsymbol{\sigma} I_j) = i\epsilon_{ikj} n_k, \quad \text{(E.10)}$$

$$Tr(\sigma_i \boldsymbol{n} \cdot \boldsymbol{\sigma} I_j \boldsymbol{n} \cdot \boldsymbol{\sigma}) = Tr(\sigma_i \sigma_k I_j \sigma_l n_k n_l)$$

$$= \frac{1}{2} Tr\left((\delta_{ik} I + i\epsilon_{ikm} \sigma_m)(\delta_{jl} I + i\epsilon_{jln} \sigma_n) n_k n_l\right)$$

$$= \delta_{ik}\delta_{jl} n_k n_l - \epsilon_{ikm}\epsilon_{jlm} n_k n_l$$

$$= \delta_{ij} n_i n_j + (2 - \delta_{ij}) n_i n_j - \delta_{ij} = 2 n_i n_j - \delta_{ij}. \quad \text{(E.11)}$$

Substituting (E.11) and (E.10) back into (E.9), we obtain

$$\begin{aligned}
\mathcal{R}(U)_{ij} &= \cos^2 \tfrac{\theta}{2}\delta_{ij} + \sin^2 \tfrac{\theta}{2}(2 n_i n_j - \delta_{ij}) \\
&\quad -\epsilon_{ijk} n_k \sin \tfrac{\theta}{2} \cos \tfrac{\theta}{2} + \epsilon_{ikj} n_k \sin \tfrac{\theta}{2} \cos \tfrac{\theta}{2} \\
&= \cos\theta\, \delta_{ij} + (1 - \cos\theta) n_i n_j + \sin\theta\, \epsilon_{ikj} n_k,
\end{aligned} \quad \text{(E.12)}$$

showing that $\mathcal{R}(U)_{ij}$ is *real*. We also claim that $\mathcal{R}(U)$ is a rotation about the axis $\boldsymbol{n}$ with angle θ in the three dimensional Euclidean space by comparing it with the standard formula of a rotation matrix. Because every matrix in SO(3) can be regarded as a rotation about a certain axis with a certain angle, that matrix is now shown to be an image of some $U \in$ SU(2). Thus the mapping is surjective.

Finally, we want to show that this mapping from SU(2) to SO(3) is a homomorphism and to investigate the multiplicity. Let U and T be two matrices in SU(2), $\boldsymbol{v}$ be an arbitrary vector on the Bloch sphere, and $\boldsymbol{v}' = \mathcal{R}(U)\boldsymbol{v}$, $\boldsymbol{v}'' = \mathcal{R}(T)\boldsymbol{v}' = \mathcal{R}(T)\mathcal{R}(U)\boldsymbol{v}$. Using (E.2) repeatedly, we obtain

$$\boldsymbol{v}'' \cdot A = T(\boldsymbol{v}' \cdot A)T^\dagger = TU(\boldsymbol{v} \cdot A)U^\dagger T^\dagger = \mathcal{R}(TU)\boldsymbol{v} \cdot A,$$

thus $\boldsymbol{v}'' = \mathcal{R}(TU)\boldsymbol{v}$. Because $\boldsymbol{v}$ is arbitrary, $\mathcal{R}(VU) = \mathcal{R}(V)\mathcal{R}(U)$, implying that the mapping is a homomorphism.

For the multiplicity of this mapping, we investigate the kernel of $\mathcal{R}$, $\ker(\mathcal{R})$, which is an invariant subgroup of SU(2), and the quotient group SU(2)$/\ker(\mathcal{R})$ will then be isomorphic to SO(3). Suppose that $U \in$ SU(2) and $\mathcal{R}(U) = I_3$, the identity matrix in SO(3). Then for any $\boldsymbol{v}$ on the Bloch sphere or $\mathbb{R}^3$, we have equality $\boldsymbol{v} \cdot A = U(\boldsymbol{v} \cdot A)U^\dagger$, since $\boldsymbol{v} = I\boldsymbol{v}$. Multiplying both sides with U from the right and using (D.2), we have

$$\boldsymbol{v} \cdot A(\cos \tfrac{\theta}{2} I - i \sin \tfrac{\theta}{2} \boldsymbol{n} \cdot \boldsymbol{\sigma}) = (\cos \tfrac{\theta}{2} I - i \sin \tfrac{\theta}{2} \boldsymbol{n} \cdot \boldsymbol{\sigma})\boldsymbol{v} \cdot A.$$

Subtract from both sides the term $\boldsymbol{v} \cdot A \cos \tfrac{\theta}{2} I$,

$$(\boldsymbol{v} \cdot A)\,(\boldsymbol{n} \cdot \boldsymbol{\sigma}) \sin \tfrac{\theta}{2} = (\boldsymbol{n} \cdot \boldsymbol{\sigma})\,(\boldsymbol{v} \cdot A) \sin \tfrac{\theta}{2}.$$

Assuming that $\sin \tfrac{\theta}{2} \neq 0$, we can divide this factor from both sides to obtain

$$(\boldsymbol{v} \cdot A)\,(\boldsymbol{n} \cdot \boldsymbol{\sigma}) = (\boldsymbol{n} \cdot A)\,(\boldsymbol{v} \cdot \boldsymbol{\sigma}).$$

Let $v = [1, 0, 0]^T$. Using properties of Pauli matrices, we obtain

$$n_2 I_z - n_3 I_y = 0.$$

The above is possible only when $n_2 = n_3 = 0$. Trying other different v leads to $n_1 = 0$, too. This is a contradiction, because we know that n is a unit vector. Thus we need $\sin \frac{\theta}{2} = 0$. In this case $U = I$ or $U = -I$ and it is easy to verify that these two are really mapped to the identity in $\mathbf{SO}(3)$. Now, we can conclude that the mapping we defined by (E.4) is a *two-to-one homomorphism* from $\mathbf{SU}(2)$ to $\mathbf{SO}(3)$ with kernel $\ker(\mathcal{R}) = \{I, -I\}$. The mapping is also surjective, so it defines an isomorphism from the quotient group $\mathbf{SU}(2)/\ker(\mathcal{R})$ to $\mathbf{SO}(3)$. The two elements in the kernel, $\pm I$, are in fact the same transformation for quantum systems because only the relative phase matters for a quantum system. For any $O \in \mathbf{SO}(3)$, the two elements in $\mathcal{R}^{-1}(O)$, U and $-U$ for some $U \in \mathbf{SU}(2)$, represent the same transformation, too. Thus, nothing is lost if we employ $\mathbf{SO}(3)$ to represent the transformations of a 1-spin quantum system.

References

[1] Abramowitz and I. Stegun, *Handbook of Mathematical Functions*, 9th printing, Dover, New York, 1970.

[2] G. Burkard, D. Loss, and D. DiVincenzo, Coupled quantum dots as quantum gates, *Phys. Rev. B* **59** (1999), 2070–2078.

[3] C. Darwin, The diamagnetism of the free electron, *Proc. Cambridge Phil. Soc.* **27** (1930), 86.

[4] V. Fock, Bermerkung zur Quantelung des harmonischen Oszillators in Magnetfeld, *Zeit. Physik* **47** (1928), 446.

[5] I.S. Gradshteyn and I.M. Ryzhik, *Table of Integrals, Series and Products*, 6th edition, Academic Press, New York, 2000.

[6] M.A. Nielsen and I.L. Chuang, *Quantum Computation and Quantum Information*, Cambridge University Press, Cambridge, U.K., 2000.

[7] J. Normand, *A Lie Group: Rotations in Quantum Mechanics*, North-Holland, New York, 1980.

[8] M.O. Scully, N. Erez, and E.S. Fry, *Phys. Lett. A* **347** (2005), 56–61.

[9] J.C. Slater, *Quantum Theory of Molecules and Solids*, Vol. 1, McGraw-Hill, New York, 1963.

[10] E.P. Wigner, *Am. J. Phys.* **38** (1970), 1005.

Index

Authors

Goong Chen was born in Kaohsiung, Taiwan in 1950. He received his BSc
(Math) from the National Tsing Hua University in Hsichu, Taiwan in 1972
and PhD (Math) from the University of Wisconsin at Madison in 1977. He
has taught at the Southern Illinois University at Carbondale (1977–1978),
and the Pennsylvania State University at University Park (1978–1987).
Since 1987, he has been Professor of Mathematics and Aerospace Engineering, and (since 2000) a member of the Institute for Quantum Studies
at Texas A&M University in College Station, Texas. He has also held short-
term visiting positions at INRIA in Rocquencourt, France, the Université
de Montréal, the Technical University of Denmark in Lyngby, Denmark,
the National University of Singapore, and National Tsing Hua University
in Hsinchu, Taiwan.

Chen has research interests in control theory for partial differential equations, boundary element methods and numerical solutions of PDEs, chaotic
dynamics, quantum computation, and molecular quantum mechanics. He
has written over 100 papers, 5 advanced texts/monographs, and co-edited

3 books. He is Editor-in-Chief of the Chapman & Hall/CRC Press Applied Mathematics and Nonlinear Science Series, and has served on several editorial boards, including the *SIAM Journal on Control and Optimization*, *Journal of Mathematical Analysis and Applications*, the *International Journal on Quantum Information*, and the *Electronic Journal of Differential Equations*. He is also a co-holder of a U.S. patent on certain quantum circuit design. He has memberships in the American Mathematical Society (AMS) and the Society for Industrial and Applied Mathematics (SIAM).

David A. Church is from the Great North Woods, born in Berlin, New Hampshire, USA in 1939. He received his BA (Physics) from Dartmouth College in 1961 and PhD (Physics) from the University of Washington, Seattle, in 1969. He held temporary positions at the University of Bonn (Germany) 1969, University of Mainz (Germany) 1969-1971, University of Arizona, 1971-1972, Lawrence Berkeley Laboratory 1972-1975, and was Assistant Professor, Texas A&M University 1975-1981. He was also a Visiting Professor at the University of Tennessee (1984-1985). From 1981 to the present, he has held permanent positions, first as Associate Professor and then Professor at Texas A&M University in College Station, Texas.

He has had research collaborations with the Oak Ridge National Laboratory (1982-1985), Brookhaven National Laboratory (1985-1991), Stanford Synchrotron Radiation Laboratory (1986-1987), Argonne National Laboratory (1989), Lawrence Livermore National Laboratory (1991-2001), NIST Gaithersberg (1995-1996), University of Nevada-Reno (1997-1999), Lawrence Berkeley Laboratory (2003-2005), and the Jet Propulsion Laboratory (2001-present).

His research interests include cold ion confinement, g-factor of the free electron, coherence in fast ion excitation, recoil ion production, energies and confinement, polarization of x-rays in fast ion-atom and ion-foil collisions, excitation, synchrotron radiation photoionization of ions and slow multi-charged ion production and confinement, electron capture in low energy ion-atom collisions, mean lives of metastable ion levels, resonant laser excitation of ions in beams for hyperfine structure and mean life measurements, confinement and cooling of low-energy highly-charged ions, g-factor of the bound electron in high-Z one-electron ions, and quantum computing with confined particles.

He is a Fellow and life member of the American Physical Society (APS) since 1986.

Berthold-Georg Englert was born in Urach, Germany, in 1953. He received his doctoral degree in Tübingen, Germany in 1981, his postdoctoral training at UCLA from 1981 to 1985, and his habilitation in Munich, Germany in 1990. He has taught at the University of Munich, Texas A&M University, the Technical University of Vienna, the University of Ulm, the University of Milan, and is now Professor of Physics at the National University of Singapore. He was a permanent guest of the quantum optics institute in Garching, Germany for many years and made extended research visits at the University of Paris-Nord and the Institut Nonlineaire de Nice. He is an Associate Editor of *Zeitschrift für Naturforschung*, Managing Editor of the *International Journal of Quantum Information*, and a member of the Editorial board for the Lecture Notes in Physics book series (Springer-Verlag).

Englert has contributed to the refinement of semiclassical methods in atomic physics (Springer-Verlag monograph, 1988), the development of algebraic methods for handling quantum-optical master equations bearing particularly on the theory of the micromaser, the technical formulation of the principle of complementarity, the understanding of the quantitative aspects of wave-particle duality, and most recently has co-invented novel schemes for quantum cryptography and direct quantum communication. He has just published undergraduate-level *Lectures on Quantum Mechanics* in three companion volumes (World Scientific, Singapore, 2006).

Carsten Henkel was born in Ratzeburg, West Germany in 1967. He studied in Kiel, Germany and Paris, France and graduated in theoretical physics in 1993 at Ecole Normale Supérieure, Paris, and Université de Paris-Sud Orsay, France. He received his PhD (Docteur en Sciences) from the Université de Paris-Sud Orsay in 1996. He has been doing research and teaching at Universität Potsdam, Germany since 1997. In 2004, he completed his Habilitation and was appointed Privatdozent (assistant professor) at Universität Potsdam. He has also held short-term visiting positions at the Université de Toulouse III, France, the Ecole Supérieure d'Optique in Orsay, France, and the Ecole Centrale Paris, France, among others.

His research is oriented towards atom optics and nano optics, with a particular emphasis on the physical implementations of quantum information processing systems, coherence and decoherence theory, and the mechanical effects of radiation. He has written about 40 articles and review papers and co-edited two special issues. He is a member of the German Physical Society (DPG), the French Society of Optics (SFO), and is a member of the European Physical Society (EPS).

Bernd Rohwedder was born in Buenos Aires, Argentina, in 1967. He studied physics at the Ludwig-Maximilians Universität in München, Germany
and received his diploma in November 1992. His diploma thesis on semi-
classical quantization in momentum space was supervised by Prof. B.-G. Englert. He then shifted his main research interests to quantum optics and
worked in the group lead by Prof. M. Orszag at Pontificia Universidad
Católica de Chile (Santiago de Chile), where he received his PhD in 1997.
During his postdoctoral appointment with the quantum optics group at the
Federal University of Rio de Janeiro, Brazil, he focused on atom optics, with
an increasing interest in theoretical optics in general and diffraction phenomena near caustics and quasiperiodical structures in particular. He has
been a visiting scientist at the Institut Laue-Langevin (Grenoble, France),
Massachusetts Institute of Technology (Cambridge, Massachusetts, USA),
the Australian National University (Canberra, Australia), and the National
University of Singapore (Singapore), where he developed his current interests in quantum information theory. Rohwedder is a member of the German
Physical Society (DPG) and the American Physical Society (APS).

Marlan O. Scully was born in Casper, Wyoming in 1939. He received his BSc (Engineering Physics) from the University of Wyoming in 1961 and PhD (Physics) from Yale University in 1966. He was on the faculty of Yale University (1966-1967), the Massachusetts Institute of Technology (1968-1971), the University of Arizona (1969-1980), the University of New Mexico (1980-1992), Max-Planck Institut für Quantenoptik in Garching, Germany (1980-present), and Texas A&M University in College Station, Texas (1992-present). Since 1997, he has been the Herschel Burgess Chair at Texas A&M University. He was elected to the National Academy of Sciences in 2001, and earlier also to the Academia Europa and the Max Planck Society. He has received numerous awards, including the Arthur Schawlow prize of the American Physical Society, the Charles H. Townes Award of the Optical Society of America, the Quantum Electronics Award of IEEE, the Elliott Cresson Medal of the Franklin Institute, the Adolph E. Lomb Medal of the OSA, a Guggenheim Fellowship, and the Alexander von Humboldt Distinguished Faculty Prize. He has a joint appointment between Texas A&M and Princeton Universities.

Scully has made pioneering contributions to laser physics, including the first fully quantum mechanical theory of laser that predicted the photon statistics and the laser linewidth of a laser. His work on the applications of atomic coherence effects in lasing systems and precision spectroscopy is seminal, as his study on the foundations of quantum mechanics such as the introduction of the quantum eraser concept is fundamental. He has also co-authored two major textbooks on laser physics and quantum optics.

M. Suhail Zubairy was born in Lahore, Pakistan in 1952. He received his PhD from the University of Rochester in 1978. He held research and teaching appointments at the Optical Sciences Center of the University of Arizona and the Center for Advanced Studies at the University of New Mexico before joining the Quaid-i-Azam University, Islamabad, Pakistan in 1984. He was Professor of Electronics and the founding chairman of the Department of Electronics at the Quaid-i-Azam University. In 2000 he joined Texas A&M University, where he is presently Professor of Physics and Associate Director of the Institute of Quantum Studies.

Zubairy is a Fellow of the American Physical Society, the Optical Society of America, and the Pakistan Academy of Sciences. He has received the Orders of Hilal-e-Imtiaz and Sitara-e-Imtiaz from the President of Pakistan, the COMSTECH Prize for Physics, the Abdus Salam Prize for Physics and the International Khwarizmi Award.

His main research interests are quantum optics and laser physics. His recent work includes quantum optical applications to quantum computing and quantum informatics, quantum state measurement of the radiation field and sub-wavelength atom localization. His other interests include coherent atomic effects and quantum thermodynamics. *Quantum Optics* (Cambridge University Press, 1997), coauthored with Marlan O. Scully, is a major tome on the topic.

Printed and bound by CPI Group (UK) Ltd, Croydon, CR0 4YY

25/10/2024

01779529-0001